Library of Congress Cataloging-in-Publication Data

Rotman, Joseph J.
    A first course in abstract algebra/Joseph J. Rotman -- 2nd ed.
      p. cm.
    Includes bibliographical references and index.
    ISBN 0-13-011584-3
    1. Algebra, Abstract/ I. Title.
QA162.R68   2000
512'.02--dc21
                                    99-088794
                                    CIP

Acquisitions Editor: *George Lobell*
Production Editor: *Bayani Mendoza de Leon*
Assistant Vice President of Production and Manufacturing: *David W. Riccardi*
Executive Managing Editor: *Kathleen Schiaparelli*
Senior Managing Editor: *Linda Mihatov Behrens*
Manufacturing Buyer: *Alan Fischer*
Manufacturing Manager: *Trudy Pisciotti*
Marketing Manager: *Melody Marcus*
Marketing Assistant: *Vince Jansen*
Director of Marketing: *John Tweeddale*
Editorial Assistant: *Gale Epps*
Art Director: *Jayne Conte*
Cover Designer: *Bruce Kenselaar*
Cover Image: *The original painting of 10×10 orthogonal Latin squares by Emi Kasia*
hangs in the office of the Mathematics Department of the University
of Illinois at Urbana-Champaign.

-011584-3

RNATIONAL (UK) LIMITED, LONDON
STRALIA PTY. LIMITED, SYDNEY
A, INC., TORONTO
AMERICANA, S.A., MEXICO
PRIVATE LIMITED, NEW DELHI
INC., TOKYO
PTE. LTD.
BRASIL, LTDA., RIO DE JANEIRO

The
Library
Mansfield
College
Oxford

A FIRST COURSE
IN ABSTRACT ALGEBRA

Second Edition

JOSEPH J. ROTMAN
University of Illinois
at Urbana-Champaign

All
be re
witho

Printed i
10 9 8

ISBN 0-13

PRENTICE-HALL INTE
PRENTICE-HALL OF A
PRENTICE-HALL CANA
PRENTICE-HALL HISPAN
PRENTICE-HALL OF INDIA
PRENTICE-HALL OF JAPAN,
PEARSON EDUCATION ASIA
EDITORA PRENTICE-HALL DO

PRENTICE HALL, Up

To my two wonderful kids,

Danny and Ella,

whom I love very much

Contents

Preface to the First Edition

A First Course in Abstract Algebra introduces three related topics: number theory (division algorithm, greatest common divisors, unique factorization into primes, and congruences), group theory (permutations, Lagrange's theorem, homomorphisms, and quotient groups), and commutative ring theory (domains, fields, polynomial rings, homomorphisms, quotient rings, and finite fields). The final chapter combines the preceding chapters to solve some classical problems: angle trisection, squaring the circle, doubling the cube, construction of regular n-gons, and impossibility of generalizing the quadratic, cubic, and quartic formulas to polynomials of higher degree. Such results make it clear that mathematics is, indeed, one subject whose various areas do bear one on the other.

A complicating factor, permeating introductory courses, is that this may be one of the first times students are expected to read and write proofs. This book is my attempt to cover the required topics, to give models of proofs, and to make it all enjoyable.

There is enough material here for a two-semester course, even though many readers may be interested in only one semester's worth. All the "usual suspects" are assembled here, however, and I hope that instructors will be able to find those theorems and examples they believe to be appropriate for a first course. When teaching a one-semester course, one must skip parts of the text; however, it is often possible simply to state and use theorems whose proofs have been omitted. For example, if the discussion of generalized associativity is omitted, one can safely cite the laws of exponents; if the proof of Gauss's lemma is omitted, one can quote it and still derive irreducibility criteria for polynomials in $\mathbb{Q}[x]$. ...

I do not enjoy reading introductory chapters of books that consist wholly of "tools" needed for understanding subsequent material. By the Golden Rule, I do not inflict such greetings on my readers. Rather than beginning with a discussion of logic, sets, Boolean operations, functions, equivalence relations, and so forth, I introduce such tools as they are needed. For example, functions and bijections are introduced with permutation groups; equivalence relations are introduced in Chapter 3 to construct fraction fields of domains (I recognize that this late entry of equivalence relations and equivalence classes may annoy those who prefer introducing quotient groups with them; however, I feel that readers first meeting cosets and quotient groups do not need the extra baggage of an earlier discussion of equivalence classes). The first section of Chapter 1 does introduce an essential tool, induction, but induction also serves there as a vehicle to introduce more interesting topics such as primes and De Moivre's theorem.

Several results that are not usually included in a first course have been included just because they are interesting and accessible applications; they should not be presented in class because they are designed for curious readers only. In Chapter 1 on number theory, congruences are used to find on which day of the week a given date falls. In Chapter 2 on groups, the group of motions of the plane is used to describe symmetry of planar figures, the affine group is used to prove theorems of plane geometry, and a counting lemma is applied to solve some difficult combinatorial problems. In Chapter 3 on rings, we construct finite fields, and then we use them to construct complete sets of orthogonal Latin squares. The (fourth) chapter is both a dessert and an appetizer. After a short discussion of vector spaces and dimension (which reinforces the categorical viewpoint of objects and morphisms), we show how modern algebra solves several classical problems of geometry. After giving the quadratic, cubic, and quartic formulas, we present an analogy between symmetry groups of figures and Galois groups, and we prove the theorem of Abel and Ruffini that there is no generalization of the classical formulas to higher degree polynomials. This discussion can serve as an introduction to Galois Theory.

Since Birkhoff and Mac Lane created this course half a century ago, there has been mild controversy about the order of presentation: should the exposition of groups precede that of rings, or should rings be done first (Birkhoff and Mac Lane do rings first). There are arguments on both sides and, after being a rings first man for a long time, I have come to believe that it is more reasonable to do groups first. The definition of group is very simple, and permutation groups offer an immediate nontrivial example. Many elementary properties of rings are much simpler once one has studied groups. Indeed, the very definition of a ring is more palatable once one has seen groups. As a second example, the quotient group construction can be used to construct quotient rings (since rings are additive abelian groups and ideals are normal subgroups), but the quotient ring construction cannot be used directly in constructing quotient groups. Thus, discussing groups first is more efficient than the alternative. Finally, whenever I have taught rings first, I have found an initial confusion in the class about the relation of general rings to the particular ring $\mathbb{Z}$ of integers. There is a need to develop some arithmetic properties of $\mathbb{Z}$, and bouncing back and forth between commutative rings and $\mathbb{Z}$ creates an unnecessary difficulty for many students. In particular, students become unsure about which properties of $\mathbb{Z}$ may be assumed and which need proof. The organization here avoids this problem by separating these two subjects by group theory.

Giving the etymology of mathematical terms is rarely done. Let me explain, with an analogy, why I have included derivations of many terms. There are many variations of standard poker games and, in my poker group, the dealer announces the game of his choice by naming it. Now some names are better than others. For example, "Little Red" is a game in which one's smallest red card is wild; this is a good name because it reminds the players of its distinctive feature. On the other hand, "Aggravation" is not such a good name, for though it is, indeed, suggestive, the name does not distinguish this particular game from several others. Most terms in mathematics have been well chosen; there are more red names than aggravating ones. An example of a good name is *even* permutation, for a permutation is even if it is a product of an even number of transpositions. Another example of a good term is the *parallelogram law* describing vector addition. But many good names, clear when they were chosen, are now obscure because their roots are either in another language or in another discipline. The term *mathematics* is obscure only because most of us do not know that it comes from the classical Greek word meaning "to learn." The term *corollary* is doubly obscure; it comes from the Latin word meaning "flower," but what do flowers have to do with theorems? A plausible explanation is that it was common, in ancient Rome, to give flowers as gifts, and so a corollary is a gift bequeathed by a theorem. The term *theorem* comes from the Greek word meaning "to watch" or "to contemplate" (*theatre* has the same root); it was used by Euclid with its present meaning. The term *lemma* comes from the Greek word meaning "taken" or "received;" it is a statement that is taken for granted (for it has already been proved) in the course of proving a theorem. On the other hand, I am not too fond of the mathematical terms *normal* and *regular* for, in themselves, they convey no specific meaning. Since the etymology of terms often removes unnecessary obscurity, it is worthwhile (and interesting!) to do so.

It is a pleasure to thank Dan Grayson, Heini Halberstam, David G. Poole, Ed Reingold, and John Wetzel for their suggestions. I also thank the Hebrew University of Jerusalem for the hospitality given me as I completed my manuscript. I thank the several reviewers who carefully read my manuscript and made valuable suggestions. They are Daniel D. Anderson, University of Iowa; Michael J. J. Barry, Allegheny College; Brad Shelton, University of Oregon; Warren M. Sinnott, Ohio State University; and Dalton Tarwater, Texas Tech University. And I thank George Lobell, who persuaded me to develop and improve my first manuscript into the present text.

Joseph J. Rotman

Preface to the Second Edition

I was reluctant to accept Prentice Hall's offer to write a second edition of this book. When I wrote the first edition several years ago, I assumed the usual assumption: All first courses in algebra have essentially the same material, and so it is not necessary to ask *what* is in such a book, but rather *how* it is in it. I think that most people accept this axiom, at least tacitly, and so their books are almost all clones of one another, differing only in the quality of the writing. Looking at the first version of my book, I now see many flaws; there were some interesting ideas in it, but the book was not significantly different from others. I could improve the text I had written, but I saw no reason to redo it if I were to make only cosmetic changes.

I then thought more carefully about what an introduction to algebra ought to be. When Birkhoff and Mac Lane wrote their pioneering *A Survey of Modern Algebra* about 60 years ago, they chose the topics that they believed were most important, both for students with a strong interest in algebra and those with other primary interests in which algebraic ideas and methods are used. Birkhoff and Mac Lane were superb mathematicians, and they chose the topics for their book very well. Indeed, their excellent choice of topics is what has led to the clone producing assumption I have mentioned above. But times have changed; indeed, Mac Lane himself has written a version of *A Survey of Modern Algebra* from a categorical point of view. [I feel it is too early to mention categories explicitly in this book, for I believe one learns from the particular to the general, but categories are present implicitly in the almost routine way homomorphisms are introduced as soon as possible after introducting algebraic systems.] Whereas emphasis on rings and groups is still fundamental, there are today major directions which either did not exist in 1940 or were not then recognized to be so important. These new directions involve algebraic geometry, computers, homology, and representations. One may view this new edition as the first of a two volume sequence. This book, the first volume, is designed for students beginning their study of algebra. The sequel, designed for beginning graduate students, is designed to be independent of this one. Hence, the sequel will have a substantial overlap with this book, but it will go on to discuss some of the basic results which lead to the most interesting contemporary topics. Each generation should survey algebra to make it serve the present time.

When I was writing this second edition, I was careful to keep the pace of the exposition at its original level; one should not rush at the beginning. Besides rewriting and rearranging theorems, examples, and exercises that were present in the first edition, I have added new material. For example, there is a short subsection on euclidean rings which contains a proof of Fermat's Two-Squares Theorem; and the Fundamental Theorem of Galois Theory is stated and used to prove the Fundamental Theorem of Algebra: the complex numbers are algebraically closed.

I have also added two new chapters, one with more group theory and one with more commutative rings, so that the book is now more suitable for a one-year course (one can also base a one-semester course on the first three chapters). The new chapter on groups proves the Sylow theorems, the Jordan-Hölder theorem, and the fundamental theorem of finite abelian groups, and it introduces free groups and presentations by generators and relations. The new chapter on rings discusses prime and maximal ideals, unique factorization in polynomial rings in several variables, noetherian rings, varieties, and Gröbner bases. Finally, a new section contains hints for most of the exercises (and an instructor's solution manual contains complete solutions for all the exercises in the first four chapters).

In addition to thanking again those who helped me with the first edition, it is a pleasure to thank Daniel D. Anderson, Aldo Brigaglia, E. Graham Evans, Daniel Flath, William Haboush, Dan Grayson, Christopher Heil, Gerald J. Janusz, Jennifer D. Key, Steven L. Kleiman, Emma Previato, Juan Jorge Schaffer, and Thomas M. Songer for their valuable suggestions for this book.

And so here is edition two; my hope is that it makes modern algebra accessible to beginners, and that it will make its readers want to pursue algebra further.

Joseph J. Rotman

1

Number Theory

1.1 INDUCTION

There are many styles of proof, and mathematical induction is one of them. We begin by saying what mathematical induction is not. In the natural sciences, *inductive reasoning* is the assertion that a freqently observed phenomenon will always occur. Thus, one says that the Sun will rise tomorrow morning because, from the dawn of time, the Sun has risen every morning. This is not a legitimate kind of proof in mathematics, for even though a phenomenon has been observed many times, it need not occur forever.

Inductive reasoning is valuable in mathematics because seeing patterns in data often helps in guessing what may be true in general. On the other hand, inductive reasoning is not adequate for proving theorems. Before we see examples, let us make sure that we agree on the meaning of some standard terms.

Definition. An *integer* is one of $0, 1, -1, 2, -2, 3, \ldots$.

Definition. An integer d is a *divisor* of an integer n if $n = da$ for some integer a. An integer $n \geq 2$ is called *prime*[1] if its only positive divisors are 1 and n; otherwise, n is called *composite*.

---

[1] One reason the number 1 is not called a prime is that many theorems involving primes would otherwise be more complicated to state.

1

An integer $n \geq 2$ is composite if it has a factorization $n = ab$, where $a < n$ and $b < n$ are positive integers; the inequalities are present to eliminate the uninteresting factorization $n = n \times 1$. The first few primes are 2, 3, 5, 7, 11, 13, 17, 19, 23, 29, 31, 37, 41, ...; that this sequence never ends is proved in Corollary 1.27.

Consider the assertion, for n a positive integer, that

$$f(n) = n^2 - n + 41$$

is always prime. Evaluating $f(n)$ for $n = 1, 2, 3, \ldots, 40$ gives the numbers

41, 43, 47, 53, 61, 71, 83, 97, 113, 131,

151, 173, 197, 223, 251, 281, 313, 347, 383, 421,

461, 503, 547, 593, 641, 691, 743, 797, 853, 911,

971, 1033, 1097, 1163, 1231, 1301, 1373, 1447, 1523, 1601.

It is tedious, but not very difficult, to show that every one of these numbers is prime (see Proposition 1.3). Inductive reasoning predicts that *all* the numbers of the form $f(n)$ are prime. But the next number, $f(41) = 1681$, is not prime, for $f(41) = 41^2 - 41 + 41 = 41^2$, which is, obviously, composite. Thus, inductive reasoning is not appropriate for mathematical proofs.

Here is an even more spectacular example (which I first saw in an article by W. Sierpinski). Recall that **perfect squares** are numbers of the form n^2, where n is an integer; the first few perfect squares are 1, 4, 9, 16, 25, 36, ... For each $n \geq 1$, consider the statement

$S(n)$: $991n^2 + 1$ is not a perfect square.

The nth statement, $S(n)$, is true for many n; in fact, the smallest number n for which $S(n)$ is false is

$$n = 12, 055, 735, 790, 331, 359, 447, 442, 538, 767$$
$$\approx 1.2 \times 10^{28}.$$

(The original equation, $m^2 = 991n^2 + 1$, is an example of **Pell's equation**— an equation of the form $m^2 = pn^2 + 1$, where p is prime—and there is a way of calculating all possible solutions of it. An even more spectacular example of Pell's equation involves the prime $p = 1{,}000{,}099$; the smallest n for which $1{,}000{,}099n^2 + 1$ is a perfect square has 1116 digits.) The most generous estimate of the age of the earth is 10 billion (10,000,000,000) years, or 3.65×10^{12} days, a number insignificant when compared to 1.2×10^{28}, let alone 10^{1115}. If, starting from the Earth's very first day, one verified statement $S(n)$ on the nth day, then there would be today as much evidence of the general truth of these statements as there is that the Sun will rise tomorrow morning. And yet some of the statements $S(n)$ are false!

As a final example, let us consider the following statement, known as **Goldbach's conjecture**: Every even number $m \geq 4$ is a sum of two primes. (It would be foolish to demand that all odd numbers be sums of two primes. For example, let us show that 27 is not the sum of two primes. Otherwise, $27 = p + q$, where p and q are primes. Now one of the summands must be even (for the sum of two odds is even); as $p = 2$ is the only even prime, it follows that $q = 25$, which is not prime.)

No one has ever found a counterexample to Goldbach's Conjecture, but neither has anyone ever proved it. At present, the conjecture has been verified for all even numbers $m < 10^{13}$ by H. J. J. te Riele and J.-M. Deshouillers. It has been proved by J.-R. Chen (with a simplification by P. M. Ross) that every sufficiently large even number m can be written as $p + q$, where p is prime and q is "almost" a prime; that is, q is either a prime or a product of two primes. Even with all of this positive evidence, however, no mathematician will say that Goldbach's Conjecture must, therefore, be true for all even m.

We have seen what (mathematical) induction is not; let us now discuss what induction is. Suppose one has a list of statements

$$S(1), S(2), \ldots, S(n), \ldots,$$

one for each positive integer n. Having determined that many statements on this list are true, one may guess that every $S(n)$ is true. Induction is a technique of proving that *all* the statements $S(n)$ on the list are, indeed, true. For example, the reader may check that $2^n > n$ for many values of n, but is this inequality true for every positive integer n? We will soon prove, using induction, that this is so.

Our discussion is based on the following property of positive integers (usually called the *Well Ordering Principle*).

Least Integer Axiom. There is a smallest integer in every nonempty collection C of positive integers.

Saying that C is *nonempty* merely means that there is at least one integer in the collection C. Although this axiom cannot be proved (it arises in analyzing what integers are), it is certainly plausible. Consider the following procedure. Check whether 1 belongs to C; if it does, then it is the smallest integer in C. Otherwise, check whether 2 belongs to C; if it does, then 2 is the smallest integer; if not, check 3. Continue this procedure until one bumps into C; this will occur eventually because C is nonempty.

Remark. The Least Integer Axiom holds for the set of nonnegative integers as well as for the set of positive integers: any nonempty collection C of the nonnegative integers contains a smallest integer. If C contains 0, then 0 is the smallest integer in C; otherwise, C is actually a nonempty collection of positive integers, and the original axiom now applies to C. ◄

We begin by recasting the Least Integer Axiom.

Proposition 1.1 (Least Criminal). *Let $S(1), S(2), \ldots, S(n), \ldots$ be statements, one for each integer $n \geq 1$. If some of these statements are false, then there is a first false statement.*

Proof. Let C be the collection of all those positive integers n for which $S(n)$ is false; by hypothesis, C is nonempty. The Least Integer Axiom provides a smallest integer m in C, and $S(m)$ is the first false statement. •

This seemingly innocuous proposition is useful.

Theorem 1.2. *Every integer $n \geq 2$ is either a prime or a product of primes.*

Proof. Were this not so, there would be "criminals," that is, integers $n \geq 2$ neither prime nor a product of primes; a least criminal m is the smallest such integer. Since m is not a prime, it is composite; there is thus a factorization $m = ab$ with $2 \leq a < m$ and $2 \leq b < m$ (since a is an integer, $1 < a$ implies $2 \leq a$). Since m is the least criminal, both a and b are "honest," i.e.,

$$a = pp'p'' \cdots \text{ and } b = qq'q'' \cdots,$$

where the factors p and q are primes. Therefore,

$$m = ab = pp'p'' \cdots qq'q'' \cdots$$

is a product of (at least two) primes, which is a contradiction. •

Proposition 1.3.   *If $m \geq 2$ is a positive integer which is not divisible by any prime p with $p \leq \sqrt{m}$, then m is a prime.*

Proof.[2]   If m is not prime, then $m = ab$, where $a < m$ and $b < m$ are positive integers. If $a > \sqrt{m}$ and $b > \sqrt{m}$, then $m = ab > \sqrt{m}\sqrt{m} = m$, a contradiction. Therefore, we may assume that $a \leq \sqrt{m}$. By Theorem 1.2, a is either a prime or a product of primes, and any (prime) divisor p of a is also a divisor of m. Thus, if m is not prime, then it has a "small" prime divisor p; i.e., $p \leq \sqrt{m}$. The contrapositive says that if m has no small prime divisor, then m is prime.   •

Proposition 1.3 can be used to show that 991 is a prime. It suffices to check whether 991 is divisible by some prime p with $p \leq \sqrt{991} \approx 31.48$; if 991 is not divisible by 2, 3, 5, ..., or 29, then it is prime. There are 10 such primes, and one checks (by long division) that none of them is a divisor of 991. One can check that 1,000,099 is a prime in the same way, but it is a longer enterprise because its square root is a bit over 1000. It is also tedious, but not difficult, to see that the numbers $f(n) = n^2 - n + 41$, for $1 \leq n \leq 40$, are all prime.

Mathematical induction is a version of least criminal that is more convenient to use. The key idea is just this. Imagine a stairway to the sky. If its first step is white and if the next step above a white step is also white, then all the steps of the stairway must be white. (One can trace this idea back to Levi ben Gershon in 1321. There is an explicit description of induction, cited by Pascal, written by Francesco Maurolico in 1557.) For example, the statement "$2^n > n$ for all $n \geq 1$" can be regarded as an infinite sequence of statements (a stairway to the sky):

$$2^1 > 1; \ 2^2 > 2; \ 2^3 > 3; \ 2^4 > 4; \ 2^5 > 5; \cdots.$$

Certainly, $2^1 = 2 > 1$. If $2^{100} > 100$, then $2^{101} = 2 \times 2^{100} > 2 \times 100 = 100 + 100 > 101$. There is nothing magic about the exponent 100; once we have reached any stair, we can climb up to the next one. This argument will be formalized in Proposition 1.6.

---

[2]The ***contrapositive*** of an implication "P implies Q" is the implication "(not Q) implies (not P)." For example, the contrapositive of "If a series $\sum a_n$ converges, then $\lim_{n\to\infty} a_n = 0$" is "If $\lim_{n\to\infty} a_n \neq 0$, then $\sum a_n$ diverges." If an implication is true, then so is its contrapositive; conversely, if the contrapositive is true, then so is the original implication. The strategy of this proof is to prove the contrapositive of the original implication. Although a statement and its contrapositive are logically equivalent, it is sometimes more convenient to prove the contrapositive. This method is also called ***indirect proof*** or ***proof by contradiction***.

Theorem 1.4 (Mathematical Induction).   *Given statements* $S(n)$, *one for each integer* $n \geq 1$, *suppose that*

(i) $S(1)$ *is true, and*

(ii) *if* $S(n)$ *is true, then* $S(n + 1)$ *is true.*

Then $S(n)$ *is true for all integers* $n \geq 1$.

Proof.   We must show that the collection C of all those positive integers n for which the statement $S(n)$ is false is empty.

If, on the contrary, C is nonempty, then there is a first false statement $S(m)$. Since $S(1)$ is true, by (i), we must have $m \geq 2$. This implies that $m - 1 \geq 1$, and so there is a statement $S(m - 1)$ [there is no statement $S(0)$]. As m is the least criminal, $m - 1$ must be honest; that is, $S(m - 1)$ is true. But now (ii) says that $S(m) = S([m - 1] + 1)$ is true, and this is a contradiction. We conclude that C is empty and, hence, that all the statements $S(n)$ are true. $\bullet$

Before we illustrate how to use (mathematical) induction, let us make sure that we can manipulate inequalities. We know that if two real numbers a and b are both positive, then ab, $a + b$, and $1/a$ are also positive. On the other hand, the product of a positive number and a negative number is negative. (In Chapter 3, we will see why the product of two negatives is positive.)

Definition.   For any two real numbers a and A, define

$$a < A \quad (\text{ also written } A > a)$$

to mean that $A - a$ is positive. We write $a \leq A$ to mean either $a < A$ or $a = A$.

Notice that if $a > b$ and $b > c$, then $a > c$ [for $a - c = (a - b) + (b - c)$ is a sum of positive numbers and, hence, is itself positive]. One often abbreviates these two inequalities as $a > b > c$. The reader may check that if $a > b \geq c$, then $a > c$.

Proposition 1.5.   *Assume that* $b < B$ *are real numbers.*

(i) *If* m *is positive, then* $mb < mB$; *if* m *is negative, then* $mb > mB$.

(ii) *For any number* N, *positive, negative, or zero, we have*

$$N + b < N + B \quad and \quad N - b > N - B.$$

(iii) *Let* a *and* A *be positive numbers. If* $a < A$, *then* $1/a > 1/A$, *and, conversely, if* $1/A < 1/a$, *then* $A > a$.

Proof.   (i) By hypothesis, $B - b > 0$. If $m > 0$, then the product of positive numbers being positive implies that $m(B - b) = mB - mb$ is positive; that is, $mb < mB$. If $m < 0$, then the product $m(B - b) = mB - mb$ is negative; that is, $mB < mb$.

(ii) The difference $(N + B) - (N + b)$ is positive, for it equals $B - b$. For the other inequality, $(N - b) - (N - B) = -b + B$ is positive, and, hence, $N - b > N - B$.

(iii) If $a < A$, then $A - a$ is positive. Hence, $1/a - 1/A = (A - a)/Aa$ is positive, being the product of the positive numbers $A - a$ and $1/Aa$ (by hypothesis, both A and a are positive). Therefore, $1/a > 1/A$. Conversely, if $1/A < 1/a$, then part (i) gives $a = Aa(1/A) < Aa(1/a) = A$; that is, $A > a$.   •

We now show how to use induction.

Proposition 1.6. $2^n > n$ *for all integers* $n \geq 1$.

Proof.   The nth statement $S(n)$ is

$$S(n) : 2^n > n.$$

Two steps are required for induction, corresponding to the two hypotheses in Theorem 1.4.

   Base step. The initial statement

$$S(1) : 2^1 > 1$$

is true, for $2^1 = 2 > 1$.

   Inductive step. If $S(n)$ were true, then $S(n + 1)$ would also be true; that is, using the ***inductive hypothesis*** $S(n)$, we must prove

$$S(n + 1) : 2^{n+1} > n + 1.$$

If $2^n > n$ were true, then multiplying both sides of its inequality by 2 would give, by Proposition 1.5, the valid inequality:

$$2^{n+1} = 2 \times 2^n > 2n.$$

Now $2n = n + n \geq n + 1$ (because $n \geq 1$), and hence $2^{n+1} > 2n \geq n + 1$, as desired.

   Having verified both the base step and the inductive step, we conclude that $2^n > n$ for all $n \geq 1$.   •

Induction is plausible in the same sense that the Least Integer Axiom is plausible. Suppose that a given list $S(1)$, $S(2)$, $S(3)$, ... of statements has the property that $S(n + 1)$ is true whenever $S(n)$ is true. If, in addition, $S(1)$ is true, then $S(2)$ is true; the truth of $S(2)$ now gives the truth of $S(3)$; the truth of $S(3)$ now gives the truth of $S(4)$; and so forth. Induction replaces the phrase *and so forth* by the inductive step; this guarantees, for every n, that there is never an obstruction in the passage from any statement $S(n)$ to the next one $S(n + 1)$.

Here are two comments before we give more illustrations of induction. First, one must verify both the base step and the inductive step; verification of only one of them is inadequate. For example, consider the statements $S(n) : n^2 = n$. The base step is true, but one cannot prove the inductive step (of course, these statements are false for all $n > 1$). Another example is given by the statements $S(n) : n = n + 1$. It is easy to see that the inductive step is true: if $n = n + 1$, then Proposition 1.5(ii) says that adding 1 to both sides gives $n + 1 = (n + 1) + 1 = n + 2$, which is the next statement $S(n + 1)$. But the base step is false (of course, all these statements are false).

Second, when first seeing induction, many people suspect that the inductive step is circular reasoning: one is using $S(n)$, and this is what one wants to prove! A closer analysis shows that this is not at all what is happening. The inductive step, by itself, does not prove that $S(n + 1)$ is true. Rather, it says that *if* $S(n)$ is true, *then* $S(n + 1)$ is also true. In other words, the inductive step proves that the *implication* "If $S(n)$ is true, then $S(n + 1)$ is true" is correct. The truth of this implication is not the same thing as the truth of its conclusion. For example, consider the two statements: "Your grade on every exam is 100%" and "Your grade in the course is A." The implication "If all your exams are perfect, then you will get the highest grade for the course" is true. Unfortunately, this does not say that it is inevitable that your grade in the course will be A. Our discussion above gives a mathematical example: the implication "If $n = n + 1$, then $n + 1 = n + 2$" is correct, but the conclusion "$n + 1 = n + 2$" is false.

Proposition 1.7. $1 + 2 + \cdots + n = \frac{1}{2}n(n + 1)$ *for every integer* $n \geq 1$.

Proof.   The proof is by induction.

  Base step. If $n = 1$, then the left side is 1 and the right side is $\frac{1}{2}1(1 + 1) = 1$, as desired.

  Inductive step. It is always a good idea to write the $(n + 1)$st statement

$S(n + 1)$ so one can see what has to be proved. We must prove

$$S(n + 1): \quad 1 + 2 + \cdots + n + (n + 1) = \tfrac{1}{2}(n + 1)(n + 2).$$

By the inductive hypothesis, i.e., using $S(n)$, the left side is

$$[1 + 2 + \cdots + n] + (n + 1) = \tfrac{1}{2}n(n + 1) + (n + 1),$$

and high school algebra shows that $\tfrac{1}{2}n(n + 1) + (n + 1) = \tfrac{1}{2}(n + 1)(n + 2)$. By induction,[3] the formula holds for all $n \geq 1$. •

There is a story told about Gauss as a boy. One of his teachers asked the students to add up all the numbers from 1 to 100, thereby hoping to get some time for himself for other tasks. But Gauss quickly volunteered that the answer was 5050. Here is what he did (without induction). Let s denote the sum of all the numbers from 1 to 100; $s = 1 + 2 + \cdots + 99 + 100$. Of course, $s = 100 + 99 + \cdots + 2 + 1$. Arrange these nicely:

$$
\begin{aligned}
s &= 1 + 2 + \cdots + 99 + 100 \\
s &= 100 + 99 + \cdots + 2 + 1,
\end{aligned}
$$

add:

$$2s = 101 + 101 + \cdots + 101 + 101 \quad (100 \text{ times}),$$

and solve: $s = \tfrac{1}{2}(101 \times 100) = 5050$. The same argument works for any number n in place of 100. Not only does this give a new proof of Proposition 1.7, it also shows how the formula could have been discovered.[4]

It is not always the case, in an inductive proof, that the base step is very simple. In fact, all possibilities can occur: both steps are easy; both are difficult; one is harder than the other.

---

[3] *Induction*, having a Latin root meaning "to lead," came to mean "prevailing upon to do something" or "influencing." This is an apt name here, for the nth statement influences the $(n + 1)$st one.

[4] Actually, this formula goes back at least a thousand years; Alhazen (Ibn al-Haytham) (965-1039), found a geometric way to add

$$1^k + 2^k + \cdots + n^k$$

for any fixed integer $k \geq 1$ [see Exercise 1.10(ii)].

Proposition 1.8.   *If one assumes the product rule for derivatives, $D(fg) = (Df)g + f(Dg)$, then*

$$D(x^n) = nx^{n-1} \text{ for all integers } n \geq 1.$$

Proof.   We proceed by induction.

Base step. If $n = 1$, then we ask whether $D(x) = x^0 = 1$. Now

$$D(f(x)) = \lim_{h \to 0}(1/h)[f(x+h) - f(x)].$$

When $f(x) = x$, therefore,

$$D(x) = \lim_{h \to 0}(1/h)[x + h - x] = \lim_{h \to 0} h/h = 1.$$

Inductive step. We must prove that $D(x^{n+1}) = (n+1)x^n$. It is permissible to use the inductive hypothesis, $D(x^n) = nx^{n-1}$, as well as $D(x) = 1$, for the base step has already been proved. Since $x^{n+1} = x^n x$, the product rule gives

$$D(x^{n+1}) = D(x^n x) = D(x^n)x + x^n D(x)$$
$$= x(nx^{n-1}) + x^n 1 = (n+1)x^n.$$

We conclude that $D(x^n) = nx^{n-1}$ is true for all $n \geq 1$.   •

The base step of an induction may be an integer other than 1. For example, consider the statements

$$S(n) : 2^n > n^2.$$

This is not true for small values of n: if $n = 2$ or 4, then there is equality, not inequality; if $n = 3$, the left side, 8, is smaller than the right side, 9. However, $S(5)$ is true, for $32 > 25$.

Proposition 1.9. $2^n > n^2$ *is true for all integers $n \geq 5$.*

Proof.   We have just checked the base step $S(5)$. In proving

$$S(n+1) : 2^{n+1} > (n+1)^2,$$

we are allowed to assume that $n \geq 5$ (actually, we will need only $n \geq 3$) as well as the inductive hypothesis. Multiply both sides of $2^n > n^2$ by 2 to get

$$2^{n+1} = 2 \times 2^n > 2n^2 = n^2 + n^2 = n^2 + nn.$$

Since $n \geq 5$, we have $n \geq 3$, and so

$$nn \geq 3n = 2n + n \geq 2n + 1.$$

Therefore,

$$2^{n+1} > n^2 + nn \geq n^2 + 2n + 1 = (n + 1)^2. \quad \bullet$$

We have seen that the base step of an induction can begin at $n = 1$ or $n = 5$. Indeed, the base step of an induction can begin at any integer k; of course, the conclusion is that the statements are true for all $n \geq k$. Assuming that there is a statement $S(0)$, one may also start an induction with base step $n = 0$ (we remarked just before Proposition 1.1 that the Least Integer Axiom also applies to nonempty subsets of the nonnegative integers); occasionally, one even runs across inductions with base step $n = -1$.

There is another version of induction, usually called the *second form of induction*, that is often convenient to use.

Definition.   The ***predecessors*** of an integer $n \geq 2$ are the positive integers k with $k < n$, namely, $1, 2, \ldots, n - 1$.

Theorem 1.10 (Second Form of Induction).   *Let $S(n)$ be a family of statements, one for each integer $n \geq 1$, and suppose that*

(i) *$S(1)$ is true, and*

(ii) *if $S(k)$ is true for all predecessors k of n, then $S(n)$ is itself true.*

Then $S(n)$ is true for all integers $n \geq 1$.

Proof.   It suffices to show that there are no integers n for which $S(n)$ is false; that is, the collection C of all positive integers n for which $S(n)$ is false is empty.

If, on the contrary, C is nonempty, then there is a least criminal m; that is, there is a first false statement $S(m)$. Since $S(1)$ is true, by (i), we must have $m \geq 2$. As m is the *least* criminal, k must be honest for all $k < m$; that is, $S(k)$ is true for all the predecessors of m. By (ii), $S(m)$ is true, and this is a contradiction. We conclude that C is empty and, hence, that all the statements $S(n)$ are true. $\bullet$

The second form of induction can be used to give a second proof of Theorem 1.2 (as with the first form, the base step need not occur at 1).

Theorem 1.11 (= Theorem 1.2). *Every integer $n \geq 2$ is either a prime or a product of primes.*

Proof.[5]   Base step. The statement is true when $n = 2$ because 2 is a prime.

   Inductive step. If $n \geq 2$ is a prime, we are done. Otherwise, $n = ab$, where $2 \leq a < n$ and $2 \leq b < n$. As a and b are predecessors of n, each of them is either prime or a product of primes:

$$a = pp'p'' \cdots \quad \text{and} \quad b = qq'q'' \cdots,$$

and so $n = pp'p'' \cdots qq'q'' \cdots$ is a product of (at least two) primes.   •

   The reason the second form of induction is more convenient here is that it is more natural to use $S(a)$ and $S(b)$ than to use $S(n-1)$; indeed, it is not at all clear how to use $S(n-1)$.

   Here is a notational remark. When using the second form of induction, we speak of n and its predecessors, not $n + 1$ and its predecessors. If one wants to compare the two forms of induction, one can say that the first form uses $S(n-1)$ to prove $S(n)$, whereas the second form uses any or all of the earlier statements $S(1), S(2), \dots, S(n-1)$ to prove $S(n)$.

   The next result says that one can always factor out a largest power of 2 from any integer.

Proposition 1.12. *Every integer $n \geq 1$ has a unique factorization $n = 2^k m$, where $k \geq 0$ and $m \geq 1$ is odd.*

Proof.   We use the second form of induction on $n \geq 1$ to prove the existence of k and m.

   Base step. If $n = 1$, take $k = 0$ and $m = 1$.

   Inductive step. If $n \geq 1$, then n is either odd or even. If n is odd, then take $k = 0$ and $m = n$. If n is even, then $n = 2b$. Because $b < n$, it is a predecessor of n, and so the inductive hypothesis allows us to assume $S(b) : b = 2^\ell m$, where $\ell \geq 0$ and m is odd. The desired factorization is $n = 2b = 2^{\ell+1} m$.

   The word *unique* means "exactly one." We prove uniqueness by showing that if $n = 2^k m = 2^t m'$, where both k and t are nonnegative and both m and m' are odd, then $k = t$ and $m = m'$. We may assume that $k \geq t$. If $k > t$, then canceling 2^t from both sides gives $2^{k-t} m = m'$. Since $k - t > 0$, the left side is even while the right side is odd; this contradiction shows that $k = t$. We may thus cancel 2^k from both sides, leaving $m = m'$.   •

---

[5]The similarity of the proofs of Theorems 1.2 and 1.11 indicates that the second form of induction is merely a variation of least criminal.

Definition.   The *Fibonacci sequence* $F_0, F_1, F_2, \ldots$ is defined as follows:

$$F_0 = 0, \quad F_1 = 1, \quad \text{and} \quad F_n = F_{n-1} + F_{n-2} \quad \text{for all integers } n \geq 2.$$

The Fibonacci sequence begins: $0, 1, 1, 2, 3, 5, 8, 13, \ldots$. There is an intimate connection between the Fibonacci sequence and the **golden ratio** $\alpha = \frac{1}{2}(1 + \sqrt{5})$ (which one finds when studying *recurrence relations* in general). Notice that α is a root of $x^2 - x - 1$ [as is $\beta = \frac{1}{2}(1 - \sqrt{5})$]. The ancient Greeks thought a rectangular figure most pleasing if its edges a and b were in the proportion

$$a : b = b : a + b.$$

It follows that $b^2 = a(a + b)$, so that $b^2 - ab - a^2 = 0$, and the quadratic formula gives $b = \frac{1}{2}\left(a \pm \sqrt{a^2 + 4a^2}\right) = a\frac{1}{2}(1 \pm \sqrt{5})$. Therefore,

$$b/a = \alpha \quad \text{or} \quad b/a = \beta.$$

Theorem 1.13.   *If F_n denotes the nth term of the Fibonacci sequence, then for all $n \geq 0$,*

$$F_n = \tfrac{1}{\sqrt{5}}(\alpha^n - \beta^n),$$

where $\alpha = \frac{1}{2}(1 + \sqrt{5})$ and $\beta = \frac{1}{2}(1 - \sqrt{5})$.

Proof.   We are going to use the second form of induction [the second form is the appropriate induction here, for the equation $F_n = F_{n-1} + F_{n-2}$ suggests that proving $S(n)$ will involve not only $S(n - 1)$ but $S(n - 2)$ as well].

Base step.   The formula is true for $n = 0 : \frac{1}{\sqrt{5}}(\alpha^0 - \beta^0) = 0 = F_0$. The formula is also true for $n = 1$:

$$\begin{aligned}
\tfrac{1}{\sqrt{5}}(\alpha^1 - \beta^1) &= \tfrac{1}{\sqrt{5}}(\alpha - \beta) \\
&= \tfrac{1}{\sqrt{5}}\left[\tfrac{1}{2}(1 + \sqrt{5}) - \tfrac{1}{2}(1 - \sqrt{5})\right] \\
&= \tfrac{1}{\sqrt{5}}(\sqrt{5}) = 1 = F_1.
\end{aligned}$$

(We have mentioned both $n = 0$ and $n = 1$ because the inductive step will use two predecessors.)

Inductive step. If $n \geq 2$, then

$$
\begin{aligned}
F_n &= F_{n-1} + F_{n-2} \\
&= \tfrac{1}{\sqrt{5}}(\alpha^{n-1} - \beta^{n-1}) + \tfrac{1}{\sqrt{5}}(\alpha^{n-2} - \beta^{n-2}) \\
&= \tfrac{1}{\sqrt{5}}\left[(\alpha^{n-1} + \alpha^{n-2}) - (\beta^{n-1} + \beta^{n-2})\right] \\
&= \tfrac{1}{\sqrt{5}}\left[\alpha^{n-2}(\alpha + 1) - \beta^{n-2}(\beta + 1)\right] \\
&= \tfrac{1}{\sqrt{5}}\left[\alpha^{n-2}(\alpha^2) - \beta^{n-2}(\beta^2)\right] \\
&= \tfrac{1}{\sqrt{5}}(\alpha^n - \beta^n),
\end{aligned}
$$

because $\alpha + 1 = \alpha^2$ and $\beta + 1 = \beta^2$.   •

It is curious that the integers F_n are expressed in terms of the irrational number $\sqrt{5}$.

Corollary 1.14.   *If* $\alpha = \tfrac{1}{2}\left(1 + \sqrt{5}\right)$, *then*

$$
F_n > \alpha^{n-2}
$$

for all integers $n \geq 3$.

Remark.   If $n = 2$, then $F_2 = 1 = \alpha^0$, and so there is equality, not inequality.   ◄

Proof.   *Base step.* If $n = 3$, then $F_3 = 2 > \alpha$, for $\alpha \approx 1.618$.

    Inductive step.   We must show that $F_{n+1} > \alpha^{n-1}$. By the inductive hypothesis,

$$
\begin{aligned}
F_{n+1} = F_n + F_{n-1} &> \alpha^{n-2} + \alpha^{n-3} \\
&= \alpha^{n-3}(\alpha + 1) = \alpha^{n-3}\alpha^2 = \alpha^{n-1}. \quad •
\end{aligned}
$$

One can also use induction to give definitions. For example, we can define ***n factorial***,[6] denoted by $n!$, by induction on $n \geq 0$. Define $0! = 1$, and if $n!$ is known, then define

$$
(n + 1)! = n!(n + 1).
$$

One reason for defining $0! = 1$ will be apparent in the next section.

---

[6]The term *factor* comes from the Latin "to make" or "to contribute"; the term *factorial* recalls that $n!$ has many factors.

EXERCISES

1.1 Find a formula for $1 + 3 + 5 + \cdots + (2n - 1)$, and use mathematical induction to prove that your formula is correct. (Inductive reasoning is used in mathematics to help guess what might be true. Once a guess has been made, it must still be proved, perhaps using mathematical induction, perhaps by some other method.)

1.2 Find a formula for $1 + \sum_{j=1}^{n} j! \, j$, and use induction to prove that your formula is correct.

1.3 (i) For any $n \geq 0$ and any $r \neq 1$, prove that

$$1 + r + r^2 + r^3 + \cdots + r^n = (1 - r^{n+1})/(1 - r).$$

(ii) Prove that

$$1 + 2 + 2^2 + \cdots + 2^n = 2^{n+1} - 1.$$

1.4 Show, for all $n \geq 1$, that 10^n leaves remainder 1 after dividing by 9.

1.5 Prove that if $0 \leq a \leq b$, then $a^n \leq b^n$ for all $n \geq 0$.

1.6 Prove that $1^2 + 2^2 + \cdots + n^2 = \frac{1}{6}n(n + 1)(2n + 1) = \frac{1}{3}n^3 + \frac{1}{2}n^2 + \frac{1}{6}n$.

1.7 Prove that $1^3 + 2^3 + \cdots + n^3 = \frac{1}{4}n^4 + \frac{1}{2}n^3 + \frac{1}{4}n^2$.

1.8 Prove that $1^4 + 2^4 + \cdots + n^4 = \frac{1}{5}n^5 + \frac{1}{2}n^4 + \frac{1}{3}n^3 - \frac{1}{30}n$.

1.9 Derive the formula for $\sum_{i=1}^{n} i$ by computing the area $(n + 1)^2$ of a square with sides of length $n + 1$ using Figure 1.1.

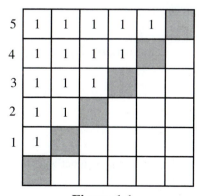

Figure 1.1

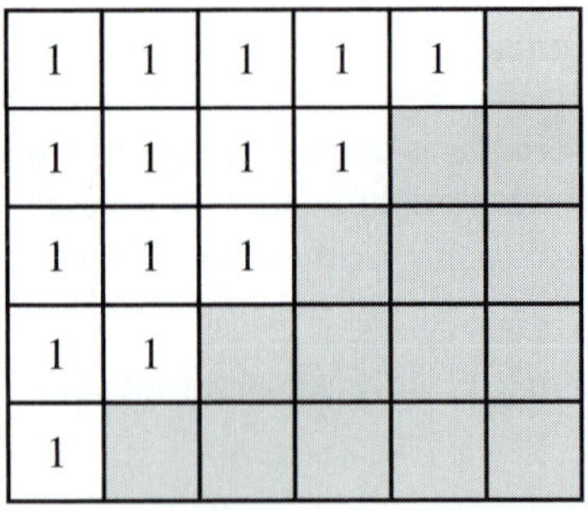

Figure 1.2

1.10   (i) Derive the formula for $\sum_{i=1}^{n} i$ by computing the area $n(n+1)$ of a rectangle with height $n+1$ and base n, as pictured in Figure 1.2.

(ii) (**Alhazen**) For fixed $k \geq 1$, use Figure 1.3 to prove

$$(n+1)\sum_{i=1}^{n} i^k = \sum_{i=1}^{n} i^{k+1} + \sum_{i=1}^{n}\left(\sum_{p=1}^{i} p^k\right).$$

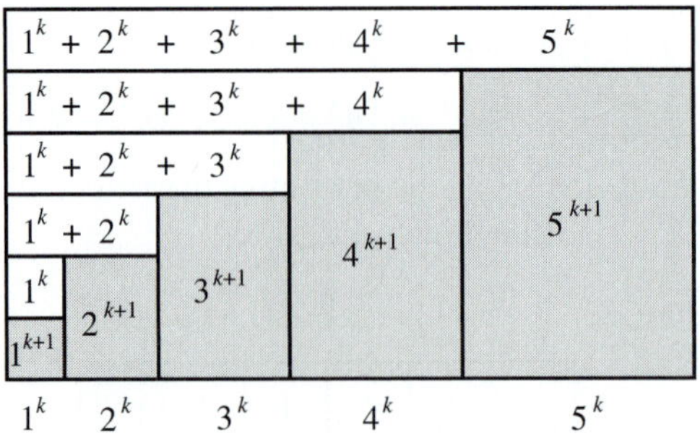

Figure 1.3

(iii) Given the formula $\sum_{i=1}^{n} i = \frac{1}{2}n(n+1)$, use part (ii) to derive the formula for $\sum_{i=1}^{n} i^2$.

1.11   (i) Prove that $2^n > n^3$ for all $n \geq 10$.

(ii) Prove that $2^n > n^4$ for all $n \geq 17$.

1.12 Let $g_1(x), \ldots, g_n(x)$ be differentiable functions, and let $f(x)$ be their product: $f(x) = g_1(x)\cdots g_n(x)$. Prove, for all integers $n \geq 2$, that

$$Df(x) = \sum_{i=1}^{n} g_1(x)\cdots g_{i-1}(x)Dg_i(x)g_{i+1}(x)\cdots g_n(x).$$

1.13 Prove that $(1 + x)^n \geq 1 + nx$ if $1 + x > 0$.

1.14 Prove that every positive integer a has a unique factorization $a = 3^k m$, where $k \geq 0$ and m is not a multiple of 3.

1.15 Prove that $F_n < 2^n$ for all $n \geq 0$, where $F_0, F_1, F_2, \ldots$ is the Fibonacci sequence.

1.16 If F_n denotes the nth term of the Fibonacci sequence, prove that

$$\sum_{n=1}^{m} F_n = F_{m+2} - 1.$$

1.17 Prove that $4^{n+1} + 5^{2n-1}$ is divisible by 21 for all $n \geq 1$.

1.18 For any integer $n \geq 2$, prove that there are n consecutive composite numbers. Conclude that the gap between consecutive primes can be arbitrarily large.

1.19 (**Double Induction**) Let $S(m, n)$ be a doubly indexed family of statements, one for each $m \geq 1$ and $n \geq 1$. Suppose that

  (i) $S(1, 1)$ is true;
  (ii) if $S(m, 1)$ is true, then $S(m + 1, 1)$ is true;
  (iii) if $S(m, n)$ is true for all m, then $S(m, n + 1)$ is true for all m.

Prove that $S(m, n)$ is true for all $m \geq 1$ and $n \geq 1$.

1.20 Use double induction to prove that

$$(m + 1)^n > mn$$

for all $m, n \geq 1$.

1.2 BINOMIAL COEFFICIENTS

What is the pattern of the coefficients in the formulas for the powers $(1 + x)^n$ of the binomial $1 + x$? The first few such formulas are:

$$(1 + x)^0 = 1$$
$$(1 + x)^1 = 1 + 1x$$
$$(1 + x)^2 = 1 + 2x + 1x^2$$
$$(1 + x)^3 = 1 + 3x + 3x^2 + 1x^3$$
$$(1 + x)^4 = 1 + 4x + 6x^2 + 4x^3 + 1x^5.$$

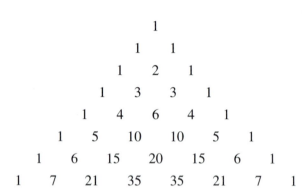

Figure 1.4

Figure 1.4, called ***Pascal's triangle***, after B. Pascal (1623–1662), displays an arrangement of the first few coefficients. Figure 1.5, a picture from China in the year 1303, shows that the pattern of coefficients had been recognized long before Pascal was born.

The expansion of $(1 + x)^n$ is an expression of the form

$$c_0 + c_1 x + c_2 x^2 + \cdots + c_n x^n.$$

The coefficients c_r are called ***binomial coefficients***;[7] L. Euler (1707–1783) introduced the notation $\left(\frac{n}{r}\right)$ for them; this symbol evolved into $\binom{n}{r}$, which is generally accepted nowadays:

$$\binom{n}{r} = \text{coefficient } c_r \text{ of } x^r \text{ in } (1 + x)^n.$$

Hence,

$$(1 + x)^n = \sum_{r=0}^{n} \binom{n}{r} x^r.$$

The number $\binom{n}{r}$ is pronounced "*n* choose *r*" because it also arises in counting problems, as we shall see later in this section.

---

[7]*Binomial*, coming from the Latin *bi*, meaning "two," and *nomen*, meaning "name" or "term," describes expressions of the form $a + b$. Similarly, *trinomial* describes expressions of the form $a + b + c$, and *monomial* describes expressions with a single term. The word *polynomial* is a hybrid, coming from the Greek *poly* meaning "many" and the Latin *nomen*; polynomials are certain expressions having many terms.

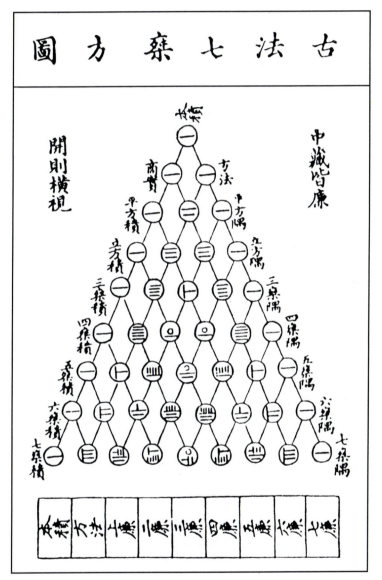

Figure 1.5

Observe, in Figure 1.4, that an inside number (i.e., not a 1 on the border) of the $(n + 1)$th row can be computed by going up to the nth row and adding the two neighboring numbers above it. For example, the inside numbers in row 4 can be computed from row 3 as follows:

$$1 \quad 3 \quad 3 \quad 1$$
$$1 \quad 4 \quad 6 \quad 4 \quad 1$$

$4 = 1 + 3, 6 = 3 + 3$, and $4 = 3 + 1$. Let us prove that this observation always holds.

Lemma 1.15. *For all integers $n \geq 1$ and r, where $0 < r < n + 1$,*

$$\binom{n + 1}{r} = \binom{n}{r - 1} + \binom{n}{r}.$$

Proof. We prove the lemma by induction on $n \geq 1$.

Base step. If $n = 1$, then

$$2 = \binom{2}{1} = 1 + 1 = \binom{1}{0} + \binom{1}{1}.$$

Inductive Step. We must show that if

$$(1 + x)^n = c_0 + c_1 x + c_2 x^2 + \cdots + c_n x^n,$$

then the coefficient of x^r in $(1 + x)^{n+1}$ is $c_{r-1} + c_r$. Since $c_0 = 1$,

$$\begin{aligned}
(1 + x)^{n+1} &= (1 + x)(1 + x)^n \\
&= (1 + x)^n + x(1 + x)^n \\
&= (c_0 + c_1 x + c_2 x^2 + \cdots + c_n x^n) \\
&\quad + x(c_0 + c_1 x + c_2 x^2 + \cdots + c_n x^n) \\
&= (c_0 + c_1 x + c_2 x^2 + \cdots + c_n x^n) \\
&\quad + c_0 x + c_1 x^2 + c_2 x^3 + \cdots + c_n x^{n+1} \\
&= 1 + (c_0 + c_1)x + (c_1 + c_2)x^2 + (c_2 + c_3)x^3 + \cdots .
\end{aligned}$$

Thus $\binom{n+1}{r}$, the coefficient of x^r in $(1 + x)^{n+1}$, is

$$c_{r-1} + c_r = \binom{n}{r - 1} + \binom{n}{r}. \quad \bullet$$

Proposition 1.16 (Pascal). *For all $n \geq 0$ and all r with $0 \leq r \leq n$,*

$$\binom{n}{r} = \frac{n!}{r!(n-r)!}.$$

Proof. We prove the theorem by induction on $n \geq 0$.

    Base step.[8] If $n = 0$, then

$$\binom{0}{0} = 0!/0!0! = 1.$$

Inductive step. We must prove

$$\binom{n+1}{r} = \frac{(n+1)!}{r!(n+1-r)!}.$$

If $r = 0$, then $\binom{n+1}{0} = 1 = (n+1)!/0!(n+1-0)!$; if $r = n+1$, then $\binom{n+1}{n+1} = 1 = (n+1)!/(n+1)!0!$; if $0 < r < n+1$, we use Lemma 1.15:

$$\binom{n+1}{r} = \binom{n}{r-1} + \binom{n}{r}$$

$$= \frac{n!}{(r-1)!(n-r+1)!} + \frac{n!}{r!(n-r)!}$$

$$= \frac{n!}{(r-1)!(n-r)!}\left(\frac{1}{(n-r+1)} + \frac{1}{r}\right)$$

$$= \frac{n!}{(r-1)!(n-r)!}\left(\frac{r+n-r+1}{r(n-r+1)}\right)$$

$$= \frac{n!}{(r-1)!(n-r)!}\left(\frac{n+1}{r(n-r+1)}\right)$$

$$= \frac{(n+1)!}{r!(n+1-r)!}. \quad \bullet$$

Corollary 1.17. *For any number x and for all integers $n \geq 0$,*

$$(1+x)^n = \sum_{r=0}^{n}\binom{n}{r}x^r = \sum_{r=0}^{n}\frac{n!}{r!(n-r)!}x^r.$$

Proof. The first equation is the definition of the binomial coefficients, and the second equation replaces $\binom{n}{r}$ by the value given in Pascal's theorem. $\bullet$

---

[8]This is one reason why 0! is defined to be 1.

Corollary 1.18 (Binomial Theorem). *For all numbers a and b and for all integers $n \geq 1$,*

$$(a+b)^n = \sum_{r=0}^{n} \binom{n}{r} a^{n-r} b^r = \sum_{r=0}^{n} (n!/r!(n-r)!)\, a^{n-r} b^r.$$

Proof. The result is trivially true when $a = 0$ (if we agree that $0^0 = 1$). If $a \neq 0$, set $x = b/a$ in Corollary 1.17, and observe that

$$\left(1 + \frac{b}{a}\right)^n = \left(\frac{a+b}{a}\right)^n = \frac{(a+b)^n}{a^n}.$$

Therefore,

$$(a+b)^n = a^n \left(1 + \frac{b}{a}\right)^n = a^n \sum_{r=0}^{n} \binom{n}{r} \frac{b^r}{a^r} = \sum_{r=0}^{n} \binom{n}{r} a^{n-r} b^r. \quad \bullet$$

Remark. The binomial theorem can be proved without first proving Corollary 1.17; just prove the formula for $(a+b)^n$ by induction on $n \geq 0$. We have chosen the proof above for clearer exposition. ◀

Here is a combinatorial interpretation of the binomial coefficients. Given a set X, an **r-subset** is a subset of X with exactly r elements. If X has n elements, denote the number of its r-subsets by

$$[n, r];$$

that is, $[n, r]$ is the number of ways one can choose r things from a box of n things.

We compute $[n, r]$ by considering a related question. Given an "alphabet" with n (distinct) letters and a number r with $1 \leq r \leq n$, an **r-anagram** is a sequence of r of the letters with no repetitions. For example, the 2-anagrams on the alphabet a, b, c are

$$ab, ba, ac, ca, bc, cb$$

(note that aa, bb, cc are not on this list). How many r-anagrams are there on an alphabet with n letters? We count the number of such anagrams in two ways.

(1) There are n choices for the first letter; since no letter is repeated, there are only $n - 1$ choices for the second letter, only $n - 2$ choices for the third letter, and so forth. Thus, the number of r-anagrams is

$$n(n - 1)(n - 2) \cdots (n - [r - 1]) = n(n - 1)(n - 2) \cdots (n - r + 1).$$

Note the special case $n = r$: the number of n-anagrams on n letters is $n!$.

(2) Here is a second way to count these anagrams. First choose an r-subset of the alphabet (consisting of r letters); there are $[n, r]$ ways to do this, for this is exactly what the symbol $[n, r]$ means. For each chosen r-subset, there are $r!$ ways to arrange the r letters in it (this is the special case of (1) when $n = r$). The number of r-anagrams is thus

$$r![n, r].$$

We conclude that

$$r![n, r] = n(n - 1)(n - 2) \cdots (n - r + 1),$$

from which it follows, by Pascal's formula, that

$$[n, r] = n(n - 1)(n - 2) \cdots (n - r + 1)/r! = \binom{n}{r}.$$

This fact is the reason one often pronounces the binomial coefficient $\binom{n}{r}$ as n choose r.

As an example, how many ways are there to choose 2 hats from a closet containing 14 different hats? (One of my friends does not like the phrasing of this question. After all, one can choose 2 hats with one's left hand, with one's right hand, with one's teeth, ... ; but I continue the evil tradition.) The answer is $\binom{14}{2}$, and Pascal's formula allows us to compute this as $(14 \times 13)/2 = 91$.

Our first interpretation of the binomial coefficients $\binom{n}{r}$ was *algebraic*; that is, as coefficients of polynomials which can be calculated by Pascal's formula; our second interpretation is *combinatorial*; that is, as n choose r. Quite often, each interpretation can be used to prove a desired result. For example, here is a combinatorial proof of Lemma 1.15. Let X be a set with $n + 1$ elements, and let us color one of its elements red and the other n elements blue. Now $\binom{n+1}{r}$ is the number of r-subsets of X. There are two possibilities for an r-subset Y: either it contains the red element or it is all blue. If Y contains the red element, then Y consists of the red element and $r - 1$ blue elements, and so the number of such Y is the same as the number of all blue $(r-1)$-subsets, namely, $\binom{n}{r-1}$. The other possibility is that Y is all blue, and there are $\binom{n}{r}$ such r-subsets. Therefore, $\binom{n+1}{r} = \binom{n}{r-1} + \binom{n}{r}$, as desired.

We are now going to apply the binomial theorem to trigonometry. Recall that the **modulus** $|z|$ of a complex number $z = a + ib$ is defined to be

$$|z| = \sqrt{a^2 + b^2}.$$

If we identify a complex number $z = a + ib$ with the point (a, b) in the plane, then its modulus $|z|$ is the distance from z to the origin. It follows that every complex number z of modulus 1 corresponds to a point on the unit circle, and so it has coordinates $(\cos\theta, \sin\theta)$ for some angle θ.

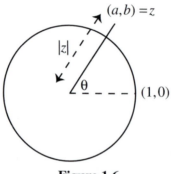

Figure 1.6

Proposition 1.19 (Polar Decomposition).   *Every complex number z has a factorization*

$$z = r(\cos\theta + i\sin\theta),$$

where $r = |z| \geq 0$ and $0 \leq \theta < 2\pi$.

Proof.   If $z = 0$, then $|z| = 0$ and any choice of θ works. If $z = a + bi \neq 0$, then $|z| \neq 0$. Now $z/|z| = (a/|z|, b/|z|)$ has modulus 1, for $(a/|z|)^2 + (b/|z|)^2 = (a^2 + b^2)/|z|^2 = 1$. Therefore, there is an angle θ with

$$\frac{z}{|z|} = \cos\theta + i\sin\theta,$$

and so $z = |z|(\cos\theta + i\sin\theta) = r(\cos\theta + i\sin\theta)$. $\bullet$

If $z = a + ib = r(\cos\theta + i\sin\theta)$, then (r, θ) are the **polar coordinates** of z; this is the reason Proposition 1.19 is called the polar decomposition of z.

The trigonometric addition formulas for $\cos(\theta + \psi)$ and $\sin(\theta + \psi)$ have a lovely translation in the language of complex numbers.

Proposition 1.20 (Addition Theorem).   *If*

$$z = \cos\theta + i\sin\theta \text{ and } w = \cos\psi + i\sin\psi,$$

then

$$zw = \cos(\theta + \psi) + i\sin(\theta + \psi).$$

Proof.

$$zw = (\cos\theta + i\sin\theta)(\cos\psi + i\sin\psi)$$
$$= (\cos\theta\cos\psi - \sin\theta\sin\psi) + i(\sin\theta\cos\psi + \cos\theta\sin\psi).$$

The trigonometric addition formulas show that

$$zw = \cos(\theta + \psi) + i\sin(\theta + \psi). \quad \bullet$$

The addition theorem gives a geometric interpretation of complex multiplication: if $z = r(\cos\theta + i\sin\theta)$ and $w = s(\cos\psi + i\sin\psi)$, then

$$zw = rs[\cos(\theta + \psi) + i\sin(\theta + \psi)],$$

and the polar coordinates of zw are

$$(rs, \theta + \psi).$$

In particular, if $|z| = 1 = |w|$, then $|zw| = 1$; that is, the product of two complex numbers on the unit circle also lies on the unit circle.

Corollary 1.21.   *If z and w are complex numbers, then*

$$|zw| = |z|\,|w|.$$

Proof.   If the polar decompositions of z and w are $z = r(\cos\theta + i\sin\theta)$ and $w = s(\cos\psi + i\sin\psi)$, respectively, then we have just seen that $|z| = r$, $|w| = s$, and $|zw| = rs$. $\quad \bullet$

In 1707, A. De Moivre (1667–1754) proved the following elegant result.

Theorem 1.22 (De Moivre). *For every real number x and every positive integer n,*

$$\cos(nx) + i\sin(nx) = (\cos x + i\sin x)^n.$$

Proof. We prove De Moivre's theorem by induction on $n \geq 1$. The base step $n = 1$ is obviously true. For the inductive step,

$$
\begin{aligned}
(\cos x + i\sin x)^{n+1} &= (\cos x + i\sin x)^n(\cos x + i\sin x) \\
&= (\cos(nx) + i\sin(nx))(\cos x + i\sin x) \\
&\qquad \text{(inductive hypothesis)} \\
&= \cos(nx + x) + i\sin(nx + x) \\
&\qquad \text{(addition formula)} \\
&= \cos([n+1]x) + i\sin([n+1]x). \quad \bullet
\end{aligned}
$$

Example 1.1.
Find the value of $(\cos 3° + i\sin 3°)^{40}$.
    By De Moivre's theorem,

$$(\cos 3° + i\sin 3°)^{40} = \cos 120° + i\sin 120° = -\tfrac{1}{2} + i\tfrac{\sqrt{3}}{2}. \quad \blacktriangleleft$$

Corollary 1.23.

(i) $\cos(2x) = \cos^2 x - \sin^2 x = 2\cos^2 x - 1$
         $\sin(2x) = 2\sin x \cos x$.

(ii) $\cos(3x) = \cos^3 x - 3\cos x \sin^2 x = 4\cos^3 x - 3\cos x$
        $\sin(3x) = 3\cos^2 x \sin x - \sin^3 x = 3\sin x - 4\sin^3 x$.

Proof. (i)

$$
\begin{aligned}
\cos(2x) + i\sin(2x) &= (\cos x + i\sin x)^2 \\
&= \cos^2 x + 2i\sin x \cos x + i^2\sin^2 x \\
&= \cos^2 x - \sin^2 x + i(2\sin x \cos x).
\end{aligned}
$$

Equating real and imaginary parts gives both double angle formulas.

(ii) De Moivre's theorem gives

$$\cos(3x) + i\sin(3x) = (\cos x + i\sin x)^3$$
$$= \cos^3 x + 3i\cos^2 x \sin x + 3i^2 \cos x \sin^2 x + i^3 \sin^3 x$$
$$= \cos^3 x - 3\cos x \sin^2 x + i(3\cos^2 x \sin x - \sin^3 x).$$

Equality of the real parts gives $\cos(3x) = \cos^3 x - 3\cos x \sin^2 x$; the second formula for $\cos(3x)$ follows by replacing $\sin^2 x$ by $1 - \cos^2 x$. Equality of the imaginary parts gives $\sin(3x) = 3\cos^2 x \sin x - \sin^3 x = 3\sin x - 4\sin^3 x$; the second formula arises by replacing $\cos^2 x$ by $1 - \sin^2 x$.   ●

Corollary 1.23 can be generalized. If $f_2(x) = 2x^2 - 1$, then

$$\cos(2x) = 2\cos^2 x - 1 = f_2(\cos x),$$

and if $f_3(x) = 4x^3 - 3x$, then

$$\cos(3x) = 4\cos^3 x - 3\cos x = f_3(\cos x).$$

Proposition 1.24.   *For all $n \geq 1$, there is a polynomial $f_n(x)$ having all coefficients integers such that*

$$\cos(nx) = f_n(\cos x).$$

Proof.   By De Moivre's theorem,

$$\cos(nx) + i\sin(nx) = (\cos x + i\sin x)^n$$
$$= \sum_{r=0}^{n} \binom{n}{r} \cos^{n-r} x \, i^r \sin^r x.$$

The real part of the left side, $\cos(nx)$, must be equal to the real part of the right side. Now i^r is real if and only if r is even, and so

$$\cos(nx) = \sum_{r \text{ even}} \binom{n}{r} \cos^{n-r} x \, i^r \sin^r x.$$

If $r = 2k$, then $i^r = i^{2k} = (-1)^k$, and

$$\cos(nx) = \sum_{k=0} (-1)^k \binom{n}{2k} \cos^{n-2k} x \, \sin^{2k} x.$$

But $\sin^{2k} x = (\sin^2 x)^k = (1 - \cos^2 x)^k$, which is a polynomial in $\cos x$. This completes the proof.   ●

It is not difficult to show that $f_n(x)$ begins with $2^{n-1}x^n$. A sine version of Proposition 1.24 can be found in Exercise 1.27.

We are now going to present a beautiful formula discovered by Euler, but we begin by recalling some power series formulas from calculus to see how it arises. For every real number x,

$$e^x = 1 + x + \frac{x^2}{2!} + \cdots + \frac{x^n}{n!} + \cdots ,$$

$$\cos x = 1 - \frac{x^2}{2!} + \frac{x^4}{4!} - \cdots + \frac{(-1)^n x^{2n}}{(2n)!} + \cdots ,$$

and

$$\sin x = x - \frac{x^3}{3!} + \frac{x^5}{5!} - \cdots + \frac{(-1)^{n-1} x^{2n+1}}{(2n+1)!} + \cdots .$$

Euler's Theorem. *For all real numbers x,*

$$e^{ix} = \cos x + i \sin x.$$

Proof. (*Sketch*) As n varies over $0, 1, 2, 3, \ldots$, the powers of i repeat every four steps: that is, i^n takes values

$$1, i, -1, -i, 1, i, -1, -i, 1, \ldots .$$

It follows, for every real number x, that $(ix)^n$ takes values

$$1, ix, -x^2, -ix^3, x^4, ix^5, -x^6, -ix^7, x^8, \ldots .$$

One can define convergence of power series $\sum_{n=0}^{\infty} c_n z^n$ for z and c_n complex numbers, and one can show that the series

$$1 + z + \frac{z^2}{2!} + \cdots + \frac{z^n}{n!} + \cdots$$

converges for every complex number z. The ***complex exponential*** e^z is defined to be the sum of this series. In particular, the series for e^{ix} converges for all real numbers x, and

$$e^{ix} = 1 + ix + \frac{(ix)^2}{2!} + \cdots + \frac{(ix)^n}{n!} + \cdots .$$

The even powers of ix do not involve i, whereas the odd powers do. Collecting terms, one has $e^{ix} =$ even terms $+$ odd terms. But

$$\text{even terms} \quad = \quad 1 + \frac{(ix)^2}{2!} + \frac{(ix)^4}{4!} + \cdots$$

$$= \quad 1 - \frac{x^2}{2!} + \frac{1x^4}{4!} - \cdots = \cos x$$

and

$$\text{odd terms} \quad = \quad ix + \frac{(ix)^3}{3!} + \frac{(ix)^5}{5!} + \cdots.$$

$$= \quad i\left(x - \frac{x^3}{3!} + \frac{x^5}{5!} - \cdots\right) = i \sin x.$$

Therefore, $e^{ix} = \cos x + i \sin x$.   •

It is said that Euler was especially pleased with the equation

$$e^{\pi i} = -1;$$

indeed, this formula is inscribed on his tombstone.

As a consequence of Euler's theorem, the polar decomposition can be rewritten in exponential form: every complex number z has a factorization

$$z = re^{i\theta},$$

where $r \geq 0$ and $0 \leq \theta < 2\pi$.

The addition theorem and De Moivre's theorem can be restated in complex exponential form. The first becomes

$$e^{ix}e^{iy} = e^{i(x+y)};$$

the second becomes

$$(e^{ix})^n = e^{inx}.$$

Definition.   If $n \geq 1$ is an integer, then an ***nth root of unity*** is a complex number ζ with $\zeta^n = 1$.

The geometric interpretation of complex multiplication is particularly interesting when z and w lie on the unit circle, so that $|z| = 1 = |w|$. Given a positive integer n, let $\theta = 2\pi/n$ and let $\zeta = e^{i\theta}$. The polar coordinates of ζ are $(1, \theta)$, the polar coordinates of ζ^2 are $(1, 2\theta)$, the polar coordinates of ζ^3 are $(1, 3\theta)$, ... , the polar coordinates of ζ^{n-1} are $(1, (n-1)\theta)$, and the polar coordinates of $\zeta^n = 1$ are $(1, n\theta) = (1, 0)$. Thus, the nth roots of unity are equally spaced around the unit circle. Figure 1.7 shows the 8th roots of unity (here, $\theta = 2\pi/8 = \pi/4$).

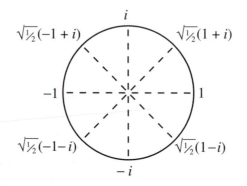

Figure 1.7

Corollary 1.25.   *Every nth root of unity is equal to*

$$e^{2\pi i k/n} = \cos(2\pi k/n) + i \sin(2\pi k/n),$$

where $k = 0, 1, 2, \ldots , n - 1$.

Proof.   Note that $e^{2\pi i} = \cos 2\pi + i \sin 2\pi = 1$. By De Moivre's theorem, if $\zeta = e^{2\pi i/n} = \cos(2\pi/n) + i \sin(2\pi/n)$, then

$$\zeta^n = (e^{2\pi i/n})^n = e^{2\pi i} = 1,$$

so that ζ is an nth root of unity (ζ is called a ***primitive nth root of unity***). Since $\zeta^n = 1$, it follows that $(\zeta^k)^n = (\zeta^n)^k = 1^k = 1$ for all $k = 0, 1, 2, \ldots,$ $n - 1$, so that $\zeta^k = e^{2\pi i k/n}$ is also an nth root of unity. We have exhibited n distinct nth roots of unity; there can be no others, for it will be proved (Theorem 3.33) that a polynomial of degree n with rational coefficients, e.g., $x^n - 1$, has at most n complex roots.   ●

Just as there are two square roots of a number a, namely, $\sqrt{a}$ and $-\sqrt{a}$, there are n different nth roots of a, namely, $e^{2\pi i/k}\sqrt[n]{a}$ for $k = 0, 1, \ldots, k-1$.

For example, the cube roots of unity are 1,

$$\zeta = \cos 120° + i \sin 120° = -\tfrac{1}{2} + i\tfrac{\sqrt{3}}{2}$$

and

$$\zeta^2 = \cos 240° + i \sin 240° = -\tfrac{1}{2} - i\tfrac{\sqrt{3}}{2}.$$

There are 3 cube roots of 2, namely, $\sqrt[3]{2}$, $\zeta\sqrt[3]{2}$, and $\zeta^2\sqrt[3]{2}$.

Why is a root so called? Just as the Greeks called the bottom side of a triangle its base (as in the area formula $\frac{1}{2}$ altitude × base), they also called the bottom side of a square its base. A natural question for the Greeks was: Given a square of area A, what is the length of its side? Of course, the answer is $\sqrt{A}$. Were we inventing a word for $\sqrt{A}$, we might have called it the *base* of A or the *side* of A. Similarly, were we seeking a term for the analogous three-dimensional question: What is the length of a side of a cube of volume V, we might have called $\sqrt[3]{V}$ the *cube base* of V and $\sqrt{A}$ the *square base* of A. Why, then, do we call these numbers cube *root* and square *root*? What has any of this to do with plants?

Since tracing the etymology of words is not a simple matter, we only suggest the following explanation. Through about the fourth and fifth centuries, most mathematics was written in Greek, but, by the fifth century, India had become a center of mathematics, and important mathematical texts were also written in Sanskrit. The Sanskrit term for square root is *pada*. Both Sanskrit and Greek are Indo-European languages, and the Sanskrit word *pada* is a cognate of the Greek word *podos*; both mean *base* in the sense of the foot of a pillar or, as above, the bottom of a square. In both languages, however, there is a secondary meaning: the root of a plant. In translating from Sanskrit, Arab mathematicians chose the secondary meaning, perhaps in error (Arabic is not an Indo-European language), perhaps for some unknown reason. For example, the influential book by al-Khwarizmi, *Al-jabr w'al muqabala*,[9] which appeared in the year 830, used the Arabic word *jidhr*, meaning root of a plant. (The word *algebra* is a European version of the first word in the title of this book; the author's name has also come into the English language, as the word

---

[9]One can translate this title from Arabic, but the words already had a technical meaning: both *jabr* and *muqabala* refer to certain operations akin to subtracting the same number from both sides of an equation.

algorithm.) This mistranslation has since been handed down through the centuries; the term *jidhr* became standard in Arabic mathematical writings, and European translations from Arabic into Latin used the word *radix* (meaning root, as in radish or radical). The notation $r2$ for $\sqrt{2}$ occurs in European writings from about the twelfth century (but the square root symbol did not arise from the letter r; it evolved from an old dot notation). However, there was a competing notation in use at the same time, for some scholars who translated directly from the Greek denoted $\sqrt{2}$ by $l2$, where l abbreviates the Latin word *latus*, meaning side. Finally, with the invention of logarithms in the 1500s, r won out over l, for the notation $l2$ was then commonly used to denote $\log 2$. The passage from square root to cube root to the root of a polynomial equation other than $x^2 - a$ and $x^3 - a$ is a natural enough generalization. Thus, as pleasant as it would be, there seems to be no botanical connection with roots of equations.

As long as we are discussing etymology, where do the names of the trigonometric functions come from? The circle in Figure 1.8 is the unit circle, and

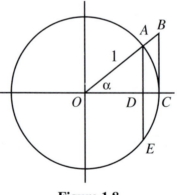

Figure 1.8

so the coordinates of the point A are $(\cos \alpha, \sin \alpha)$; that is, $|OD| = \cos \alpha$ and $|AD| = \sin \alpha$. The reader may show that $|BC| = \tan \alpha$ (the Latin word *tangere* means "to touch," and a *tangent* is a line which touches the circle in only one point), and that $|OB| = \sec \alpha$ (the Latin word *secare* means "to cut," and a *secant* is a line that cuts a circle). The *complement* of an acute angle α is $90° - \alpha$, and so the name *cosine* arises from that of sine because of the identity $\cos \alpha = \sin(90° - \alpha)$.

The reason for the term *sine* is more amusing. We see in Figure 1.8 that

$$\sin \alpha = |AD| = \tfrac{1}{2}|AE|;$$

that is, $\sin \alpha$ is half the length of the chord AE. The fifth century Indian mathematician Aryabhata called the sine *ardha-jya* (half chord) in Sanskrit, which was later abbreviated to *jya*. A few centuries later, books in Arabic transliterated *jya* as *jiba*. In Arabic script, there are letters and diacritical marks; roughly speaking, the letters correspond to our consonants, while the diacritical marks correspond to our vowels. It is customary to suppress dia-critical marks in writing; for example, the Arabic version of *jiba* is written *jb* (using Arabic characters, of course). Now *jiba*, having no other meaning in Arabic, eventually evolved into *jaib*, which is an Arabic word, meaning "bosom of a dress" (a fine word, but having absolutely nothing to do with half-chord). Finally, Gherardo of Cremona, ca. 1150, translated *jaib* into its Latin equivalent, *sinus*. And this is why sine is so called, for sine means bosom!

EXERCISES

1.21 Prove that the binomial theorem holds for complex numbers: If u and v are complex numbers, then

$$(u + v)^n = \sum_{r=0}^{n} \binom{n}{r} u^{n-r} v^r.$$

1.22 Show that the binomial coefficients are "symmetric":

$$\binom{n}{r} = \binom{n}{n-r}$$

for all r with $0 \leq r \leq n$.

1.23 Show, for every n, that the sum of the binomial coefficients is 2^n:

$$\binom{n}{0} + \binom{n}{1} + \binom{n}{2} + \cdots + \binom{n}{n} = 2^n.$$

1.24 (i) Show, for every $n \geq 1$, that the "alternating sum" of the binomial coeffi-cients is zero:

$$\binom{n}{0} - \binom{n}{1} + \binom{n}{2} - \cdots \pm \binom{n}{n} = 0.$$

(ii) Use part (i) to prove, for a given n, that the sum of all the binomial coefficients $\binom{n}{r}$ with r even is equal to the sum of all those $\binom{n}{r}$ with r odd.

1.25 Prove that if $n \geq 2$, then

$$\sum_{r=1}^{n} (-1)^{r-1} r \binom{n}{r} = 0.$$

1.26 Let $\varepsilon_1, \ldots, \varepsilon_n$ be complex numbers with $|\varepsilon_j| = 1$ for all j, where $n \geq 2$.

(i) Prove that

$$\left| \sum_{j=1}^{n} \varepsilon_j \right| \leq \sum_{j=1}^{n} |\varepsilon_j| = n.$$

(ii) Prove that there is equality,

$$\left| \sum_{j=1}^{n} \varepsilon_j \right| = n,$$

if and only if all the ε_j are equal.

1.27 For all odd $n \geq 1$, prove that there is a polynomial $g_n(x)$, all of whose coefficients are integers, such that

$$\sin(nx) = g_n(\sin x).$$

1.28 (**Star of David**) Prove, for all $n > r \geq 1$, that

$$\binom{n-1}{r-1}\binom{n}{r+1}\binom{n+1}{r} = \binom{n-1}{r}\binom{n}{r-1}\binom{n+1}{r+1}.$$

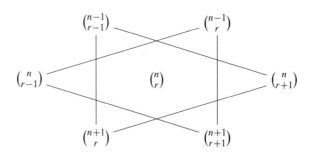

1.29   (i) What is the coefficient of x^{16} in $(1+x)^{20}$?

      (ii) How many ways are there to choose 4 colors from a palette containing paints of 20 different colors?

1.30 Give at least two different proofs that a set X with n elements has exactly 2^n subsets.

1.31 A weekly lottery asks you to select 5 numbers between 1 and 45. At the week's end, 5 such numbers are drawn at random, and you win the jackpot if all your numbers match the drawn numbers. What is your chance of winning?

1.32 Assume that "term-by-term" differentiation holds for power series: If $f(x) = c_0 + c_1 x + c_2 x^2 + \cdots + c_n x^n + \cdots$, then the power series for the derivative $Df(x)$ is

$$Df(x) = c_1 + 2c_2 x + 3c_3 x^2 + \cdots + n c_n x^{n-1} + \cdots.$$

  (i) Prove that $f(0) = c_0$.

  (ii) Prove, for all $n \geq 0$, that

$$D^n f(x) = n! c_n + (n+1)! c_{n+1} x + x^2 g_n(x),$$

  where $g_n(x)$ is some power series ($D_0 f(x)$ is defined to be $f(x)$).

  (iii) Prove that $c_n = D^n f(x)(0)/n!$ for all $n \geq 0$, where $D^n f(0)$ denotes the nth derivative of $f(x)$ evaluated at $x = 0$. (Of course, this is Taylor's formula.)

1.33 (*Leibniz*) A function $f : \mathbb{R} \to \mathbb{R}$ is called a C^∞-*function* if it has a kth derivative $D^k f$ for every $k \geq 0$ ($D^0 f$ is defined to be f). Prove that if f and g are C^∞-functions, then

$$D^n(fg) = \sum_{k=0}^{n} \binom{n}{k} D^k f \cdot D^{n-k} g.$$

1.34   (i) If $z = a + ib \neq 0$, prove that

$$\frac{1}{z} = \frac{a}{a^2 + b^2} - i \frac{b}{a^2 + b^2}.$$

  (ii) Find $\sqrt{i}$.

1.35   (i) If $z = r[\cos\theta + i\sin\theta]$, show that

$$w = \sqrt[n]{r}\,[\cos(\theta/n) + i\sin(\theta/n)]$$

  is an nth root of z.

  (ii) Show that every nth root of z has the form $\zeta^k w$, where ζ is a primitive nth root of unity and $k = 0, 1, 2, \ldots, n-1$.

1.36   (i) Find $\sqrt{8 + 15i}$.

  (ii) Find all the fourth roots of $8 + 15i$.

1.37 If $z = q + ip$, where $q, p \in \mathbb{Z}$, prove that $|z^2|$ is a perfect square in $\mathbb{Z}$.

1.3 GREATEST COMMON DIVISORS

This is an appropriate time to introduce notation for some popular sets of numbers.

$\mathbb{N} =$ all natural numbers, that is, 0, 1, 2, ...[10]

$\mathbb{Z} =$ all integers, positive, negative, and zero (after the German word *Zahlen* meaning numbers)

$\mathbb{Q} =$ all rational numbers (or fractions), that is, all numbers of the form a/b, where a and b are integers and $b \neq 0$ (after the word *quotient*)

$\mathbb{R} =$ all real numbers

$\mathbb{C} =$ all complex numbers

Long division involves dividing an integer b by a nonzero integer a, giving

$$\frac{b}{a} = q + \frac{r}{a},$$

where q is an integer and $0 \leq r/a < 1$. We clear denominators to get a statement wholly in $\mathbb{Z}$.

Theorem 1.26 (Division Algorithm).   *Given integers a and b with a $\neq$ 0, there exist unique integers q and r with*

$$b = qa + r \quad and \quad 0 \leq r < |a|.$$

Proof.   We will prove the theorem in the special case in which $a > 0$ and $b \geq 0$; Exercise 1.38 asks the reader to complete the proof. Long division involves finding the largest integer q with $qa \leq b$, which is the same thing as finding the smallest nonnegative integer of the form $b - qa$. We formalize this.

The set C of all nonnegative integers of the form $b - na$, where $n \geq 0$, is not empty because it contains $b = b - 0a$ (we are assuming that $b \geq 0$). By the Least Integer Axiom, C contains a smallest element, say, $r = b - qa$ (for some $q \geq 0$); of course, $r \geq 0$, by its definition. If $r \geq a$, then

$$b - (q + 1)a = b - qa - a = r - a \geq 0.$$

---

[10]Some mathematicians do not call 0 a natural number.

Hence, $r - a = b - (q + 1)a$ is an element of C that is smaller than r, contradicting r being the smallest integer in C. Therefore, $0 \leq r < a$.

It remains to prove the uniqueness of q and r. Suppose that $b = q'a + r'$, where $0 \leq r' < a$, so that

$$(q - q')a = r' - r.$$

We may assume that $r' \geq r$, so that $r' - r \geq 0$ and hence $q - q' \geq 0$. If $q \neq q'$, then $q - q' \geq 1$ (for $q - q'$ is an integer); thus, since $a > 0$,

$$(q - q')a \geq a.$$

On the other hand, since $r' < a$, Proposition 1.5(ii) gives

$$r' - r < a - r \leq a.$$

Therefore, $(q - q')a \geq a$ and $r' - r < a$, contradicting the given equation $(q - q')a = r' - r$. We conclude that $q = q'$ and hence $r = r'$.   •

Definition.   If a and b are integers with $a \neq 0$, then the integers q and r occurring in the division algorithm are called the *quotient* and the *remainder* after dividing b by a.

For example, there are only two possible remainders after dividing by 2, namely, 0 and 1. A number m is even if the remainder is 0; m is odd if the remainder is 1. Thus, either $m = 2q$ or $m = 2q + 1$.

Warning! The division algorithm makes sense, in particular, when b is negative. A careless person may assume that b and $-b$ leave the same remainder after dividing by a, and this is usually false. For example, let us divide 60 and -60 by 7.

$$60 = 7 \cdot 8 + 4 \quad \text{and} \quad -60 = 7 \cdot (-9) + 3.$$

Thus, the remainders after dividing 60 and -60 by 7 are different.

The next result shows that there is no largest prime.

Corollary 1.27.   *There are infinitely many primes.*

Proof. (Euclid)   Suppose, on the contrary, that there are only finitely many primes. If $p_1, p_2, \ldots, p_k$ is the complete list of all the primes, define $M = (p_1 \cdots p_k) + 1$. By Theorem 1.2, M is either a prime or a product of primes.

But M is neither a prime ($M > p_i$ for every i) nor does it have any prime divisor p_i, for dividing M by p_i gives remainder 1 and not 0. For example, dividing M by p_1 gives $M = p_1(p_2 \cdots p_k) + 1$, so that the quotient and remainder are $q = p_2 \cdots p_k$ and $r = 1$; dividing M by p_2 gives $M = p_2(p_1 p_3 \cdots p_k) + 1$, so that $q = p_1 p_3 \cdots p_k$ and $r = 1$; and so forth. This contradiction proves that there cannot be only finitely many primes, and so there must be an infinite number of them. ●

An algorithm solving a problem is a set of directions which gives the correct answer after a finite number of steps, never at any stage leaving the user in doubt as to what to do next. The division algorithm is an algorithm in this sense: one starts with a and b and ends with q and r. We are now going to treat algorithms more formally, using **pseudocodes**, which are general directions that can easily be translated into a programming language. The basic building blocks of a pseudocode are *assignments*, *looping structures*, and *branching structures*.

An **assignment** is an instruction written in the form

$$\langle \text{variable} \rangle := \langle \text{expression} \rangle.$$

This instruction evaluates the expression on the right, using any stored values for the variables appearing in it; this value is then stored on the left. Thus, the assignment replaces the variable on the left by the new value on the right.

Example 1.2.
Consider the following pseudocode for the division algorithm.

```
1: Input: a, b
2: Output: q, r
3: q := 0;   r := b
4: WHILE r ≥ 0 DO
5:    q := r - a
6:    q := q + 1
7: END WHILE
```

The meaning of the first two lines is clear; line 3 has two assignments giving initial values to the variables q and r. Let us explain the **looping structure** WHILE . . . DO before considering assignments 5 and 6. The general form is

$$\text{WHILE} \; \langle \text{condition} \rangle \; \text{DO}$$
$$\langle \text{action} \rangle.$$

Here, *action* means a sequence of instructions. The loop repeats the action as long as the condition holds, but it stops either when the condition is no longer valid or when it is told to end. In the example above, one begins with $r = b$ and $q = 0$; since $b \geq 0$, the condition holds, and so assignment 5 replaces $r = b$ by $r = b - a$. Similarly, assignment 6 replaces $q = 0$ by $q = 1$. If $r = b - a \geq 0$, this loop repeats this action using the new values of r and q just obtained.

This pseudocode is not a substitute for a proof of the existence of a quotient and a remainder. Had we begun with it, we would still have been obliged to prove two things: first, that the loop does stop eventually; second, that the output q and r satisfies the desiderata of the division algorithm: $b = qa + r$ and $0 \leq r < |a|$. ◄

Example 1.3.
Another popular looping structure is REPEAT, written

$$\text{REPEAT } \langle\text{action}\rangle \text{ UNTIL } \langle\text{condition}\rangle.$$

In WHILE, the condition tells when to proceed, whereas in REPEAT, the condition tells when to stop. Another difference is that WHILE may not do a single step, for the condition is checked before acting; REPEAT always does at least one step, for it checks that the condition holds only after it acts.

For example, consider Newton's method for finding a real root of a polynomial $f(x)$. Recall that one begins with a guess x_1 for a root of $f(x)$ and, inductively, defines

$$x_{n+1} = x_n - \frac{f(x_n)}{f'(x_n)}.$$

If the sequence $\{x_n\}$ converges (and it may not), then its limit is a root of $f(x)$. The following pseudocode finds a real root of $f(x) = x^3 + x^2 - 36$ with error at most .0001.

```
Input: x
Output: x, y, y'
REPEAT
    y := x³ + x² - 36
    y' := 3x² + 2x
    x := x - y/y'
UNTIL y < .0001  ◄
```

Example 1.4.

Our last example of a looping structure is FOR, written

$$\text{FOR each } k \text{ in } K \text{ DO } \langle \text{action} \rangle.$$

Here, a (finite) set K is given, and the action consists in performing the action on every element in K.

For example,

> Input: n
> Output: f
> FOR $0 \le n \le 41$ DO
> $\quad f := n^2 - n + 41$
> END FOR   ◄

Example 1.5.

An example of a ***branching structure*** is

$$\text{IF } \langle \text{condition} \rangle \text{ THEN } \langle \text{action \#1} \rangle \text{ ELSE } \langle \text{action \#2} \rangle.$$

When this structure is reached and the condition holds, then action #1 is taken (only once), but if the condition does not hold when this structure is reached, then action #2 is taken (only once). One can omit ELSE $\langle$action #2$\rangle$, in which case the directions are

$$\text{IF } \langle \text{condition} \rangle \text{ THEN } \langle \text{action \#1} \rangle \text{ do nothing.}   ◄$$

Let us return to number theory.

Definition.   If a and b are integers, then a is a ***divisor*** of b if there is an integer d with $b = ad$. We also say that a ***divides*** b or that b is a ***multiple*** of a, and we denote this by

$$a \mid b.$$

Note that $3 \mid 6$ because $6 = 3 \times 2$, but that $3 \nmid 5$, for even though $5 = 3 \times \frac{5}{3}$, the fraction $\frac{5}{3}$ is not an integer. For every number b, we have $1 \mid b$, $-1 \mid b$, $b \mid b$, $-b \mid b$, and $b \mid 0$ (because $0 = b \times 0$); on the other hand, if $0 \mid b$, then $b = 0$ (because there is some d with $b = 0 \times d = 0$).

If a and m are positive integers with $a \mid m$, say, $m = ab$, we claim that $a \leq m$. Since $0 < b$, we have $1 \leq b$, because b is an integer, and so $a \leq ab = m$.

If a and b are integers with $a \neq 0$, then a is a divisor of b *if and only if*[11] the remainder r given by the division algorithm is 0. If a is a divisor of b, then the remainder r given by the division algorithm is 0; conversely, if the remainder r is 0, then a is a divisor of b.

There is going to be a shift in viewpoint. When first learning long division, one emphasizes the quotient q; the remainder r is merely the fragment left over. Here, we are interested in whether or not a given number b is a multiple of a number a, but we are not so interested in which multiple it may be. Hence, from now on, we will emphasize the remainder.

Definition.   A *common divisor* of integers a and b is an integer c with $c \mid a$ and $c \mid b$. The *greatest common divisor* (gcd) of a and b, denoted by (a, b), is defined by

$$(a, b) = \begin{cases} 0 \text{ if } a = 0 = b \\ \text{the largest common divisor of } a \text{ and } b \text{ otherwise.} \end{cases}$$

Note that gcd's exist and are always nonnegative: if c is a common divisor of a and b, then so is $-c$, and one of $\pm c$ is nonnegative. It is easy to check that $(0, b) = |b|$.

Proposition 1.28.   *If p is a prime and b is any integer, then*

$$(p, b) = \begin{cases} p \ \textit{if } p \mid b \\ 1 \ \textit{otherwise.} \end{cases}$$

Proof.   A common divisor c of p and a is, of course, a divisor of p. But the only positive divisors of p are p and 1, and so $(p, a) = p$ or 1; it is p if $p \mid a$, and it is 1 otherwise.   •

---

[11] The *converse* of an implication "If P is true, then Q is true" is the implication "If Q is true, then P is true." An implication may be true without its converse being true. For example, "If $a = b$, then $a^2 = b^2$." The phrase *if and only if* means that both the statement and its converse are true.

not know irrationals." But the Greeks knew irrational ratios very well. ... That they did not consider $\sqrt{2}$ as a number was not a result of ignorance, but of strict adherence to the definition of number. *Arithomos* means quantity, therefore whole number. Their logical rigor did not even allow them to admit fractions; they replaced them by ratios of integers.

For the Babylonians, every segment and every area simply represented a number. ... When they could not determine a square root exactly, they calmly accepted an approximation. Engineers and natural scientists have always done this. But the Greeks were concerned with exact knowledge, with "the diagonal itself," as Plato expresses it, not with an acceptable approximation.

In the domain of numbers (positive integers), the equation $x^2 = 2$ cannot be solved, not even in that of ratios of numbers. But it is solvable in the domain of segments; indeed the diagonal of the unit square is a solution. Consequently, in order to obtain exact solutions of quadratic equations, we have to pass from the domain of numbers (positive integers) to that of geometric magnitudes. Geometric algebra is valid also for irrational segments and is nevertheless an exact science. It is therefore logical necessity, not the mere delight in the visible, which compelled the Pythagoreans to transmute their algebra into a geometric form.

Even though the Pythagorean definition of number is no longer popular, the Pythagorean dichotomy still persists. For example, almost all American high schools teach one year of algebra followed by one year of geometry, instead of two years in which both subjects are developed together. The problem of defining *number* has arisen several times since the classical Greek era. In the 1500s, mathematicians had to deal with negative numbers and with complex numbers (see our discussion of cubic polynomials in Chapter 4); the description of real numbers generally accepted today dates from the late 1800s. There are echos of Pythagoras in our time. L. Kronecker (1823–1891) wrote,

> Die ganzen Zahlen hat der liebe Gott gemacht, alles andere ist Menschenwerk.

(God created the integers; everything else is the work of Man), and even today some logicians argue for a new definition of number.

Our discussion of gcd's is incomplete. What is the gcd $(12327, 2409)$? To ask the question another way, is the expression $2409/12327$ in lowest terms? The next result not only enables one to compute gcd's, it also allows one to compute integers s and t expressing the gcd as a linear combination. Before giving the theorem, consider the following example. Since $(2, 3) = 1$, there are integers s and t with $1 = 2s + 3t$. A moment's thought gives $s = -1$ and $t = 1$; but another moment's thought gives $s = 2$ and $t = -1$. We conclude that the coefficients s and t expressing the gcd as a linear combination are not uniquely determined. The algorithm below, however, always picks out a particular pair of coefficients.

Theorem 1.37 (Euclidean Algorithm).   *Let a and b be positive integers. There is an algorithm that finds the gcd $d = (a, b)$, and there is an algorithm that finds a pair of integers s and t with $d = sa + tb$.*

Remark.   The general case for arbitrary a and b follows from this, for

$$(-a, b) = (-a, -b) = (a, -b) = (a, b). \quad \blacktriangleleft$$

Proof.   The idea is to keep repeating the division algorithm (we will show where this idea comes from after the proof is completed). Let us set $b = r_0$ and $a = r_1$. Repeated application of the division algorithm gives a series of equations:

$$
\begin{aligned}
b &= q_1 r_1 + r_2, & r_2 &< r_1 = a \\
r_1 &= q_2 r_2 + r_3, & r_3 &< r_2 \\
r_2 &= q_3 r_3 + r_4, & r_4 &< r_3 \\
&\;\;\vdots & &\;\;\vdots \\
r_{n-3} &= q_{n-2} r_{n-2} + r_{n-1}, & r_{n-1} &< r_{n-2} \\
r_{n-2} &= q_{n-1} r_{n-1} + r_n, & r_n &< r_{n-1} \\
r_{n-1} &= q_n r_n
\end{aligned}
$$

(remember that all q_j and r_j are explicitly known from the division algorithm). We use Proposition 1.30 to show that the last remainder $d = r_n$ is the gcd. First, notice that there is a last remainder; the procedure stops because the remainders form a strictly decreasing sequence of nonnegative integers (indeed, the number of steps needed is less than a, but see Proposition 1.38). Second, d is a common divisor of b and a: the number $d = r_n$ divides r_{n-1}, and so the $(n-1)$st equation $r_{n-2} = q_{n-1}r_{n-1} + r_n$ shows that $d \mid r_{n-2}$. Working up the list ultimately gives $d \mid r_1 = a$ and $d \mid r_0 = b$. Third, if c is a common divisor of a and b, start at the top of the list and work down: $c \mid b$ and $c \mid a$ implies $c \mid r_1$, and so forth; eventually, $c \mid r_n = d$. Therefore, $d = r_n$ is the gcd (a, b).

We find coefficients s and t with $d = sa + tb$ by working from the bottom of the list upward. Thus,

$$d = r_n = r_{n-2} - q_{n-1}r_{n-1}$$

is a linear combination of r_{n-2} and r_{n-1}. Combining this with the equation immediately above it, $r_{n-1} = r_{n-3} - q_{n-2}r_{n-2}$, gives

$$d = r_n = r_{n-2} - q_{n-1}(r_{n-3} - q_{n-2}r_{n-2})$$
$$= (1 + q_{n-1}q_{n-2})r_{n-2} - q_{n-1}r_{n-3},$$

a linear combination of r_{n-2} and r_{n-3}. This procedure ends with $d = sa + tb$.   •

We say that n is the **number of steps** in the euclidean algorithm, for one does not know whether r_n in the $(n-1)$st step $r_{n-2} = q_{n-1}r_{n-1} + r_n$ is the gcd until the division algorithm is applied to r_{n-1} and r_n.

Here is a pseudocode implementing the euclidean algorithm.

```
Input: a, b
Output: d
d := a;   s := b
WHILE s ≥ 0 DO
    rem := remainder (d, s)
    d := s
    s := rem
END WHILE
```

Example 1.6.
Find the gcd $(326, 78)$ and express it as a linear combination. Write $78/326$

in lowest terms.

$$\boxed{326} = 4 \times \boxed{78} + \boxed{14}$$
$$\boxed{78} = 5 \times \boxed{14} + \boxed{8}$$
$$\boxed{14} = 1 \times \boxed{8} + \boxed{6}$$
$$\boxed{8} = 1 \times \boxed{6} + \boxed{2}$$
$$\boxed{6} = 3 \times \boxed{2}.$$

The euclidean algorithm gives $(326, 78) = 2$.

We now express 2 as a linear combination of 326 and 78, using the equations above.

$$2 = \boxed{8} - 1\boxed{6} \quad \text{by (4)}$$
$$= \boxed{8} - 1\left(\boxed{14} - 1\boxed{8}\right) \quad \text{by (3)}$$
$$= 2\boxed{8} - 1\boxed{14}$$
$$= 2\left(\boxed{78} - 5\boxed{14}\right) - 1\boxed{14} \quad \text{by (2)}$$
$$= 2\boxed{78} - 11\boxed{14}$$
$$= 2\boxed{78} - 11\left(\boxed{326} - 4\boxed{78}\right) \quad \text{by (1)}$$
$$= 46\boxed{78} - 11\boxed{326};$$

thus, $s = 46$ and $t = -11$.

Dividing numerator and denominator by the gcd $= 2$, we have $78/326 = 39/163$, and the last expression is in lowest terms. ◄

The Greeks called the euclidean algorithm *antanairesis*, meaning "reciprocal subtraction." It rests on the simple observation that if $b \geq a$, then any common divisor of a and b is also a common divisor of a and $b - a$, so that $(a, b) = (a, b - a)$. Repeat for the pair a and $b - a$, for the pair a and $b - 2a$, and keep repeating until one reaches a pair a and $b - qa$ (for some q) with $b - qa < a$. Thus, if $b = qa + r$, where $0 \leq r < a$, then

$$(a, b) = (a, b - a) = (a, b - 2a) = \cdots = (a, b - qa) = (a, r).$$

Now, repeat the procedure beginning with the pair $(a, r) = (r, a)$; eventually one reaches the pair d and 0.

For example, antanairesis computes the gcd (326, 78) as follows:

$$(326, 78) = (248, 78) = (170, 78) = (92, 78) = (14, 78).$$

So far, we have been subtracting 78 from the other larger numbers. At this point, we now subtract 14 (this is the "reciprocal" aspect of antanairesis), for $78 > 14$.

$$(78, 14) = (64, 14) = (50, 14) = (36, 14) = (22, 14) = (8, 14).$$

Again we "reciprocate":
$$(14, 8) = (6, 8).$$

Reciprocate once again to get $(8, 6) = (2, 6)$, and reciprocate one last time to get

$$(6, 2) = (4, 2) = (2, 2) = (0, 2) = 2.$$

Thus, gcd $(326, 78) = 2$.

The division algorithm (which is just iterated subtraction!) is a more efficient way of performing antanairesis. There are four steps in the passage from (326, 78) to (14, 78); the division algorithm expresses this as

$$326 = 4 \times 78 + 14.$$

There are then five steps in the passage from (78, 14) to (8, 14); the division algorithm expresses this as

$$78 = 5 \times 14 + 8.$$

There is one step in the passage from (14, 8) to (6, 8): that is,

$$14 = 1 \times 8 + 6.$$

There is one step in the passage from (8, 6) to (2, 6); that is,

$$8 = 1 \times 6 + 2,$$

and there are three steps from (6, 2) to (0, 2) $= 2$; that is,

$$6 = 3 \times 2.$$

These are the equations in the euclidean algorithm.

Here is a result whose proof involves the Fibonacci sequence

$$F_0 = 0, F_1 = 1, F_2 = 1, F_3 = 2, \ldots , F_n = F_{n-1} + F_{n-2}, \ldots .$$

Proposition 1.38 (Lamé's[13] Theorem).   *Let $a \geq b$ be positive integers, and let β be the number of digits in the decimal expression of b. If n is the number of steps in the euclidean algorithm computing $\gcd(a, b)$, then*

$$n \leq 5\beta.$$

Proof.   Let us denote a by r_0 and b by r_1, so that every equation appearing in the euclidean algorithm has the form

$$r_j = r_{j+1} q_{j+1} + r_{j+2}$$

except the last one, which is

$$r_{n-1} = r_n q_n.$$

Note that $q_n \geq 2$: if $q_n = 0$, then $r_{n-1} = 0$, and the algorithm would have ended a step earlier; if $q_n = 1$, then $r_{n-1} = r_n$, contradicting $r_n < r_{n-1}$. Similarly, all $q_1, q_2, \dots, q_{n-1} \geq 1$: otherwise $q_j = 0$ for some $j \leq n - 1$, and $r_{j-1} = r_{j+1}$, contradicting the given strict inequalities $r_n < r_{n-1} < \cdots < r_1 = b$.

   Now

$$r_n \geq 1 = F_2$$

and, since $q_n \geq 2$,

$$r_{n-1} = r_n q_n \geq 2r_n \geq 2F_2 \geq 2 = F_3.$$

More generally, let us prove by induction on $j \geq 0$ that

$$r_{n-j} \geq F_{j+2}.$$

The inductive step is

$$
\begin{aligned}
r_{n-j-1} = r_{n-j} q_{n-j} + r_{n-j+1} & \\
\geq r_{n-j} + r_{n-j+1} \quad & \text{(since } q_{n-j} \geq 1) \\
\geq F_{j+2} + F_{j+1} = F_{j+3}. &
\end{aligned}
$$

---

[13]This is an example in which a theorem's name is not that of its discoverer. Lamé's proof appeared in 1844. The earliest estimate for the number of steps in the euclidean algorithm can be found in a rare book by Simon Jacob, published around 1564. There were also estimates by T. F. de Lagny in 1733, A-A-L Reynaud in 1821, E. Léger in 1837, and P-J-E Finck in 1841. (This earlier work is described in articles of P. Shallit and P. Schreiber, respectively, in the journal *Historica Mathematica*.)

We conclude that $b = r_1 = r_{n-(n-1)} \geq F_{n-1+2} = F_{n+1}$. By Corollary 1.14, $F_{n+1} > \alpha^{n-1}$, where $\alpha = \frac{1}{2}(1 + \sqrt{5})$, and so

$$b > \alpha^{n-1}.$$

Now $\log_{10} \alpha \approx .208 > \frac{1}{5}$, so that

$$\log_{10} b > (n-1) \log_{10} \alpha > (n-1)/5;$$

that is,

$$n - 1 < 5 \log_{10} b < 5\beta,$$

because $\beta = \lfloor \log_{10} b \rfloor + 1$, and so $n \leq 5\beta$ because β, hence 5β, is an integer.   ●

For example, Lamé's theorem guarantees there are at most 10 steps needed to compute $(326, 78)$ (actually, there are 5 steps).

The usual notation for the integer 5754 is an abbreviation of

$$5 \times 10^3 + 7 \times 10^2 + 5 \times 10 + 4.$$

The next result shows that there is nothing magic about the number 10.

Proposition 1.39. *If $b \geq 2$ is an integer, then every positive integer m has an expression in **base b**: there are integers d_i with $0 \leq d_i < b$ such that*

$$m = d_k b^k + d_{k-1} b^{k-1} + \cdots + d_0;$$

moreover, this expression is unique if $d_k \neq 0$.

Remark. The numbers $d_k, \ldots, d_0$ are called the **b-adic digits** of m.   ◄

Proof. Let m be a positive integer; since $b \geq 2$, there are powers of b larger than m. We prove, by induction on $k \geq 0$, that if $b^k \leq m < b^{k+1}$, then m has an expression

$$m = d_k b^k + d_{k-1} b^{k-1} + \cdots + d_0$$

in base b.

If $k = 0$, then $1 = b^0 \leq m < b^1 = b$, and we may define $d_0 = m$. If $k > 0$, then the division algorithm gives integers d_k and r with $m = d_k b^k + r$, where $0 \leq r < b^k$. Notice that $d_k < b$ (lest $m \geq b^{k+1}$) and $0 < d_k$ (lest $m < b^k$). If $r = 0$, define $d_0 = \cdots = d_{k-1} = 0$, and $m = d_k b^k$ is an expression in base *b.* If $r > 0$, then the inductive hypothesis shows that r and, hence, m has an expression in base b.

Before proving uniqueness of the b-adic digits d_i, we first observe that if $0 \leq d_i < b$ for all i, then $\sum_{i=0}^{k} d_i b^i < b^{k+1}$:

$$\sum_{i=0}^{k} d_i b^i \leq \sum_{i=0}^{k} (b-1) b^i$$
$$= \sum_{i=0}^{k} b^{i+1} - \sum_{i=0}^{k} b^i$$
$$= b^{k+1} - 1$$
$$< b^{k+1}.$$

We now prove, by induction on $k \geq 0$, that if $b^k \leq m < b^{k+1}$, then the b-adic digits d_i in the expression $m = \sum_{i=0}^{k} d_i b^i$ are uniquely determined by m. Let $m = \sum_{i=0}^{k} d_i b^i = \sum_{i=0}^{k} c_i b^i$, where $0 \leq d_i < b$ and $0 \leq c_i < b$ for all i. Subtracting, we obtain

$$0 = \sum_{i=0}^{k} (d_i - c_i) b^i.$$

If some of the coefficients $d_i - c_i$ are nonzero, then we may rewrite this equation so that all coefficients are positive and the index sets I and J are disjoint:

$$L = \sum_{i \in I} (d_i - c_i) b^i = \sum_{j \in J} (c_j - d_j) b^j = R.$$

Let p be the largest index in I and let q be the largest index in J. Since I and J are disjoint, we may assume that $q < p$. As the left side L involves b^p with a nonzero coefficient, $L \geq b^p$; but our observation above shows that the right side $R < b^{q+1} \leq b^p$, a contradiction. Therefore, the b-adic digits are uniquely determined. •

Example 1.7.
Let us follow the steps in the proof of Proposition 1.39 to write 12345 in base 7. First write the powers of 7 until 12345 is exceeded: $7; 7^2 = 40; 7^3 = 343;$

$7^4 = 2401; 7^5 = 16807$. Repeated use of the division algorithm gives

$$12345 = 5 \times 7^4 + 340 \quad \text{and} \quad 340 < 7^4 = 2401;$$
$$340 = 0 \times 7^3 + 340 \quad \text{and} \quad 340 < 7^3 = 343;$$
$$340 = 6 \times 7^2 + 46 \quad \text{and} \quad 46 < 7^2 = 49;$$
$$46 = 6 \times 7 + 4 \quad \text{and} \quad 4 < 7;$$
$$4 = 4 \times 1.$$

The 7-adic digits of 12345 are thus 50664. ◄

The most popular bases are $b = 10$ (giving everyday *decimal* digits), $b = 2$ (giving *binary* digits, useful because a computer can interpret 1 as "on" and 0 as "off"), and $b = 16$ (*hexadecimal*, also for computers), but let us see now that other bases can also be useful.

Example 1.8.
Here is a problem of Bachet de Méziriac from 1624. A merchant had a 40-pound weight that broke into 4 pieces. When the pieces were weighed, it was found that each piece was a whole number of pounds and that the four pieces could be used to weigh every integral weight between 1 and 40 pounds. What were the weights of the pieces?

Weighing means using a balance scale having two pans, with weights being put on either pan. Thus, given weights of 1 and 3 pounds, one can weigh a 2-pound weight □ by putting 1 and □ on one pan and 3 on the other pan.

A solution to Bachet's problem is 1, 3, 9, 27. If □ denotes a given integral weight, let us write the weights on one pan to the left of the semicolon and the weights on the other pan to the right of the semicolon.

1	1 ; □	9	9 ; □
2	3 ; 1, □	10	9, 1 ; □
3	3 ; □	11	9, 3 ; 1, □
4	3, 1 ; □	12	9, 3 ; □
5	9 ; 3, 1, □	13	9, 3, 1 ; □
6	9 ; 3, □	14	27 ; 9, 3, 1, □
7	9, 1 ; 3, □	15	27 ; 9, 3, □
8	9 ; 1, □		

The reader may complete this table for □ ≤ 40. ◄

Example 1.9.

Given a balance scale, the weight of any person can be found using only six lead weights. (This is not quite accurate, for we cannot weigh a very heavy person.)

We begin by proving that every positive integer m can be written

$$m = e_k 3^k + e_{k-1} 3^{k-1} + \cdots + 3e_1 + e_0,$$

where $e_i = -1, 0,$ or 1.

The idea is to modify the 3-adic expansion

$$m = d_k 3^k + d_{k-1} 3^{k-1} + \cdots + 3d_1 + d_0.$$

where $d_i = 0, 1, 2,$ by "carrying." If $d_0 = 0$ or 1, set $e_0 = d_0$ and leave d_1 alone. If $d_0 = 2$, set $e_0 = -1$, and replace d_1 by $d_1 + 1$ (we have merely substituted $3 - 1$ for 2). Now $1 \le d_1 + 1 \le 3$. If $d_1 + 1 = 1$, set $e_1 = 1$, and leave d_2 alone; if $d_1 + 1 = 2$, set $e_1 = -1$, and replace d_2 by $d_2 + 1$; if $d_1 + 1 = 3$, define $e_1 = 0$ and replace d_2 by $d_2 + 1$. Continue in this way (the ultimate expansion of m may begin with either $e_k 3^k$ or $e_{k+1} 3^{k+1}$). Here is a table of the first few numbers in this new expansion (we write $\bar{1}$ instead of -1).

1	1	9	100
2	$1\bar{1}$	10	101
3	10	11	$11\bar{1}$
4	11	12	110
5	$1\bar{1}\bar{1}$	13	111
6	$1\bar{1}1$	14	$1\bar{1}\bar{1}\bar{1}$
7	$10\bar{1}$	15	$1\bar{1}\bar{1}0$
8	$10\bar{1}$		

The reader should now understand Example 1.8. If $\square$ weighs m pounds, write $m = \sum e_i 3^i$, where $e_i = 1, 0,$ or -1, and then transpose those terms having negative coefficients. Those weights with $e_i = -1$ go on the pan with $\square$, while those weights with $e_i = 1$ go on the other pan.

The solution to the current weighing problem involves choosing as weights $1, 3, 9, 27, 81,$ and 243 pounds. One can find the weight of anyone under 365 pounds, because $1 + 3 + 9 + 27 + 81 = 364$. ◀

EXERCISES

1.38 Given integers a and b (perhaps negative) with $a \neq 0$, prove that there exist unique integers q and r with $b = qa + r$ and $0 \leq r < |a|$.

1.39 Prove that $\sqrt{2}$ is irrational using Proposition 1.12 instead of Euclid's lemma.

1.40 Prove the converse of Euclid's lemma: an integer $p \geq 2$, which, whenever it divides a product necessarily divides one of the factors, must be a prime.

1.41 Let $p_1, p_2, p_3, \ldots$ be the list of the primes in ascending order: $p_1 = 2$, $p_2 = 3$, $p_3 = 5, \ldots$ Define $f_k = p_1 p_2 \cdots p_k + 1$ for $k \geq 1$. Find the smallest k for which f_k is not a prime.

1.42 Prove that if d and d' are nonzero integers, each of which divides the other, then $d' = \pm d$.

1.43 Show that every positive integer m can be written as a sum of distinct powers of 2; show, moreover, that there is only one way in which m can so be written.

1.44 Find the b-adic digits of 1000 for $b = 2, 3, 4, 5$, and 20.

1.45 (i) Prove that if n is **squarefree** (i.e., n is not divisible by the square of any prime), then $\sqrt{n}$ is irrational.

   (ii) Prove that $\sqrt[3]{2}$ is irrational.

1.46 (i) Find the gcd $d = (12327, 2409)$, find integers s and t with $d = 12327s + 2409t$, and put the fraction 2409/12327 in lowest terms.

   (ii) Find the gcd $d = (7563, 526)$, and express it as a linear combination of 7563 and 526.

1.47 Find gcd $(7404621, 73122)$ and write it as a linear combination; that is, find integers s and t with $7404621s + 73122t$ for integers s and t.

1.48 Let a and b be integers, and let $sa + tb = 1$ for s, t in $\mathbb{Z}$. Prove that a and b are relatively prime.

1.49 Prove that if $(r, m) = 1 = (r', m)$, then $(rr', m) = 1$.

1.50 Assume that $d = sa + tb$ is a linear combination of integers a and b. Find infinitely many pairs of integers (s_k, t_k) with

$$d = s_k a + t_k b.$$

1.51 If a and b are relatively prime and if each divides an integer n, then their product ab also divides n.

1.52 If $a > 0$, prove that $a(b, c) = (ab, ac)$. [One must assume that $a > 0$ lest $a(b, c)$ be negative.]

1.53 If F_n denotes the nth term of the Fibonacci sequence $0, 1, 1, 2, 3, 5, 8, \ldots$, prove that

$$\gcd(F_{n+1}, F_n) = 1 \text{ for all } n \geq 1.$$

Definition.   A *common divisor* of integers $a_1, a_2, \ldots, a_n$ is an integer c with $c \mid a_i$ for all i; the largest of the common divisors, denoted by $(a_1, a_2, \ldots, a_n)$, is called the *greatest common divisor*.

1.54   (i) Show that if d is the greatest common divisor of $a_1, a_2, \ldots, a_n$, then $d = \sum t_i a_i$, where t_i is in $\mathbb{Z}$ for $1 \leq i \leq n$.

    (ii) Prove that if c is a common divisor of $a_1, a_2, \ldots, a_n$, then $c \mid d$.

1.55   (i) Show that (a, b, c), the gcd of a, b, c, is equal to $(a, (b, c))$.

    (ii) Compute $(120, 168, 328)$.

1.56 A *Pythagorean triple* is a triple (a, b, c) of positive integers for which

$$a^2 + b^2 = c^2;$$

it is called *primitive* if the gcd $(a, b, c) = 1$.

    (i) Consider a complex number $z = q + ip$, where $q > p$ are positive integers. Prove that

$$(q^2 - p^2, 2qp, q^2 + p^2)$$

is a Pythagorean triple by showing that $|z^2| = |z|^2$. [One can prove that every *primitive* Pythagorean triple (a, b, c) is of this type.]

    (ii) Show that the Pythagorean triple $(9, 12, 15)$ (which is not primitive) is not of the type given in part (i).

    (iii) Using a calculator which can find square roots but which can display only 8 digits, prove that

$$(19597501, 28397460, 34503301)$$

is a Pythagorean triple by finding q and p.

Definition.   A *common multiple* of $a_1, a_2, \ldots, a_n$ is an integer m with $a_i \mid m$ for all i. The *least common multiple*, written lcm and denoted by $[a_1, a_2, \ldots, a_n]$, is the smallest positive common multiple if all $a_i \neq 0$, and it is 0 otherwise.

1.57 Prove that an integer $M \geq 0$ is the lcm of $a_1, a_2, \ldots, a_n$ if and only if it is a common multiple of $a_1, a_2, \ldots, a_n$ which divides every other common multiple.

1.58   (i) Prove that $[a, b](a, b) = ab$, where $[a, b]$ is the least common multiple of a and b.

    (ii) Find $[1371, 123]$.

1.4 THE FUNDAMENTAL THEOREM OF ARITHMETIC

We have already seen, in Theorem 1.2, that every integer $a \geq 2$ is either a prime or a product of primes. We are now going to generalize Proposition 1.12 on page 12 by showing that the primes in such a factorization and the number of times each of them occurs are uniquely determined by a.

Theorem 1.40 (Fundamental Theorem of Arithmetic). *Assume that an integer $a \geq 2$ has factorizations*

$$a = p_1 \cdots p_m \ \text{and} \ a = q_1 \cdots q_n,$$

where the p's and q's are primes. Then $n = m$ and the q's may be reindexed so that $q_i = p_i$ for all i.

Proof.   We prove the theorem by induction on ℓ, the larger of m and n.

Base step. If $\ell = 1$, then the given equation is $a = p_1 = q_1$, and the result is obvious.

Inductive step.   The equation gives $p_m \mid q_1 \cdots q_n$. By Theorem 1.32, Euclid's lemma, there is some i with $p_m \mid q_i$. But q_i, being a prime, has no positive divisors other than 1 and itself, so that $q_i = p_m$. Reindexing, we may assume that $q_n = p_m$. Canceling, we have $p_1 \cdots p_{m-1} = q_1 \cdots q_{n-1}$. By the inductive hypothesis, $n - 1 = m - 1$ and the q's may be reindexed so that $q_i = p_i$ for all i.   •

Corollary 1.41.   *If $a \geq 2$ is an integer, then there are unique distinct primes p_i and unique integers $e_i > 0$ with*

$$a = p_1^{e_1} \cdots p_n^{e_n}.$$

Proof.   Just collect like terms in a prime factorization.   •

Here is an impractical code. Make a list of the 52 English letters (lower case and upper case) together with a space and the 11 punctuation marks

$$, \quad . \quad ; \quad : \quad ! \quad ? \quad - \quad ' \quad " \quad (\quad)$$

Assign a to 1, b to 2, c to 3; in general, assign the ith item on the list of 64 to the number i. One can now encode the King James translation of the Bible "In the beginning, ... " as one integer, in the following way: "I" = 35 (first come the 26 lower case letters, and "I" is the ninth upper case letter), "n" = 14, "space" $= 53$, "t" $= 20$, and so forth. The Biblical text now corresponds to the number $B = 2^{35}3^{14}5^{53}7^{20} \cdots$ (since there are infinitely many primes, there are as many primes as one needs). The fundamental theorem of arithmetic says that one can decode; given the huge number B, one can rewrite the Biblical text in ordinary characters (assuming that one could factor B).

One can write an algorithm giving the prime factorization of an integer m: enumerate all the primes $\leq m$, and then see which of them are factors (use the division algorithm to see whether a remainder is 0). We should mention that actually factoring large numbers is a lengthy task. Even with high-speed computers, it may take months to factor $m = pq$ if p and q are unknown sufficiently large primes; indeed, this is the basis of the construction of "public access codes" in which the sender and receiver know primes p and q, whereas only the product pq appears in the message.

It is often convenient, when comparing two numbers a and b, to allow zero exponents, thus introducing "dummy" factors equal to 1 in factorizations. By this device, we may assume that the same family of primes occurs in the factorizations of a and b. For example, $168 = 2^3 3^1 7^1$ and $60 = 2^2 3^1 5^1$ may be rewritten as $168 = 2^3 3^1 5^0 7^1$ and $60 = 2^2 3^1 5^1 7^0$.

Corollary 1.42. *Every positive rational number $r \neq 1$ has a unique factorization*

$$r = p_1^{g_1} \cdots p_n^{g_n}$$

where the p_i are distinct primes and the g_i are nonzero integers. Moreover, r is an integer if and only if $g_i > 0$ for all i.

Proof. There are positive integers a and b with $r = a/b$. If $a = p_1^{e_1} \cdots p_n^{e_n}$ and $b = p_1^{f_1} \cdots p_n^{f_n}$, then $r = p_1^{g_1} \cdots p_n^{g_n}$, where $g_i = e_i - f_i$ (we may assume that the same primes appear in both factorizations by allowing zero exponents). The desired factorization is obtained if one deletes those factors $p_i^{g_i}$, if any, with $g_i = 0$.

Suppose there were another such factorization

$$r = p_1^{h_1} \cdots p_n^{h_n}$$

(by allowing zero exponents, we may again assume that the same primes occur in each factorization). Suppose that $g_i \neq h_i$ for some i; reindexing if necessary, we may assume that $i = 1$ and that $g_1 > h_1$. Therefore,

$$p_1^{g_1 - h_1} p_2^{g_2} \cdots p_n^{g_n} = p_2^{h_2} \cdots p_n^{h_n}.$$

This is an equation of rational numbers, for some of the exponents may be negative. Cross-multiplying gives an equation in $\mathbb{Z}$ whose left side involves the prime p_1 and whose right side does not; this contradicts the fundamental theorem of arithmetic.

If all the exponents in the factorization of r are positive, then r is an integer because it is a product of integers. Conversely, if r is an integer, then it has a prime factorization in which all exponents are positive.   •

We can now give a new description of gcd's and lcm's.

Lemma 1.43. *Let the prime factorization of positive integers a and b be*

$$a = p_1^{e_1} \cdots p_n^{e_n} \text{ and } b = p_1^{f_1} \cdots p_n^{f_n},$$

where $e_i, f_i \geq 0$ for all i. Then $a \mid b$ if and only if $e_i \leq f_i$ for all i.

Proof. If $e_i \leq f_i$ for all i, then $b = ac$, where $c = p_1^{f_1-e_1} \cdots p_n^{f_n-e_n}$. The number c is an integer because $f_i - e_i \geq 0$ for all i. Therefore, $a \mid b$.

Conversely, if $b = ac$, let the prime factorization of c be $c = p_1^{g_1} \cdots p_n^{g_n}$, where $g_i \geq 0$ for all i. It follows from the Fundamental Theorem of Arithmetic that $e_i + g_i = f_i$ for all i, and so $f_i - e_i = g_i \geq 0$ for all i.   •

Proposition 1.44. *Let $a = p_1^{e_1} \cdots p_n^{e_n}$ and let $b = p_1^{f_1} \cdots p_n^{f_n}$, where e_i, $f_i \geq 0$ for all i; define*

$$m_i = \min\{e_i, f_i\} \quad \text{and} \quad M_i = \max\{e_i, f_i\}.$$

Then

$$\gcd(a, b) = p_1^{m_1} \cdots p_n^{m_n} \quad \text{and} \quad \mathrm{lcm}[a, b] = p_1^{M_1} \cdots p_n^{M_n}.$$

Proof. Define $d = p_1^{m_1} \cdots p_n^{m_n}$. Lemma 1.43 shows that d is a (positive) common divisor of a and b; moreover, if c is any (positive) common divisor, then $c = p_1^{g_1} \cdots p_n^{g_n}$, where $0 \leq g_i \leq \min\{e_i, f_i\} = m_i$ for all i. Therefore, $c \mid d$.

A similar argument shows that $D = p_1^{M_1} \cdots p_n^{M_n}$ is a common multiple that divides every other such.   •

For example, $168 = 2^3 3^1 5^0 7^1$ and $60 = 2^2 3^1 5^1 7^0$, and so $(168, 60) = 2^2 3^1 5^0 7^0 = 12$ and $[168, 60] = 2^3 3^1 5^1 7^1 = 840$.

For small numbers a and b, using their prime factorizations is a more efficient way to compute their gcd than using the euclidean algorithm.

EXERCISES

1.59   (i) Find the gcd (210, 48) using factorizations into primes.

(ii) Find (1234, 5678).

1.60 Use Proposition 1.44 to give another proof of Exercise 1.58(i):

$$[a, b](a, b) = ab.$$

1.61   (i) Prove that an integer $m \geq 2$ is a perfect square if and only if each of its prime factors occurs an even number of times.

(ii) Prove that if m is a positive integer for which $\sqrt{m}$ is rational, then m is a perfect square.

1.62 If a and b are positive integers with $(a, b) = 1$, and if ab is a square, prove that both a and b are squares.

1.63 Let $n = p^r m$, where p is a prime not dividing an integer $m \geq 1$. Prove that

$$p \nmid \binom{n}{p^r}.$$

1.64 Definition.   If p is a prime, define the **p-adic norm** of a rational number a as follows: $\|0\|_p = 0$; if $a \neq 0$, then $a = p^e p_1^{e_1} \cdots p_n^{e_n}$, where $p, p_1, \ldots, p_n$ are distinct primes, and we set $\|a\|_p = p^{-e}$.

(i) For all rationals a and b, prove that

$$\|ab\|_p = \|a\|_p \|b\|_p \text{ and } \|a + b\|_p \leq \max\{\|a\|_p, \|b\|_p\}.$$

(ii) Define $\delta_p(a, b) = \|a - b\|_p$.

(a) For all rationals a, b, prove $\delta_p(a, b) \geq 0$ and $\delta_p(a, b) = 0$ if and only if $a = b$;

(b) For all rationals a, b, prove that $\delta_p(a, b) = \delta_p(b, a)$;

(c) For all rationals a, b, c, prove $\delta_p(a, b) \leq \delta_p(a, c) + \delta_p(c, b)$.

(iii) If a and b are integers and $p^n \mid a - b$, then $\delta_p(a, b) \leq p^{-n}$. (Thus, a and b are "close" if $a - b$ is divisible by a "large" power of n.)

1.65 Let a and b be in $\mathbb{Z}$. Prove that if $\delta_p(a, b) \leq p^{-n}$, then a and b have the same first n p-adic digits, $d_0, \ldots, d_{n-1}$.

1.5 CONGRUENCES

Two integers a and b are said to have the **same parity** if they are both even or both odd. If a and b have the same parity, then $a - b$ is even: this is surely true if a and b are both even; if a and b are both odd, then $a = 2m + 1$, $b = 2n + 1$, and $a - b = 2(m - n)$ is even. Conversely, if $a - b$ is even, then we cannot have one of them even and the other odd lest $a - b$ be odd. The next definition generalizes this notion of parity, letting any positive integer m play the role of 2.

Definition.   Let $m \geq 0$ be fixed. Then integers a and b are **congruent modulo** m, denoted by

$$a \equiv b \bmod m,$$

if $m \mid a - b$.

   Usually, one assumes that the **modulus** $m \geq 2$ because the cases $m = 0$ and $m = 1$ are not very interesting: if a and b are integers, then $a \equiv b \bmod 0$ if and only if $0 \mid a - b$, that is, $a = b$; the congruence $a \equiv b \bmod 1$ is true for every pair of integers a and b because $1 \mid a - b$ always.

   The word "modulo" is usually abbreviated to "mod." The Latin root of this word means a standard of measure; thus, the term *modular unit* is used today in architecture: a fixed length m is chosen, say, $m = 1$ foot, and plans are drawn so that the dimensions of every window, door, wall, etc., are integral multiples of m.

   London time is 6 hours later than Chicago time. What time is it in London if it is 10:00 A.M. in Chicago? Since clocks are set up with 12 hour cycles, this is really a problem about congruence mod 12. To solve it, note that

$$10 + 6 = 16 \equiv 4 \bmod 12,$$

and so it is 4:00 P.M. in London. A similar question involves dates. If today is May 26, what is the date two weeks from today? There are 31 days in May, and so we compute

$$26 + 14 = 40 \equiv 9 \bmod 31.$$

The answer is June 9.

   The next theorem shows that congruence mod m behaves very much like equality.

Proposition 1.45.   *If $m \geq 0$ is a fixed integer, then for all integers a, b, c,*

(i) $a \equiv a \bmod m$;

(ii) *if $a \equiv b \bmod m$, then $b \equiv a \bmod m$;*

(iii) *if $a \equiv b \bmod m$ and $b \equiv c \bmod m$, then $a \equiv c \bmod m$.*

Remark.   (i) says that congruence is **reflexive,** (ii) says it is **symmetric,** and (iii) says it is **transitive.**   ◀

Proof.   (i) Since $m \mid a - a = 0$, we have $a \equiv a \bmod m$.

(ii) If $m \mid a - b$, then $m \mid -(a - b) = b - a$ and so $b \equiv a \bmod m$.

(iii) If $m \mid a - b$ and $m \mid b - c$, then $m \mid (a - b) + (b - c) = a - c$, and so $a \equiv c \bmod m$.   •

We now generalize the observation that $a \equiv 0 \bmod m$ if and only if $m \mid a$.

Proposition 1.46.   *Let $m \geq 0$ be a fixed integer.*

(i) *If $a = qm + r$, then $a \equiv r \bmod m$.*

(ii) *If $0 \leq r' < r < m$, then r and r' are not congruent $\bmod m$; in symbols, $r \not\equiv r' \bmod m$.*

(iii) *$a \equiv b \bmod m$ if and only if a and b leave the same remainder after dividing by m.*

Proof.   (i) The equation $a - r = qm$ shows that $m \mid a - r$.

(ii) If $r \equiv r' \bmod m$, then $m \mid r - r'$ and $m \leq r - r'$. But $r - r' \leq r < m$, a contradiction.

(iii) If $a = qm + r$ and $b = q'm + r'$, where $0 \leq r < m$ and $0 \leq r' < m$, then $a - b = (q - q')m + (r - r')$; that is,

$$a - b \equiv r - r' \bmod m.$$

Therefore, if $a \equiv b \bmod m$, then $a - b \equiv 0 \bmod m$, hence $r - r' \equiv 0 \bmod m$, and $r \equiv r' \bmod m$; by (ii), $r = r'$.

Conversely, if $r = r'$, then $a = qm + r$ and $b = q'm + r$, so that $a - b = (q' - q)m$ and $a \equiv b \bmod m$.   •

Corollary 1.47. *Given $m \geq 2$, every integer a is congruent* mod m *to exactly one of* $0, 1, \ldots, m - 1$.

Proof. The division algorithm says that $a \equiv r$ mod m, where $0 \leq r < m$; that is, r is an integer on the list $0, 1, \ldots, m - 1$. If a were congruent to two integers on the list, say, r and r', then $r \equiv r'$ mod m, contradicting part (ii) of the theorem. Therefore, a is congruent to a unique such r.   •

We know that every integer a is either even or odd; that is, a has the form $2k$ or $2k + 1$. We now see that if $m \geq 2$, then every integer a has exactly one of the forms $mk, mk + 1, mk + 2, \ldots, mk + (m - 1)$; we have generalized the even – odd dichotomy from $m = 2$ to $m \geq 2$. Notice how we continue to focus on the remainder in the division algorithm and not upon the quotient.

Congruence is compatible with addition and multiplication.

Proposition 1.48. *Let $m \geq 0$ be a fixed integer.*

(i) *If $a \equiv a'$ mod m and $b \equiv b'$ mod m, then*

$$a + b \equiv a' + b' \text{ mod } m.$$

(i') *If $a_i \equiv a_i'$ mod m for $i = 1, 2, \ldots, n$, then*

$$a_1 + \cdots + a_n \equiv a_1' + \cdots + a_n' \text{ mod } m.$$

(ii) *If $a \equiv a'$ mod m and $b \equiv b'$ mod m, then*

$$ab \equiv a'b' \text{ mod } m.$$

(ii') *If $a_i \equiv a_i'$ mod m for $i = 1, 2, \ldots, n$, then*

$$a_1 \cdots a_n \equiv a_1' \cdots a_n' \text{ mod } m.$$

(iii) *If $a \equiv b$ mod m, then $a^n \equiv b^n$ mod m for all $n \geq 1$.*

Proof. (i) If $m \mid a - a'$ and $m \mid b - b'$, then $m \mid a - a' + b - b' = (a + b) - (a' + b')$.

(i') Induction on n, using (i).

(ii) We must show that if $m \mid a - a'$ and $m \mid b - b'$, then $m \mid ab - a'b'$, and this follows from the identity

$$ab - a'b' = (ab - a'b) + (a'b - a'b')$$
$$= (a - a')b + a'(b - b').$$

(ii′) Induction on n, using (ii).

(iii) This is the special case of (ii′) when $a_i = a$ and $a'_i = b$ for all i.   •

Let us repeat a warning given when we proved the division algorithm. A number and its negative usually have different remainders after being divided by the same number. For example, $60 = 7 \cdot 8 + 4$ and $-60 = 7 \cdot (-9) + 3$. In terms of congruences,

$$60 \equiv 4 \bmod 7$$

while

$$-60 \equiv 3 \bmod 7.$$

In light of Proposition 1.46(i), if the remainder after dividing b by m is r and the remainder after dividing $-b$ by m is s, then $b \equiv r \bmod m$ and $-b \equiv s \bmod m$. Therefore, Proposition 1.48(i) gives

$$r + s \equiv b - b \equiv 0 \bmod m.$$

Thus, $r + s = m$, for $0 \le r, s < m$. For example, the remainders after dividing 60 and -60 by 7 are 4 and 3, respectively.

The next example shows how one can use congruences. In each case, the key idea is to solve a problem by replacing numbers by their remainders.

Example 1.10.
(i) Prove that if a is in $\mathbb{Z}$, then $a^2 \equiv 0$, 1, or 4 mod 8.

If a is an integer, then $a \equiv r \bmod 8$, where $0 \le r \le 7$; moreover, by Proposition 1.48(iii), $a^2 \equiv r^2 \bmod 8$, and so it suffices to look at the squares of the remainders.

r	0	1	2	3	4	5	6	7
r^2	0	1	4	9	16	25	36	49
$r^2 \bmod 8$	0	1	4	1	0	1	4	1

Table 1.1.   Squares mod 8

We see in the table that only 0, 1, or 4 can be a remainder after dividing a perfect square by 8.

(ii) Prove that 1003456789 is not a perfect square.

Dividing $n = 1003456789$ by 8 leaves remainder 5; that is, $n \equiv 5 \bmod 8$. Were n a perfect square, then $n \equiv 0$, 1, or 4 mod 8.

(iii) If m and n are positive integers, are there any perfect squares of the form $3^m + 3^n + 1$?

Again, let us look at remainders mod 8. Now $3^2 = 9 \equiv 1 \bmod 8$, and so we can evaluate $3^m \bmod 8$ as follows: if $m = 2k$, then $3^m = 3^{2k} = 9^k \equiv 1 \bmod 8$; if $m = 2k + 1$, then $3^m = 3^{2k+1} = 9^k \cdot 3 \equiv 3 \bmod 8$. Thus,

$$3^m \equiv \begin{cases} 1 \bmod 8 & \text{if } m \text{ is even;} \\ 3 \bmod 8 & \text{if } m \text{ is odd.} \end{cases}$$

Replacing numbers by their remainders after dividing by 8, we have the following possibilities for the remainder of $3^m + 3^n + 1$, depending on the parities of m and n:

$$3 + 1 + 1 \equiv 5 \bmod 8$$
$$3 + 3 + 1 \equiv 7 \bmod 8$$
$$1 + 1 + 1 \equiv 3 \bmod 8$$
$$1 + 3 + 1 \equiv 5 \bmod 8.$$

In no case is the remainder 0, 1, or 4, and so no number of the form $3^m + 3^n + 1$ can be a perfect square, by part (i).   ◄

The next result shows how congruence can simplify complicated expressions.

Proposition 1.49.   *If p is a prime and a and b are integers, then*

$$(a + b)^p \equiv a^p + b^p \bmod p.$$

Proof.   The binomial theorem gives

$$(a + b)^p = a^p + b^p + \sum_{r=1}^{p-1} \binom{p}{r} a^{p-r} b^r.$$

But Proposition 1.33 gives $\binom{p}{r} \equiv 0 \bmod p$ for $0 < r < p$, and so Proposition 1.48 gives $(a + b)^p \equiv a^p + b^p \bmod p$.   •

Theorem 1.50 (Fermat).

(i) *If p is a prime, then*
$$a^p \equiv a \bmod p$$

for every a in $\mathbb{Z}$.

(ii) *If p is a prime, then*

$$a^{p^k} \equiv a \bmod p$$

for every a in $\mathbb{Z}$ and every integer $k \geq 1$.

Proof.   (i) Assume first that $a \geq 0$; we proceed by induction on a. The base step $a = 0$ is plainly true. For the inductive step, observe that

$$(a + 1)^p \equiv a^p + 1 \bmod p,$$

by Proposition 1.49. The inductive hypothesis gives $a^p \equiv a \bmod p$, and so $(a + 1)^p \equiv a^p + 1 \equiv a + 1 \bmod p$, as desired.

Now consider $-a$, where $a \geq 0$. If $p = 2$, then $-a \equiv a$; hence, $(-a)^2 = a^2 \equiv a \equiv -a \bmod 2$. If p is an odd prime, then $(-a)^p = (-1)^p a^p \equiv (-1)^p a \equiv -a \bmod p$, as desired.

(ii) A straightforward induction on $k \geq 1$; the base step is part (i).   •

Corollary 1.51.   *A positive integer a is divisible by 3 if and only if the sum of its (decimal) digits is divisible by 3.*

Proof.   If the decimal form of a is $d_k \ldots d_1 d_0$, then

$$a = d_k 10^k + \cdots + d_1 10 + d_0.$$

Now $10 \equiv 1 \bmod 3$, so that Proposition 1.48(iii) gives $10^i \equiv 1^i = 1 \bmod 3$ for all i; thus Proposition 1.48(i') gives $a \equiv d_k + \cdots + d_1 + d_0 \bmod m$. Therefore, a is divisible by 3 if and only if $a \equiv 0 \bmod 3$ if and only if $d_k + \cdots + d_1 + d_0 \equiv 0 \bmod 3$.   •

Remark.   Since $10 \equiv 1 \bmod 9$, the same result holds if we replace 3 by 9 (it is often called *casting out 9's*): A positive integer a is divisible by 9 if and only if the sum of its (decimal) digits is divisible by 9.   ◄

Corollary 1.52.   *Let p be a prime and let n be a positive integer. If $m \geq 0$ and if Σ is the sum of the p-adic digits of m, then*

$$n^m \equiv n^\Sigma \bmod p.$$

Proof.   Let $m = d_k p^k + \cdots + d_1 p + d_0$ be the expression of m in base p. By Fermat's theorem, Theorem 1.50(ii), $n^{p^i} \equiv n \bmod p$ for all i; thus, $n^{d_i p^i} = (n^{d_i})^{p^i} \equiv n^{d_i} \bmod p$. Therefore,

$$
\begin{aligned}
n^m &= n^{d_k p^k + \cdots + d_1 p + d_0} \\
&= n^{d_k p^k} n^{d_{k-1} p^{k-1}} \cdots n^{d_1 p} n^{d_0} \\
&\equiv n^{d_k} n^{d_{k-1}} \cdots n^{d_1} n^{d_0} \bmod p \\
&\equiv n^{d_k + \cdots + d_1 + d_0} \bmod p \\
&\equiv n^\Sigma \bmod p. \quad \bullet
\end{aligned}
$$

Example 1.11.
What is the remainder after dividing 3^{12345} by 7? By Example 1.7, the 7-adic digits of 12345 are 50664. Therefore, $3^{12345} \equiv 3^{21} \bmod 7$ (because $5+0+6+6+4 = 21$). The 7-adic digits of 21 are 30 (because $21 = 3 \times 7$), and so $3^{21} \equiv 3^3 \bmod 7$ (because $3 + 0 = 3$). We conclude that $3^{12345} \equiv 3^3 = 27 \equiv 6 \bmod 7$. ◄

Theorem 1.53.   *If $(a, m) = 1$, then, for every integer b, the congruence*

$$ax \equiv b \bmod m$$

can be solved for x; in fact, $x = sb$, where $sa \equiv 1 \bmod m$. Moreover, any two solutions are congruent mod m.

Remark.   We consider the case $(a, m) \neq 1$ in Exercise 1.76. ◄

Proof.   Since $(a, m) = 1$, there is an integer s with $as \equiv 1 \bmod m$ (because there is a linear combination $1 = sa + tm$). It follows that $asb \equiv b \bmod m$, so that $x = sb$ is a solution. (Note that Proposition 1.46(i) allows us to take s with $1 \leq s < m$.)

If y is another solution, then $ax \equiv ay \bmod m$, and so $m \mid a(x - y)$. Since $(a, m) = 1$, Corollary 1.34 gives $m \mid x - y$; that is, $x \equiv y \bmod m$. $\bullet$

Corollary 1.54.   *If p is prime, the congruence* $ax \equiv b \bmod p$ *is always solvable if a is not divisible by p.*

Proof.   Since p is a prime, $p \nmid a$ implies $(a, p) = 1$.   •

If $(a, m) = 1$, then there are integers s and t with $sa + tm = 1$ (so that $sa \equiv 1 \bmod m$), and s can be found by the euclidean algorithm. However, when m is small, it is easier to find s by trying each of $r = 2, 3, \ldots, m - 1$ in turn, at each step checking whether $ra \equiv 1 \bmod m$. For example, if $m = 13$ and $a = 2$, such trials eventually lead to $7 \times 2 = 14 \equiv 1 \bmod 13$. Therefore, $s = 7$.

There are problems solved in ancient Chinese manuscripts that involve simultaneous congruences with relatively prime moduli.

Theorem 1.55 (Chinese Remainder Theorem).   *If m and m' are relatively prime, then the two congruences*

$$x \equiv b \bmod m$$
$$x \equiv b' \bmod m'$$

have a common solution, and any two solutions are congruent mod mm'.

Proof.   Every solution of the first congruence has the form $x = b + km$ for some integer k; hence, we must find k such that $b + km \equiv b' \bmod m'$; that is, $km \equiv b' - b \bmod m'$. Since $(m, m') = 1$, however, Theorem 1.53 applies at once to show that such an integer k does exist.

If y is another common solution, then both m and m' divide $x - y$; by Exercise 1.51, $mm' \mid x - y$, and so $x \equiv y \bmod mm'$.   •

Example 1.12.
Solve the simultaneous congruences

$$x \equiv 2 \bmod 5$$
$$3x \equiv 5 \bmod 13.$$

Every solution to the first congruence has the form $x = 5k + 2$ for k in $\mathbb{Z}$. Substituting into the second congruence, we have

$$3(5k + 2) \equiv 5 \bmod 13.$$

Therefore,

$$15k + 6 \equiv 5 \bmod 13$$
$$2k \equiv -1 \bmod 13.$$

Now $7 \times 2 \equiv 1 \bmod 13$, and so multiplying by 7 gives

$$k \equiv -7 \equiv 6 \bmod 13.$$

By the Chinese Remainder Theorem, all the simultaneous solutions x have the form

$$x \equiv 5k + 2 \equiv 5 \cdot 6 + 2 = 32 \bmod 65;$$

that is, the solutions are

$$\ldots, -98, -33, 32, 97, 162, \ldots . \quad \blacktriangleleft$$

Example 1.13 (A Mayan Calendar).
A congruence arises whenever there is cyclic behavior. For example, suppose we choose some particular Sunday as time zero and enumerate all the days according to the time elapsed since then. Every date now corresponds to some integer (which is negative if it occurred before time zero), and, given two days t_1 and t_2, we ask for the number $x = t_2 - t_1$ of days from one to the other. If, for example, t_1 falls on a Thursday and t_2 falls on a Tuesday, then $t_1 \equiv 4 \bmod 7$ and $t_2 \equiv 2 \bmod 7$, and so $x = t_2 - t_1 = -2 \equiv 5 \bmod 7$. Thus, $x = 7k + 5$ for some k.

About 2500 years ago, the Maya of Central America and Mexico developed three calendars (each having a different use) . Their religious calendar, called *tzolkin*, consisted of 20 "months," each having 13 days (so that the tzolkin "year" had 260 days). The months were

1. Imix	6. Cimi	11. Chuen	16. Cib
2. Ik	7. Manik	12. Eb	17. Caban
3. Akbal	8. Lamat	13. Ben	18. Etznab
4. Kan	9. Muluc	14. Ix	19. Cauac
5. Chicchan	10. Oc	15. Men	20. Ahau

Let us describe a tzolkin date by an ordered pair $\{m, d\}$, where $1 \leq m \leq 20$ and $1 \leq d \leq 13$ (thus, m denotes the month and d denotes the day). Instead of enumerating as we do (so that Imix 1 is followed by Imix 2, then by Imix 3, and so forth), the Maya let both month and day cycle simultaneously; that is, the days proceed as follows:

Imix 1, Ik 2, Akbal 3, . . . , Ben 13, Ix 1, Men 2, . . . ,
Cauac 6, Ahau 7, Imix 8, Ik 9, ...

We now ask how many days have elapsed between Oc 11 and Etznab 5. More generally, let us find the number x of days that have elapsed from tzolkin $\{m, d\}$ to tzolkin $\{m', d'\}$. As we remarked at the beginning of this example, the cyclic behavior of the days gives the congruence

$$x \equiv d' - d \bmod 13$$

(e.g., there are 13 days between Imix 1 and Ix 1; here, $x \equiv 0 \bmod 13$), while the cyclic behavior of the months gives the congruence

$$x \equiv m' - m \bmod 20$$

(e.g., there are 20 days between Imix 1 and Imix 8; here, $x \equiv 0 \bmod 20$). To answer the original question, Oc 11 corresponds to the ordered pair $\{10, 11\}$ and Etznab 5 corresponds to $\{18, 5\}$. The simultaneous congruences are thus

$$x \equiv -6 \bmod 13$$
$$x \equiv \ \ \ 8 \bmod 20.$$

Since $(13, 20) = 1$, we can solve this system as in the proof of the Chinese Remainder Theorem. The first congruence gives

$$x = 13k - 6,$$

and the second gives

$$13k - 6 \equiv \ \ 8 \bmod 20;$$

that is,

$$13k \equiv 14 \bmod 20.$$

Since $13 \times 17 = 221 \equiv 1 \bmod 20$,[14] we have $k \equiv 17 \times 14 \bmod 20$, that is,

$$k \equiv 18 \bmod 20,$$

---

[14]One finds 17 either by trying each number between 1 and 19 or by using the euclidean algorithm.

and so the Chinese Remainder Theorem gives

$$x = 13k - 6 \equiv 13 \times 18 - 6 \equiv 228 \text{ mod } 260.$$

It is not clear whether Oc 11 precedes Etznab 5 in a given year (one must look). If it does, then there are 228 days between them; otherwise, there are $32 = 260 - 228$ days between them. ◀

EXERCISES

1.66 Find all solutions x to each of the following congruences:

(i) $3x \equiv 2 \text{ mod } 5$.

(ii) $7x \equiv 4 \text{ mod } 10$.

(iii) $243x + 17 \equiv 101 \text{ mod } 725$.

(iv) $4x + 3 \equiv 4 \text{ mod } 5$.

(v) $6x + 3 \equiv 4 \text{ mod } 10$.

(vi) $6x + 3 \equiv 1 \text{ mod } 10$.

1.67 Let m be a positive integer, and let m' be an integer obtained from m by rearranging its (decimal) digits (e.g., take $m = 314159$ and $m' = 539114$). Prove that $m - m'$ is a multiple of 9.

1.68 Prove that a positive integer n is divisible by 11 if and only if the alternating sum of its digits is divisible by 11 (if the digits of a are $d_k \ldots d_2 d_1 d_0$, then their *alternating sum is $d_0 - d_1 + d_2 - \cdots$*).

1.69 What is the remainder after dividing 10^{100} by 7? (The huge number 10^{100} is called a *googol*[15] in children's stories.)

1.70 (i) Prove that $10q + r$ is divisible by 7 if and only if $q - 2r$ is divisible by 7.

(ii) Given an integer a with decimal digits $d_k d_{k-1} \ldots d_0$, define

$$a' = d_k d_{k-1} \cdots d_1 - 2d_0.$$

Show that a is divisible by 7 if and only if some one of $a', a'', a''', \ldots$ is divisible by 7. (For example, if $a = 65464$, then $a' = 6546 - 8 = 6538$, $a'' = 653 - 16 = 637$, and $a''' = 63 - 14 = 49$; we conclude that 65464 is divisible by 7.)

1.71 (i) Show that $1000 \equiv -1 \text{ mod } 7$.

(ii) Show that if $a = r_0 + 1000r_1 + 1000^2 r_2 + \cdots$, then a is divisible by 7 if and only if $r_0 - r_1 + r_2 - \cdots$ is divisible by 7.

---

[15]This word was invented by a 9-year-old boy when his uncle asked him to think up a name for the number 1 followed by a hundred zeros. At the same time, the boy suggested *googolplex* for a 1 followed by a googol zeros.

Remark.   Exercises 1.70 and 1.71 combine to give an efficient way to determine whether large numbers are divisible by 7. If $a = 33456789123987$, for example, then $a \equiv 0 \bmod 7$ if and only if $987 - 123 + 789 - 456 + 33 = 1230 \equiv 0 \bmod 7$. By Exercise 1.70, $1230 \equiv 123 \equiv 6 \bmod 7$, and so a is not divisible by 7.   ◄

1.72 Prove that there are no integers x, y, and z such that

$$x^2 + y^2 + z^2 = 999.$$

1.73 Prove that there is no perfect square a^2 whose last two digits are 35.

1.74 If x is an odd number not divisible by 3, prove that $x^2 \equiv 1 \bmod 4$.

1.75 Prove that if p is a prime and if $a^2 \equiv 1 \bmod p$, then $a \equiv \pm 1 \bmod p$.

1.76 Consider the congruence $ax \equiv b \bmod m$ when $(a, m) = d$. Show that $ax \equiv b \bmod m$ has a solution if and only if $d \mid b$.

1.77 Solve the congruence $x^2 \equiv 1 \bmod 21$.

1.78 Solve the simultaneous congruences:
  (i) $x \equiv 2 \bmod 5$ and $3x \equiv 1 \bmod 8$;
  (ii) $3x \equiv 2 \bmod 5$ and $2x \equiv 1 \bmod 3$.

1.79 How many days are there between Akbal 13 and Muluc 8 in the Mayan tzolkin calendar?

1.80   (i) Show that $(a + b)^n \equiv a^n + b^n \bmod 2$ for all a and b and for all $n \geq 1$.
  (ii) Show that $(a + b)^2 \not\equiv a^2 + b^2 \bmod 3$.

1.81 On a desert island, five men and a monkey gather coconuts all day, then sleep. The first man awakens and decides to take his share. He divides the coconuts into five equal shares, with one coconut left over. He gives the extra one to the monkey, hides his share, and goes to sleep. Later, the second man awakens and takes his fifth from the remaining pile; he too finds one extra and gives it to the monkey. Each of the remaining three men does likewise in turn. Find the minimum number of coconuts originally present.

1.6 DATES AND DAYS

Congruences can be used to determine on which day of the week a given date falls. For example, on what day of the week was July 4, 1776?

A *year* is the amount of time it takes the Earth to make one complete orbit around the Sun; a *day* is the amount of time it takes the Earth to make a complete rotation about the axis through its north and south poles. There is no reason why the number of days in a year should be an integer, and it is not; a year is approximately 365.2422 days long. In 46 B.C., Julius Caesar (and his scientific advisors) compensated for this by creating the ***Julian calendar***, containing a ***leap year*** every 4 years; that is, every fourth year has an extra day, namely, February 29, and so it contains 366 days (a ***common year*** is a year that is not a leap year). This would be fine if the year were exactly 365.25 days long, but it has the effect of making the year $365.25 - 365.2422 = .0078$ days (about 11 minutes and 14 seconds) too long. After 128 years, a full day was added to the calendar. In the year 1582, the vernal equinox (the Spring day on which there are exactly 12 hours of daylight and 12 hours of night) occurred on March 11 instead of on March 21. Pope Gregory XIII (and his scientific advisors) then installed the ***Gregorian calendar*** by erasing 10 days that year; the day after October 4, 1582 was October 15, 1582, and this caused confusion and fear among the people. The Gregorian calendar modified the Julian calendar as follows. Call a year y ending in 00 a ***century year***. If a year y is not a century year, then it is a leap year if it is divisible by 4; if y is a century year, it is a leap year only if it is divisible by 400. For example, 1900 is not a leap year, but 2000 is a leap year. The Gregorian calendar is the one in common use today, but it was not uniformly adopted throughout Europe. For example, the British did not accept it until 1752, when 11 days were erased, and the Russians did not accept it until 1918, when 13 days were erased (thus, the Russians called their 1917 revolution the October Revolution, even though it occurred in November of the Gregorian calendar).

The true number of days in 400 years is about

$$400 \times 365.2422 = 146096.88 \text{ days.}$$

The Julian calendar has

$$400 \times 365 + 100 = 146100 \text{ days}$$

in this period, while the Gregorian calendar has 146097 days (it eliminated 3 leap years from this time period). Thus, the Gregorian calendar gains only about 0.12 days (about 2 hours and 53 minutes) every 400 years.

A little arithmetic shows that there are 1628 years from 46 B.C. to 1582, which is approximately $13 \times 128 = 1662$ years. Why didn't Gregory have to erase 13 days? The first Nicene Council, meeting in the year 325, was interested in setting up the dates for Easter, which depend on the vernal equinox. The vernal equinox that year occurred on March 21, and so they decreed that the vernal equinox fall on March 21 of every year. The discrepancy observed in 1582 was thus the result of 1257 years of the Julian calendar: approximately 10 days.

Let us now seek a calendar formula. For easier calculation, we choose 0000 as our reference year, even though there was no year zero! Assign a number to each day of the week, according to the following scheme:

Sun	Mon	Tues	Wed	Thurs	Fri	Sat
0	1	2	3	4	5	6

In particular, March 1, 0000, has some number a, where $0 \le a \le 6$. Now March 1, 0001, has number $a + 1 \pmod 7$, for 365 days have elapsed from March 1, 0000, to March 1, 0001, and

$$365 = 52 \times 7 + 1 \equiv 1 \bmod 7.$$

Similarly, March 1, 0002, has number $a + 2$, and March 1, 0003, has number $a + 3$. However, March 1, 0004, has number $a + 5$, for February 29, 0004, fell between March 1, 0003, and March 1, 0004, and so $366 \equiv 2 \bmod 7$ days had elapsed since the previous March 1. We see, therefore, that every common year adds 1 to the previous number for March 1, while each leap year adds 2. Thus, if March 1, 0000, has number a, then the number a' of March 1, year y, is

$$a' \equiv a + y + L \bmod 7,$$

where L is the number of leap years from year 0000 to year y. To compute L, count all those years divisible by 4, then throw away all the century years, and then put back those century years that are leap years. Thus,

$$L = \lfloor y/4 \rfloor - \lfloor y/100 \rfloor + \lfloor y/400 \rfloor,$$

where $\lfloor x \rfloor$ denotes the *greatest integer* or *floor* in x (for example, $\lfloor 5 \rfloor = 5$ and $\lfloor \pi \rfloor = 3$). Therefore, we have

$$a' \equiv a + y + L$$
$$\equiv a + y + \lfloor y/4 \rfloor - \lfloor y/100 \rfloor + \lfloor y/400 \rfloor \bmod 7.$$

We can actually find a' by looking at a calendar. Since March 1, 1994, fell on a Tuesday,

$$2 \equiv a + 1994 + \lfloor 1994/4 \rfloor - \lfloor 1994/100 \rfloor + \lfloor 1994/400 \rfloor$$
$$\equiv a + 1994 + 498 - 19 + 4 \bmod 7$$

and so

$$a \equiv -2475 \equiv -4 \equiv 3 \bmod 7$$

(that is, March 1, year 0000, fell on Wednesday). One can now determine the day of the week on which March 1 will fall in any year $y > 0$, namely, the day corresponding to

$$3 + y + \lfloor y/4 \rfloor - \lfloor y/100 \rfloor + \lfloor y/400 \rfloor \bmod 7.$$

There is a reason we have been discussing March 1. Life would have been simpler had Julius Caesar decreed that the extra day of a leap year be December 32 instead of February 29.[16] Let us now analyze February 28. For example, suppose that February 28, 1600, has number b. As 1600 is a leap year, February 29, 1600, occurs between February 28, 1600, and February 28, 1601; hence, 366 days have elapsed between these two February 28's, so that February 28, 1601, has number $b + 2$. February 28, 1602, has number $b + 3$, February 28, 1603, has number $b + 4$, February 28, 1604, has number $b + 5$, but February 28, 1605, has number $b + 7$ (for there was a February 29 in 1604).

Let us compare the pattern of behavior of February 28, 1600, namely, b, $b+2, b+3, b+4, b+5, b+7, \ldots$, with that of some date in 1599. If May 26, 1599, has number c, then May 26, 1600, has number $c + 2$, for February 29, 1600, comes between these two May 26's, and so there are $366 \equiv 2 \bmod 7$ intervening days. The numbers of the next few May 26's, beginning with May 26, 1601, are $c + 3, c + 4, c + 5, c + 7$. We see that the pattern of the days for February 28, starting in 1600, is exactly the same as the pattern of the days for May 26, starting in 1599; indeed, the same is true for any date in January or February. Thus, the pattern of the days for any date in January or February of a year y is the same as the pattern for a date occurring in the preceding year $y - 1$: a year preceding a leap year adds 2 to the number for such a date, whereas all other years add 1. Therefore, we revert to the ancient calendar by making New Year's Day fall on March 1; any date in January or February is treated as if it had occurred in the previous year.

How do we find the day corresponding to a date other than March 1? Since March 1, 0000, has number 3 (as we have seen above), April 1, 0000, has number 6, for March has 31 days and $3 + 31 \equiv 6 \bmod 7$. Since April has 30 days, May 1, 0000, has number $6 + 30 \equiv 1 \bmod 7$. Here is the table giving

---

[16] Actually, March 1 was the first day of the year in the old Roman calendar. This explains why the leap day was added onto February and not onto some other month. It also explains why months 9, 10, 11, and 12, namely, September, October, November, and December, are so named; originally, they were months 7, 8, 9, and 10.

the number of the first day of each month in year 0000:

March 1, 0000, has number	3
April 1	6
May 1	1
June 1	4
July 1	6
August 1	2
September 1	5
October 1	0
November 1	3
December 1	5
January 1	1
February 1	4

Remember that we are pretending that March is month 1, April is month 2, etc. Let us denote these numbers by $1 + j(m)$, where $j(m)$, for $m = 1, 2, \ldots, 12$, is defined by

$$j(m) : \ 2, 5, 0, 3, 5, 1, 4, 6, 2, 4, 0, 3.$$

It follows that month m, day 1, year y, has number

$$1 + j(m) + g(y) \bmod 7,$$

where

$$g(y) = y + \lfloor y/4 \rfloor - \lfloor y/100 \rfloor + \lfloor y/400 \rfloor.$$

Proposition 1.56 (Calendar[17] Formula).   *The date with month m, day d, year y has number*

$$d + j(m) + g(y) \bmod 7,$$

where

$$j(m) = 2, 5, 0, 3, 5, 1, 4, 6, 2, 4, 0, 3,$$

(March corresponds to m = 1, April to m = 2, ..., February to m = 12) and

$$g(y) = y + \lfloor y/4 \rfloor - \lfloor y/100 \rfloor + \lfloor y/400 \rfloor,$$

provided that dates in January and February are treated as having occurred in the previous year.

---

[17] The word *calendar* comes from the Greek "to call," which evolved into the Latin for the first day of a month (when accounts were due).

Proof. We have seen that the number mod 7 corresponding to month m, day 1, year y, is $1 + j(m) + g(y)$. It follows that $2 + j(m) + g(y)$ corresponds to month m, day 2, year y, and, more generally, that $d + j(m) + g(y)$ corresponds to month m, day d, year y. •

Example 1.14.
Let us use the calendar formula to find the day of the week on which July 4, 1776, fell. Here $m = 5$, $d = 4$, and $y = 1776$. Substituting in the formula, we obtain the number

$$4 + 5 + 1776 + 444 - 17 + 4 = 2217 \equiv 4 \bmod 7;$$

therefore, July 4, 1776, fell on a Thursday. ◀

Most of us need paper and pencil (or a calculator) to use the calendar formula in the theorem. Here are some ways to simplify the formula so that one can do the calculation in one's head and amaze one's friends.

One mnemonic for $j(m)$ is given by

$$j(m) = \lfloor 2.6m - 0.2 \rfloor, \quad \text{where } 1 \le m \le 12.$$

Another mnemonic for $j(m)$ is the sentence:

My Uncle Charles has eaten a cold supper; he eats nothing hot.

| 2 | 5 | $7 \equiv 0$ | 3 | 5 | 1 | 4 | 6 | 2 | 4 | $7 \equiv 0$ | 3 |

Corollary 1.57. *The date with month m, day d, year $y = 100C + N$, where $0 \le N \le 99$, has number*

$$d + j(m) + N + \lfloor N/4 \rfloor + \lfloor C/4 \rfloor - 2C \bmod 7,$$

provided that dates in January and February are treated as having occurred in the previous year.

Proof. If we write a year $y = 100C + N$, where $0 \le N \le 99$, then

$$y = 100C + N \equiv 2C + N \bmod 7,$$

$$\lfloor y/4 \rfloor = 25C + \lfloor N/4 \rfloor \equiv 4C + \lfloor N/4 \rfloor \bmod 7,$$

$$\lfloor y/100 \rfloor = C, \text{ and } \lfloor y/400 \rfloor = \lfloor C/4 \rfloor.$$

Therefore,

$$y + \lfloor y/4 \rfloor - \lfloor y/100 \rfloor + \lfloor y/400 \rfloor \equiv N + 5C + \lfloor N/4 \rfloor + \lfloor C/4 \rfloor \bmod 7$$
$$\equiv N + \lfloor N/4 \rfloor + \lfloor C/4 \rfloor - 2C \bmod 7. \quad \bullet$$

This formula is simpler than the first one. For example, the number corresponding to July 4, 1776, is now obtained as

$$4 + 5 + 76 + 19 + 4 - 34 = 74 \equiv 4 \bmod 7,$$

agreeing with our previous calculation in Example 1.14. The reader may now discover the day of his or her birth.

Example 1.15.
The birthday of Amalia, the grandmother of Danny and Ella, is December 5, 1906; on what day of the week was she born?

If A is the number of the day, then

$$A \equiv 5 + 4 + 6 + \lfloor 6/4 \rfloor + \lfloor 19/4 \rfloor - 38$$
$$\equiv -18 \bmod 7$$
$$\equiv 3 \bmod 7.$$

Amalia was born on a Wednesday.   ◄

Does every year y contain a Friday 13? We have

$$5 \equiv 13 + j(m) + g(y) \bmod 7.$$

The question is answered positively if the numbers $j(m)$, as m varies from 1 through 12, give all the remainders 0 through 6 mod 7. And this is what happens. The sequence of remainders mod 7 is

$$2, 5, \boxed{0}, \boxed{3}, \boxed{5}, \boxed{1}, \boxed{4}, \boxed{6}, \boxed{2}, 4, 0, 3.$$

Indeed, we see that there must be a Friday 13 occurring between May and November. No number occurs three times on the list, but it is possible that there are three Friday 13's in a year because January and February are viewed as having occurred in the previous year; for example, there were three Friday

13s in 1987 (see Exercise 1.85). Of course, we may replace Friday by any other day of the week, and we may replace 13 by any number between 1 and 28.

J. H. Conway has found an even simpler calendar formula. In his system, he calls **doomsday** of a year that day of the week on which the last day of February occurs. For example, doomsday 1900, corresponding to February 28, 1900 (1900 is not a leap year), is Wednesday = 3, while doomsday 2000, corresponding to February 29, 2000, is Tuesday = 2, as we know from Corollary 1.57.

Knowing the doomsday of a century year $100C$, one can find the doomsday of any other year $y = 100C + N$ in that century, as follows. Since $100C$ is a century year, the number of leap years from $100C$ to y does not involve the Gregorian alteration. Thus, if D is doomsday $100C$ (of course, $0 \leq D \leq 6$), then doomsday $100C + N$ is congruent to

$$D + N + \lfloor N/4 \rfloor \bmod 7.$$

For example, since doomsday 1900 is Wednesday = 3, we see that doomsday 1994 is Monday = 1, for

$$3 + 94 + 23 = 120 \equiv 1 \bmod 7.$$

Proposition 1.58 (Conway). *Let D be doomsday $100C$, and let $0 \leq N \leq 99$. If $N = 12q + r$, where $0 \leq r < 12$, then the formula for doomsday $100C + N$ is*

$$D + q + r + \lfloor r/4 \rfloor \bmod 7.$$

Proof.

$$
\begin{aligned}
\text{Doomsday } (100C + N) &\equiv D + N + \lfloor N/4 \rfloor \\
&\equiv D + 12q + r + \lfloor (12q + r)/4 \rfloor \\
&\equiv D + 15q + r + \lfloor r/4 \rfloor \\
&\equiv D + q + r + \lfloor r/4 \rfloor \bmod 7. \quad \bullet
\end{aligned}
$$

For example, $94 = 12 \times 7 + 10$, so that doomsday 1994 is $3 + 7 + 10 + 2 \equiv 1 \bmod 7$; that is, doomsday 1994 is Monday, as we saw above.

Once one knows doomsday of a particular year, one can use various tricks (e.g., my Uncle Charles) to pass from doomsday to any other day in the year.

Conway observes that some other dates falling on the same day of the week as the doomsday are

April 4, June 6, August 8, October 10, December 12,

May 9, July 11, September 5, and November 7;

it is easier to remember these dates using the notation

4/4, 6/6, 8/8, 10/10, 12/12 and 5/9, 7/11, 9/5, and 11/7,

where m/d denotes month/day (we now return to the usual counting having January as the first month: $1 =$ January). Since doomsday corresponds to the last day of February, we are now within a few weeks of any date in the calendar, and we can easily interpolate to find the desired day.

EXERCISES

1.82 A suspect said that he had spent the Easter holiday April 21, 1893, with his ailing mother; Sherlock Holmes challenged his veracity at once. How could the great detective have been so certain?

1.83 How many times in 1900 did the first day of the month fall on a Tuesday?

1.84 On what day of the week did February 29, 1896, fall? Conclude from your method of solution that no extra fuss is needed to find leap days.

1.85   (i) Show that 1987 had three Friday 13's.

    (ii) Show, for any year $y > 0$, that $g(y) - g(y-1) = 1$ or 2, where $g(y) = y + \lfloor y/4 \rfloor - \lfloor y/100 \rfloor + \lfloor y/400 \rfloor$.

    (iii) Can there be a year with exactly one Friday 13?

2

Groups I

2.1 FUNCTIONS

We are going to study algebraic systems called *groups*, which involve objects that can be "multiplied"; in the next chapter we will study *rings*, which involve objects that can be multiplied and added. Since the best examples of groups are built from permutations, which are special kinds of functions, we begin by discussing functions. (Of course, functions arise in all parts of mathematics.)

A set X is a collection of elements (numbers, points, herring, etc.); one writes

$$x \in X$$

to denote x belonging to X. The terms *set*, *element*, and *belongs to* are undefined terms (there have to be such in any language), and they are used so that a set is determined by the elements in it. Thus, we define two sets X and Y to be *equal*, denoted by

$$X = Y,$$

if they are comprised of exactly the same elements; for every element x, we have $x \in X$ if and only if $x \in Y$.

A *subset* of a set X is a set S each of whose elements also belongs to X: if $s \in S$, then $s \in X$. One denotes S being a subset of X by

$$S \subset X;$$

synonyms for this are S is **contained** in X and S is **included** in X. (Note that $X \subset X$ is always true; we say that a subset S of X is a **proper subset** of X if $S \subset X$ and $S \neq X$.) It follows from the definitions that two sets X and Y are equal if and only if each is a subset of the other:

$$X = Y \qquad \text{if and only if} \qquad X \subset Y \text{ and } Y \subset X.$$

Because of this remark, many proofs showing that two sets are equal break into two parts, each half showing that one of the sets is a subset of the other. For example, let

$$X = \{a \in \mathbb{R} : a \geq 0\} \qquad \text{and} \qquad Y = \{r^2 : r \in \mathbb{R}\}.$$

If $a \in X$, then $a \geq 0$ and $a = r^2$, where $r = \sqrt{a}$; hence, $a \in Y$ and $X \subset Y$. For the reverse inclusion, choose $r^2 \in Y$. If $r \geq 0$, then $r^2 \geq 0$; if $r < 0$, then $r = -s$, where $s > 0$, and $r^2 = (-1)^2 s^2 = s^2 \geq 0$. In either case, $r^2 \geq 0$ and $r^2 \in X$. Therefore, $Y \subset X$, and so $X = Y$.

The idea of a *function* occurs in calculus (and earlier); examples are x^2, $\sin x$, $\sqrt{x}$, $1/x$, $x + 1$, e^x, etc. Calculus books define a function $f(x)$ as a "rule" that assigns, to each number a, exactly one number, namely, $f(a)$. Thus, the squaring function assigns the number 81 to the number 9; the square root function assigns the number 3 to the number 9. Notice that there are two candidates for $\sqrt{9}$, namely, 3 and -3. In order that there be exactly one number assigned to 9, one must select one of the two possible values ± 3; everyone has agreed that $\sqrt{x} \geq 0$ whenever $x \geq 0$, and so $\sqrt{x}$ is a function.

The calculus definition of function is certainly in the right spirit, but it has a defect: What is a rule? To ask this question another way, when are two rules the same? For example, consider the functions

$$f(x) = (x + 1)^2 \qquad \text{and} \qquad g(x) = x^2 + 2x + 1.$$

Is $f(x) = g(x)$? The evaluation procedures are certainly different: for example, $f(6) = (6+1)^2 = 7^2$, while $g(6) = 6^2 + 2 \cdot 6 + 1 = 36 + 12 + 1$. Since the term *rule* has not been defined, it is ambiguous, and our question cannot be answered. Surely the calculus description is inadequate if one cannot decide whether these two functions are the same.

To find a reasonable definition, let us return to examples of what we seek to define. Each of the functions x^2, $\sin x$, etc., has a *graph*, namely, the subset of the plane consisting of all those points of the form $(a, f(a))$. For example, the graph of $f(x) = x^2$ is the parabola consisting of all the points of the form (a, a^2).

A graph is a concrete thing, and the upcoming formal definition of a function amounts to saying that a function *is* its graph. The informal calculus definition of a function as a rule remains, but we will have avoided the problem of saying what a rule is. In order to give the definition, we first need an analog of the plane (for we will want to use functions $f(x)$ whose argument x does not vary over numbers).

Definition. If X and Y are (not necessarily distinct) sets, then their ***cartesian***[1] ***product*** $X \times Y$ is the set of all ordered pairs (x, y), where $x \in X$ and $y \in Y$.

The plane is $\mathbb{R} \times \mathbb{R}$.
The only thing one needs to know about ordered pairs is that

$$(x, y) = (x', y') \quad \text{if and only if} \quad x = x' \text{ and } y = y'$$

(see Exercise 2.12).
Observe that if X and Y are finite sets, say, $|X| = m$ and $|Y| = n$ (we denote the number of elements in a finite set X by $|X|$), then $|X \times Y| = mn$.

Definition. Let X and Y be (not necessarily distinct) sets. A ***function*** f from X to Y, denoted by

$$f : X \to Y,$$

is a subset $f \subset X \times Y$ such that, for each $a \in X$, there is a unique $b \in Y$ with $(a, b) \in f$.

For each $a \in X$, the unique element $b \in Y$ for which $(a, b) \in f$ is called the ***value*** of f at a, and b is denoted by $f(a)$. Thus, f consists of all those points in $X \times Y$ of the form $(a, f(a))$. When $f : \mathbb{R} \to \mathbb{R}$, then f *is* the graph of $f(x)$.

Some other examples of functions are: the ***identity function*** on a set X, denoted by $1_X : X \to X$, defined by $1_X(x) = x$ for every $x \in X$ [when $X = \mathbb{R}$, the graph of the identity function is the 45° line consisting of all points of the form (a, a)]; ***constant functions***: if $y_0 \in Y$, then $f(x) = y_0$ for all $x \in X$ (when $X = \mathbb{R} = Y$, then the graphs of constant functions are horizontal lines).

From now on, we depart from the calculus notation; we denote a function by f and not by $f(x)$; the latter notation is reserved for the value of f at an

---

[1] This term honors R. Descartes, one of the founders of analytic geometry.

element x (there are a few exceptions; we will continue to write the familiar functions, e.g., polynomials, $\sin x$, e^x, $\sqrt{x}$, $\log x$, as usual). Here are some more words. Call X the **domain** of f, call Y the **target** (or *codomain*) of f, and define the **image** (or *range*) of f, denoted by im f, to be the subset of Y consisting of all the values of f. For example, the domain of $\sin x$ is $\mathbb{R}$, its target is usually $\mathbb{R}$, and its image is $[-1, 1]$. The domain of $1/x$ is the set of all nonzero reals, and the domain of the square root function is the set $\{x \in \mathbb{R} : x \geq 0\}$ of all nonnegative reals.

Definition.   Two functions $f : X \to Y$ and $g : X' \to Y'$ are **equal** if $X = X'$, $Y = Y'$, and the subsets $f \subset X \times Y$ and $g \subset X' \times Y'$ are equal.

A function $f: X \to Y$ has three ingredients: its domain X, its target Y, and its graph, and we are saying that two functions are equal if and only if they have the same domains, the same targets, and the same graphs. It is plain that the domain and the graph are essential parts of a function, and some reasons for caring about the target are given in a remark at the end of this section.

Here are more examples of functions.

Definition.   If $f: X \to Y$ is a function, and if S is a subset of X, then the **restriction** of f to S is the function $f|S : S \to Y$ defined by $(f|S)(s) = f(s)$ for all $s \in S$.

If S is a subset of a set X, define the **inclusion** $i : S \to X$ to be the function defined by $i(s) = s$ for all $s \in S$.

If S is a proper subset of X, then the inclusion i is not the identity function 1_S because its target is X, not S; it is not the identity function 1_X because its domain is S, not X.

Proposition 2.1.   *Let $f : X \to Y$ and $g : X \to Y$ be functions. Then $f = g$ if and only if $f(a) = g(a)$ for every $a \in X$.*

Remark.   This proposition resolves the problem raised by the ambiguous term *rule*. If $f, g : \mathbb{R} \to \mathbb{R}$ are given by $f(x) = (x + 1)^2$ and $g(x) = x^2 + 2x + 1$, then $f = g$ because $f(a) = g(a)$ for every number a. ◄

Proof.   Assume that $f = g$. Functions are subsets of $X \times Y$, and so $f = g$ means that each of f and g is a subset of the other (informally, we are saying that f and g have the same graph). If $a \in X$, then $(a, f(a)) \in f = g$, and so $(a, f(a)) \in g$. But there is only one ordered pair in g with first coordinate a, namely, $(a, g(a))$ [because the definition of function says that g gives a unique value to a]. Therefore, $(a, f(a)) = (a, g(a))$, and equality of ordered pairs gives $f(a) = g(a)$, as desired.

Conversely, assume that $f(a) = g(a)$ for every $a \in X$. To see that $f = g$, it suffices to show that $f \subset g$ and $g \subset f$. Each element of f has the form $(a, f(a))$. Since $f(a) = g(a)$, we have $(a, f(a)) = (a, g(a))$, and hence $(a, f(a)) \in g$. Therefore, $f \subset g$. The reverse inclusion $g \subset f$ is proved similarly. •

Let us make the contrapositive explicit: If $f, g : X \to Y$ are functions that disagree at even one point, i.e., if there is some $a \in X$ with $f(a) \neq g(a)$, then $f \neq g$.

We continue to regard a function f as a rule sending $x \in X$ to $f(x) \in Y$, but the precise definition is now available whenever we need it, as in Proposition 2.1. However, to reinforce our wanting to regard functions $f : X \to Y$ as dynamic things sending points in X to points in Y, we often write

$$f : x \mapsto y$$

instead of $f(x) = y$. For example, we may write $f : x \mapsto x^2$ instead of $f(x) = x^2$, and we may describe the identity function by $x \mapsto x$ for all x.

The special case when the image is the whole target has a name.

Definition.   A function $f : X \to Y$ is **surjective** (or *onto*) if

$$\operatorname{im} f = Y.$$

Thus, f is surjective if, for each $y \in Y$, there is some $x \in X$ (probably depending on y) with $y = f(x)$.

Example 2.1.
(i) Of course, identity functions are surjections.

(ii) The sine function $\mathbb{R} \to \mathbb{R}$ is not surjective, for its image is $[-1, 1]$ which is a proper subset of its target $\mathbb{R}$.

(iii) The functions x^2 and e^x have domain $\mathbb{R}$ and target $\mathbb{R}$. Now $\operatorname{im} x^2$ consists of the nonnegative reals and $\operatorname{im} e^x$ consists of the positive reals, and so neither $x^2 : \mathbb{R} \to \mathbb{R}$ nor $e^x : \mathbb{R} \to \mathbb{R}$ is surjective.

(iv) Let $f : \mathbb{R} \to \mathbb{R}$ be defined by

$$f(a) = 6a + 4.$$

To see whether f is a surjection, we ask whether every $b \in \mathbb{R}$ has the form $b = f(a)$ for some a; that is, given b, can one find a so that

$$6a + 4 = b?$$

One can always solve this equation for a, obtaining $a = \frac{1}{6}(b - 4)$. Therefore, f is a surjection.

(v) Let $f : \mathbb{R} \to \mathbb{R}$ be defined by

$$f(a) = \frac{6a + 4}{2a - 3}.$$

To see whether f is a surjection, we seek a solution a for a given b: can one always solve

$$\frac{6a + 4}{2a - 3} = b?$$

One is led to the equation $a(6 - 2b) = -3b - 4$, which can be solved for a if $6 - 2b \neq 0$. On the other hand, it suggests that there is no solution when $b = 3$ and, indeed, there is not: if $(6a + 4)/(2a - 3) = 3$, cross multiplying gives the false equation $6a + 4 = 6a - 9$. Thus, $3 \notin \operatorname{im} f$, and f is not a surjection.   ◀

Instead of saying that values of a function f are unique, one sometimes says that f is **single-valued** or that f is *well-defined* . For example, if $\mathbb{R}^+$ denotes the set of nonnegative reals, then $\sqrt{} : \mathbb{R}^+ \to \mathbb{R}^+$ is a function because we have agreed that $\sqrt{a} \geq 0$ for every positive number a. On the other hand, $f(a) = \pm\sqrt{a}$ is not single-valued, and hence it is not a function.

The simplest way to verify whether an alleged function f is single-valued is to rephrase uniqueness of values; the contrapositive states

$$\text{if } a = a', \text{ then } f(a) = f(a').$$

Does the formula $g(a/b) = ab$ define a function $g : \mathbb{Q} \to \mathbb{Q}$? There are many ways to write a fraction; since $\frac{1}{2} = \frac{3}{6}$, we see that $g(\frac{1}{2}) = 1 \cdot 2 \neq 3 \cdot 6 = g\left(\frac{3}{6}\right)$, and so g is not a function. Had we said that the formula $g(a/b) = ab$ holds whenever a/b is in lowest terms, then g would be a function. As another example, the formula $f(a/b) = 3a/b$ does define a function $f : \mathbb{Q} \to \mathbb{Q}$ for it is single-valued: if $a/b = a'/b'$, we show that

$$f(a/b) = 3a/b = 3a'/b' = f(a'/b'):$$

$a/b = a'/b'$ gives $ab' = a'b$, so that $3ab' = 3a'b$ and $3a/b = 3a'/b'$. Thus, f is a bona fide function.

The following definition gives another important property a function may have.

Definition.   A function $f : X \rightarrow Y$ is **injective** (or *one-to-one*) if, whenever a and a' are distinct elements of X, then $f(a) \neq f(a')$. Equivalently, (the contrapositive states that) f is injective if, for every pair $a, a' \in X$, we have

$$f(a) = f(a') \text{ implies } a = a'.$$

The reader should note that being injective is the converse of being single-valued: f is single-valued if $a = a'$ implies $f(a) = f(a')$; f is injective if $f(a) = f(a')$ implies $a = a'$.

Most functions are neither injective nor surjective. For example, the squaring function $f : \mathbb{R} \rightarrow \mathbb{R}$, defined by $f(x) = x^2$, is neither.

Example 2.2.
(i) Identity functions 1_X are injective.

(ii) Let $f : \mathbb{R} \rightarrow \mathbb{R}$ be defined by

$$f(a) = \frac{6a + 4}{2a - 3}.$$

To check whether f is injective, suppose that $f(a) = f(b)$:

$$\frac{6a + 4}{2a - 3} = \frac{6b + 4}{2b - 3}.$$

Cross multiplying yields

$$12ab + 8b - 18a - 12 = 12ab + 8a - 18b - 12,$$

which simplifies to $26a = 26b$ and hence $a = b$. We conclude that f is injective. (We saw, in Example 2.1(v), that f is not surjective.)

(iii) Consider $f : \mathbb{R} \rightarrow \mathbb{R}$ given by $f(x) = x^2 - 2x - 3$. If we try to check whether f is an injection by looking at the consequences of $f(a) = f(b)$, as in part (i), we arrive at the equation $a^2 - 2a = b^2 - 2b$; it is not instantly clear whether this forces $a = b$. Instead, we seek the roots of $f(x)$, which are 3 and -1. It follows that f is not injective, for $f(3) = 0 = f(-1)$; there are two distinct numbers having the same value.   ◄

Sometimes there is a way of "multiplying" two functions to form their *composite*.

Definition. If $f : X \to Y$ and $g : Y \to Z$ are functions (note that the target of f is equal to the domain of g), then their *composite*, denoted by $g \circ f$, is the function $X \to Z$ given by

$$g \circ f : x \mapsto g(f(x));$$

that is, first evaluate f on x and then evaluate g on $f(x)$.

Composition is thus a two-step process: $x \mapsto f(x) \mapsto g(f(x))$. For example, the function $h : \mathbb{R} \to \mathbb{R}$, defined by $h(x) = e^{\cos x}$, is the composite $g \circ f$, where $f(x) = \cos x$ and $g(x) = e^x$. This "factorization" is plain as soon as one tries to evaluate, say, $h(\pi)$; one must first evaluate $f(\pi) = \cos \pi = -1$ and then evaluate $g(f(\pi)) = g(-1) = e^{-1}$ in order to evaluate $h(\pi)$. The chain rule in calculus is a formula for computing the derivative $D(g \circ f)$ in terms of Dg and Df (see Exercise 2.3).

If $\mathbb{N}$ denotes the *natural numbers*,

$$\mathbb{N} = \{n \in \mathbb{Z} : n \geq 0\},$$

consider $f : \mathbb{N} \to \mathbb{N}$, given by $f(n) = n!$ and $g : \mathbb{N} \to \mathbb{R}$, given by $g(n) = \sqrt{n}$. The composite $g \circ f : n \mapsto \sqrt{n!}$ is defined, but the composite $f \circ g$ is not defined (what is $\pi!$?). Of course, if $f : X \to Y$ and $g : Y \to X$, then both composites $g \circ f$ and $f \circ g$ are defined. On the other hand, they need not be equal. For example, define $f, g : \mathbb{N} \to \mathbb{N}$ by $f : n \mapsto n^2$ and $g : n \mapsto 3n$; then $g \circ f : 2 \mapsto g(4) = 12$ and $f \circ g : 2 \mapsto f(6) = 36$. Hence, $g \circ f \neq f \circ g$.

Given a set X, let

$$\mathcal{F}(X) = \{\text{all functions } X \to X\}.$$

We have just seen that the composite of two functions in $\mathcal{F}(X)$ is always defined; moreover, the composite is again a function in $\mathcal{F}(X)$. We may thus regard $\mathcal{F}(X)$ as being equipped with a kind of multiplication. As we have just seen, this multiplication is not *commutative*; that is, $f \circ g$ and $g \circ f$ need not be equal. Let us now show that composition is always *associative*.

Lemma 2.2. *Composition of functions is associative: if*

$$f : X \to Y, \quad g : Y \to Z, \quad and \quad h : Z \to W$$

are functions, then
$$h \circ (g \circ f) = (h \circ g) \circ f.$$

Proof.   We show that the value of either composite on an element $a \in X$ is just $w = h(g(f(a)))$. If $x \in X$, then

$$h \circ (g \circ f) : x \mapsto (g \circ f)(x) = g(f(x)) \mapsto h(g(f(x))) = w,$$

and

$$(h \circ g) \circ f : x \mapsto f(x) \mapsto (h \circ g)(f(x)) = h(g(f(x))) = w.$$

It follows from Proposition 2.1 that the composites are equal.[2]   •

The next result implies that 1_X behaves for the multiplication in $\mathcal{F}(X)$ just as the number one does for ordinary multiplication of numbers.

Lemma 2.3.   *If $f : X \to Y$, then $1_Y \circ f = f = f \circ 1_X$.*

Proof.   If $x \in X$, then

$$1_Y \circ f : x \mapsto f(x) \mapsto f(x)$$

and

$$f \circ 1_X : x \mapsto x \mapsto f(x).   \bullet$$

Are there "reciprocals" in $\mathcal{F}(X)$; that is, are there any functions f for which there is $g \in \mathcal{F}(X)$ with $f \circ g = 1_X$ and $g \circ f = 1_X$?

Definition.   A function $f : X \to Y$ is **bijective** (or a *one-one correspondence*) if it is both injective and surjective.

Example 2.3.
(i) Identity functions are always bijections.
(ii) Let $X = \{1, 2, 3\}$ and define $f : X \to X$ by

$$f(1) = 2, \quad f(2) = 3, \quad f(3) = 1.$$

It is easy to see that f is a bijection.   ◄

---

[2]The reader should now understand why one may be tempted to omit parentheses and write fx instead of $f(x)$.

Example 2.4.

Our definitions allow us to treat a degenerate case. If X is a set, what are the functions $X \to \varnothing$? Note first that an element of $X \times \varnothing$ is an ordered pair (x, y) with $x \in X$ and $y \in \varnothing$; since there is no $y \in \varnothing$, there are no such ordered pairs, and so $X \times \varnothing = \varnothing$. Now a function $X \to \varnothing$ is a subset of $X \times \varnothing$ of a certain type; but $X \times \varnothing = \varnothing$, so there is only one subset, namely $\varnothing$, and hence at most one function, namely, $f = \varnothing$. The definition of function $X \to \varnothing$ says that, for each $x \in X$, there exists a unique $y \in \varnothing$ with $(x, y) \in f$. If $X \neq \varnothing$, then there exists $x \in X$ for which no such y exists (there are no elements y at all in $\varnothing$), and so f is not a function. Thus, if $X \neq \varnothing$, there are no functions from X to $\varnothing$. On the other hand, if $X = \varnothing$, then $f = \varnothing$ is a function. Otherwise, the negation of the statement "f is a function" would be true: "there exists $x \in \varnothing$, etc." We need not go on; since $\varnothing$ has no elements in it, there is no way to complete the sentence so that it is a true statement. We conclude that $f = \varnothing$ is a function $\varnothing \to \varnothing$, and we declare it to be the identity function $1_\varnothing$. It follows that $1_\varnothing$ is a bijection, for $1_\varnothing \circ 1_\varnothing = 1_\varnothing$. ◄

We can draw a picture of a function in the special case when X and Y are finite sets. Let $X = \{1, 2, 3, 4, 5\}$, let $Y = \{a, b, c, d, e\}$, and define $f : X \to Y$ by

$$f(1) = b; \qquad f(2) = e; \qquad f(3) = a; \qquad f(4) = b; \qquad f(5) = c.$$

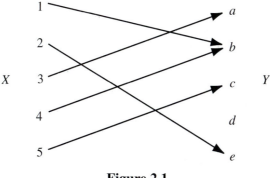

Figure 2.1

We see that f is not injective because $f(1) = b = f(4)$, and f is not surjective because there is no $x \in X$ with $f(x) = d$. Can one reverse the arrows to get a function $g : Y \to X$? There are two reasons why one cannot.

First, there is no arrow going to d, and so $g(d)$ is not defined. Second, what is $g(b)$? Is it 1 or 4? The first problem is that the domain of g is not all of Y, and it arises because f is not surjective; the second problem is that g is not single-valued, and it arises because f is not injective (this reflects the fact that being single-valued is the converse of being injective). Therefore, neither problem arises when f is a bijection.

Definition.   A function $f : X \to Y$ has an *inverse* if there is a function $g : Y \to X$ with both composites $g \circ f$ and $f \circ g$ being identity functions.

We do not say that every function f has an inverse; on the contrary, we have just analyzed the reasons that most functions do not have an inverse. Notice that if an inverse function g does exist, then it "reverses the arrows" in Figure 2.1. If $f(a) = y$, then there is an arrow from a to y. Now $g \circ f$ being the identity says that $a = (g \circ f)(a) = g(f(a)) = g(y)$; therefore $g : y \mapsto a$, and so the picture of g is obtained from the picture of f by reversing arrows. If f twists something, then its inverse g untwists it.

Lemma 2.4.   *If $f : X \to Y$ and $g : Y \to X$ are functions such that $g \circ f = 1_X$, then f is injective and g is surjective.*

Proof.   Suppose that $f(a) = f(a')$; apply g to obtain $g(f(a)) = g(f(a'))$; that is, $a = a'$ [because $g(f(a)) = a$], and so f is injective. If $y \in Y$, then $y = g(f(y))$, so that $y \in \text{im } g$; hence g is surjective.   ●

Lemma 2.5.   *A function $f : X \to Y$ has an inverse $g : Y \to X$ if and only if it is a bijection.*

Proof.   If f has an inverse g, then Lemma 2.4 shows that f is injective and surjective, for both composites $g \circ f$ and $f \circ g$ are identities.

Assume that f is a bijection. For each $y \in Y$, there is $a \in X$ with $f(a) = y$, since f is surjective, and this element a is unique because f is injective. Defining $g(y) = a$ thus gives a (single-valued) function whose domain is Y, and it is plain that g is the inverse of f; that is, $f(g(y)) = f(a) = y$ for all $y \in Y$ and $g(f(a)) = g(y) = a$ for all $a \in X$.   ●

Notation.   The inverse of a bijection f is denoted by f^{-1} (Exercise 2.4 says that a function cannot have two inverses). This is the same notation used for inverse trigonometric functions in calculus; e.g., $\sin^{-1} x = \arcsin x$ satisfies $\sin(\arcsin(x)) = x$ and $\arcsin(\sin(x)) = x$.

Example 2.5.
Here is an example of two functions f and g one of whose composites $g \circ f$ is the identity while the other composite $f \circ g$ is not the identity; thus, f and g are not inverse functions.
   Define $f, g : \mathbb{N} \to \mathbb{N}$ as follows:

$$f(n) = n + 1;$$

$$g(n) = \begin{cases} 0 & \text{if } n = 0 \\ n - 1 & \text{if } n \geq 1. \end{cases}$$

The composite $g \circ f = 1_{\mathbb{N}}$, for $g(f(n)) = g(n+1) = n$ (because $n+1 \geq 1$). On the other hand, $f \circ g \neq 1_{\mathbb{N}}$ because $f(g(0)) = f(0) = 1 \neq 0$. ◀

Example 2.6.
If a is a real number, then ***multiplication by*** a is the function $\mu_a : \mathbb{R} \to \mathbb{R}$ defined by $r \mapsto ar$ for all $r \in \mathbb{R}$. If $a \neq 0$, then μ_a is a bijection whose inverse function is ***division by*** a, namely, $\delta_a : \mathbb{R} \to \mathbb{R}$, defined by $r \mapsto \frac{1}{a}r$. If $a = 0$, however, then $\mu_a = \mu_0$ is the constant function $\mu_0 : r \mapsto 0$ for all $r \in \mathbb{R}$, which has no inverse function because it is not a bijection. ◀

   Two strategies are now available to determine whether or not a given function is a bijection: use the definitions of injective and surjective; find an inverse. For example, if $\mathbb{R}^+$ denotes the positive real numbers, let us show that the exponential function $f : \mathbb{R} \to \mathbb{R}^+$, defined by $f(x) = e^x = \sum x^n/n!$, is a bijection. It is simplest to use the (natural) logarithm $g(y) = \ln y = \int_1^y dt/t$. The usual formulas $e^{\ln y} = y$ and $\ln e^x = x$ say that both composites $f \circ g$ and $g \circ f$ are identities, and so f and g are inverse functions. Therefore, f is a bijection, for it has an inverse. (A direct proof that f is an injection would require showing that if $e^a = e^b$, then $a = b$; a direct proof showing that f is surjective would involve showing that every positive real number c has the form e^a for some a.)
   Let us summarize the results of this section.

Proposition 2.6.   *If the set of all the bijections from a set X to itself is denoted by S_X, then composition of functions satisfies the following properties:*

   (i) *if $f, g \in S_X$, then $f \circ g \in S_X$;*

   (ii) *$h \circ (g \circ f) = (h \circ g) \circ f$ for all $f, g, h \in S_X$;*

   (iii) *the identity 1_X lies in S_X, and $1_X \circ f = f = f \circ 1_X$ for every $f \in S_X$;*

(iv) *for every $f \in S_X$, there is $g \in S_X$ with $g \circ f = 1_X = f \circ g$.*

Proof.   We have merely restated results of Exercise 2.9(iii), Lemmas 2.2, 2.3, and 2.5.   •

Here is one interesting use of bijections. It is easy to prove (see Exercise 2.7) that two finite sets X and Y have the same number of elements if and only if there is a bijection $f : X \rightarrow Y$. This suggests the following definition.

Definition.   Two (possibly infinite) sets X and Y have the *same number of elements*, denoted by $|X| = |Y|$, if there exists a bijection $f : X \rightarrow Y$.

A set X is called *countable* if either X is finite or X has the same number of elements as the natural numbers $\mathbb{N}$.

G. Cantor (1845–1918) proved that $\mathbb{R}$ is *uncountable*; that is, $\mathbb{R}$ is not countable. Thus, there are different sizes of infinity; in fact, there are infinitely many different sizes of infinity. One can define *cardinal numbers* which describe the number of elements in infinite sets in the same way that natural numbers describe the number of elements in finite sets. For example, an infinite countable set is said to have cardinal $\aleph_0$ ($\aleph$ is the first letter of the Hebrew alphabet, and it is pronounced *aleph*); the cardinal of $\mathbb{R}$ is usually denoted by c (abbreviating *continuum*). The distinction between cardinals can be useful. For example, one calls a real number z *algebraic* if it is a root of some polynomial $f(x) = q_0 + q_1 x + \cdots + q_n x^n$, all of whose coefficients $q_0, q_1, \ldots, q_n$ are rational; one calls z *transcendental* if it is not algebraic. Of course, every rational r is algebraic, for it is a root of $x - r$. But irrational algebraic numbers do exist; for example, $\sqrt{2}$ is algebraic, being a root of $x^2 - 2$. Are there any transcendental numbers? One can prove that there are only countably many algebraic numbers, and so it follows from Cantor's theorem, the uncountability of $\mathbb{R}$, that not every real number is algebraic.

Remark.   Why should we care about the target of a function when its image is more important?

As a practical matter, when first defining a function, one usually does not know its image. For example, we say that $f : \mathbb{R} \rightarrow \mathbb{R}$, given by $f(x) = x^2 + 3x - 8$, is a real-valued function, and we then analyze f to find its image. But if targets have to be images, then we could not even write down $f : X \rightarrow Y$ without having first found the image of f (and finding the precise image is often very difficult, if not impossible); thus, targets are convenient to use.

Part of the definition of equality of functions is that their targets are equal; changing the target changes the function. Suppose we do not do this. Consider a function $f : X \to Y$ that is not surjective, let $Y' = \text{im } f$, and define $g : X \to Y'$ by $g(x) = f(x)$ for all $x \in X$. The functions f and g have the same domain and the same values (i.e., the same graph); they differ only in their targets. Now g is surjective. Had we decided that targets are not a necessary ingredient in the definition of a function, then we would not be able to distinguish between f, which is not surjective, and g, which is. It would then follow that every function is a surjection (this would not shake the foundations of mathematics, but it would force us into using cumbersome circumlocutions).

The next, more advanced, example may not be intelligible now, but it does give a convincing reason why targets are necessary in a function's definition.

In linear algebra, one considers a vector space V and its *dual space* $V^* = \{$all linear functionals on $V\}$ (which is also a vector space). Moreover, every linear transformation $T : V \to W$ defines a linear transformation

$$T^* : W^* \to V^*,$$

and the domain of T^*, being W^*, is determined by the target W of T. (In fact, if a matrix for T is A, then a matrix for T^* is A^t, the transpose of A.) Thus, changing the target of T changes the domain of T^*, and so T^* is changed in an essential way.

The moral of this story is that the target is an essential part of the definition of function, but the underlying reasons why it is essential may not be clear until later. ◄

EXERCISES

2.1 Let X and Y be sets, and let $f : X \to Y$ be a function. If S is a subset of X, prove that the restriction $f|S$ is equal to the composite $f \circ i$, where $i : S \to X$ is the inclusion map.

2.2 If $f : X \to Y$ has an inverse g, show that g is a bijection.

2.3 If f and g are differentiable functions, show that the chain rule may be written

$$D(g \circ f) = [(Dg) \circ f] \cdot Df.$$

2.4 Show that if $f : X \to Y$ is a bijection, then it has exactly one inverse.

2.5 Show that $f : \mathbb{R} \to \mathbb{R}$, defined by $f(x) = 3x + 5$, is a bijection, and find its inverse.

2.6 Determine whether $f : \mathbb{Q} \times \mathbb{Q} \to \mathbb{Q}$, given by

$$f(a/b, c/d) = (a + c)/(b + d)$$

is a function.

2.7 Let $X = \{x_1, \ldots, x_m\}$ and $Y = \{y_1, \ldots, y_n\}$ be finite sets. Show that there is a bijection $f : X \to Y$ if and only if $|X| = |Y|$; that is, $m = n$.

2.8 If X and Y are finite sets with the same number of elements, show that the following conditions are equivalent for a function $f : X \to Y$:

    (i) f is injective;

    (ii) f is bijective;

    (iii) f is surjective.

2.9 Let $f : X \to Y$ and $g : Y \to Z$ be functions.

    (i) If both f and g are injective, then $g \circ f$ is injective.

    (ii) If both f and g are surjective, then $g \circ f$ is surjective.

    (iii) If both f and g are bijective, then $g \circ f$ is bijective.

    (iv) If $g \circ f$ is a bijection, prove that f is an injection and g is a surjection.

2.10   (i) If $f : (-\pi/2, \pi/2) \to \mathbb{R}$ is defined by $a \mapsto \tan a$, then f has an inverse function g; indeed, $g = \arctan$.

    (ii) Show that the other "inverse trigonometric functions" of calculus are inverse functions as defined in this section. (Domains and targets must be chosen with care.)

2.11 If A and B are subsets of a set X, define

$$A - B = \{a \in A : a \notin B\}.$$

Prove that $A - B = A \cap B'$, where $B' = X - B$ is the **complement** of B; that is, $B' = \{x \in X : x \notin B\}$.

2.12 Let A and B be sets, and let $a \in A$ and $b \in B$. Define their **ordered pair** as follows:

$$(a, b) = \{a, \{a, b\}\}.$$

If $a' \in A$ and $b' \in B$, prove that $(a', b') = (a, b)$ if and only if $a' = a$ and $b' = b$.

2.13 Let $\Delta = \{(x, x) : x \in \mathbb{R}\}$; thus, Δ is the line in the plane which passes through the origin and which makes an angle of $45°$ with the x-axis.

    (i) If $P = (a, b)$ is a point in the plane with $a \neq b$, prove that Δ is the perpendicular bisector of the segment PP' having endpoints $P = (a, b)$ and $P' = (b, a)$.

(ii) If $f : \mathbb{R} \to \mathbb{R}$ is a bijection whose graph consists of certain points (a, b) [of course, $b = f(a)$], prove that the graph of f^{-1} is

$$\{(b, a) : (a, b) \in f\}.$$

2.14 (i) Prove that an infinite set X is countable if and only if there is a sequence $x_0, x_1, x_2, \ldots, x_n, \ldots$ of *all* the elements of X which has no repetitions.

(ii) Prove that every subset S of a countable set X is itself countable.

(iii) Prove that if there is a sequence $x_0, x_1, x_2, \ldots, x_n, \ldots$ of all the elements of a set X, possibly with repetitions, then X is countable.

(iv) If X is countable and $f : X \to Y$ is a surjection, prove that Y is countable.

2.15 Prove that if $X_0, X_1, \ldots, X_n, \ldots$ are countable sets, then $\bigcup_{n=0}^{\infty} X_n$ is also countable.

2.2 PERMUTATIONS

Definition.   A *permutation* of a set X is a bijection from X to itself.

In high school mathematics, a permutation of a set X is defined as a rearrangement of its elements. For example, there are six rearrangements of $X = \{1, 2, 3\}$:

$$123; \quad 132; \quad 213; \quad 231; \quad 312; \quad 321.$$

Now let $X = \{1, 2, \ldots, n\}$. A *rearrangement* is a list, with no repetitions, of all the elements of X. All we can do with such lists is count them, and there are exactly $n!$ permutations of the n-element set X.

Now a rearrangement $i_1, i_2, \ldots, i_n$ of X determines a function $\alpha : X \to X$, namely, $\alpha(1) = i_1, \alpha(2) = i_2, \ldots, \alpha(n) = i_n$. For example, the rearrangement 213 determines the function α with $\alpha(1) = 2, \alpha(2) = 1$, and $\alpha(3) = 3$. We use a two-rowed notation to denote the function corresponding to a rearrangement; if $\alpha(j)$ is the jth item on the list, then

$$\alpha = \begin{pmatrix} 1 & 2 & \cdots & j & \cdots & n \\ \alpha(1) & \alpha(2) & \cdots & \alpha(j) & \cdots & \alpha(n) \end{pmatrix}.$$

That a list contains *all* the elements of X says that the corresponding function α is surjective, for the bottom row is $\operatorname{im} \alpha$; that there are no repetitions on the list says that distinct points have distinct values; that is, α is injective. Thus, each list determines a bijection $\alpha : X \to X$; that is, each rearrangement determines a permutation. Conversely, every permutation α determines a rearrangement, namely, the list $\alpha(1)$, $\alpha(2)$, ..., $\alpha(n)$ displayed as the bottom row. Therefore, rearrangement and permutation are simply different ways of describing the same thing.[3] The advantage of viewing permutations as functions, however, is that they can now be composed and, by Exercise 2.9(iii), their composite is also a permutation.

The results in this section first appeared in an article of Cauchy in 1815.

Definition.   The family of all the permutations of a set X, denoted by S_X, is called the **symmetric group** on X. When $X = \{1, 2, \ldots, n\}$, S_X is usually denoted by S_n, and it is called the **symmetric group on n letters**.

Notice that composition in S_3 is not commutative. If

$$\alpha = \begin{pmatrix} 1 & 2 & 3 \\ 2 & 3 & 1 \end{pmatrix} \quad \text{and} \quad \beta = \begin{pmatrix} 1 & 2 & 3 \\ 2 & 1 & 3 \end{pmatrix},$$

then their composites are

$$\alpha \circ \beta = \begin{pmatrix} 1 & 2 & 3 \\ 3 & 2 & 1 \end{pmatrix} \quad \text{and} \quad \beta \circ \alpha = \begin{pmatrix} 1 & 2 & 3 \\ 1 & 3 & 2 \end{pmatrix},$$

so that $\alpha \circ \beta \neq \beta \circ \alpha$ [for example, $\alpha \circ \beta$: $1 \mapsto \alpha(\beta(1)) = \alpha(2) = 3$ while $\beta \circ \alpha$: $1 \mapsto 2 \mapsto 1$].

On the other hand, some permutations do commute; for example,

$$\begin{pmatrix} 1 & 2 & 3 & 4 \\ 2 & 1 & 3 & 4 \end{pmatrix} \quad \text{and} \quad \begin{pmatrix} 1 & 2 & 3 & 4 \\ 1 & 2 & 4 & 3 \end{pmatrix}$$

commute, as the reader may check.

Composition in S_X satisfies the **cancellation law**:

$$\text{if } \gamma \circ \alpha = \gamma \circ \beta, \text{ then } \alpha = \beta.$$

---

[3] In Example 2.4, we saw that there is exactly one function $\varnothing \to \varnothing$, namely, $1_\varnothing$, and it is a permutation. This is another justification for defining $0! = 1$.

To see this,

$$\alpha = 1_X \circ \alpha$$
$$= (\gamma^{-1} \circ \gamma) \circ \alpha$$
$$= \gamma^{-1} \circ (\gamma \circ \alpha)$$
$$= \gamma^{-1} \circ (\gamma \circ \beta)$$
$$= (\gamma^{-1} \circ \gamma) \circ \beta$$
$$= 1_X \circ \beta = \beta.$$

A similar argument shows that

$$\alpha \circ \gamma = \beta \circ \gamma \text{ implies } \alpha = \beta.$$

Aside from being cumbersome, there is a major problem with the two-rowed notation for permutations. It hides the answers to elementary questions such as: Do two permutations commute? Is the square of a permutation the identity? The special permutations introduced below will remedy this defect.

Let us first simplify notation by writing $\beta\alpha$ instead of $\beta \circ \alpha$ and (1) instead of 1_X.

Definition. If $\alpha \in S_n$ and $i \in \{1, 2, \ldots, n\}$, then α **fixes** i if $\alpha(i) = i$, and α **moves** i if $\alpha(i) \neq i$.

Definition. Let $i_1, i_2, \ldots, i_r$ be distinct integers in $\{1, 2, \ldots, n\}$. If $\alpha \in S_n$ fixes the other integers (if any) and if

$$\alpha(i_1) = i_2, \alpha(i_2) = i_3, \ldots, \alpha(i_{r-1}) = i_r, \alpha(i_r) = i_1,$$

then α is called an **r-cycle**. One also says that α is a cycle of **length** r.

A 2-cycle interchanges i_1 and i_2 and fixes everything else; 2-cycles are also called **transpositions**. A 1-cycle is the identity, for it fixes every i; thus, all 1-cycles are equal: $(i) = (1)$ for all i.

Consider the permutation

$$\alpha = \begin{pmatrix} 1 & 2 & 3 & 4 & 5 \\ 4 & 3 & 1 & 5 & 2 \end{pmatrix}.$$

The two-rowed notation does not help us recognize that α is, in fact, a 5-cycle: $\alpha(1) = 4$, $\alpha(4) = 5$, $\alpha(5) = 2$, $\alpha(2) = 3$, and $\alpha(3) = 1$. We now introduce new notation: an r-cycle α, as in the definition, shall be denoted by

$$\alpha = (i_1 \; i_2 \; \ldots \; i_r).$$

For example, the 5-cycle α above will be written $\alpha = (1\ 4\ 5\ 2\ 3)$. The reader may check that

$$\begin{pmatrix} 1 & 2 & 3 & 4 \\ 2 & 3 & 4 & 1 \end{pmatrix} = (1\ 2\ 3\ 4),$$

$$\begin{pmatrix} 1 & 2 & 3 & 4 & 5 \\ 5 & 1 & 4 & 2 & 3 \end{pmatrix} = (1\ 5\ 3\ 4\ 2), \quad \text{and}$$

$$\begin{pmatrix} 1 & 2 & 3 & 4 & 5 \\ 2 & 3 & 1 & 4 & 5 \end{pmatrix} = (1\ 2\ 3).$$

Notice that

$$\beta = \begin{pmatrix} 1 & 2 & 3 & 4 \\ 2 & 1 & 4 & 3 \end{pmatrix}$$

is not a cycle; in fact, $\beta = (1\ 2)(3\ 4)$.

The term *cycle* comes from the Greek word for circle. One can picture the cycle $(i_1\ i_2\ \dots\ i_r)$ as a clockwise rotation of the circle (see Figure 2.2).

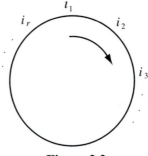

Figure 2.2

Any i_j can be taken as the "starting point," and so there are r different cycle notations for any r-cycle:

$$(i_1\ i_2\ \dots\ i_r) = (i_2\ i_3\ \dots\ i_r\ i_1) = \cdots = (i_r\ i_1\ i_2\ \dots\ i_{r-1}).$$

Let us now give an ***algorithm*** to factor a permutation into a product of cycles.

For example, take

$$\alpha = \begin{pmatrix} 1 & 2 & 3 & 4 & 5 & 6 & 7 & 8 & 9 \\ 6 & 4 & 7 & 2 & 5 & 1 & 8 & 9 & 3 \end{pmatrix}.$$

Begin by writing "(1." Now $\alpha : 1 \mapsto 6$, so write "(1 6." Next, $\alpha : 6 \mapsto 1$, and so the parentheses close: α begins "(1 6)." The first number not having appeared is 2, and so we write "(1 6)(2." Now $\alpha : 2 \mapsto 4$, so we write "(1 6)(2 4." Since $\alpha : 4 \mapsto 2$, the parentheses close once again, and we write "(1 6)(2 4)." The smallest remaining number is 3; now $3 \mapsto 7, 7 \mapsto 8$, $8 \mapsto 9$, and $9 \mapsto 3$; this gives the 4-cycle (3 7 8 9 5). Finally, $\alpha(5) = 5$; we claim that

$$\alpha = (1\ 6)(2\ 4)(3\ 7\ 8\ 9)(5).$$

Since multiplication in S_n is composition of functions, our claim is that

$$\alpha(i) = [(1\ 6)(2\ 4)(3\ 7\ 8\ 9)(5)](i)$$

for every i between 1 and n (after all, two functions f and g are equal if and only if $f(i) = g(i)$ for every i in their common domain). The right side is the composite $\beta \gamma \delta$, where $\beta = (1\ 6)$, $\gamma = (2\ 4)$, and $\delta = (3\ 7\ 8\ 9)$ (actually, there is also the 1-cycle (5), which we may ignore when we are evaluating, for (5) is the identity function). Now $\alpha(1) = 6$; multiplication of permutations views the permutations as functions and then takes their composite. For example, if $i = 1$, then

$$\begin{aligned} \beta \gamma \delta(1) &= \beta(\gamma(\delta(1))) \\ &= \beta(\gamma(1)) \quad \text{δ fixes 1} \\ &= \beta(1) \quad\quad \text{γ fixes 1} \\ &= 6. \end{aligned}$$

Factorizations into cycles are very convenient for multiplication of permutations. For example, in S_5, what is the product

$$\sigma = (1\ 2)(1\ 3\ 4\ 2\ 5)(2\ 5\ 1\ 3)?$$

First, we find the two-rowed notation for σ; we evaluate, starting with the cycle on the right.

$$\sigma : 1 \mapsto 3 \mapsto 4 \mapsto 4;$$
$$\sigma : 2 \mapsto 5 \mapsto 1 \mapsto 2;$$
$$\sigma : 3 \mapsto 2 \mapsto 5 \mapsto 5;$$
$$\sigma : 4 \mapsto 4 \mapsto 2 \mapsto 1;$$
$$\sigma : 5 \mapsto 1 \mapsto 3 \mapsto 3.$$

Thus,[4]

$$\sigma = \begin{pmatrix} 1 & 2 & 3 & 4 & 5 \\ 4 & 2 & 5 & 1 & 3 \end{pmatrix}.$$

The algorithm given earlier, when applied to this two-rowed notation for σ, now gives

$$\sigma = (1\ 4)(2)(3\ 5).$$

In the factorization of a permutation into cycles, given by the algorithm above, one notes that the family of cycles is *disjoint* in the following sense.

Definition.   Two permutations α, $\beta \in S_n$ are **disjoint** if every i moved by one is fixed by the other: if $\alpha(i) \neq i$, then $\beta(i) = i$, and if $\beta(j) \neq j$, then $\alpha(j) = j$. A family $\beta_1 \ldots, \beta_t$ of permutations is **disjoint** if each pair of them is disjoint.

Consider the special case of cycles. If $\alpha = (i_1\ i_2\ \ldots\ i_r)$ and $\beta = (j_1\ j_2\ \ldots\ j_s)$, then any k in the intersection $\{i_1, i_2, \ldots, i_r\} \cap \{j_1, j_2, \ldots, j_s\}$ is moved by both α and β. Thus, it is easy to see that two cycles are disjoint if and only if $\{i_1, i_2, \ldots, i_r\} \cap \{j_1, j_2, \ldots, j_s\}$ is empty; that is, $\{i_1, i_2, \ldots, i_r\}$ and $\{j_1, j_2, \ldots, j_s\}$ are disjoint sets.

When permutations α and β are disjoint, there are exactly three distinct possibilities for a number i: it is moved by α, it is moved by β, or it is moved by neither (that is, it is fixed by both).

---

[4]There are authors who multiply permutations differently, so that their $\alpha \circ \beta$ is our $\beta \circ \alpha$. This is a consequence of their putting "functions on the right": instead of writing $\alpha(i)$ as we do, they write $(i)\alpha$. Consider the composite of permutations α and β in which we first apply β and then apply α. We write $i \mapsto \beta(i) \mapsto \alpha(\beta(i))$. In the right-sided notation, $i \mapsto (i)\beta \mapsto ((i)\beta)\alpha$. Thus, the notational switch causes a switch in the order of multiplication.

Lemma 2.7.   *Disjoint permutations α, $\beta \in S_n$ commute.*

Proof.   It suffices to prove that if $1 \le i \le n$, then $\alpha\beta(i) = \beta\alpha(i)$. If β moves i, say, $\beta(i) = j \ne i$, then β also moves j [otherwise, $\beta(j) = j$ and $\beta(i) = j$ contradicts β's being an injection]; since α and β are disjoint, $\alpha(i) = i$ and $\alpha(j) = j$. Hence $\beta\alpha(i) = j = \alpha\beta(i)$. A similar argument shows that $\alpha\beta(i) = \beta\alpha(i)$ if α moves i. The last possibility is that neither α nor β moves i; in this case, $\alpha\beta(i) = i = \beta\alpha(i)$. Therefore, $\alpha\beta = \beta\alpha$, by Proposition 2.1.   ●

Of course, cycles are permutations, and so disjoint cycles commute.

It is possible for permutations that are not disjoint to commute; for example, the reader may check that $(1\ 2\ 3)(4\ 5)$ and $(1\ 3\ 2)(6\ 7)$ do commute.

Proposition 2.8.   *Every permutation $\alpha \in S_n$ is either a cycle or a product of disjoint cycles.*

Proof.   The proof is by induction on the number k of points moved by α. The base step $k = 0$ is true, for now α is the identity, which is a 1-cycle.

If $k > 0$, let i_1 be a point moved by α. Define $i_2 = \alpha(i_1)$, $i_3 = \alpha(i_2)$, ..., $i_{r+1} = \alpha(i_r)$, where r is the smallest integer for which $i_{r+1} \in \{i_1, i_2, \ldots, i_r\}$ (the list $i_1, i_2, i_3, \ldots, i_k, \ldots$ must eventually have a repetition, because there are only n possible values). We claim that $\alpha(i_r) = i_1$. Otherwise, $\alpha(i_r) = i_j$ for some $j \ge 2$; but $\alpha(i_{j-1}) = i_j$, and this contradicts the hypothesis that α is an injection. Let σ be the r-cycle $(i_1\ i_2\ i_3\ \ldots\ i_r)$. If $r = n$, then $\alpha = \sigma$. If $r < n$, then σ fixes each point in Y, where Y consists of the remaining $n - r$ points, while $\alpha(Y) = Y$. Define α' to be the permutation with $\alpha'(i) = \alpha(i)$ for $i \in Y$ which fixes all $i \notin Y$, and note that

$$\alpha = \sigma\alpha'.$$

The inductive hypothesis gives $\alpha' = \beta_1 \cdots \beta_t$, where $\beta_1, \ldots, \beta_t$ are disjoint cycles. Since σ and α' are disjoint, $\alpha = \sigma\beta_1 \cdots \beta_t$ is a product of disjoint cycles, as desired.   ●

Usually one suppresses the 1-cycles in this factorization [for 1-cycles equal the identity (1)]. However, a factorization of α containing one 1-cycle for each i fixed by α, if any, will arise several times in the sequel.

Definition.   A *complete factorization* of a permutation α is a factorization of α into disjoint cycles that contains one 1-cycle (i) for every i fixed by α.

The factorization algorithm always yields a complete factorization. For example, suppose that

$$\alpha = \begin{pmatrix} 1 & 2 & 3 & 4 & 5 \\ 1 & 3 & 4 & 2 & 5 \end{pmatrix}.$$

Now $\alpha = (1)(2\ 3\ 4)(5)$ is a complete factorization. However, if one suppresses 1-cycles, the factorizations

$$\alpha = (2\ 3\ 4) = (1)(2\ 3\ 4) = (2\ 3\ 4)(5)$$

are not complete factorizations. In a complete factorization $\alpha = \beta_1 \cdots \beta_t$, every symbol i between 1 and n occurs in exactly one of the β's.

There is a relation between an r-cycle $\beta = (i_1\ i_2\ \ldots\ i_r)$ and its **powers** β^k, where β^k denotes the composite of β with itself k times. Note that $i_2 = \beta(i_1)$, $i_3 = \beta(i_2) = \beta(\beta(i_1)) = \beta^2(i_1)$, $i_4 = \beta(i_3) = \beta(\beta^2(i_1)) = \beta^3(i_1)$, and, more generally,

$$i_{k+1} = \beta^k(i_1)$$

for all $k < r$.

Lemma 2.9.

(i) Let $\alpha = \beta_1 \cdots \beta_t$ be a factorization into disjoint permutations. If β_1 moves i, then $\alpha^k(i) = \beta_1^k(i)$ for all $k \geq 1$.

(ii) If β and γ are cycles both of which move some i, and if $\beta^k(i) = \gamma^k(i)$ for all $k \geq 1$, then $\beta = \gamma$.

Proof. (i) Denote $\beta_2 \cdots \beta_t$ by δ, so that $\alpha = \beta_1\delta$; note that β_1 and δ are disjoint. Since β_1 moves i, it follows that δ fixes i; indeed, every power of δ fixes i. Disjointness of β_1 and δ implies that they commute, by Lemma 2.7, and so Exercise 2.24(i) gives $(\beta_1\delta)^k(i) = \beta_1^k(\delta^k(i)) = \beta_1^k(i)$, as desired.

(ii) As we saw above, if $\beta = (i\ i_2\ \ldots\ i_r)$, then $i_{k+1} = \beta^k(i)$ for all $k < r$. Similarly, if $\gamma = (i\ j_2 \ldots j_s)$, then $j_{k+1} = \gamma^k(i)$ for $k < s$. We may assume that $r \leq s$ so that $i_2 = j_2, \ldots, i_r = j_r$. Since $j_{r+1} = \gamma^r(i) = \beta^r(i) = i$, it follows that $s = r$ and $j_k = i_k$ for all k, and so $\beta = (i\ i_2\ \ldots\ i_r) = \gamma$. •

Theorem 2.10. *Let $\alpha \in S_n$ and let $\alpha = \beta_1 \cdots \beta_t$ be a complete factorization into disjoint cycles. This factorization is unique except for the order in which the cycles occur.*

Proof.   Since every complete factorization of α has exactly one 1-cycle for each i fixed by α, it suffices to consider cycles of length ≥ 2. Let $\alpha = \gamma_1 \cdots \gamma_s$ be a second complete factorization of α into disjoint cycles.

We prove the theorem by induction on ℓ, the larger of t and s. The base step is true, for when $\ell = 1$, the hypothesis is $\beta_1 = \alpha = \gamma_1$.

To prove the inductive step, note first that if β_t moves i_1, then $\beta_t^k(i_1) = \alpha^k(i_1)$ for all $k \geq 1$, by Lemma 2.9(i). Now some γ_j must move i_1; since disjoint cycles commute, we may assume that γ_s moves i_1, and, as above, $\gamma_s^k(i_1) = \alpha^k(i_1)$ for all k. It follows from Lemma 2.9(ii) that $\beta_t = \gamma_s$, and the cancellation law gives $\beta_1 \cdots \beta_{t-1} = \gamma_1 \cdots \gamma_{s-1}$. By the inductive hypothesis, $s = t$ and the γ's can be reindexed so that $\gamma_1 = \beta_1, \ldots, \gamma_{t-1} = \beta_{t-1}$.   •

Every permutation is a bijection; how do we find its inverse? In the pictorial representation of a cycle β as a clockwise rotation of a circle, then the inverse β^{-1} is just a counterclockwise rotation.

Proposition 2.11.

(i) *The inverse of the cycle* $\alpha = (i_1 \ i_2 \ \ldots \ i_r)$ *is the cycle* $(i_r \ i_{r-1} \ldots \ i_1)$:

$$(i_1 \ i_2 \ \ldots \ i_r)^{-1} = (i_r \ i_{r-1} \ldots \ i_1).$$

(ii) *If* $\gamma \in S_n$ *and* $\gamma = \beta_1 \cdots \beta_k$, *then*

$$\gamma^{-1} = \beta_k^{-1} \cdots \beta_1^{-1}$$

(note that the order of the factors in γ^{-1} *has been reversed).*

Proof.   (i) If $\alpha \in S_n$, we show that both composites are equal to (1). Now $(i_1 \ i_2 \ \ldots \ i_r)(i_r \ i_{r-1} \ldots \ i_1)$ fixes each integer between 1 and n, if any, other than $i_1, \ldots, i_r$. The composite also sends $i_1 \mapsto i_r \mapsto i_1$ while it acts on i_j, for $j \geq 2$, by $i_j \mapsto i_{j-1} \mapsto i_j$. Thus, each integer between 1 and n is fixed by the composite, and so it is (1). A similar argument proves that the composite in the other order is also equal to (1), from which it follows that

$$(i_1 \ i_2 \ \ldots \ i_r)^{-1} = (i_r \ i_{r-1} \ldots \ i_1).$$

(ii) The proof is by induction on $k \geq 2$. For the base step $k = 2$, we have

$$(\beta_1 \beta_2)(\beta_2^{-1} \beta_1^{-1}) = \beta_1(\beta_2 \beta_2^{-1})\beta_1^{-1} = \beta_1(1)\beta_1^{-1} = \beta_1 \beta_1^{-1} = (1).$$

Similarly, $(\beta_2^{-1} \beta_1^{-1})(\beta_1 \beta_2) = (1)$.

For the inductive step, let $\delta = \beta_1 \cdots \beta_k$, so that $\beta_1 \cdots \beta_k \beta_{k+1} = \delta \beta_{k+1}$. Then

$$(\beta_1 \cdots \beta_k \beta_{k+1})^{-1} = (\delta \beta_{k+1})^{-1}$$
$$= \beta_{k+1}^{-1} \delta^{-1}$$
$$= \beta_{k+1}^{-1} (\beta_1 \cdots \beta_k)^{-1}$$
$$= \beta_{k+1}^{-1} \beta_k^{-1} \cdots \beta_1^{-1}. \quad \bullet$$

Thus, $(1\ 2\ 3\ 4)^{-1} = (4\ 3\ 2\ 1) = (1\ 4\ 3\ 2)$ and $(1\ 2)^{-1} = (2\ 1) = (1\ 2)$ (every transposition is equal to its own inverse).

Example 2.7.
The result in Proposition 2.11 holds, in particular, if the factors are disjoint cycles (in which case the reversal of the order of the factors is unnecessary because they commute with one another). Thus, if

$$\alpha = \begin{pmatrix} 1 & 2 & 3 & 4 & 5 & 6 & 7 & 8 & 9 \\ 6 & 4 & 7 & 2 & 5 & 1 & 8 & 9 & 3 \end{pmatrix},$$

then $\alpha = (1\ 6)(2\ 4)(3\ 7\ 8\ 9)(5)$ and

$$\alpha^{-1} = (5)(9\ 8\ 7\ 3)(4\ 2)(6\ 1)$$
$$= (1\ 6)(2\ 4)(3\ 7\ 8\ 7). \quad \blacktriangleleft$$

Definition.   Two permutations $\alpha, \beta \in S_n$ have the **same cycle structure** if their complete factorizations have the same number of r-cycles for each r.

According to Exercise 2.18, there are

$$(1/r)[n(n-1) \cdots (n-r+1)]$$

r-cycles in S_n. This formula can be used to count the number of permutations having any given cycle structure if one is careful about factorizations having several cycles of the same length. For example, the number of permutations in S_4 of the form $(a\ b)(c\ d)$ is

$$\tfrac{1}{2}\left[\tfrac{1}{2}(4 \times 3)\right] \times \left[\tfrac{1}{2}(2 \times 1)\right] = 3,$$

the extra factor $\tfrac{1}{2}$ occurring so that we do not count $(a\ b)(c\ d) = (c\ d)(a\ b)$ twice.

Example 2.8.
Let us count the permutations in $G = S_4$.

Cycle Structure	Number
(1)	1
(1 2)	6
(1 2 3)	8
(1 2 3 4)	6
(1 2)(3 4)	3
	$\overline{24}$

Table 2.1.   Permutations in S_4 ◄

Example 2.9.
Let us count the permutations in $G = S_5$.

Cycle Structure	Number
(1)	1
(1 2)	10
(1 2 3)	20
(1 2 3 4)	30
(1 2 3 4 5)	24
(1 2)(3 4 5)	20
(1 2)(3 4)	15
	$\overline{120}$

Table 2.2.   Permutations in S_5 ◄

Here is a computational aid.

Proposition 2.12.   *If $\gamma, \alpha \in S_n$, then $\alpha\gamma\alpha^{-1}$ has the same cycle structure as γ. In more detail, if the complete factorization of γ is*

$$\gamma = \beta_1\beta_2 \cdots (\ldots \ i \ j \ \ldots) \cdots \beta_t,$$

then $\alpha\gamma\alpha^{-1}$ is the permutation which is obtained from γ by applying α to the symbols in the cycles of γ.

Remark.   For example, if $\gamma = (1 \ 3)(2 \ 4 \ 7)(5)(6)$ and $\alpha = (2 \ 5 \ 6)(1 \ 4 \ 3)$, then

$$\alpha\gamma\alpha^{-1} = (\alpha 1 \ \alpha 3)(\alpha 2 \ \alpha 4 \ \alpha 7)(\alpha 5)(\alpha 6) = (4 \ 1)(5 \ 3 \ 7)(6)(2). \quad ◄$$

Proof. Let σ denote the permutation defined in the statement.

If γ fixes i, then σ fixes $\alpha(i)$, for the definition of σ says that $\alpha(i)$ lives in a 1-cycle in the factorization of σ. On the other hand, $\alpha\gamma\alpha^{-1}$ also fixes $\alpha(i)$:

$$\alpha\gamma\alpha^{-1}(\alpha(i)) = \alpha\gamma(i) = \alpha(i),$$

because γ fixes i.

Assume that γ moves i, say, $\gamma(i) = j$, so that one of the cycles in the complete factorization of γ is

$$(\ldots\ i\ j\ \ldots).$$

By definition, one of the cycles in σ is

$$(\ldots\ k\ \ell\ \ldots),$$

where $\alpha(i) = k$ and $\alpha(j) = \ell$; hence, $\sigma : k \mapsto \ell$. But $\alpha\gamma\alpha^{-1} : k \mapsto i \mapsto j \mapsto \ell$, and so $\alpha\gamma\alpha^{-1}(k) = \sigma(k)$. Therefore, σ and $\alpha\gamma\alpha^{-1}$ agree on all symbols of the form $k = \alpha(i)$. Since α is surjective, every k is of this form, and so $\sigma = \alpha\gamma\alpha^{-1}$, as desired.   •

There is another useful factorization of a permutation.

Proposition 2.13. *If $n \geq 2$, then every $\alpha \in S_n$ is a product of transpositions.*

Proof. By Proposition 2.8, it suffices to factor an r-cycle β into a product of transpositions. This is done as follows. If $r = 1$, then β is the identity, and $\beta = (1\ 2)(1\ 2)$. If $r \geq 2$, then

$$\beta = (1\ 2\ \ldots\ r) = (1\ r)(1\ r-1)\cdots(1\ 3)(1\ 2).$$

[One checks that this is an equality by evaluating each side. For example, the left side β sends $1 \mapsto 2$; each of $(1\ r), (1\ r-1), \ldots, (1\ 3)$ fixes 2, and so the right side also sends $1 \mapsto 2$.]   •

Every permutation can thus be realized as a sequence of interchanges. Such a factorization is not as nice as the factorization into disjoint cycles. First of all, the transpositions occurring need not commute: $(1\ 2\ 3) = (1\ 3)(1\ 2) \neq (1\ 2)(1\ 3)$; second, neither the factors themselves nor the number of factors are uniquely determined. For example, here are some factorizations of (1 2 3) in S_4:

$$
\begin{aligned}
(1\ 2\ 3) &= (1\ 3)(1\ 2) \\
&= (2\ 3)(1\ 3) \\
&= (1\ 3)(4\ 2)(1\ 2)(1\ 4) \\
&= (1\ 3)(4\ 2)(1\ 2)(1\ 4)(2\ 3)(2\ 3).
\end{aligned}
$$

Is there any uniqueness at all in such a factorization? We now prove that the parity of the number of factors is the same for all factorizations of a permutation α; that is, the number of transpositions is always even or always odd [as is suggested by the factorizations of $\alpha = (1\ 2\ 3)$ displayed above].

Example 2.10.
The *15-puzzle* consists of a *starting position*, which is a 4×4 array of the numbers between 1 and 15 and a symbol # (which we interpret as "blank"), and *simple moves*. For example, consider the starting position shown below.

3	15	4	8
10	11	1	9
2	5	13	12
6	7	14	#

A *simple move* interchanges the blank with a symbol adjacent to it; for example, there are two beginning simple moves for this starting position: either interchange # and 14 or interchange # and 12. One wins the game if, after a sequence of simple moves, the starting position is transformed into the standard array 1, 2, 3, ..., 15, #.

To analyze this game, note that the given array is really a permutation α of $\{1, 2, \ldots, 15, 16\}$ (if we now call the blank 16 instead of #); that is, $\alpha \in S_{16}$. More precisely, if the spaces are labeled 1 through 16, then $\alpha(i)$ is the symbol occupying the ith square. For example, the starting position given above is

$$
\begin{pmatrix}
1 & 2 & 3 & 4 & 5 & 6 & 7 & 8 & 9 & 10 & 11 & 12 & 13 & 14 & 15 & 16 \\
3 & 15 & 4 & 8 & 10 & 11 & 1 & 9 & 2 & 5 & 13 & 12 & 6 & 7 & 14 & 16
\end{pmatrix}.
$$

Each simple move is a special kind of transposition, namely, one that moves 16 (remember that the blank is now 16). Moreover, performing a simple move (corresponding to a special transposition τ) from a position (corresponding to a permutation β) yields a new position corresponding to the permutation $\tau\beta$. For example, if α is the position above and τ is the transposition interchanging 14 and 16, then $\tau\alpha(16) = \tau(16) = 14$ and $\tau\alpha(15) = \tau(14) = 16$, while $\tau\alpha(i) = i$ for all other i. That is, the new configuration has all the numbers in their original positions except for 14 and 16 being interchanged. Therefore, to win the game, one needs special transpositions $\tau_1, \tau_2, \ldots, \tau_m$ so that

$$\tau_m \cdots \tau_2 \tau_1 \alpha = (1).$$

It turns out that there are some choices of α for which the game can be won, but there are others for which it cannot be won, as we shall see in Example 2.11. ◄

Definition.   A permutation $\alpha \in S_n$ is **even** if it can be factored into a product of an even number of transpositions; otherwise, α is **odd**. The **parity** of a permutation is whether it is even or odd.

It is easy to see that $\alpha = (1\ 2\ 3)$ is even, for there is a factorization $\alpha = (1\ 3)(1\ 2)$ with two transpositions. On the other hand, we do not yet have any examples of odd permutations! If α is a product of an odd number of transpositions, perhaps it also has some other factorization into an even number of transpositions. The definition of odd permutation α, after all, says that there is no factorization of α into an even number of transpositions.

Lemma 2.14.   *If* $k, \ell \geq 0$ *and the letters* a, b, c_i, d_j *are all distinct, then*

$$(a\ b)(a\ c_1\ \ldots\ c_k\ b\ d_1\ \ldots\ d_\ell) = (a\ c_1\ \ldots\ c_k)(b\ d_1\ \ldots\ d_\ell)$$

and

$$(a\ b)(a\ c_1\ \ldots\ c_k)(b\ d_1\ \ldots\ d_\ell) = (a\ c_1\ \ldots\ c_k\ b\ d_1\ \ldots\ d_\ell).$$

Proof.   The left side sends

$$a \mapsto c_1 \mapsto c_1;$$
$$c_i \mapsto c_{i+1} \mapsto c_{i+1} \text{ if } i < k;$$
$$c_k \mapsto b \mapsto a;$$
$$b \mapsto d_1 \mapsto d_1;$$
$$d_j \mapsto d_{j+1} \mapsto d_{j+1} \text{ if } j < \ell;$$
$$d_\ell \mapsto a \mapsto b.$$

Similar evaluation of the right side shows that both permutations are equal. For the second equation, just multiply both sides of the first equation by $(a\ b)$ on the left:

$$(a\ b)(a\ c_1\ \ldots\ c_k)(b\ d_1\ \ldots\ d_\ell) = (a\ b)(a\ b)(a\ c_1\ \ldots\ c_k\ b\ d_1\ \ldots\ d_\ell)$$
$$= (a\ c_1\ \ldots\ c_k)(b\ d_1\ \ldots\ d_\ell).\quad \bullet$$

An illustration of the lemma is

$$(1\ 2)(1\ 3\ 4\ 2\ 5\ 6\ 7) = (1\ 3\ 4)(2\ 5\ 6\ 7).$$

Definition.   If $\alpha \in S_n$ and $\alpha = \beta_t \cdots \beta_t$ is a complete factorization into disjoint cycles, then ***signum***[5] α is defined by

$$\text{sgn}(\alpha) = (-1)^{n-t}.$$

Theorem 2.10 shows that sgn is a (single-valued) function, for the number t is uniquely determined by α. Notice that $\text{sgn}(\varepsilon) = 1$ for every 1-cycle ε because $t = n$. If τ is a transposition, then it moves two numbers, and it fixes each of the $n - 2$ other numbers; therefore, $t = (n - 2) + 1 = n - 1$, and so $\text{sgn}(\tau) = (-1)^{n-(n-1)} = -1$.

Lemma 2.15.   *If $\alpha, \tau \in S_n$, where τ is a transposition, then*

$$\text{sgn}(\tau\alpha) = -\,\text{sgn}(\alpha).$$

Proof.   Let $\alpha = \beta_1 \cdots \beta_t$ be a complete factorization of α into disjoint cycles, and let $\tau = (a\ b)$. If a and b occur in the same β, say, in β_1, then $\beta_1 = (a\ c_1 \ldots c_k\ b\ d_1 \ldots d_\ell)$, where $k, \ell \geq 0$. By Lemma 2.13,

$$\tau\beta_1 = (a\ c_1 \ldots c_k)(b\ d_1 \ldots d_\ell),$$

and so the complete factorization $\tau\alpha = (\tau\beta_1)\beta_2 \cdots \beta_t$ has acquired an extra cycle ($\tau\beta_1$ splits into two disjoint cycles). Therefore, $\text{sgn}(\tau\alpha) = (-1)^{n-(t+1)} = -\,\text{sgn}(\alpha)$.

The other possibility is that a and b occur in different cycles, say, $\beta_1 = (a\ c_1 \ldots c_k)$ and $\beta_2 = (b\ d_1 \ldots d_\ell)$, where $k, \ell \geq 0$. But now $\tau\alpha = (\tau\beta_1\beta_2)\beta_3 \cdots \beta_t$, and Lemma 2.14 gives

$$\tau\beta_1\beta_2 = (a\ c_1 \ldots c_k\ b\ d_1 \ldots d_\ell).$$

---

[5] *Signum* is the Latin word for "mark" or "token"; of course, it has become the word *sign*.

Therefore $\tau\alpha$ has a complete factorization with one cycle fewer than does α, and so $\mathrm{sgn}(\tau\alpha) = (-1)^{n-(t-1)} = -\mathrm{sgn}(\alpha)$, as desired. ●

Theorem 2.16. *For all $\alpha, \beta \in S_n$,*

$$\mathrm{sgn}(\alpha\beta) = \mathrm{sgn}(\alpha)\,\mathrm{sgn}(\beta).$$

Proof. Assume that $\alpha \in S_n$ is given and that α has a factorization as a product of m transpositions: $\alpha = \tau_1 \cdots \tau_m$. We prove, by induction on m, that $\mathrm{sgn}(\alpha\beta) = \mathrm{sgn}(\alpha)\,\mathrm{sgn}(\beta)$ for every $\beta \in S_n$. The base step $m = 1$ is precisely Lemma 2.15, for $m = 1$ says that α is a transposition. If $m > 1$, then the inductive hypothesis applies to $\tau_2 \cdots \tau_m$, and so

$$
\begin{aligned}
\mathrm{sgn}(\alpha\beta) &= \mathrm{sgn}(\tau_1 \cdots \tau_m \beta) \\
&= -\mathrm{sgn}(\tau_2 \cdots \tau_m \beta) && \text{(Lemma 2.15)} \\
&= -\mathrm{sgn}(\tau_2 \cdots \tau_m)\,\mathrm{sgn}(\beta) && \text{(by induction)} \\
&= \mathrm{sgn}(\tau_1 \cdots \tau_m)\,\mathrm{sgn}(\beta) && \text{(Lemma 2.15)} \\
&= \mathrm{sgn}(\alpha)\,\mathrm{sgn}(\beta). && ●
\end{aligned}
$$

It follows by induction on $k \geq 2$ that

$$\mathrm{sgn}(\alpha_1 \alpha_2 \cdots \alpha_k) = \mathrm{sgn}(\alpha_1)\,\mathrm{sgn}(\alpha_2) \cdots \mathrm{sgn}(\alpha_k).$$

Theorem 2.17.

(i) *Let $\alpha \in S_n$; if $\mathrm{sgn}(\alpha) = 1$, then α is even, and if $\mathrm{sgn}(\alpha) = -1$, then α is odd.*

(ii) *A permutation α is odd if and only if it has a factorization into an odd number of transpositions.*

Proof. (i) We have seen that $\mathrm{sgn}(\tau) = -1$ for every transposition τ. Therefore, if $\alpha = \tau_1 \cdots \tau_q$ is a factorization of α into transpositions, then Proposition 2.16 gives $\mathrm{sgn}(\alpha) = \mathrm{sgn}(\tau_1) \cdots \mathrm{sgn}(\tau_q) = (-1)^q$. Thus, if $\mathrm{sgn}(\alpha) = 1$, then q must always be even, and if $\mathrm{sgn}(\alpha) = -1$, then q must always be odd.

(ii) If α is odd, then α is not even, and so $\mathrm{sgn}(\alpha) \neq 1$; that is, $\mathrm{sgn}(\alpha) = -1$. Now $\alpha = \tau_1 \cdots \tau_q$, where the τ_i are transpositions, so that $\mathrm{sgn}(\alpha) = -1 =$

$(-1)^q$; hence, q is odd (we have proved more; every factorization of α into transpositions has an odd number of factors). Conversely, if $\alpha = \tau_1 \cdots \tau_q$ is a product of transpositions with q odd, then $\mathrm{sgn}(\alpha) = -1$; therefore, α is not even and, hence, α is odd.   •

Corollary 2.18.   *Let α, $\beta \in S_n$. If α and β have the same parity, then $\alpha\beta$ is even, while if α and β have distinct parity, then $\alpha\beta$ is odd.*

Proof.   If α and β have the same parity, then $\mathrm{sgn}(\alpha) = \mathrm{sgn}(\beta)$, and so

$$\mathrm{sgn}(\alpha\beta) = \mathrm{sgn}(\alpha)\,\mathrm{sgn}(\beta) = (\pm 1)^2 = 1;$$

hence, $\alpha\beta$ is even. On the other hand, if α and β have different parity, then $\mathrm{sgn}(\alpha) = -\mathrm{sgn}(\beta)$, and

$$\mathrm{sgn}(\alpha\beta) = \mathrm{sgn}(\alpha)\,\mathrm{sgn}(\beta) = -1;$$

hence, $\alpha\beta$ is odd.   •

Example 2.11.
An analysis of the 15-puzzle in Example 2.10 shows that if $\alpha \in S_{16}$ is the starting position, then the game can be won if and only if α is an even permutation that fixes 16. For a proof of this, we refer the reader to McCoy and Janusz, *Introduction to Modern Algebra*. The proof in one direction is fairly clear, however. The blank 16 starts in position 16. Each simple move takes 16 up, down, left, or right. Thus, the total number m of moves is $u+d+l+r$, where u is the number of up moves, etc. If 16 is to return home, each one of these must be undone: there must be the same number of up moves as down moves, i.e., $u = d$, and the same number of left moves as right moves, i.e., $r = l$. Thus, the total number of moves is even: $m = 2u + 2r$. That is, if $\tau_m \cdots \tau_1 \alpha = (1)$, then m is even; hence, $\alpha = \tau_1 \cdots \tau_m$ (because $\tau^{-1} = \tau$ for every transposition τ), and so α is an even permutation. Armed with this theorem, one sees that the starting position α in Example 2.10 is, in cycle notation,

$$\alpha = (1\ 3\ 4\ 8\ 9\ 2\ 15\ 14\ 7)(5\ 10)(6\ 11\ 13).$$

Now $\mathrm{sgn}(\alpha) = (-1)^{16-3} = -1$, so that α is an odd permutation; therefore, the game starting with α cannot be won.   ◄

EXERCISES

2.16 Find sgn(α) and α^{-1}, where

$$\alpha = \begin{pmatrix} 1 & 2 & 3 & 4 & 5 & 6 & 7 & 8 & 9 \\ 9 & 8 & 7 & 6 & 5 & 4 & 3 & 2 & 1 \end{pmatrix}.$$

2.17 If $\sigma \in S_n$ fixes some j, where $1 \le j \le n$ (that is, $\sigma(j) = j$), define $\sigma' \in S_{n-1}$ by $\sigma'(i) = \sigma(i)$ for all $i \ne j$. Prove that

$$\mathrm{sgn}(\sigma') = \mathrm{sgn}(\sigma).$$

2.18 If $1 \le r \le n$, show that there are

$$\tfrac{1}{r}[n(n-1)\cdots(n-r+1)]$$

r-cycles in S_n.

2.19   (i) If α is an r-cycle, show that $\alpha^r = (1)$.
   (ii) If α is an r-cycle, show that r is the smallest positive integer k such that $\alpha^k = (1)$.

2.20 Show that an r-cycle is an even permutation if and only if r is odd.

2.21 Given $X = \{1, 2, \dots, n\}$, let us call a permutation τ of X an **adjacency** if it is a transposition of the form $(i\ i+1)$ for $i < n$. If $i < j$, prove that $(i\ j)$ is a product of an odd number of adjacencies.

2.22 Define $f : \{0, 1, 2, \dots, 10\} \to \{0, 1, 2, \dots, 10\}$ by

$$f(n) = \text{the remainder after dividing } 4n^2 - 3n^7 \text{ by } 11.$$

   (i) Show that f is a permutation.
   (ii) Compute the parity of f.
   (iii) Compute the inverse of f.

2.23 If α is an r-cycle, is α^k an r-cycle?

2.24   (i) Prove that if α and β are (not necessarily disjoint) permutations that commute, then $(\alpha\beta)^k = \alpha^k \beta^k$ for all $k \ge 1$.
   (ii) Give an example of two permutations α and β for which $(\alpha\beta)^2 \ne \alpha^2\beta^2$.

2.25 Prove the converse of Proposition 2.12: If $\beta, \gamma \in S_n$ have the same cycle structure, then there is $\alpha \in S_n$ with $\beta = \alpha\gamma\alpha^{-1}$.

2.26   (i) Prove, for all i, that $\alpha \in S_n$ moves i if and only if α^{-1} moves i.
   (ii) Prove that if $\alpha, \beta \in S_n$ are disjoint and if $\alpha\beta = (1)$, then $\alpha = (1)$ and $\beta = (1)$.

2.27 Prove that the number of even permutations in S_n is $\frac{1}{2}n!$.

2.28   (i) How many permutations in S_5 commute with (1 2 3), and how many *even* permutations in S_5 commute with (1 2 3)?

   (ii) Same questions for (1 2)(3 4).

2.29 Give an example of $\alpha, \beta, \gamma \in S_5$ with $\alpha\beta = \beta\alpha$, $\alpha\gamma = \gamma\alpha$, but with $\beta\gamma \neq \gamma\beta$.

2.30 If $n \geq 3$, show that if $\alpha \in S_n$ commutes with every $\beta \in S_n$, then $\alpha = (1)$.

2.3 GROUPS

Generalizations of the quadratic formula for finding the roots of cubic and quartic polynomials were discovered in the early 1500s. Over the next three centuries, many tried to find analogous formulas for the roots of higher-degree polynomials, but in 1824, N. H. Abel (1802–1829) proved that there is no such formula giving the roots of any polynomial of degree 5. (P. Ruffini (1765–1822) had essentially proved this result, in 1799, but his proof had gaps, was difficult to read, and was not accepted by his contemporaries.) In 1831, E. Galois (1811–1832) determined all those polynomials $f(x)$, of any degree, for which such a formula exists. First, he examined certain permutations of the roots of $f(x)$, and he saw that they comprise an algebraic system which he called a *group* [nowadays called the *Galois group* of $f(x)$]; that is, the composite of two such was another permutation of the same kind. In order to use this group of permutations, Galois found that he had to understand its "structure"; he needed the notions of *subgroup*, *index*, and *normal subgroup*. He could then define *solvability* of a group, and he proved that there is a formula for the roots of a polynomial $f(x)$ if and only if its group is solvable. Since Galois's time, groups have arisen in many other areas of mathematics, for they are also the way to describe the notion of symmetry.

The essence of a "product" is that two things are combined to form a third thing of the same kind. For example, ordinary multiplication, addition, and subtraction combine two numbers to give another number, while composition combines two permutations to give another permutation.

Definition.   An *operation* on a set G is a function

$$* : G \times G \to G.$$

In more detail, an operation assigns an element $*(x, y)$ in G to each ordered pair (x, y) of elements in G. It is more natural to write $x * y$ instead of

$*(x, y)$; thus, composition of functions is the function $(g, f) \mapsto g \circ f$; multiplication, addition, and subtraction are, respectively, the functions $(x, y) \mapsto xy$, $(x, y) \mapsto x + y$, and $(x, y) \mapsto x - y$. The examples of composition and subtraction show why we want ordered pairs, for $x * y$ and $y * x$ may be distinct. Like any function, an operation is single-valued; when one says this explicitly, it is usually called the *law of substitution*:

$$\text{If } x = x' \text{ and } y = y', \text{ then } x * y = x' * y'.$$

Definition. A *group* is a set G which is equipped with an operation $*$ and a special element $e \in G$, called the *identity*, such that

(i) the *associative law* holds: for every $x, y, z \in G$,

$$x * (y * z) = (x * y) * z;$$

(ii) $e * x = x = x * e$ for all $x \in G$;

(iii) for every $x \in G$, there is $x' \in G$ with $x * x' = e = x' * x$.

By Proposition 2.6, the set S_X of all permutations of a set X, with composition as the operation and (1) as the identity, is a group (the *symmetric group* on X).

We are now at the precise point when algebra becomes *abstract* algebra. In contrast to the concrete group S_n consisting of all the permutations of $\{1, 2, \ldots, n\}$, we have passed to groups whose elements are unspecified. Moreover, products of elements are not explicitly computable but are, instead, merely subject to certain rules. It will be seen that this approach is quite fruitful, for theorems now apply to many different groups, and it is more efficient to prove theorems once for all instead of proving them anew for each group encountered. In addition to this obvious economy, it is often simpler to work with the "abstract" viewpoint even when dealing with a particular concrete group. For example, we will see that certain properties of S_n are simpler to treat without recognizing that the elements in question are permutations (see Example 2.16).

Definition. A group G is called *abelian*[6] if it satisfies the *commutative law*: $x * y = y * x$ holds for every $x, y \in G$.

---

[6]This term honors N. H. Abel who proved a theorem, in 1827, equivalent to there being a formula for the roots of a polynomial if its Galois group is commutative. This theorem is virtually forgotten today, because it was superseded by work of Galois around 1830.

The groups S_n, for $n \geq 3$, are not abelian because (1 2) and (1 3) are elements of S_n that do not commute: (1 2)(1 3) = (1 3 2) and (1 3)(1 2) = (1 2 3).

We prove a basic lemma before giving more examples of groups.

Lemma 2.19.   *Let G be a group.*

(i) *The **cancellation laws** hold: if either $x * a = x * b$ or $a * x = b * x$, then $a = b$.*

(ii) *The element e is the unique element in G with $e * x = x = x * e$ for all $x \in G$.*

(iii) *Given $x \in G$, there is a unique $x' \in G$ with $x * x' = e = x' * x$ (x' is called the **inverse** of x, and it is denoted by x^{-1}).*

(iv) *$(x^{-1})^{-1} = x$ for all $x \in G$.*

Proof.   (i) Choose x' with $x' * x = e = x * x'$; then

$$a = e * a = (x' * x) * a = x' * (x * a)$$
$$= x' * (x * b) = (x' * x) * b = e * b = b.$$

A similar proof works when x is on the right.

(ii) Let $e_0 \in G$ satisfy $e_0 * x = x = x * e_0$ for all $x \in G$. In particular, setting $x = e$ in the second equation gives $e = e * e_0$; on the other hand, the defining property of e gives $e * e_0 = e_0$, so that $e = e_0$.

(iii) Assume that $x'' \in G$ satisfies $x * x'' = e = x'' * x$. Multiply the equation $e = x * x'$ on the left by x'' to obtain

$$x'' = x'' * e = x'' * (x * x') = (x'' * x) * x' = e * x' = x'.$$

(iv) By definition, $(x^{-1})^{-1} * x^{-1} = e = x^{-1} * (x^{-1})^{-1}$. But $x * x^{-1} = e = x^{-1} * x$, so that $(x^{-1})^{-1} = x$, by (iii).   •

From now on, we will usually denote the product $x * y$ in a group by xy (we have already abbreviated $\alpha \circ \beta$ to $\alpha\beta$ in symmetric groups), and we will denote the identity by 1 instead of by e. When a group is abelian, however, we will often use the ***additive notation*** $x + y$; in this case, we will denote the identity by 0, and we will denote the inverse of an element x by $-x$ instead of by x^{-1}.

We now give too many examples of groups (and there are more!). Glance over the list and choose one or two that look interesting to you.

Example 2.12.

(i) The set $\mathbb{Z}$ of all integers is an additive abelian group with $a * b = a + b$, with identity $e = 0$, and with the inverse of an integer n being $-n$. Similarly, one can see that $\mathbb{Q}$, $\mathbb{R}$, and $\mathbb{C}$ are additive abelian groups.

(ii) The set $\mathbb{Q}^\times$ of all nonzero rationals is an abelian group, where $*$ is ordinary multiplication, the number 1 is the identity, and the inverse of $r \in \mathbb{Q}^\times$ is $1/r$. Similarly, $\mathbb{R}^\times$ is a multiplicative abelian group. We show, in the next example, that $\mathbb{C}^\times$ is also a multiplicative group.

Note that $\mathbb{Z}^\times$ is not a group, for none of its elements (aside from ± 1) has a multiplicative inverse.

(iii) The nonzero complex numbers $\mathbb{C}^\times$ form a group under multiplication. It is easy to see that multiplication is an associative operation and that 1 is the identity. Here is the simplest way to find inverses. If $z = a + ib \in \mathbb{C}$, where $a, b \in \mathbb{R}$, define its **complex conjugate** $\bar{z} = a - ib$. Note that $z\bar{z} = a^2 + b^2$, so that $z \neq 0$ if and only if $z\bar{z} \neq 0$. If $z \neq 0$, then

$$z^{-1} = 1/z = \bar{z}/z\bar{z} = (a/z\bar{z}) - (b/z\bar{z})i.$$

(iv) The circle S^1 of radius 1 with center the origin can be made into a multiplicative group if we regard its points as complex numbers of modulus 1. The **circle group** is defined by

$$S^1 = \{z \in \mathbb{C} : |z| = 1\},$$

where the operation is multiplication of complex numbers; this is an operation on S^1, for the addition formulas for sine and cosine give

$$(\cos x + i \sin x)(\cos y + i \sin y) = \cos(x + y) + i \sin(x + y).$$

Of course, complex multiplication is associative, the identity is 1 (which has modulus 1), and the inverse of any complex number of modulus 1 is its complex conjugate, which also has modulus 1. Therefore, S^1 is a group.

(v) For any positive integer n, let

$$\Gamma_n = \left\{ \zeta^k : 0 \leq k < n \right\}$$

be the set of all the nth roots of unity, where

$$\zeta = e^{2\pi i/n} = \cos(2\pi/n) + i \sin(2\pi/n).$$

The reader may use De Moivre's theorem to see that Γ_n is a group with operation multiplication of complex numbers; moreover, the inverse of any root of unity is its complex conjugate.

(vi) The plane $\mathbb{R} \times \mathbb{R}$ is a group with operation vector addition; that is, if $\alpha = (x, y)$ and $\alpha' = (x', y')$, then $\alpha + \alpha' = (x + x', y + y')$. The identity is the origin $O = (0, 0)$, and the inverse of (x, y) is $(-x, -y)$.   ◄

Example 2.13.
The **parity group** $\mathcal{P}$ has two elements, the words "even" and "odd," with operation
$$\text{even} + \text{even} = \text{even} = \text{odd} + \text{odd}$$
and
$$\text{even} + \text{odd} = \text{odd} = \text{odd} + \text{even}.$$

The reader may show that $\mathcal{P}$ is an abelian group.   ◄

Example 2.14.
Let X be a set. If U and V are subsets of X, define
$$U - V = \{x \in U : x \notin V\}.$$

The **Boolean group** $\mathcal{B}(X)$ [named after the logician G. Boole (1815–1864)] is the family of all the subsets of X equipped with addition given by **symmetric difference** $A + B$, where
$$A + B = (A - B) \cup (B - A);$$

symmetric difference is pictured in Figure 2.3.

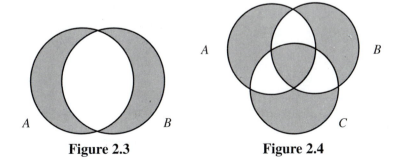

Figure 2.3   Figure 2.4

It is plain that $A+B = B+A$, so that symmetric difference is commutative. The identity is $\varnothing$, the empty set, and the inverse of A is A itself, for $A + A = \varnothing$. The reader may verify associativity by showing that both $(A + B) + C$ and $A + (B + C)$ are described by Figure 2.4. ◀

Example 2.15.
(i) A (2 × 2 real) *matrix*[7] A is

$$A = \begin{bmatrix} a & c \\ b & d \end{bmatrix},$$

where $a, b, c, d \in \mathbb{R}$. If

$$B = \begin{bmatrix} w & y \\ x & z \end{bmatrix},$$

then the *product* AB is defined by

$$AB = \begin{bmatrix} a & c \\ b & d \end{bmatrix}\begin{bmatrix} w & y \\ x & z \end{bmatrix} = \begin{bmatrix} aw + cx & ay + cz \\ bw + dx & by + dz \end{bmatrix}.$$

The elements a, b, c, d are called the *entries* of A. Call (a, c) the first *row* of A and call (b, d) the second row; call (a, b) the first *column* of A and call (c, d) the second column. Thus, each entry of the product AB is a dot product of a row of A with a column of B. The *determinant* of A, denoted by $\det(A)$, is the number $ad - bc$, and a matrix A is called *nonsingular* if $\det(A) \neq 0$. The reader may calculate that

$$\det(AB) = \det(A)\det(B),$$

from which it follows that the product of nonsingular matrices is itself non-singular. The set $GL(2, \mathbb{R})$ of all nonsingular matrices, with operation matrix multiplication, is a (nonabelian) group, called the *general linear group*: the identity is the *identity matrix*

$$E = \begin{bmatrix} 1 & 0 \\ 0 & 1 \end{bmatrix}$$

---

[7]The word *matrix* (derived from the word meaning "mother") means "womb" in Latin; more generally, it means something that contains the essence of a thing. Its mathematical usage arises because a 2 × 2 matrix, which is an array of four numbers, completely describes a certain type of function $\mathbb{R}^2 \to \mathbb{R}^2$ called a linear transformation (more generally, larger matrices contain the essence of linear transformations between higher-dimensional spaces).

and the inverse of a nonsingular matrix A is

$$A^{-1} = \begin{bmatrix} d/\Delta & -c/\Delta \\ -b/\Delta & a/\Delta \end{bmatrix},$$

where $\Delta = ad - bc = \det(A)$. (The proof of associativity is routine, though tedious; a "clean" proof of associativity can be given once one knows the relation between matrices and linear transformations [see Corollary 4.16].)

(ii) The previous example can be modified in two ways. First, we may allow the entries to lie in $\mathbb{Q}$ or in $\mathbb{C}$, giving the groups GL(2, $\mathbb{Q}$) or GL(2, $\mathbb{C}$). We may even allow the entries to be in $\mathbb{Z}$, in which case GL(2, $\mathbb{Z}$) is defined to be the set of all such matrices with determinant ± 1 (one wants all the entries of A^{-1} to be in $\mathbb{Z}$).

(iii) All ***special***[8] ***orthogonal*** matrices, that is, all matrices of the form

$$A = \begin{bmatrix} \cos\alpha & -\sin\alpha \\ \sin\alpha & \cos\alpha \end{bmatrix}$$

form a group, denoted by $SO(2, \mathbb{R})$ and called the 2×2 ***special orthogonal group***. Let us show that matrix multiplication is an operation on $SO(2, \mathbb{R})$. The product

$$\begin{bmatrix} \cos\alpha & -\sin\alpha \\ \sin\alpha & \cos\alpha \end{bmatrix} \begin{bmatrix} \cos\beta & -\sin\beta \\ \sin\beta & \cos\beta \end{bmatrix}$$

is

$$\begin{bmatrix} \cos\alpha\cos\beta - \sin\alpha\sin\beta & -[\cos\alpha\sin\beta + \sin\alpha\cos\beta] \\ \sin\alpha\cos\beta + \cos\alpha\sin\beta & \cos\alpha\cos\beta - \sin\alpha\sin\beta \end{bmatrix}.$$

The addition theorem for sine and cosine shows that this product is again a special orthogonal matrix, for it is

$$\begin{bmatrix} \cos(\alpha+\beta) & -\sin(\alpha+\beta) \\ \sin(\alpha+\beta) & \cos(\alpha+\beta) \end{bmatrix}.$$

In fact, this calculation shows that $SO(2, \mathbb{R})$ is abelian. It is clear that the identity matrix is special orthogonal, and we let the reader check that the inverse of a special orthogonal matrix (which exists because special orthogonal matrices have determinant 1) is also special orthogonal.

---

[8]The adjective *special* applied to a matrix usually means its determinant is 1.

(iv) The ***affine*[9] *group*** Aff(1, $\mathbb{R}$) consists of all functions $f: \mathbb{R} \to \mathbb{R}$ (called *affine maps*) of the form

$$f(x) = ax + b,$$

where a and b are fixed real numbers with $a \neq 0$. Let us check that Aff(1, $\mathbb{R}$) is a group under composition. If $g : \mathbb{R} \to \mathbb{R}$ has the form $g(x) = cx + d$, where $c \neq 0$, then

$$fg(x) = f(g(x)) = f(cx + d) = a(cx + d) + b = acx + (ad + b).$$

Since $ac \neq 0$, the composite fg is an affine map. The identity function $1 : \mathbb{R} \to \mathbb{R}$ is an affine map (set $a = 1$ and $b = 0$), while the inverse of f is easily seen to be $x \mapsto a^{-1}x - a^{-1}b$.

Similarly, replacing $\mathbb{R}$ by $\mathbb{Q}$ or $\mathbb{C}$ gives the groups Aff(1, $\mathbb{Q}$) and Aff(1, $\mathbb{C}$). ◀

An operation allows one to multiply two elements at a time; how does one multiply three elements? There is a choice. Given the expression $2 \times 3 \times 4$, for example, one can first multiply $2 \times 3 = 6$ and then multiply $6 \times 4 = 24$; or, one can first multiply $3 \times 4 = 12$ and then multiply $2 \times 12 = 24$; of course, the answers agree, for multiplication of numbers is associative. Not all operations are associative, however. For example, subtraction is not associative: if $c \neq 0$, then

$$a - (b - c) \neq (a - b) - c.$$

Definition.   If G is a group and if $a \in G$, define the ***powers***[10] a^n, for $n \geq 1$, inductively:

$$a^1 = a \quad \text{and} \quad a^{n+1} = aa^n.$$

---

[9]Projective geometry involves enlarging the plane (and higher-dimensional spaces) by adjoining "points at infinity." The enlarged plane is called the projective plane, and the original plane is called an affine (or finite) plane. Affine functions are special functions between affine planes.

[10]The terminology x square and x cube for x^2 and x^3 is, of course, geometric in origin. Usage of the word *power* in this context goes back to Euclid, who wrote, "The power of a line is the square of the same line" (from the first English translation of Euclid, in 1570, by H. Billingsley). "Power" was the standard European rendition of the Greek *dunamis* (from which dynamo derives). However, contemporaries of Euclid, e.g., Aristotle and Plato, often used *dunamis* to mean amplification, and this seems to be a more appropriate translation, for Euclid was probably thinking of a 1-dimensional line sweeping out a 2-dimensional square. (I thank Donna Shalev for informing me of the classical usage of *dunamis*.)

Define $a^0 = 1$ and, if n is a positive integer, define

$$a^{-n} = (a^{-1})^n.$$

We let the reader prove that $(a^{-1})^n = (a^n)^{-1}$; this is a special case of the equation in Exercise 2.31.

There is a hidden complication here. The first and second powers are fine: $a^1 = a$ and $a^2 = aa$. There are two possible cubes: we have defined $a^3 = aa^2 = a(aa)$, but there is another reasonable contender: $(aa)a = a^2a$. If one assumes associativity, then these are equal:

$$a^3 = aa^2 = a(aa) = (aa)a = a^2a.$$

There are several possible products of a with itself four times; assuming that the operation is associative, is it obvious that $a^4 = a^3a = a^2a^2$? And what about higher powers?

An **expression** $a_1a_2 \cdots a_n$ is an n-tuple in $G \times \cdots \times G$ (n factors), and it yields many elements of G. Choose two adjacent a's, multiply them, and obtain an expression with $n - 1$ factors: the new product just formed and $n - 2$ original factors. In this shorter new expression, choose two adjacent factors (either an original pair or an original one together with the new product from the first step) and multiply them. Repeat this procedure until there is an expression with only two factors; multiply them and obtain an element of G; call this an **ultimate product** derived from the expression. For example, consider the expression $abcd$. We may first multiply ab, obtaining $(ab)cd$, an expression with three factors, namely, ab, c, d. We may now choose either the pair c, d or the pair ab, c; in either case, multiply these, obtaining expressions with two factors: $(ab)(cd)$ having factors ab and cd or $[(ab)c]d$ having factors $(ab)c$ and d. The two factors in either of these last expressions can now be multiplied to give an ultimate product from $abcd$. Other ultimate products derived from the expression $abcd$ arise by multiplying bc or cd as the first step. It is not obvious whether the ultimate products derived from a given expression are all equal.

Definition.   An expression $a_1a_2 \cdots a_n$ **needs no parentheses** if all the ultimate products derived from it are equal — that is, no matter what choices are made of adjacent factors to multiply, all the resulting products in G are equal.

Theorem 2.20 (Generalized Associativity).   *If G is a group and $a_1, a_2, \ldots, a_n \in G$, then the expression $a_1a_2 \cdots a_n$ needs no parentheses.*

Remark.   This result holds in greater generality, for neither the identity element nor inverses will be used in the proof.   ◄

Proof.   The proof is by (the second form of) induction. The base step $n = 3$ follows from associativity. For the inductive step, consider two elements of G obtained from an expression $a_1 a_2 \cdots a_n$ after two series of choices:

$$(a_1 \cdots a_i)(a_{i+1} \cdots a_n) \quad \text{and} \quad (a_1 \cdots a_j)(a_{j+1} \cdots a_n);$$

the parentheses indicate the last two factors before obtaining ultimate products from $a_1 a_2 \cdots a_n$; there are many parentheses inside each of these shorter expressions. We may assume that $i \leq j$. Since each of the four expressions in parentheses has fewer than n factors, the inductive hypothesis says that each needs no parentheses. It follows that the two products displayed above give the same elements of G if $i = j$. If $i < j$, then the inductive hypothesis allows the first expression to be rewritten

$$(a_1 \cdots a_i) \left([a_{i+1} \cdots a_j][a_{j+1} \cdots a_n] \right)$$

and the second to be rewritten

$$\left([a_1 \cdots a_i][a_{i+1} \cdots a_j] \right) (a_{j+1} \cdots a_n),$$

where each of the expressions $a_1 \cdots a_i$, $a_{i+1} \cdots a_j$, and $a_{j+1} \cdots a_n$ needs no parentheses. Thus, these expressions yield unique elements A, B, and C of G, respectively. The first expression yields $A(BC)$, the second yields $(AB)C$, and these two expression give the same element of G, by associativity.   ●

Corollary 2.21.   *If G is a group and $a, b \in G$, then*

$$(ab)^{-1} = b^{-1}a^{-1}.$$

Proof.   By Lemma 2.19(iii), it suffices to prove $(ab)(b^{-1}a^{-1}) = 1$ and $(b^{-1}a^{-1})(ab) = 1$. Using generalized associativity,

$$(ab)(b^{-1}a^{-1}) = [a(bb^{-1})]a^{-1} = (a1)a^{-1} = aa^{-1} = 1.$$

A similar argument proves the other equation. (We have been very careful about parentheses because we want to show that generalized associativity is being used. On the other hand, there is no need to be so fussy any more.)   ●

Corollary 2.22.   *If G is a group, if $a \in G$, and if $m, n \geq 1$, then*

$$a^{m+n} = a^m a^n \quad and \quad (a^m)^n = a^{mn}.$$

Proof.   In the first instance, both elements arise from the expression having $m + n$ factors each equal to a; in the second instance, both elements arise from the expression having mn factors each equal to a.   ◉

It follows that any two powers of an element a in a group commute:

$$a^m a^n = a^{m+n} = a^{n+m} = a^n a^m.$$

The proofs of the various statements in the next proposition, while straight-forward, are not short.

Proposition 2.23 (Laws of Exponents).   *Let G be a group, let $a, b \in G$, and let m and n be (not necessarily positive) integers.*

(i) *If a and b commute, then $(ab)^n = a^n b^n$.*

(ii) *$(a^n)^m = a^{mn}$.*

(iii) *$a^m a^n = a^{m+n}$.*

Proof.   Exercises for the reader.   ●

The notation a^n is the natural way to denote $a * a * \cdots * a$ if a appears n times. However, if the operation is +, then it is more natural to denote $a + a + \cdots + a$ by na. Let G be a group written additively; if $a, b \in G$ and m and n are (not necessarily positive) integers, then Proposition 2.23 is usually rewritten:

(i) $n(a + b) = na + nb$.

(ii) $m(na) = (mn)a$.

(iii) $ma + na = (m + n)a$.

Example 2.16.
Suppose a deck of cards is shuffled, so that the order of the cards has changed from $1, 2, 3, 4, \ldots, 52$ to $2, 1, 4, 3, \ldots, 52, 51$. If we shuffle again in the same way, then the cards return to their original order. But a similar thing happens for any permutation α of the 52 cards: if one repeats α sufficiently often, the deck is eventually restored to its original order. One way to see this

uses our knowledge of permutations. Write α as a product of disjoint cycles, say, $\alpha = \beta_1 \beta_2 \cdots \beta_t$, where β_i is an r_i-cycle. Now $\beta_i^{r_i} = (1)$ for every i, by Exercise 2.19, and so $\beta_i^k = (1)$, where k is the least common multiple of the r_i. Therefore,

$$\alpha^k = (\beta_1 \cdots \beta_t)^t = \beta_1^k \cdots \beta_t^k,$$

by Exercise 2.24.

Here is a more general result with a simpler proof (abstract algebra can be easier than algebra): If G is a finite group and $a \in G$, then $a^k = 1$ for some $k \geq 1$. Consider the subset

$$\{1, a, a^2, \ldots, a^n, \ldots\}.$$

Since G is finite, there must be a repetition occurring on this infinite list: there are integers $m > n$ with $a^m = a^n$, and hence $1 = a^m a^{-n} = a^{m-n}$. We have shown that there is some positive power of a equal to 1. (Our original argument that $\alpha^n = (1)$ for a permutation α of 52 cards is not worthless, for it gives an algorithm computing n.) ◄

Definition. Let G be a group and let $a \in G$. If $a^k = 1$ for some $k \geq 1$, then the smallest such exponent $k \geq 1$ is called the **order** of a; if no such power exists, then one says that a has **infinite order**.

The additive group of integers, $\mathbb{Z}$, is a group, and 1 is an element in it having infinite order.

The argument given above shows that every element in a finite group has finite order. In any group G, the identity has order 1, and it is the only element in G of order 1; an element has order 2 if and only if it is equal to its own inverse.

Lemma 2.24. *Let G be a group and assume that $a \in G$ has finite order k. If $a^n = 1$, then $k \mid n$. In fact, $\{n \in \mathbb{Z} : a^n = 1\}$ is the set of all the multiples of k.*

Proof. It is easy to see that $I = \{n \in \mathbb{Z} : a^n = 1\} \subset \mathbb{Z}$ satisfies the hypotheses of Corollary 1.31: $a^0 = 1$; if $a^n = 1$ and $a^m = 1$, then $a^{n-m} = a^n a^{-m} = 1$; if $a^n = 1$ and if q is any integer, then $a^{qn} = (a^n)^q = 1$. Therefore, I consists of all the multiples of k, where k is the smallest positive integer in I. But the smallest positive k in I is, by definition, the order of a. Therefore, if $a^n = 1$, then $n \in I$, and so n is a multiple of k. •

What is the order of a permutation in S_n?

Proposition 2.25.   *Let $\alpha \in S_n$.*

(i) *If α is an r-cycle, then α has order r.*

(ii) *If $\alpha = \beta_1 \cdots \beta_t$ is a product of disjoint r_i-cycles β_i, then α has order $m = \mathrm{lcm}\{r_1, \ldots, r_t\}$.*

(iii) *If p is a prime, then α has order p if and only if it is a p-cycle or a product of disjoint p-cycles.*

Proof.   (i) This is Exercise 2.19(ii).

(ii) Each β_i has order r_i, by (i). Suppose that $\alpha^M = (1)$. Since the β_i commute, $(1) = \alpha^M = (\beta_1 \cdots \beta_t)^M = \beta_1^M \cdots \beta_t^M$. By Exercise 2.26(ii), disjointness of the β's implies that $\beta_i^M = (1)$ for each i, so that Lemma 2.24 gives $r_i \mid M$ for all i; that is, M is a common multiple of $r_1, \ldots, r_t$. On the other hand, if $m = \mathrm{lcm}\{r_1, \ldots, r_t\}$, then it is easy to see that $\alpha^m = (1)$. Therefore, α has order m.

(iii) Write α as a product of disjoint cycles and use (ii).   •

For example, a permutation in S_n has order 2 if and only if it is either a transposition or a product of disjoint transpositions.

We can now augment the table in Example 2.9.

Cycle Structure	Number	Order	Parity
(1)	1	1	Even
(1 2)	10	2	Odd
(1 2 3)	20	3	Even
(1 2 3 4)	30	4	Odd
(1 2 3 4 5)	24	5	Even
(1 2)(3 4 5)	20	6	Odd
(1 2)(3 4)	15	2	Even
	120		

Table 2.3.   Permutations in S_5

Symmetry

We now present a connection between groups and symmetry.

Definition. A *motion* is a bijection[11] $\varphi : \mathbb{R}^2 \to \mathbb{R}^2$ that is *distance preserving*: for all points $P = (a, b)$ and $Q = (c, d)$ in $\mathbb{R}^2$,

$$\|\varphi(P) - \varphi(Q)\| = \|P - Q\|,$$

where

$$\|P - Q\| = \sqrt{(a - c)^2 + (b - d)^2}.$$

We prove that the set $\mathcal{M}$ of all motions of the plane is a group under composition by showing that $\mathcal{M}$ is a subgroup of $S_{\mathbb{R}^2}$, the group of all permutations of $\mathbb{R}^2$. The identity function 1 is clearly a motion. If φ' and φ are motions, then, for all P and Q, we have

$$\|\varphi'(\varphi(P)) - \varphi'(\varphi(Q))\| = \|\varphi(P) - \varphi(Q)\| = \|P - Q\|,$$

and so the composite $\varphi'\varphi$ is also a motion. Finally, the inverse of a motion φ (which exists because motions are bijections) is a motion:

$$\|P - Q\| = \|\varphi(\varphi^{-1}(P)) - \varphi(\varphi^{-1}(Q))\| = \|\varphi^{-1}(P) - \varphi^{-1}(Q)\|.$$

Therefore, $\mathcal{M}$ is a group.

If P and Q are points in the plane, we denote the line segment with endpoints P and Q by PQ.

Here are three kinds of motions.

(i) Given an angle θ, *rotation* R_θ about the origin O is defined as follows: $R_\theta(O) = O$; if $P \neq O$, draw the line OP in Figure 2.5, rotate it θ (counterclockwise if θ is positive, clockwise if θ is negative) to OP', and define $R_\theta(P) = P'$. Of course, one can rotate about any point in the plane.

(ii) *Reflection* ρ_L in a line L fixes each point in L; if $P \notin L$, then $\rho_L(P) = P'$, as in Figure 2.6 (L is the perpendicular bisector of PP'). If one pretends that L is a mirror, then P' is the mirror image of P.

(iii) Given a point V, *translation*[12] *by* V is the function $\tau_V : \mathbb{R}^2 \to \mathbb{R}^2$ defined by $\tau_V(U) = U + V$ (see Figure 2.7).

---

[11] It can be proved that a distance-preserving function must be a bijection.

[12] The word *translation* comes from the Latin word meaning "to transfer." It usually means passing from one language to another, but here it means a special way of moving each point to another.

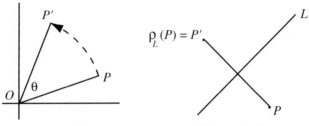

Figure 2.5   **Figure 2.6**

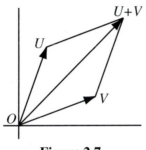

Figure 2.7

It can be proved that every motion is a composite of motions of these three special types. It follows that motions preserve various geometric figures (although there are simpler direct proofs for the following facts).

(i) If φ is a motion and PQ is the line segment with endpoints P and Q, then $\varphi(PQ)$ is the line segment with endpoints $\varphi(P)$ and $\varphi(Q)$.

(ii) If Ω is a polygon with vertices $v_1, v_2, \ldots, v_n$, then $\varphi(\Omega)$ is a polygon with vertices $\varphi(v_1), \varphi(v_2), \ldots, \varphi(v_n)$ and Ω and $\varphi(\Omega)$ are congruent.

Example 2.17.
As we have just remarked, if Δ is a triangle with vertices P, Q, U and if φ is a motion, then $\varphi(\Delta)$ is the triangle with vertices $\varphi(P)$, $\varphi(Q)$, $\varphi(U)$. If we assume further that $\varphi(\Delta) = \Delta$, then φ permutes the vertices P, Q, U (see Figure 2.8). Assume that the center of Δ is O. If Δ is isosceles (with equal sides PQ and PU), and if ρ_{PO} is the reflection in PO, then $\rho_{PO}(\Delta) = \Delta$ (we can thus describe ρ_{PO} by the transposition $(Q\ U)$, for it fixes P and interchanges Q and U); on the other hand, if Δ is not isosceles, then $\rho_{PO}(\Delta) \neq \Delta$. If Δ is equilateral, then $\rho_{QO}(\Delta) = \Delta$ and $\rho_{UO}(\Delta) = \Delta$ [we can describe these

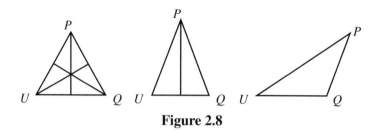

Figure 2.8

reflections by the transpositions $(P\ U)$ and $(P\ Q)$, respectively]; these re-flections do not carry Δ into itself when Δ is only isosceles. Moreover, the rotation about O by $120°$ and $240°$ also carry Δ into itself [these rotations can be described by the 3-cycles $(P\ Q\ U)$ and $(P\ U\ Q)$]. We see that an equilateral triangle is "more symmetric" than an isosceles triangle, and that an isosceles triangle is "more symmetric" than a triangle Δ that is not even isosceles [for such a triangle, $\varphi(\Delta) = \Delta$ implies that $\varphi = 1$].   ◄

Definition.   The *symmetry group* $\Sigma(\Omega)$ of a figure Ω in the plane is the set of all motions φ with $\varphi(\Omega) = \Omega$. The elements of $\Sigma(\Omega)$ are called *symmetries* of Ω.

It is easy to see that $\Sigma(\Omega)$ is a subgroup of $\mathcal{M}$, so that it is a group. The example of triangles indicates that one figure is "more symmetric" than another figure if it has a "larger" symmetry group: if π_3 is an equilateral triangle, then $|\Sigma(\pi_3)| = 6$; if Δ is only an isosceles triangle, then $|\Sigma(\Delta)| = 2$; if Δ is a triangle that is not even isosceles, then $|\Sigma(\Delta)| = \{1\}$. Thus, groups can describe the amount of symmetry present in a figure.

Example 2.18.
(i) Let π_4 be a square having sides of length 1 and vertices $\{v_1, v_2, v_3, v_4\}$; draw π_4 in the plane so that its center is at the origin O and its sides are parallel to the axes. It is easy to see that every $\varphi \in \Sigma(\pi_4)$ permutes the vertices; indeed, a symmetry φ of π_4 is determined by $\{\varphi(v_i) : 1 \le i \le 4\}$, and so there are at most $24 = 4!$ possible symmetries. Not every permutation in S_4 arises from a symmetry of π_4, however. If v_i and v_j are adjacent, then $\|v_i - v_j\| = 1$, but $\|v_1 - v_3\| = \sqrt{2} = \|v_2 - v_4\|$; it follows that φ must preserve adjacency (for motions preserve distance). The reader may now check that there are only eight symmetries of π_4. Aside from the identity and the three rotations about O by $90°$, $180°$, and $270°$, there are four reflections,

respectively, in the lines $v_1 v_3$, $v_2 v_4$, the x-axis, and the y-axis (for a generalization to come, note that the y-axis is Om_1, where m_1 is the midpoint of $v_1 v_2$, and the x-axis is Om_2, where m_2 is the midpoint of $v_2 v_3$). The group $\Sigma(\pi_4)$ is called the **dihedral group** with 8 elements, and it is denoted by D_8.

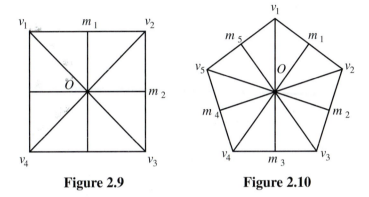

Figure 2.9   **Figure 2.10**

(ii) The symmetry group $\Sigma(\pi_5)$ of a regular pentagon π_5 with vertices $v_1, \ldots,$ v_5 and center O has 10 elements: the rotations about the origin of $(72j)°$, where $0 \leq j \leq 4$, as well as the reflections in the lines Ov_k for $1 \leq k \leq 5$. The symmetry group $\Sigma(\pi_5)$ is called the **dihedral group** with 10 elements, and it is denoted by D_{10}.   ◄

Definition.   The symmetry group $\Sigma(\pi_n)$ of a regular polygon π_n with n vertices $v_1, v_2, \ldots, v_n$ and center at O is called the **dihedral group**[13] with $2n$ elements, and it is denoted by D_{2n}.

The dihedral group D_{2n} contains the n rotations ρ^j about the center by $(360j/n)°$, where $0 \leq j \leq n - 1$. The description of the other n elements depends on the parity of n. If n is odd (as in the case of the pentagon; see Figure 2.10), then the other n symmetries are reflections in the distinct lines Ov_i, for $i = 1, 2, \ldots, n$. If $n = 2q$ is even [see the square in Figure 2.9 or

---

[13]F. Klein was investigating those finite groups occurring as subgroups of the group of motions of $\mathbb{R}^3$. Some of these occur as symmetry groups of regular polyhedra [from the Greek *poly* meaning "many" and *hedron* meaning "two-dimensional side." He invented a degenerate polyhedron that he called a *dihedron*, from the Greek word *di* meaning "two," which consists of two congruent regular polygons of zero thickness pasted together. The symmetry group of a dihedron is thus called a *dihedral group*. For our purposes, it is more natural to describe these groups as in the text.

the regular hexagon in Figure 2.11], then each line Ov_i coincides with the line Ov_{q+i}, giving only q such reflections; the remaining q symmetries are reflections in the lines Om_i for $i = 1, 2, \ldots, q$, where m_i is the midpoint of the edge $v_i v_{i+1}$. For example, the six lines of symmetry of π_6 are Ov_1, Ov_2, and Ov_3, and Om_1, Om_2, and Om_3.

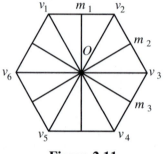

Figure 2.11

Symmetry arises in calculus when describing figures in the plane. We quote from Edwards and Penny, *Calculus and Analytic Geometry*, 3d ed., 1990, p. 456, as they describe different kinds of symmetry that might be enjoyed by a curve with equation $f(x, y) = 0$.

(i) *Symmetry about the x-axis*: the equation of the curve is unaltered when y is replaced by $-y$.

(ii) *Symmetry about the y-axis*: the equation of the curve is unaltered when x is replaced by $-x$.

(iii) *Symmetry with respect to the origin*: the equation of the curve is unaltered when x is replaced by $-x$ and y is replaced by $-y$.

(iv) *Symmetry about the 45° line $y = x$*: the equation is unaltered when x and y are interchanged.

In our language, their first symmetry is ρ_x, reflection in the x-axis, the second is ρ_y, reflection in the y-axis, the third is R_{180}, rotation by 180°, and the fourth is ρ_L, where L is the 45° line. One can now say whether an equation in two variables has symmetry. For example, a function $f(x, y)$ has the first type of symmetry if $f(x, y) = f(x, -y)$. In this case, the graph Γ of the equation $f(x, y) = 0$ [consisting of all the points (a, b) for which $f(a, b) = 0$] is symmetric about the x-axis, for $(a, b) \in \Gamma$ implies $(a, -b) \in \Gamma$. This is what physicists mean when they say that solutions to various systems of differential equations (in many variables) have certain types of symmetry.

EXERCISES

2.31 If $a_1, a_2, \ldots, a_{t-1}, a_t$ are elements in a group G, prove that

$$(a_1 a_2 \cdots a_{t-1} a_t)^{-1} = a_t^{-1} a_{t-1}^{-1} \cdots a_2^{-1} a_1^{-1}.$$

2.32 Assume that G is a set with an associative operation. Prove that $(ab)(cd) = a[(bc)d]$ without using generalized associativity.

2.33   (i) Compute the order, inverse, and parity of

$$\alpha = (1\ 2)(4\ 3)(1\ 3\ 5\ 4\ 2)(1\ 5)(1\ 3)(2\ 3).$$

   (ii) What are the respective orders of the permutations in Exercises 2.16 and 2.22?

2.34   (i) How many elements of order 2 are there in S_5 and in S_6?

   (ii) How many elements of order 2 are there in S_n?

2.35 If G is a group, prove that the only element $g \in G$ with $g^2 = g$ is 1.

2.36 Let H be a set containing an element e, and assume that there is an associative operation $*$ on H satisfying:

1. $e * x = x$ for all $x \in H$;

2. for every $x \in H$, there is $x' \in H$ with $x' * x = e$.

   (i) Prove that if $h \in H$ satisfies $h * h = h$, then $h = e$.

   (ii) For all $x \in H$, prove that $x * x' = e$.

   (iii) For all $x \in H$, prove that $x * e = x$.

   (iv) Prove that if $e' \in H$ satisfies $e' * x = x$ for all $x \in H$, then $e' = e$.

   (v) Let $x \in H$. Prove that if $x'' \in H$ satisfies $x'' * x = e$, then $x'' = x'$.

   (vi) Prove that H is a group.

2.37 Let y be a group element of order m; if $m = pt$ for some prime p, prove that y^t has order p.

2.38 Let G be a group and let $a \in G$ have order pk for some prime p, where $k \geq 1$. Prove that if there is $x \in G$ with $x^p = a$, then the order of x is $p^2 k$, and hence x has larger order than a.

2.39 Let $G = \mathrm{GL}(2, \mathbb{Q})$, and let

$$A = \begin{bmatrix} 0 & -1 \\ 1 & 0 \end{bmatrix} \quad \text{and} \quad B = \begin{bmatrix} 0 & 1 \\ -1 & 1 \end{bmatrix}.$$

Show that $A^4 = E = B^6$, but that $(AB)^n \neq E$ for all $n > 0$. Conclude that AB can have infinite order even though both factors A and B have finite order (this cannot happen in a finite group).

2.40   (i) Prove, by induction on $k \geq 1$, that

$$\begin{bmatrix} \cos\theta & -\sin\theta \\ \sin\theta & \cos\theta \end{bmatrix}^k = \begin{bmatrix} \cos k\theta & -\sin k\theta \\ \sin k\theta & \cos k\theta \end{bmatrix}.$$

(ii) Find all the elements of finite order in the special orthogonal group $SO(2, \mathbb{R})$ [see Example 2.15(iii)].

2.41 If G is a group in which $x^2 = 1$ for every $x \in G$, prove that G must be abelian. [The Boolean groups $\mathcal{B}(X)$ of Example 2.14 are such groups.]

2.42 If G is a group with an even number of elements, prove that the number of elements in G of order 2 is odd. In particular, G must contain an element of order 2.

2.43 What is the largest order of an element in S_n, where $n = 1, 2, \ldots, 10$?

2.44 The ***stochastic***[14] ***group*** $\Sigma(2, \mathbb{R})$ consists of all those matrices in $GL(2, \mathbb{R})$ whose column sums are 1; that is, $\Sigma(2, \mathbb{R})$ consists of all the nonsingular matrices

$$\begin{bmatrix} a & c \\ b & d \end{bmatrix}$$

with $a + b = 1 = c + d$. [There are also stochastic groups $\Sigma(2, \mathbb{Q})$ and $\Sigma(2, \mathbb{C})$.]

Prove that the product of two stochastic matrices is again stochastic, and that the inverse of a stochastic matrix is stochastic.

2.4 LAGRANGE'S THEOREM

Definition.   A subset H of a group G is a ***subgroup*** if

(i) $1 \in H$;

(ii) if $x, y \in H$, then $xy \in H$;

(iii) if $x \in H$, then $x^{-1} \in H$.

---

[14]The term *stochastic* comes from the Greek word meaning "to guess." Its mathematical usage occurs in statistics, and stochastic matrices first arose in the study of certain statistical problems.

If H is a subgroup of G, we write $H \leq G$. Observe that $\{1\}$ and G are always subgroups of a group G, where $\{1\}$ denotes the subset consisting of the single element 1. We call a subgroup H of G **proper**, and we write $H < G$, if $H \neq G$; we call a subgroup H of G **nontrivial** if $H \neq \{1\}$. More interesting examples will be given below.

Let us see that every subgroup $H \leq G$ is itself a group. Property (ii) shows that H is **closed**; that is, H has an operation; it is associative, for $(xy)z = x(yz)$ holds for all $x, y, z \in G$, and so this equation holds, in particular, for all $x, y, z \in H$. Finally, (i) gives the identity, and (iii) gives inverses.

It is easier to check that a subset H of a group G is a subgroup (and hence that it is a group in its own right) than to verify the group axioms for H: associativity is inherited from the operation on G and hence it need not be verified again.

Example 2.19.

(i) The four permutations

$$\mathbf{V} = \big\{(1), (1\ 2)(3\ 4), (1\ 3)(2\ 4), (1\ 4)(2\ 3)\big\}$$

form a group, because V is a subgroup of S_4 : $(1) \in V$; $\alpha^2 = (1)$ for each $\alpha \in V$, and so $\alpha^{-1} = \alpha \in V$; the product of any two distinct permutations in $V - \{(1)\}$ is the third one.

Consider what verifying associativity $a(bc) = (ab)c$ would involve: there are 4 choices for each of a, b, and c, and so there are $4^3 = 64$ equations to be checked. Of course, we may assume that none is (1), leaving us with only $3^3 = 27$ equations but, plainly, proving $\mathbf{V}$ is a group by showing it is a subgroup of S_4 is obviously the best way to proceed.

One calls $\mathbf{V}$ the **four-group** ($\mathbf{V}$ abbreviates the original German term *Vierergruppe*).

(ii) If $\mathbb{R}^2$ is the plane considered as an (additive) abelian group, then any line L through the origin is a subgroup. The easiest way to see this is to choose a point (a, b) on L and then note that L consists of all the scalar multiples (ra, rb). The reader may now verify that the axioms in the definition of subgroup do hold for L.   ◀

One can shorten the list of items needed to verify that a subset is, in fact, a subgroup.

Proposition 2.26. *A subset H of a group G is a subgroup if and only if H is nonempty and, whenever $x, y \in H$, then $xy^{-1} \in H$.*

Proof. If H is a subgroup, then it is nonempty, for $1 \in H$. If $x, y \in H$, then $y^{-1} \in H$, by part (iii) of the definition, and so $xy^{-1} \in H$, by part (ii).

Conversely, assume that H is a subset satisfying the new condition. Since H is nonempty, it contains some element, say, h. Taking $x = h = y$, we see that $1 = hh^{-1} \in H$, and so part (i) holds. If $y \in H$, then set $x = 1$ (which we can now do because $1 \in H$), giving $y^{-1} = 1y^{-1} \in H$, and so part (iii) holds. Finally, we know that $(y^{-1})^{-1} = y$, by Lemma 2.19. Hence, if $x, y \in H$, then $y^{-1} \in H$, and so $xy = x(y^{-1})^{-1} \in H$. Therefore, H is a subgroup of G. •

Note that if the operation in G is addition, then the condition in the proposition is that H is a nonempty subset such that $x, y \in H$ implies $x - y \in H$.

For Galois, a group was just a subset H of S_n that is closed under composition; that is, if $\alpha, \beta \in H$, then $\alpha \circ \beta \in H$. A. Cayley, in 1854, was the first to define an abstract group, mentioning associativity, inverses, and identity explicitly.

Proposition 2.27. *A nonempty subset H of a finite group G is a subgroup if and only if H is closed; that is, if $a, b \in H$, then $ab \in H$. In particular, a nonempty subset of S_n is a subgroup if and only if it is closed.*

Proof. Every subgroup is nonempty, by (i) in the definition of subgroup, and it is closed, by (ii).

Conversely, assume that H is a nonempty closed subset of G; thus, (ii) holds. It follows that H contains all the powers of its elements. In particular, there is some element $a \in H$, because H is nonempty, and $a^n \in H$ for all $n \geq 1$. As we saw in Example 2.16, every element in G has finite order: there is an integer m with $a^m = 1$; hence $1 \in H$ and (i) holds. Finally, if $h \in H$ and $h^m = 1$, then $h^{-1} = h^{m-1}$ (for $hh^{m-1} = 1 = h^{m-1}h$), so that $h^{-1} \in H$ and (iii) holds. Therefore, H is a subgroup of G. •

This last proposition can be false when G is an infinite group. For example, let G be the additive group $\mathbb{Z}$ and take H to be the subset

$$H = \{n \in \mathbb{Z} : n \geq 1\}.$$

Now H is closed, but it is not a subgroup of $\mathbb{Z}$.

Example 2.20.
The subset A_n of S_n, consisting of all the even permutations, is a subgroup because it is closed under multiplication: even $\circ$ even $=$ even. This subgroup $A_n \leq S_n$ is called the *alternating*[15] *group* on n letters. ◄

Definition.   If G is a group and $a \in G$, write

$$\langle a \rangle = \{a^n : n \in \mathbb{Z}\} = \{\text{all powers of } a\};$$

$\langle a \rangle$ is called the *cyclic subgroup* of G *generated* by a.

A group G is called *cyclic* if $G = \langle a \rangle$, in which case a is called a *generator* of G.

It is easy to see that $\langle a \rangle$ is, in fact, a subgroup: $1 = a^0 \in \langle a \rangle$; $a^n a^m = a^{n+m} \in \langle a \rangle$; $a^{-1} \in \langle a \rangle$. Example 2.12(v) shows, for every $n \geq 1$, that the multiplicative group Γ_n of all nth roots of unity is a cyclic group with the primitive nth root of unity $\zeta = e^{2\pi i/n}$ as a generator.

A cyclic group can have several different generators. For example, $\langle a \rangle = \langle a^{-1} \rangle$. In Exercise 2.49, it is shown that if $G = \langle a \rangle$ and if a has order n, then every element of the form a^k, where $(k, n) = 1$, is also a generator of G.

NOTATION.   If X is a finite set, let us denote the number of elements in X by $|X|$.

Proposition 2.28.   *Let G be a finite group and let $a \in G$. Then the order of a is the number of elements in $\langle a \rangle$.*

Proof.   Since G is finite, there is an integer $k \geq 1$ with $1, a, a^2, \ldots, a^{k-1}$ consisting of k distinct elements, while $1, a, a^2, \ldots, a^k$ has a repetition; hence $a^k \in \{1, a, a^2, \ldots, a^{k-1}\}$; that is, $a^k = a^i$ for some i with $0 \leq i < k$. If $i \geq 1$, then $a^{k-i} = 1$, contradicting the original list having no repetitions. Therefore, $a^k = a^0 = 1$, and k is the order of a (being the smallest positive such k).

If $H = \{1, a, a^2, \ldots, a^{k-1}\}$, then $|H| = k$; it suffices to show that $H = \langle a \rangle$. Clearly, $H \subset \langle a \rangle$. For the reverse inclusion, take $a^i \in \langle a \rangle$. By the

---

[15]The *alternating group* first arose in studying polynomials. If

$$f(x) = (x - u_1)(x - u_2) \cdots (x - u_n),$$

then the number $D = \prod_{i<j}(u_i - u_j)$ changes sign if one permutes the roots: if α is a permutation of $\{u_1, u_2, \ldots, u_n\}$, then it is easy to see that $\prod_{i<j}[\alpha(u_i) - \alpha(u_j)] = \pm D$. Thus, the sign of the product alternates as various permutations α are applied to its factors. The sign does not change for those α in the alternating group.

division algorithm, $i = qk + r$, where $0 \leq r < k$. Hence $a^i = a^{qk+r} = a^{qk}a^r = (a^k)^q a^r = a^r \in H$; this gives $\langle a \rangle \subset H$, and so $\langle a \rangle = H$.   •

Definition.   If G is a finite group, then the number of elements in G, denoted by $|G|$, is called the **order** of G.

The word "order" is used in two senses: the order of an *element* $a \in G$ and the order $|G|$ of a *group* G. Proposition 2.28 shows that the order of a group element a is equal to $| \langle a \rangle |$.

Proposition 2.29.   *The intersection $\bigcap_{i \in I} H_i$ of any family of subgroups of a group G is again a subgroup of G. In particular, if H and K are subgroups of G, then $H \cap K$ is a subgroup of G.*

Proof.   Let $D = \bigcap_{i \in I} H_i$; we prove that D is a subgroup by verifying each of the parts in the definition. Note first that $D \neq \varnothing$ because $1 \in D$ since $1 \in H_i$ for all i. Suppose that $x \in D$. Now x got into D by being in each H_i; as each H_i is a subgroup, $x^{-1} \in H_i$ for all i, and so $x^{-1} \in D$. Finally, if $x, y \in D$, then both x and y lie in every H_i, hence their product xy lies in every H_i, and so $xy \in D$.   •

Corollary 2.30.   *If X is a subset of a group G, then there is a subgroup $\langle X \rangle$ of G containing X that is **smallest** in the sense that $\langle X \rangle \leq H$ for every subgroup H of G which contains X.*

Proof.   First of all, note that there exist subgroups of G which contain X; for example, G itself contains X. Define $\langle X \rangle = \bigcap_{X \subset H} H$, the intersection of all the subgroups H of G which contain X. By Proposition 2.29, $\langle X \rangle$ is a subgroup of G; of course, $\langle X \rangle$ contains X because every H contains X. Finally, if H is any subgroup containing X, then H is one of the subgroups whose intersection is $\langle X \rangle$; that is, $\langle X \rangle \leq H$.   •

Note that there is no restriction on the subset X in the last corollary; in particular, $X = \varnothing$ is allowed. Since the empty set is a subset of every set, we have $\varnothing \subset H$ for every subgroup H of G. Thus, $\langle \varnothing \rangle$ is the intersection of *all* the subgroups of G, and so $\langle \varnothing \rangle = \{1\}$.

Definition.   If X is a subset of a group G, then $\langle X \rangle$ is called the **subgroup generated by** X.

Example 2.21.
(i) If $G = \langle a \rangle$ is a cyclic group with generator a, then G is generated by the subset $X = \{a\}$.

(ii) The dihedral group D_{2n}, the symmetry group of a regular n-gon, is generated by ρ, σ, where ρ is a rotation by $(360/n)°$ and σ is a reflection. Note that these generators satisfy the conditions $\rho^n = 1$, $\sigma^2 = 1$, and $\sigma\rho\sigma = \rho^{-1}$.

◄

Perhaps the most fundamental fact about subgroups H of a finite group G is that their orders are constrained. Certainly, we have $|H| \le |G|$, but it turns out that $|H|$ must be a divisor of $|G|$. To prove this, we introduce the notion of coset.

Definition.   If H is a subgroup of a group G and $a \in G$, then the ***coset***[16] aH is the subset aH of G, where

$$aH = \{ah : h \in H\}.$$

If we use the $*$ notation for the operation in a group G, then we denote the coset aH by $a * H$, where

$$a * H = \{a * h : h \in H.\}$$

In particular, if the operation is addition, then the coset is denoted by

$$a + H = \{a + h : h \in H\}.$$

Of course, $a = a1 \in aH$. Cosets are usually not subgroups. For example, if $a \notin H$, then $1 \notin aH$ (otherwise $1 = ah$ for some $h \in H$, and this gives the contradiction $a = h^{-1} \in H$).

Example 2.22.
(i) Consider the plane $\mathbb{R}^2$ as an (additive) abelian group and let L be a line through the origin O (see Figure 2.12 on page 140); as in Example 2.19(ii), the line L is a subgroup of $\mathbb{R}^2$. If $\beta \in \mathbb{R}^2$, then the coset $\beta + L$[17] is the line L' containing β which is parallel to L, for if $r\alpha \in L$, then the parallelogram law gives $\beta + r\alpha \in L'$.

---

[16]The cosets just defined are often called *left cosets*; there are also right cosets of H, namely, subsets of the form $Ha = \{ha : h \in H\}$; these arise in further study of groups, but we shall now work with (left) cosets only.

[17]If we revert to the original notation in which $*$ denotes the group operation, then the coset $a * H$ is $\{a * h : h \in H\}$. Since the operation in $\mathbb{R}^2$ is addition, the coset here is $\beta + L$.

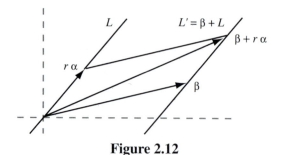

Figure 2.12

(ii) If $G = S_3$ and $H = \langle (1\ 2) \rangle$, there are exactly three cosets of H, namely:

$$
\begin{aligned}
H &= \quad \{(1), (1\ 2)\} = (1\ 2)H, \\
(1\ 3)H &= \{(1\ 3), (1\ 2\ 3)\} = (1\ 2\ 3)H, \\
(2\ 3)H &= \{(2\ 3), (1\ 3\ 2)\} = (1\ 3\ 2)H,
\end{aligned}
$$

each of which has size 2.

(iii) Recall that the dihedral group $D_{2n} = \Sigma(\pi_n)$, the group of symmetries of the regular n-gon π_n, has order $2n$ and it contains a cyclic subgroup of order n generated by a rotation ρ. The subgroup $\langle \rho \rangle$ has index $[D_{2n} : \langle \rho \rangle] = 2$. Thus, there are two cosets: $\langle \rho \rangle$ and $\sigma \langle \rho \rangle$, where σ is any reflection outside of $\langle \rho \rangle$.   ◀

Observe, in our examples, that different cosets of a given subgroup do not overlap.

Lemma 2.31.   *Let H be a subgroup of a group G, and let $a, b \in G$.*

  (i) *$aH = bH$ if and only if $b^{-1}a \in H$. In particular, $aH = H$ if and only if $a \in H$.*

  (ii) *If $aH \cap bH \neq \emptyset$, then $aH = bH$.*

  (iii) *$|aH| = |H|$ for all $a \in G$.*

Proof.   (i) Assume that $aH = bH$. Then $a = bh_0$ for some $h_0 \in H$, so that $b^{-1}a = h_0 \in H$. Conversely, suppose that $b^{-1}a \in H$, that is, $a = bh_0$ for some $h_0 \in H$. We show that $aH \subset bH$ and $bH \subset aH$. If $x \in aH$, then $x = ah = bh_0h \in bH$ (because $h_0h \in H$). For the reverse inclusion, take $y \in bH$; then $y = bh'$ for some $h' \in H$, and $y = bh' = ah_0^{-1}h' \in aH$ (because $hh_0^{-1}h' \in H$). Therefore, $aH = bH$.

The second statement follows because $H = 1H$.

(ii) If $x \in aH \cap bH$, then $x = ah = bh'$ for some $h, h' \in H$, and so $b^{-1}a = h'h^{-1} \in H$. Therefore, $aH = bH$.

(iii) The function $f : H \to aH$, given by $f(h) = ah$, is easily seen to be a bijection [its inverse $g : aH \to H$ is given by $g(x) = a^{-1}x$ for all $x \in aH$; that is, $g : ah \mapsto a^{-1}(ah) = h$]. Therefore, H and aH have the same number of elements, by Exercise 2.7.   •

Theorem 2.32 (Lagrange's Theorem).   *If H is a subgroup of a finite group G, then $|H|$ is a divisor of $|G|$.*

Proof.   Let $\{a_1 H, a_2 H, \ldots, a_t H\}$ be the family of all the distinct cosets of H in G. Then

$$G = a_1 H \cup a_2 H \cup \cdots \cup a_t H,$$

because each $g \in G$ lies in the coset gH, and $gH = a_i H$ for some i. Moreover, Lemma 2.31(ii) shows that distinct cosets $a_i H$ and $a_j H$ are disjoint. It follows that

$$|G| = |a_1 H| + |a_2 H| + \cdots + |a_t H|.$$

But $|a_i H| = |H|$ for all i, by Lemma 2.31(iii), so that $|G| = t|H|$, as desired.   •

Definition.   The *index* of a subgroup H in G, denoted by $[G : H]$, is the number of cosets of H in G.

The index $[G : H]$ is the number t in the formula $|G| = t|H|$ in the proof of Lagrange's theorem, so that

$$|G| = [G : H]|H|;$$

this formula shows that the index $[G : H]$ is also a divisor of $|G|$; moreover,

$$[G : H] = |G|/|H|.$$

Corollary 2.33.   *If G is a finite group and $a \in G$, then the order of a is a divisor of $|G|$.*

Proof.   By Proposition 2.28, the order of the element a is equal to the order of the subgroup $H = \langle a \rangle$.   •

Corollary 2.34. *If a finite group G has order m, then $a^m = 1$ for all $a \in G$.*

Proof.   By Corollary 2.33, a has order d, where $d \mid m$; that is, $m = dk$ for some integer k. Thus, $a^m = a^{dk} = (a^d)^k = 1$.   •

Corollary 2.35. *If p is a prime, then every group G of order p is cyclic.*

Proof.   Choose $a \in G$ with $a \neq 1$, and let $H = \langle a \rangle$ be the cyclic subgroup generated by a. By Lagrange's theorem, $|H|$ is a divisor of $|G| = p$. Since p is a prime and $|H| > 1$, it follows that $|H| = p = |G|$, and so $H = G$, as desired.   •

EXERCISES

2.45   (i) Define the *special linear group* by

$$SL(2, \mathbb{R}) = \{A \in GL(2, \mathbb{R}) : \det(A) = 1\}.$$

   Prove that $SL(2, \mathbb{R})$ is a subgroup of $GL(2, \mathbb{R})$.
   (ii) Prove that $GL(2, \mathbb{Q})$ is a subgroup of $GL(2, \mathbb{R})$.

2.46 Give an example of two subgroups H and K of a group G whose union $H \cup K$ is not a subgroup of G.

2.47 Let G be a finite group with subgroups H and K. If $H \leq K$, prove that

$$[G : H] = [G : K][K : H].$$

2.48 If H and K are subgroups of a group G and if $|H|$ and $|K|$ are relatively prime, prove that $H \cap K = \{1\}$.

2.49 Let $G = \langle a \rangle$ be a cyclic group of order n. Show that a^k is a generator of G (that is, $G = \langle a^k \rangle$) if and only if $(k, n) = 1$.

2.50 Prove that every subgroup S of a cyclic group $G = \langle a \rangle$ is itself cyclic.

2.51 Prove that if G is a cyclic group of order n and if $d \mid n$, then G has a subgroup of order d.

2.52 Let G be a group of order 4. Prove that either G is cyclic or $x^2 = 1$ for every $x \in G$. Conclude, using Exercise 2.41, that G must be abelian.

2.53   (i) Prove that the stochastic group $\Sigma(2, \mathbb{R})$, the set of all nonsingular 2×2 matrices whose row sums are 1, is a subgroup of $GL(2, \mathbb{R})$ (see Exercise 2.44).
   (ii) Define $\Sigma'(2, \mathbb{R})$ to be the set of all nonsingular *doubly stochastic* matrices (all row sums are 1 and all column sums are 1). Prove that $\Sigma'(2, \mathbb{R})$ is a subgroup of $GL(2, \mathbb{R})$.

2.54 Show that the symmetry group $\Sigma(C)$ of a circle C is infinite.

2.5 HOMOMORPHISMS

An important problem is determining whether two given groups G and H are somehow the same. For example, we have investigated S_3, the group of all permutations of $X = \{1, 2, 3\}$. The group S_Y of all the permutations of $Y = \{a, b, c\}$ is a group different from S_3 because permutations of $\{1, 2, 3\}$ are different than permutations of $\{a, b, c\}$. But even though S_3 and S_Y are different, they surely bear a strong resemblance to each other (see Example 2.23). The notions of homomorphism and isomorphism allow one to compare different groups, as we shall see.

Definition.   If $(G, *)$ and $(H, \circ)$ are groups (we have displayed the operation in each), then a function $f : G \to H$ is a ***homomorphism***[18] if

$$f(x * y) = f(x) \circ f(y)$$

for all $x, y \in G$. If f is also a bijection, then f is called an ***isomorphism***. Two groups G and H are called ***isomorphic***, denoted by $G \cong H$, if there exists an isomorphism $f : G \to H$ between them.

For example, let $\mathbb{R}$ be the group of all real numbers with operation addition, and let $\mathbb{R}^>$ be the group of all positive real numbers with operation multiplication. The function $f : \mathbb{R} \to \mathbb{R}^>$, defined by $f(x) = e^x$, is a bijection [its inverse function is $g(a) = \ln a$]. Moreover, f is an isomorphism, for if $x, y \in \mathbb{R}$, then

$$f(x + y) = e^{x+y} = e^x e^y = f(x) f(y).$$

Therefore, the additive group $\mathbb{R}$ is isomorphic to the multiplicative group $\mathbb{R}^>$.

As a second example, we claim that the additive group $\mathbb{C}$ of complex numbers is isomorphic to the additive group $\mathbb{R}^2$ [see Example 2.12(vi)]. Define $f : \mathbb{C} \to \mathbb{R}^2$ by

$$f : a + ib \mapsto (a, b).$$

---

[18]The word *homomorphism* comes from the Greek *homo* meaning "same" and *morph* meaning "shape" or "form." Thus, a homomorphism carries a group to another group (its image) of similar form. The word *isomorphism* involves the Greek *iso* meaning "equal," and isomorphic groups have identical form.

It is easy to check that f is a bijection; f is a homomorphism because

$$f([a + ib] + [a' + ib']) = f([a + a'] + i[b + b'])$$
$$= (a + a', b + b')$$
$$= (a, b) + (a', b')$$
$$= f(a + ib) + f(a' + ib').$$

Definition. Let $a_1, a_2, \ldots, a_n$ be a list with no repetitions of all the elements of a group G. A ***multiplication table*** for G is an $n \times n$ matrix whose ij entry is $a_i a_j$.

G	a_1	a_2	$\cdots$	a_j	$\cdots$	a_n
a_1	$a_1 a_1$	$a_1 a_2$	$\cdots$	$a_1 a_j$	$\cdots$	$a_1 a_n$
a_2	$a_2 a_1$	$a_2 a_2$	$\cdots$	$a_2 a_j$	$\cdots$	$a_2 a_n$
a_i	$a_i a_1$	$a_i a_2$	$\cdots$	$a_i a_j$	$\cdots$	$a_i a_n$
a_n	$a_n a_1$	$a_n a_2$	$\cdots$	$a_n a_j$	$\cdots$	$a_n a_n$

Let us agree, when writing a multiplication table, that the identity element is listed first; that is, $a_1 = 1$. In this case, the first row and first column of the table merely repeat the listing above, and so we usually omit them.

Consider two almost trivial examples of groups: let Γ_2 denote the multiplicative group $\{1, -1\}$, and let $\mathcal{P}$ denote the parity group (Example 2.13). Here are their multiplication tables:

$$\Gamma_2: \begin{array}{|c c|} \hline 1 & -1 \\ -1 & 1 \\ \hline \end{array}; \qquad \mathcal{P}: \begin{array}{|c c|} \hline \text{even} & \text{odd} \\ \text{odd} & \text{even} \\ \hline \end{array}.$$

It is clear that Γ_2 and $\mathcal{P}$ are distinct groups; it is equally clear that there is no significant difference between them. The notion of isomorphism formalizes this idea; Γ_2 and $\mathcal{P}$ are isomorphic, for the function $f : \Gamma_2 \to \mathcal{P}$, defined by $f(1) = $ even and $f(-1) = $ odd, is an isomorphism, as the reader can quickly check.

There are many multiplication tables for a group G of order n, one for each of the $n!$ lists of its elements. If $a_1, a_2, \ldots, a_n$ is a list of all the elements of G with no repetitions, and if $f : G \to H$ is a bijection, then $f(a_1), f(a_2), \ldots, f(a_n)$ is a list of all the elements of H with no repetitions, and so this latter list determines a multiplication table for H. That f is an isomorphism says that if we superimpose the multiplication table for G (determined by $a_1, a_2, \ldots, a_n$) upon the multiplication table for H (determined by $f(a_1), f(a_2), \ldots, f(a_n)$), then the tables match: if $a_i a_j$ is the ij entry in the given multiplication table of G, then $f(a_i) f(a_j) = f(a_i a_j)$ is the ij entry of the multiplication table of H. In this sense, isomorphic groups have the same multiplication table. Thus, isomorphic groups are essentially the same, differing only in the notation for the elements and the operations.

Example 2.23.
Here is a practical way to check whether a given bijection $f : G \to H$ between a pair of groups is actually an isomorphism: enumerate the elements $a_1, \ldots, a_n$ of G, form the multiplication table of G arising from this list, form the multiplication table for H from the list $f(a_1), \ldots, f(a_n)$, and compare the n^2 entries of the two tables.

We illustrate this for $G = S_3$, the symmetric group permuting $\{1, 2, 3\}$, and $H = S_Y$, the symmetric group of all the permutations of $Y = \{a, b, c\}$. First, enumerate G:

$$(1), \quad (1\ 2), \quad (1\ 3), \quad (2\ 3), \quad (1\ 2\ 3), \quad (1\ 3\ 2).$$

We define the obvious function $\varphi : S_3 \to S_Y$ that replaces numbers by letters:

$$(1), \quad (a\ b), \quad (a\ c), \quad (b\ c), \quad (a\ b\ c), \quad (a\ c\ b).$$

Compare the multiplication table for S_3 arising from this list of its elements with the multiplication table for S_Y arising the the corresponding list of its elements. The reader should write out the complete tables of each and superimpose one on the other to see that they match. We will only check one entry. The 3,4 position in the table for S_3 is the product $(2\ 3)(1\ 2\ 3) = (1\ 3)$, while the 3,4 position in the table for S_Y is the product $(b\ c)(a\ b\ c) = (a\ c)$.

This result is generalized in Exercise 2.55.   ◄

Example 2.24.
The formula $f(x) = x/(x - 1)$ does not give a function $\mathbb{R} \to \mathbb{R}$, because it is not defined at $x = 1$: one cannot divide by 0. However, if we define the **extended reals**, $\hat{\mathbb{R}}$, by adjoining a new element ∞ to $\mathbb{R}$, then the formula defines a function $f : \hat{\mathbb{R}} \to \hat{\mathbb{R}}$: set

$$f(1) = \infty \quad \text{and} \quad f(\infty) = 1.$$

More generally, if $g(x) = (ax+b)/(cx+d)$, define $g(y) = \infty$ if $cy+d = 0$. To evaluate $g(\infty)$, first define $h(t) = g(1/t)$, and then define $g(\infty) = h(0)$. For example, if $g(x) = 1/(1-x)$, then $g(1) = \infty$; moreover,

$$h(t) = \frac{1}{1-(1/t)} = \frac{t}{t-1},$$

and $g(\infty) = h(0) = 0$.

We claim that the set G of the following six functions $\hat{\mathbb{R}} \to \hat{\mathbb{R}}$ form a group under composition:

$$G = \left\{ x,\ 1-x,\ \frac{1}{x},\ \frac{x}{x-1},\ \frac{x-1}{x},\ \frac{1}{1-x} \right\}.$$

A multiplication table organizes the task of checking whether G is a group. One can quickly show that each $g \in G$ is a permutation of $\hat{\mathbb{R}}$, for its inverse function also lies in G. As G is finite, it suffices to show that it is closed under composition, and this is seen by displaying a multiplication table for G.

x	$1-x$	$\dfrac{1}{x}$	$\dfrac{x}{x-1}$	$\dfrac{x-1}{x}$	$\dfrac{1}{1-x}$
$1-x$	x	$\dfrac{x-1}{x}$	$\dfrac{1}{1-x}$	$\dfrac{1}{x}$	$\dfrac{x}{x-1}$
$\dfrac{1}{x}$	$\dfrac{1}{1-x}$	x	$\dfrac{x-1}{x}$	$\dfrac{x}{x-1}$	$1-x$
$\dfrac{x}{x-1}$	$\dfrac{x-1}{x}$	$\dfrac{1}{1-x}$	x	$1-x$	$\dfrac{1}{x}$
$\dfrac{x-1}{x}$	$\dfrac{x}{x-1}$	$1-x$	$\dfrac{1}{x}$	$\dfrac{1}{1-x}$	x
$\dfrac{1}{1-x}$	$\dfrac{1}{x}$	$\dfrac{x}{x-1}$	$1-x$	x	$\dfrac{x-1}{x}$

In Exercise 2.60, you are asked to prove that G is isomorphic to S_3 by comparing suitable multiplication tables. ◀

Here are some basic properties of homomorphisms.

Lemma 2.36.   *Let $f : G \to H$ be a homomorphism of groups.*

(i) $f(1) = 1$;

(ii) $f(x^{-1}) = f(x)^{-1}$;

(iii) $f(x^n) = f(x)^n$ *for all $n \in \mathbb{Z}$.*

Proof. (i) Applying f to the equation $1 \cdot 1 = 1$ in G gives the equation $f(1)f(1) = f(1)$ in H, and Exercise 2.35 gives $f(1) = 1$.

(ii) Apply f to the equations $x^{-1}x = 1 = xx^{-1}$ in G to obtain the equations $f(x^{-1})f(x) = 1 = f(x)f(x^{-1})$ in H. Uniqueness of the inverse, Lemma 2.19(iii), gives $f(x^{-1}) = f(x)^{-1}$.

(iii) It is easy to prove by induction that $f(x^n) = f(x)^n$ for all $n \geq 0$. For negative exponents,

$$f(x^{-n}) = f((x^{-1})^n) = f((x^{-1}))^n = (f(x)^{-1})^n = f(x)^{-n}. \quad \bullet$$

Example 2.25.

If G and H are cyclic groups of the same order m, then G and H are isomorphic. It follows from Corollary 2.35 that any two groups of prime order p are isomorphic.

Suppose that $G = \langle x \rangle$ and $H = \langle y \rangle$. Define $f : G \to H$ by $f(x^i) = y^i$ for $0 \leq i < m$. Now $G = \{1, x, x^2, \ldots, x^{m-1}\}$ and $H = \{1, y, y^2, \ldots, y^{m-1}\}$, and so it follows that f is a bijection. To see that f is a homomorphism (and hence an isomorphism), we must show that $f(x^i x^j) = f(x^i)f(x^j)$ for all i and j with $0 \leq i, j < m$. The desired equation clearly holds if $i + j < m$, for $f(x^{i+j}) = y^{i+j}$, and so

$$f(x^i x^j) = f(x^{i+j}) = y^{i+j} = y^i y^j = f(x^i)f(x^j).$$

If $i + j \geq m$, then $i + j = m + r$, where $0 \leq r < m$, so that

$$x^{i+j} = x^{m+r} = x^m x^r = x^r$$

(because $x^m = 1$); similarly, $y^{i+j} = y^r$ (because $y^m = 1$). Hence

$$f(x^i x^j) = f(x^{i+j}) = f(x^r)$$
$$= y^r = y^{i+j} = y^i y^j = f(x^i)f(x^j).$$

Therefore, f is an isomorphism and $G \cong H$. (See Example 2.33 for a nicer proof of this.) ◄

A property of a group G that is shared by any other group isomorphic to it is called an **invariant** of G. For example, the order $|G|$ is an invariant of G, for isomorphic groups have the same orders. Being abelian is an invariant [if a and b commute, then $ab = ba$ and

$$f(a)f(b) = f(ab) = f(ba) = f(b)f(a);$$

hence, $f(a)$ and $f(b)$ commute]. Thus, $\mathbb{R}$ and $GL(2, \mathbb{R})$ are not isomorphic, for $\mathbb{R}$ is abelian and $GL(2, \mathbb{R})$ is not. There are other invariants of a group (see Exercise 2.58); for example, the number of elements in it of any given order r, or whether or not the group is cyclic. In general, it is a challenge to decide whether two given groups are isomorphic.

Example 2.26.
We present two nonisomorphic groups of the same order.

As in Example 2.19(i), let **V** be the four-group consisting of the following four permutations:

$$\mathbf{V} = \{(1),\ (1\ 2)(3\ 4),\ (1\ 3)(2\ 4),\ (1\ 4)(2\ 3)\},$$

and let $\Gamma_4 = \langle i \rangle = \{1, i, -1, -i\}$ be the multiplicative cyclic group of fourth roots of unity, where $i^2 = -1$. If there were an isomorphism $f : \mathbf{V} \to \Gamma_4$, then surjectivity of f would provide some $x \in \mathbf{V}$ with $i = f(x)$. But $x^2 = (1)$ for all $x \in \mathbf{V}$, so that $i^2 = f(x)^2 = f(x^2) = f((1)) = 1$, contradicting $i^2 = -1$. Therefore, **V** and Γ_4 are not isomorphic.

There are other ways to prove this result. For example, Γ_4 is cyclic and **V** is not, or Γ_4 has an element of order 4 and **V** does not, or Γ_4 has a unique element of order 2, but **V** has 3 elements of order 2. At this stage, you should really believe that Γ_4 and **V** are not isomorphic! ◄

I once asked a student on an oral exam what it means when one says that groups G and H are isomorphic. He replied, "It means that there is a bijection $f : G \to H$ with $f(xy) = f(x)f(y)$ for all $x, y \in G$." "That's fine," said I. "Can you tell me whether the four-group **V** and the group Γ_4 of fourth roots of unity are isomorphic?" He wrote down the four permutations of **V** on the blackboard, then he wrote down $1, i, -1, -i$, and then he stared, first at **V**, then at Γ_4. Finally, pointing at **V**, he said, "This one is," and then, pointing at Γ_4, "but this one isn't." He did not pass. After all, suppose one replied to the question whether $\pi = \frac{22}{7}$ by saying that π is but that $\frac{22}{7}$ isn't.

Definition.   If $f : G \to H$ is a homomorphism, define

$$\textbf{\textit{kernel}}^{19} f = \{x \in G : f(x) = 1\}$$

and

$$\textbf{\textit{image}} \ f = \{h \in H : h = f(x) \text{ for some } x \in G\}.$$

We usually abbreviate kernel f to ker f and image f to im f.

Example 2.27.
(i) If $\Gamma_n = \langle \zeta \rangle$, where $\zeta = e^{2\pi i/n}$ is a primitive nth root of unity, then $f : \mathbb{Z} \to \Gamma_n$, given by $f(m) = \zeta^m$, is a surjective homomorphism with ker f all the multiples of n.

(ii) If Γ_2 is the multiplicative group $\Gamma_2 = \{\pm 1\}$, then Proposition 2.16 says that sgn : $S_n \to \Gamma_2$ is a homomorphism. The kernel of sgn is the alternating group A_n, the set of all even permutations.

(iii) Determinant is a surjective homomorphism det : $GL(2, \mathbb{R}) \to \mathbb{R}^\times$, the multiplicative group of nonzero reals, whose kernel is the special linear group $SL(2, \mathbb{R})$.   ◀

Proposition 2.37.   *Let $f : G \to H$ be a homomorphism.*

(i) ker f *is a subgroup of G and* im f *is a subgroup of H.*

(ii) *If $x \in$ ker f and if $a \in G$, then $axa^{-1} \in$ ker f.*

(iii) *f is an injection if and only if ker $f = \{1\}$.*

Proof.   (i) Lemma 2.36 shows that $1 \in$ ker f, for $f(1) = 1$. Next, if x, $y \in$ ker f, then $f(x) = 1 = f(y)$; hence, $f(xy) = f(x)f(y) = 1 \cdot 1 = 1$, and so $xy \in$ ker f. Finally, if $x \in$ ker f, then $f(x) = 1$ and so $f(x^{-1}) = f(x)^{-1} = 1^{-1} = 1$; thus, $x^{-1} \in$ ker f, and ker f is a subgroup of G.

We now show that im f is a subgroup of H. First, $1 = f(1) \in$ im f. Next, if $h = f(x) \in$ im f, then $h^{-1} = f(x)^{-1} = f(x^{-1}) \in$ im f. Finally, if $k = f(y) \in$ im f, then $hk = f(x)f(y) = f(xy) \in$ im f. Hence, im f is a subgroup of H.

(ii) If $x \in$ ker f, then $f(x) = 1$ and

$$f(axa^{-1}) = f(a)f(x)f(a)^{-1} = f(a)1f(a)^{-1} = f(a)f(a)^{-1} = 1;$$

---

[19] *Kernel* comes from the German word meaning "grain" or "seed" (*corn* comes from the same word). Its usage here indicates an important ingredient of a homomorphism.

therefore, $axa^{-1} \in \ker f$.

(iii) If f is an injection, then $x \neq 1$ implies $f(x) \neq f(1) = 1$, and so $x \notin \ker f$. Conversely, assume that $\ker f = \{1\}$ and that $f(x) = f(y)$. Then $1 = f(x)f(y)^{-1} = f(xy^{-1})$, so that $xy^{-1} \in \ker f = 1$; therefore, $xy^{-1} = 1$, $x = y$, and f is an injection.   •

Definition.   A subgroup K of a group G is called a ***normal subgroup*** if $k \in K$ and $g \in G$ imply $gkg^{-1} \in K$. If K is a normal subgroup of G, one writes $K \lhd G$.

The proposition thus says that the kernel of a homomorphism is always a normal subgroup. If G is an abelian group, then every subgroup K is normal, for if $k \in K$ and $g \in G$, then $gkg^{-1} = kgg^{-1} = k \in K$.

The cyclic subgroup $H = \langle(1\ 2)\rangle$ of S_3, consisting of the two elements (1) and $(1\ 2)$, is not a normal subgroup of S_3: if $\alpha = (1\ 2\ 3)$, then $\alpha^{-1} = (3\ 2\ 1)$, and

$$\alpha(1\ 2)\alpha^{-1} = (1\ 2\ 3)(1\ 2)(3\ 2\ 1) = (2\ 3) \notin H.$$

On the other hand, the cyclic subgroup $K = \langle(1\ 2\ 3)\rangle$ of S_3 is a normal subgroup, as the reader should verify.

It follows from Examples 2.27(ii) and 2.27(iii) that A_n is a normal subgroup of S_n and $\mathrm{SL}(2, \mathbb{R})$ is a normal subgroup of $\mathrm{GL}(2, \mathbb{R})$ (however, it is also easy to prove these facts directly).

Definition.   If G is a group and $a \in G$, then a ***conjugate*** of a is any element in G of the form

$$gag^{-1},$$

where $g \in G$.

It is clear that a subgroup $K \leq G$ is a normal subgroup if and only if K contains all the conjugates of its elements: if $k \in K$, then $gkg^{-1} \in K$ for all $g \in G$. In Proposition 2.12, we showed that any conjugate of a permutation $\alpha \in S_n$ has the same cycle structure as α (the converse is also true: see Exercise 2.67).

Remark.   In linear algebra, a linear transformation $T : V \rightarrow V$, where V is an n-dimensional vector space over $\mathbb{R}$, determines an $n \times n$ matrix A if one uses a basis of V; if one uses another basis, then one gets another matrix B from T. It turns out that A and B are *similar*; that is, there is a nonsingular matrix P with $B = PAP^{-1}$. Thus, conjugacy in $\mathrm{GL}(n, \mathbb{R})$ is similarity.   ◄

Definition.   If G is a group and $g \in G$, define *conjugation* $\gamma_g : G \to G$ by

$$\gamma_g(a) = gag^{-1}$$

for all $a \in G$.

Proposition 2.38.

(i) *If G is a group and $g \in G$, then conjugation $\gamma_g : G \to G$ is an isomorphism.*

(ii) *Conjugate elements have the same order.*

Proof.   (i) If $g, h \in G$, then $(\gamma_g \circ \gamma_h)(a) = \gamma_g(hah^{-1}) = g(hah^{-1})g^{-1} = (gh)a(gh)^{-1} = \gamma_{gh}(a)$; that is,

$$\gamma_g \circ \gamma_h = \gamma_{gh}.$$

It follows that each γ_g is a bijection, for $\gamma_g \circ \gamma_{g^{-1}} = \gamma_1 = 1 = \gamma_{g^{-1}} \circ \gamma_g$. We now show that γ_g is an isomorphism: if $a, b \in G$,

$$\gamma_g(ab) = g(ab)g^{-1} = (gag^{-1})(gbg^{-1}) = \gamma_g(a)\gamma_g(b).$$

(ii) To say that a and b are conjugate is to say that there is $g \in G$ with $b = gag^{-1}$; that is, $b = \gamma_g(a)$. But γ_g is an isomorphism, and so Exercise 2.58(ii) shows that a and $b = \gamma_g(a)$ have the same order.   •

Example 2.28.
Define the *center* of a group G, denoted by $Z(G)$, to be

$$Z(G) = \{z \in G : zg = gz \text{ for all } g \in G\};$$

that is, $Z(G)$ consists of all elements commuting with everything in G. (Note that the equation $zg = gz$ can be rewritten as $z = gzg^{-1}$, so that no other elements in G are conjugate to z.)

It is easy to see that $Z(G)$ is a subgroup of G; it is a normal subgroup because if $z \in Z(G)$ and $g \in G$, then

$$gzg^{-1} = zgg^{-1} = z \in Z(G).$$

A group G is abelian if and only if $Z(G) = G$. At the other extreme are *centerless* groups G for which $Z(G) = \{1\}$; for example, it is easy to see that $Z(S_3) = \{1\}$; indeed, all large symmetric groups are centerless, for Exercise 2.30 shows that $Z(S_n) = \{1\}$ for all $n \geq 3$.   ◄

Example 2.29.

The four-group **V** is a normal subgroup of S_4. Recall that the elements of **V** are

$$\mathbf{V} = \{(1), (1\ 2)(3\ 4), (1\ 3)(2\ 4), (1\ 4)(2\ 3)\}.$$

By Proposition 2.12, every conjugate of a product of two transpositions is another such. But we saw, in Example 2.8, that only 3 permutations in S_4 have this cycle structure, and so **V** is a normal subgroup of S_4.   ◄

Proposition 2.39.

  (i) *If H is a subgroup of index 2 in a group G, then $g^2 \in H$ for every $g \in G$.*

  (ii) *If H is a subgroup of index 2 in a group G, then H is a normal subgroup of G.*

Proof.   (i) Since H has index 2, there are exactly two cosets, namely, H and aH, where $a \notin H$. Thus, G is the disjoint union $G = H \cup aH$. Take $g \in G$ with $g \notin H$, so that $g = ah$ for some $h \in H$. If $g^2 \notin H$, then $g^2 = ah'$, where $h' \in H$. Hence,

$$g = g^{-1}g^2 = h^{-1}a^{-1}ah' = h^{-1}h' \in H,$$

and this is a contradiction.

(ii) It suffices to prove that if $h \in H$, then the conjugate $ghg^{-1} \in H$ for every $g \in G$. Since H has index 2, there are exactly two cosets, namely, H and aH, where $a \notin H$. Now, either $g \in H$ or $g \in aH$. If $g \in H$, then $ghg^{-1} \in H$, because H is a subgroup. In the second case, write $g = ax$, where $x \in H$. Then $ghg^{-1} = a(xhx^{-1})a^{-1} = ah'a^{-1}$, where $h' = xhx^{-1} \in H$ (for h' is a product of three elements in H). If $ghg^{-1} \notin H$, then $ghg^{-1} = ah'a^{-1} \in aH$; that is, $ah'a^{-1} = ay$ for some $y \in H$. Canceling a, we have $h'a^{-1} = y$, which gives the contradiction $a = y^{-1}h' \in H$. Therefore, if $h \in H$, every conjugate of h also lies in H; that is, H is a normal subgroup of G.   •

Definition.   The *quaternions*[20] is the group **Q** of order 8 consisting of the matrices in $\mathrm{GL}(2, \mathbb{C})$

$$\mathbf{Q} = \{ E, A, A^2, A^3, B, BA, BA^2, BA^3 \},$$

---

[20]The operations of addition, subtraction, multiplication, and division (by nonzero numbers) can be extended from $\mathbb{R}$ to the plane in such a way that all the usual laws of arithmetic hold; of course, the plane is usually called the complex numbers $\mathbb{C}$ in this context. W.

where E is the identity matrix,

$$A = \begin{bmatrix} 0 & 1 \\ -1 & 0 \end{bmatrix} \text{ and } B = \begin{bmatrix} 0 & i \\ i & 0 \end{bmatrix}.$$

The reader should note that the element $A \in \mathbf{Q}$ has order 4, so that $\langle A \rangle$ is a subgroup of order 4 and hence of index 2; the other coset is $B \langle A \rangle = \{B, BA, BA^2, BA^3\}$.

Example 2.30.
In Exercise 2.74, the reader will check that $\mathbf{Q}$ is a nonabelian group of order 8 having exactly one subgroup of order 2, namely, $\langle -E \rangle$. We claim that every subgroup of $\mathbf{Q}$ is normal. Lagrange's theorem says that every subgroup of Q has order a divisor of 8, and so the only possible orders of subgroups are 1, 2, 4, or 8. Clearly, the subgroup $\{1\}$ and the subgroup of order 8 (namely, $\mathbf{Q}$ itself) are normal subgroups. By Proposition 2.39(ii), any subgroup of order 4 must be normal, for it has index 2. Finally, the subgroup $\langle -E \rangle$ is normal, for if M is any matrix, then $M(\pm E) = (\pm E)M$, so that $M(\pm E)M^{-1} = (\pm E)MM^{-1} = \pm E \in \langle -E \rangle$. ◄

Example 2.30 shows that $\mathbf{Q}$ is a nonabelian group which is like abelian groups in that every subgroup is normal. This is essentially the only such example: every finite group with every subgroup normal has the form $\mathbf{Q} \times A$, where A is abelian (*direct products* will be introduced in the next section).

Lagrange's theorem states that the order of a subgroup of a finite group G must be a divisor of $|G|$. This suggests the question, given some divisor d of $|G|$, whether G must contain a subgroup of order d. The next result shows that there need not be such a subgroup.

Proposition 2.40.   *The alternating group A_4 is a group of order 12 having no subgroup of order 6.*

Proof.   First of all, $|A_4| = 12$, by Exercise 2.27. If A_4 contains a subgroup H of order 6, then H has index 2, and so $\alpha^2 \in H$ for every $\alpha \in A_4$, by

---

R. Hamilton invented a way of extending all these operations from $\mathbb{C}$ to four-dimensional space in such a way that all the usual laws of arithmetic still hold (except for commutativity of multiplication); he called the new "numbers" *quaternions* (from the Latin word meaning "four"). The multiplication is determined by knowing those products involving four particular 4-tuples 1, *i*, *j*, *k*. All the nonzero quaternions form a multiplicative group, and the group of quaternions is the smallest subgroup (which has order 8) containing these four elements. (See Exercise 2.75.)

Corollary 2.39(i). If α is a 3-cycle, however, then α has order 3, so that $\alpha = \alpha^4 = (\alpha^2)^2$. Thus, H contains every 3-cycle. This is a contradiction, for there are 8 3-cycles in A_4. $\bullet$

EXERCISES

2.55 If there is a bijection $f : X \to Y$ (that is, if X and Y have the same number of elements), then there is an isomorphism $\varphi : S_X \to S_Y$.

2.56 (i) Show that the composite of homomorphisms is itself a homomorphism.

(ii) Show that the inverse of an isomorphism is an isomorphism.

(iii) Show that two groups that are isomorphic to a third group are isomorphic to each other.

2.57 Prove that a group G is abelian if and only if the function $f : G \to G$, given by $f(a) = a^{-1}$, is a homomorphism.

2.58 This exercise gives some invariants of a group G. Let $f : G \to H$ be an isomorphism.

(i) Prove that if $a \in G$ has infinite order, then so does $f(a)$, and if a has finite order n, then so does $f(a)$. Conclude that if G has an element of some order n and H does not, then $G \ncong H$.

(ii) Prove that if $G \cong H$, then, for every divisor k of $|G|$, both G and H have the same number of elements of order k.

2.59 Find two nonisomorphic groups of order 6.

2.60 Prove that the group G in Example 2.24 is isomorphic to S_3.

2.61 (i) Find a subgroup $H \le S_4$ with $H \cong \mathbf{V}$ but with $H \ne \mathbf{V}$.

(ii) Prove that the subgroup H in part (i) is not a normal subgroup.

2.62 Show that every group G with $|G| < 6$ is abelian.

2.63 (i) If $f : G \to H$ is a homomorphism and $x \in G$ has order k, prove that $f(x) \in H$ has order m, where $m \mid k$.

(ii) If $f : G \to H$ is a homomorphism and if $(|G|, |H|) = 1$, prove that $f(x) = 1$ for all $x \in G$.

2.64 Prove that the special orthogonal group $SO(2, \mathbb{R})$ is isomorphic to the circle group S^1.

2.65 Let G be the additive group of all polynomials in x with coefficients in $\mathbb{Z}$, and let H be the multiplicative group of all positive rationals. Prove that $G \cong H$.

2.66 Show that if H is a subgroup with $bH = Hb = \{hb : h \in H\}$ for every $b \in G$, then H must be a normal subgroup.

2.67 Prove the converse of Proposition 2.12: If α and β are permutations in S_n having the same cycle structure, then α and β are conjugate.

2.68 Prove that the intersection of any family of normal subgroups of a group G is itself a normal subgroup of G.

2.69 Define $W = \langle (1\ 2)(3\ 4) \rangle$, the cyclic subgroup of S_4 generated by $(1\ 2)(3\ 4)$. Show that W is a normal subgroup of $\mathbf{V}$, but that W is not a normal subgroup of S_4. Conclude that normality is not transitive: $W \lhd \mathbf{V}$ and $\mathbf{V} \lhd G$ need not imply $W \lhd G$.

2.70 Let G be a finite abelian group written multiplicatively. Prove that if $|G|$ is odd, then every $x \in G$ has a unique square root; that is, there exists exactly one $g \in G$ with $g^2 = x$.

2.71 Give an example of a group G, a subgroup $H \le G$, and an element $g \in G$ with $[G : H] = 3$ and $g^3 \notin H$.

2.72 Show that the center of $GL(2, \mathbb{R})$ is the set of all *scalar matrices*

$$\begin{bmatrix} a & 0 \\ 0 & a \end{bmatrix}$$

with $a \neq 0$.

2.73 Let $\zeta = e^{2\pi i/n}$ be a primitive nth root of unity, and define

$$A = \begin{bmatrix} \zeta & 0 \\ 0 & \zeta^{-1} \end{bmatrix} \text{ and } B = \begin{bmatrix} 0 & 1 \\ 1 & 0 \end{bmatrix}.$$

   (i) Prove that A has order n and that B has order 2.
   (ii) Prove that $BAB = A^{-1}$.
   (iii) Prove that the matrices of the form A^i and BA^i, for $0 \le i < n$, form a multiplicative subgroup $G \le GL(2, \mathbb{C})$.
   (iv) Prove that each matrix in G has a unique expression of the form $B^i A^j$, where $i = 0, 1$ and $0 \le j < n$. Conclude that $|G| = 2n$.
   (v) Prove that $G \cong D_{2n}$.

2.74 Recall that the quaternions $\mathbf{Q}$ (defined in Example 2.30) consists of the 8 matrices in $GL(2, \mathbb{C})$

$$\mathbf{Q} = \{\ E, A, A^2, A^3, B, BA, BA^2, BA^3\ \},$$

where

$$A = \begin{bmatrix} 0 & 1 \\ -1 & 0 \end{bmatrix} \text{ and } B = \begin{bmatrix} 0 & i \\ i & 0 \end{bmatrix}.$$

   (i) Prove that $-E$ is the only element in $\mathbf{Q}$ of order 2, and that all other elements $M \neq E$ satisfy $M^2 = -E$.

(ii) Prove that **Q** is a nonabelian group with operation matrix multiplication.

(iii) Show that **Q** has a unique subgroup of order 2, and it is the center of **Q**.

2.75 Show that G, consisting of the 8 elements

$$\pm 1, \pm \mathbf{i}, \pm \mathbf{j}, \pm \mathbf{k},$$

where

$$\mathbf{i}^2 = \mathbf{j}^2 = \mathbf{k}^2 = -1, \qquad \mathbf{ij} = \mathbf{k}, \qquad \mathbf{jk} = \mathbf{i}, \qquad \mathbf{ki} = \mathbf{j},$$
$$\mathbf{ij} = -\mathbf{ji}, \qquad \mathbf{ij} = -\mathbf{ki}, \qquad \mathbf{jk} = -\mathbf{kj},$$

is a multiplicative group isomorphic to the quaternions **Q**.

2.76 Prove that the quaternions **Q** and the dihedral group D_8 are nonisomorphic groups of order 8.

2.77 Prove that A_4 is the only subgroup of S_4 of order 12. (In Exercise 2.106, this will be generalized from A_4 to A_n for all $n \geq 3$.)

2.78   (i) Let $\mathcal{A}$ be the set of all 2×2 matrices of the form

$$A = \begin{bmatrix} a & b \\ 0 & 1 \end{bmatrix},$$

where $a \neq 0$. Prove that $\mathcal{A}$ is a subgroup of $GL(2, \mathbb{R})$.

(ii) Prove that $\psi : \text{Aff}(1, \mathbb{R}) \to \mathcal{A}$, defined by $f \mapsto A$, is an isomorphism, where $f(x) = ax + b$.

2.79 Use Exercise 2.78(i) to prove that the stochastic group $\Sigma(2, \mathbb{R})$ [see Exercise 2.44] is isomorphic to the affine group $\text{Aff}(1, \mathbb{R})$ by showing that $\varphi : \Sigma(2, \mathbb{R}) \to \mathcal{A} \cong \text{Aff}(1, \mathbb{R})$, given by $\varphi(M) = QMQ^{-1}$, is an isomorphism, where

$$Q = \begin{bmatrix} 1 & 0 \\ 1 & 1 \end{bmatrix} \quad \text{and} \quad Q^{-1} = \begin{bmatrix} 1 & 0 \\ -1 & 1 \end{bmatrix}.$$

2.80 Prove that the symmetry group $\Sigma(\pi_n)$, where π_n is a regular polygon with n vertices, is isomorphic to a subgroup of S_n.

2.6 QUOTIENT GROUPS

We are now going to construct a group using congruence mod m. Once this is done, we will be able to give a proof of Fermat's theorem using group theory. This construction is the prototype of a more general way of building new groups from given groups, called *quotient groups*.

Definition.   Given $m \geq 0$ and $a \in \mathbb{Z}$, the ***congruence class*** of $a \bmod m$ is the subset $[a]$ of $\mathbb{Z}$:

$$[a] = \{b \in \mathbb{Z} : b \equiv a \bmod m\}$$
$$= \{a + km : k \in \mathbb{Z}\}$$
$$= \{\ldots, a - 2m, a - m, a, a + m, a + 2m, \ldots\};$$

the ***integers mod m***, denoted by $\mathbb{Z}_m$, is the family of all such congruence classes.

For example, if $m = 2$, then $[0] = \{b \in \mathbb{Z} : b \equiv 0 \bmod 2\}$ is the set of all even integers and $[1] = \{b \in \mathbb{Z} : b \equiv 1 \bmod 2\}$ is the set of all odd integers. Notice that $[2] = \{2 + 2k : k \in \mathbb{Z}\}$ is also the set of all even integers, so that $[2] = [0]$; indeed, $[0] = [2] = [-2] = [4] = [-4] = [6] = [-6] = \cdots$.

Remark.   Given m, we may form the cyclic subgroup $\langle m \rangle$ of $\mathbb{Z}$ generated by m. The reader should check that the congruence class $[a]$ is precisely the coset $a + \langle m \rangle$.   ◄

The notation $[a]$ is incomplete in that it does not mention the modulus m: for example, $[1]$ in $\mathbb{Z}_2$ is not the same as $[1]$ in $\mathbb{Z}_3$ (the former is the set of all odd numbers and the latter is $\{1 + 3k : k \in \mathbb{Z}\} = \{\ldots, -5, -2, 1, 4, 7, \ldots\}$). This will not cause problems, for, almost always, one works with only one $\mathbb{Z}_m$ at a time.

Proposition 2.41. $[a] = [b]$ in $\mathbb{Z}_m$ if and only if $a \equiv b \bmod m$.

Proof.   If $[a] = [b]$, then $a \in [b]$, so $a = b + km$ for some k; hence $a - b = km$, $m \mid a - b$, and $a \equiv b \bmod m$.

Conversely, if $a \equiv b \bmod m$, then $a - b = \ell m$ for some integer ℓ, and $a = b + \ell m$. If $s \in [a]$, then $s = a + km = b + (\ell + k)m \in [b]$. Hence, $[a] \subset [b]$. For the reverse inclusion, if $t \in [b]$, then $t = b + nm = a + (\ell - n)m \in [b]$. Therefore, $[b] \subset [a]$ and $[a] = [b]$.   •

In words, Proposition 2.41 says that $\mathbb{Z}_m$ converts congruence mod m between numbers into equality at the cost of replacing numbers by congruence classes.

In particular, $[a] = [0]$ in $\mathbb{Z}_m$ if and only if $a \equiv 0 \bmod m$; that is, $[a] = [0]$ in $\mathbb{Z}_m$ if and only if m is a divisor of a.

Proposition 2.42.   *Let $m \geq 2$ be given.*

(i) *If $a \in \mathbb{Z}$, then $[a] = [r]$ for some r with $0 \leq r < m$.*

(ii) *If $0 \leq r' < r < m$, then $[r'] \neq [r]$.*

(iii) $\mathbb{Z}_m$ *has exactly m elements, namely, $[0], [1], \ldots, [m-1]$.*

Proof.   (i) For each $a \in \mathbb{Z}$, the division algorithm gives $a = qm + r$, where $0 \leq r < m$; hence $a - r = qm$ and $a \equiv r \bmod m$. Therefore $[a] = [r]$, where r is the remainder after dividing a by m.

(ii) Proposition 1.46(ii) gives $r' \not\equiv r \bmod m$.

(iii) Part (i) shows that every $[a]$ in $\mathbb{Z}_m$ occurs on the list $[0], [1], [2], \ldots,$ $[m-1]$; part (ii) shows that this list of m items has no repetitions.   •

We are now going to make $\mathbb{Z}_m$ into an abelian group by equipping it with an addition.

Proposition 2.43.   *For every $m \geq 2$, $\mathbb{Z}_m$ is an (additive) cyclic group of order m and with generator $[1]$.*

Proof.   Define $\alpha : \mathbb{Z}_m \times \mathbb{Z}_m \to \mathbb{Z}_m$ by $\alpha([a], [b]) = [a+b]$. To see that α is a (single-valued) function, we must show that the law of substitution holds: if $[a] = [a']$ and $[b] = [b']$, then $\alpha([a], [b]) = \alpha([a'], [b'])$, that is, $[a+b] = [a'+b']$. But this is precisely Proposition 1.48(i).

In this proof only, we shall write $\boxplus$ for addition of congruence classes:

$$\alpha([a], [b]) = [a] \boxplus [b] = [a+b].$$

The operation $\boxplus$ is associative: since $+$ in $\mathbb{Z}$ is associative,

$$\begin{aligned}
[a] \boxplus \big([b] \boxplus [c]\big) &= [a] \boxplus [b+c] \\
&= [a + (b+c)] \\
&= [(a+b) + c] \\
&= [a+b] \boxplus [c] \\
&= \big([a] \boxplus [b]\big) \boxplus [c].
\end{aligned}$$

The operation $\boxplus$ is commutative: since ordinary addition is commutative,

$$[a] \boxplus [b] = [a+b] = [b+a] = [b] \boxplus [a].$$

The identity element is [0]: since 0 is the additive identity in $\mathbb{Z}$,

$$[0] \boxplus [a] = [0 + a] = [a].$$

The inverse of $[a]$ is $[-a]$; since $-a$ is the additive inverse of a in $\mathbb{Z}$,

$$[-a] \boxplus [a] = [-a + a] = [0].$$

Therefore, $\mathbb{Z}_m$ is an abelian group of order m; it is cyclic with generator [1], for if $0 \leq r < m$, then $[r]$ is the sum of r copies of [1].   •

We now drop the notation $\boxplus$; henceforth, we shall write

$$[a] + [b] = [a + b]$$

for the sum of congruence classes in $\mathbb{Z}_m$. The reader should notice that the group axioms in $\mathbb{Z}_m$ are "inherited" from the group axioms in $\mathbb{Z}$.

Corollary 2.44.   *Every cyclic group of order $m \geq 2$ is isomorphic to $\mathbb{Z}_m$.*

Proof.   We have already seen, in Example 2.25, that any two finite cyclic groups of the same order are isomorphic.   •

Remark.   The function $\pi : \mathbb{Z} \to \mathbb{Z}_m$, defined by $\pi(a) = [a] = a + \langle m \rangle$, is a homomorphism, for

$$\pi(a + b) = [a + b] = [a] + [b] = \pi(a) + \pi(b). \quad \blacktriangleleft$$

Proposition 2.45.

(i) *The function $\mu : \mathbb{Z}_m \times \mathbb{Z}_m \to \mathbb{Z}_m$, given by*

$$\mu([a], [b]) = [ab],$$

is an operation on $\mathbb{Z}_m$.

(ii) *This operation is associative and commutative, and [1] is an identity element.*

(iii) *Denote $\mu(a, b) = [a][b]$ by $[a][b]$. If $(a, m) = 1$, then $[a][x] = [b]$ can be solved for $[x]$ in $\mathbb{Z}_m$.*

(iv) *If p is a prime, then $\mathbb{Z}_p^\times$, the nonzero elements in $\mathbb{Z}_p$, is a multiplicative abelian group of order $p - 1$.*

Proof.   (i) To see that μ is a (single-valued) function, we must show that the law of substitution holds: if $[a] = [a']$ and $[b] = [b']$, then $\mu([a], [b]) = \mu([a'], [b'])$, that is, $[ab] = [a'b']$. But this is precisely Proposition 1.48(ii).

(ii) *In this proof only*, we are going to write $\boxtimes$ for multiplication of congruence classes:

$$\mu([a], [b]) = [a] \boxtimes [b] = [ab].$$

The operation $\boxtimes$ is associative: since ordinary multiplication in $\mathbb{Z}$ is associative,

$$
\begin{aligned}
[a] \boxtimes \big([b] \boxtimes [c]\big) &= [a] \boxtimes [bc] \\
&= [a(bc)] \\
&= [(ab)c] \\
&= [ab] \boxtimes [c] \\
&= \big([a] \boxtimes [b]\big) \boxtimes [c].
\end{aligned}
$$

The operation $\boxtimes$ is commutative: since ordinary multiplication in $\mathbb{Z}$ is commutative,

$$[a] \boxtimes [b] = [ab] = [ba] = [b] \boxtimes [a].$$

For all $a \in \mathbb{Z}$,

$$[1] \boxtimes [a] = [1a] = [a].$$

(iii) By Theorem 1.53, the congruence $ax \equiv b \bmod m$ can be solved for x if $(a, m) = 1$; that is, $[a] \boxtimes [x] = [b]$ can be solved for $[x]$ in $\mathbb{Z}_m$ when a and m are relatively prime. (Recall that if $sa + tm = 1$, then $[x] = [sb]$.)

(iv) Assume that $m = p$ is prime; if $0 < a < p$, then $(a, p) = 1$ and the equation $[a] \boxtimes [x] = [1]$ can be solved in $\mathbb{Z}_p$, by part (iii); that is, $[a]$ has an inverse in $\mathbb{Z}_p$. We have proved that $\mathbb{Z}_p^\times$ is an abelian group; its order is $p - 1$ because, as a set, it is obtained from $\mathbb{Z}_p$ by throwing away one element, namely, $[0]$.   •

Note that if $m \geq 2$ is not a prime, then $(\mathbb{Z}_m)^\times$ is not a group: if $m = ab$, where $1 < a, b < m$, then $[a], [b] \in (\mathbb{Z}_m)^\times$, but their product $[a] \boxtimes [b] = [ab] = [m] = [0] \notin (\mathbb{Z}_m)^\times$.

In Theorem 3.78 we will prove, for every prime p, that $\mathbb{Z}_p^\times$ is a cyclic group of order $p - 1$.

We now drop the notation $\boxtimes$; henceforth, we shall write

$$[a][b] = [ab]$$

for the product of congruence classes in $\mathbb{Z}_m$.

Corollary 2.46 (*Fermat*).   *If p is a prime and $a \in \mathbb{Z}$, then*

$$a^p \equiv a \bmod p.$$

Proof.   By Proposition 2.41, it suffices to show that $[a^p] = [a]$ in $\mathbb{Z}_p$. If $[a] = [0]$, then Proposition 2.45 gives $[a^p] = [a]^p = [0]^p = [0] = [a]$. If $[a] \neq [0]$, then $[a] \in \mathbb{Z}_p^\times$, the multiplicative group of nonzero elements in $\mathbb{Z}_p$. By Corollary 2.34 to Lagrange's theorem, $[a]^{p-1} = [1]$, because $|\mathbb{Z}_p^\times| = p - 1$. Multiplying by $[a]$ gives the desired result $[a^p] = [a]^p = [a]$. Therefore, $a^p \equiv a \bmod p$.   •

We are now going to give a generalization of Fermat's theorem due to Euler.

Definition.   If m is a positive integer, define the ***Euler ϕ-function*** as follows: $\phi(1) = 1$; if $m \geq 2$, then $\phi(m)$ is the number of integers r, where $1 \leq r < m$, with $(r\ m) = 1$.

If p is a prime, then $\phi(p) = p - 1$.

Theorem 2.47 (*Euler*).   *If $(r, m) = 1$, then*

$$r^{\phi(m)} \equiv 1 \bmod m.$$

Proof.   Define

$$U(\mathbb{Z}_m) = \{\, [r] \in \mathbb{Z}_m : (r, m) = 1 \}.$$

We claim that $U(\mathbb{Z}_m)$ is a multiplicative group; note that if p is prime, then $U(\mathbb{Z}_p) = \mathbb{Z}_p^\times$. By Exercise 1.49, $(r, m) = 1 = (r', m)$ implies $(rr', m) = 1$; hence $U(\mathbb{Z}_m)$ is closed under multiplication. Proposition 2.45(ii) shows that multiplication is associative and that $[1]$ is the identity. By Proposition 2.45(iii), the equation $[r][x] = [1]$ can be solved for $[x] \in \mathbb{Z}_m$. There is thus an integer x with $rx \equiv 1 \bmod m$, so that $rx + sm = 1$ for some integer s; by Exercise 1.48, $(x, m) = 1$, and so $[x] \in U(\mathbb{Z}_m)$. Therefore, each $[r] \in U(\mathbb{Z}_m)$ has an inverse, and so $U(\mathbb{Z}_m)$ is a group. The definition of the Euler ϕ-function shows that $|U(\mathbb{Z}_m)| = \phi(m)$, and Corollary 2.34 to Lagrange's theorem gives $[r]^{\phi(m)} = [1]$ for all $[r] \in U(\mathbb{Z}_m)$. In congruence notation, this says that if $(r, m) = 1$, then $r^{\phi(m)} \equiv 1 \bmod m$.   •

Example 2.31.
It is easy to see that

$$U(\mathbb{Z}_8) = \{ [1], [3], [5], [7] \} \cong \mathbf{V}$$

and

$$U(\mathbb{Z}_{10}) = \{ [1], [3], [7], [9] \} \cong \mathbb{Z}_4. \quad \blacktriangleleft$$

Theorem 2.48 (Wilson's Theorem).   *An integer p is a prime if and only if*

$$(p - 1)! \equiv -1 \bmod p.$$

Proof.   Assume that p is a prime. If $a_1, a_2, \ldots, a_n$ is a list of all the elements of a finite abelian group G, then the product $a_1 a_2 \ldots a_n$ is the same as the product of all elements a with $a^2 = 1$, for any other element cancels against its inverse. Since p is prime, Exercise 1.75 implies that $\mathbb{Z}_p^{\times}$ has only one element of order 2, namely, $[-1]$. It follows that the product of all the elements in $\mathbb{Z}_p^{\times}$, namely, $[(p - 1)!]$, is equal to $[-1]$; therefore, $(p - 1)! \equiv -1 \bmod p$.

Conversely, assume that m is composite; there are integers a and b with $m = ab$ and $1 < a \le b < m$. If $a < b$, then $m = ab$ is a divisor of $(m - 1)!$, and so $(m - 1)! \equiv 0 \bmod m$. If $a = b$, then $m = a^2$. If $a = 2$, then $(a^2 - 1)! = 3! = 6 \equiv 2 \bmod 4$ and, of course, $2 \not\equiv -1 \bmod 4$. If $2 < a$, then $2a < a^2$, and so a and $2a$ are factors of $(a^2 - 1)!$; therefore, $(a^2 - 1)! \equiv 0 \bmod a^2$, and the proof is complete.   •

Remark.   One can generalize Wilson's theorem in the same way that Euler's theorem generalizes Fermat's theorem: replace $U(\mathbb{Z}_p)$ by $U(\mathbb{Z}_n)$. For example, one can prove, for all $m \ge 3$, that $U(\mathbb{Z}_{2^m})$ has exactly 3 elements of order 2, namely, $[-1]$, $[1 + 2^{m-1}]$, and $[-(1 + 2^{m-1})]$. It now follows that the product of all the odd numbers r, where $1 \le r < 2^m$ is congruent to $1 \bmod 2^m$, because

$$(-1)(1 + 2^{m-1})(-1 - 2^{m-1}) = (1 + 2^{m-1})^2$$
$$= 1 + 2^m + 2^{2m-2} \equiv 1 \bmod 2^m. \quad \blacktriangleleft$$

The homomorphism $\pi : \mathbb{Z} \to \mathbb{Z}_m$, defined by $\pi : a \mapsto [a]$, is surjective, so that $\mathbb{Z}_m$ is equal to im π. Thus, every element of $\mathbb{Z}_m$ has the form $\pi(a)$ for some $a \in \mathbb{Z}$, and $\pi(a) + \pi(b) = \pi(a + b)$. This description of the additive group $\mathbb{Z}_m$ in terms of the additive group $\mathbb{Z}$ can be generalized to arbitrary, not necessarily abelian, groups. Suppose that $f : G \to H$ is a surjective homomorphism between groups G and H. Since f is surjective, each element of H has the form $f(a)$ for some $a \in G$, and the operation in H is given by $f(a)f(b) = f(ab)$, where $a, b \in G$. Now $K = \ker f$ is a normal subgroup of G, and we are going to reconstruct $H = \operatorname{im} f$ (as well as a surjective homomorphism $\pi : G \to H$) from G and K alone.

We begin by introducing an operation on the set

$$\mathcal{S}(G)$$

of all nonempty subsets of a group G. If $X, Y \in \mathcal{S}(G)$, define

$$XY = \{xy : x \in X \text{ and } y \in Y\}.$$

This multiplication is associative: $X(YZ)$ is the set of all $x(yz)$, where $x \in X$, $y \in Y$, and $z \in Z$, $(XY)Z$ is the set of all such $(xy)z$, and these are the same because of associativity in G.

An instance of this multiplication is the product of a one-point subset $\{a\}$ and a subgroup $K \leq G$, which is the coset aK.

As a second example, we show that if H is any subgroup of G, then

$$HH = H.$$

If $h, h' \in H$, then $hh' \in H$, because subgroups are closed under multiplication, and so $HH \subset H$. For the reverse inclusion, if $h \in H$, then $h = h1 \in HH$ (because $1 \in H$), and so $H \subset HH$.

It is possible for two subsets X and Y in $\mathcal{S}(G)$ to commute even though their constituent elements do not commute. One example has just been given; take $X = Y = H$, where H is a nonabelian subgroup of G. Here is a more interesting example: let $G = S_3$ and $K = \langle (1\ 2\ 3) \rangle$. Now $(1\ 2)$ does not commute with $(1\ 2\ 3) \in K$, but we claim that $(1\ 2)K = K(1\ 2)$. In fact, let us prove the converse of Exercise 2.66.

Lemma 2.49.   *If K is a* normal *subgroup of a group G, then*

$$bK = Kb$$

for every $b \in G$.

Proof.   Let $bk \in bK$. Since K is normal, $bkb^{-1} \in K$, say $bkb^{-1} = k' \in K$, so that $bk = (bkb^{-1})b = k'b \in Kb$, and so $bK \subset Kb$. For the reverse inclusion, let $kb \in Kb$. Since K is normal, $(b^{-1})k(b^{-1})^{-1} = b^{-1}kb \in K$, say $b^{-1}kb = k'' \in K$. Hence, $kb = b(b^{-1}kb) = bk'' \in bK$ and $Kb \subset bK$. Therefore, $bK = Kb$ when $K \lhd G$. •

A natural question is whether HK is a subgroup when both H and K are subgroups. In general, HK need not be a subgroup. For example, let $G = S_3$, let $H = \langle (1\ 2) \rangle$, and let $K = \langle (1\ 3) \rangle$. Then

$$HK = \{(1), (1\ 2), (1\ 3), (1\ 3\ 2)\}$$

is not a subgroup lest we contradict Lagrange's theorem.

Proposition 2.50.

   (i) *If H and K are subgroups of a group G, and if one of them is a normal subgroup, then HK is a subgroup of G; moreover, $HK = KH$ in this case.*

   (ii) *If both H and K are normal subgroups, then HK is a normal subgroup.*

Proof.   (i) Assume first that $K \lhd G$. We claim that $HK = KH$. If $hk \in HK$, then $k' = hkh^{-1} \in K$, because $K \lhd G$, and

$$hk = hkh^{-1}h = k'h \in KH.$$

Hence, $HK \subset KH$. For the reverse inclusion, write $kh = hh^{-1}kh = hk'' \in HK$. (Note that the same argument shows that $HK = KH$ if $H \lhd G$.)

We now show that HK is a subgroup. Since $1 \in H$ and $1 \in K$, we have $1 = 1 \cdot 1 \in HK$; if $hk \in HK$, then $(hk)^{-1} = k^{-1}h^{-1} \in KH = HK$; if $hk, h_1 k_1 \in HK$, then $h_1^{-1}kh_1 = k' \in K$ and

$$hkh_1 k_1 = hh_1(h_1^{-1}kh_1)k_1 = (hh_i)(k'k_1) \in HK.$$

Therefore, HK is a subgroup of G.

(ii) If $g \in G$, then

$$ghkg^{-1} = (ghg^{-1})(gkg^{-1}) \in HK.$$

Therefore, $HK \lhd G$ in this case.   •

Here is a fundamental construction of a new group from a given group.

Theorem 2.51. *Let G/K denote the family of all the cosets of a subgroup K of G. If K is a normal subgroup, then*

$$aKbK = abK$$

for all $a, b \in G$, and G/K is a group under this operation.

Remark. The group G/K is called the **quotient group** G mod K; when G is finite, its order $|G/K|$ is the index $[G : K] = |G|/|K|$ (presumably, this is the reason *quotient groups* are so called). ◄

Proof. The product of two cosets $(aK)(bK)$ can also be viewed as the product of 4 elements in $\mathcal{S}(G)$. Hence, associativity in $\mathcal{S}(G)$ gives

$$(aK)(bK) = a(Kb)K = a(bK)K = abKK = abK,$$

for normality of K gives $Kb = bK$ for all $b \in K$, by Lemma 2.49, while $KK = K$ because K is a subgroup. Thus, the product of two cosets of K is again a coset of K, and so an operation on G/K has been defined. Because multiplication in $\mathcal{S}(G)$ is associative, equality $X(YZ) = (XY)Z$ holds, in particular, when X, Y, and Z are cosets of K, so that the operation on G/K is associative. The identity is the coset $K = 1K$, for $(1K)(bK) = 1bK = bK = b1K = (bK)(1K)$, and the inverse of aK is $a^{-1}K$, for $(a^{-1}K)(aK) = a^{-1}aK = K = aa^{-1}K = (aK)(a^{-1}K)$. Therefore, G/K is a group. ●

Example 2.32.
We show that the quotient group G/K is precisely $\mathbb{Z}_m$ when G is the additive group $\mathbb{Z}$ and $K = \langle m \rangle$, the (cyclic) subgroup of all the multiples of a positive integer m. Since $\mathbb{Z}$ is abelian, $\langle m \rangle$ is necessarily a normal subgroup. The sets $\mathbb{Z}/\langle m \rangle$ and $\mathbb{Z}_m$ coincide because they are comprised of the same elements: the coset $a + \langle m \rangle$ is the congruence class $[a]$:

$$a + \langle m \rangle = \{a + km : k \in \mathbb{Z}\} = [a].$$

The operations also coincide: addition in $\mathbb{Z}/\langle m \rangle$ is given by

$$(a + \langle m \rangle) + (b + \langle m \rangle) = (a + b) + \langle m \rangle;$$

since $a + \langle m \rangle = [a]$, this last equation is just $[a] + [b] = [a + b]$, which is the sum in $\mathbb{Z}_m$. Therefore, $\mathbb{Z}_m$ is equal to the quotient group $\mathbb{Z}/\langle m \rangle$. ◄

We remind the reader of Lemma 2.31(i): If K is a subgroup of G, then two cosets aK and bK are equal if and only if $b^{-1}a \in K$. In particular, if $b = 1$, then $aK = K$ if and only if $a \in K$.

We can now prove the converse of Proposition 2.37(ii).

Corollary 2.52. *Every normal subgroup $K \lhd G$ is the kernel of some homomorphism.*

Proof. Define the **natural map** $\pi : G \to G/K$ by $\pi(a) = aK$. With this notation, the formula $aKbK = abK$ can be rewritten as $\pi(a)\pi(b) = \pi(ab)$; thus, π is a (surjective) homomorphism. Since K is the identity element in G/K,

$$\ker \pi = \{a \in G : \pi(a) = K\} = \{a \in G : aK = K\} = K,$$

by Lemma 2.31(i).   •

The next theorem shows that every homomorphism gives rise to an isomorphism, and that quotient groups are merely constructions of homomorphic images. It was E. Noether (1882–1935) who emphasized the fundamental importance of this fact.

Theorem 2.53 (First Isomorphism Theorem). *If $f : G \to H$ is a homomorphism, then*

$$\ker f \lhd G \text{ and } G/\ker f \cong \operatorname{im} f.$$

In more detail, if $\ker f = K$, then the function $\varphi : G/K \to \operatorname{im} f \leq H$, given by $\varphi : aK \mapsto f(a)$, is an isomorphism.

Proof. We have already seen, in Proposition 2.37(ii), that $K = \ker f$ is a normal subgroup of G. Now φ is single-valued: if $aK = bK$, then $a = bk$ for some $k \in K$, and so $f(a) = f(bk) = f(b)f(k) = f(b)$, because $f(k) = 1$.

Let us now see that φ is a homomorphism. Since f is a homomorphism and $\varphi(aK) = f(a)$,

$$\varphi(aKbK) = \varphi(abK) = f(ab) = f(a)f(b) = \varphi(aK)\varphi(bK).$$

It is clear that $\operatorname{im} \varphi \leq \operatorname{im} f$. For the reverse inclusion, note that if $y \in \operatorname{im} f$, then $y = f(a)$ for some $a \in G$, and so $y = f(a) = \varphi(aK)$. Thus, φ is surjective.

Finally, we show that φ is injective. If $\varphi(aK) = \varphi(bK)$, then $f(a) = f(b)$. Hence, $1 = f(b)^{-1}f(a) = f(b^{-1}a)$, so that $b^{-1}a \in \ker f = K$. Therefore, $aK = bK$, by LeLeLemma 2.31(i), and so φ is injective. We have proved that $\varphi : G/K \to \operatorname{im} f$ is an isomorphism.   •

Here is a minor application of the first isomorphism theorem. For any group G, the identity function $f : G \to G$ is a surjective homomorphism with $\ker f = \{1\}$. By the first isomorphism theorem, we have

$$G/\{1\} \cong G.$$

Remark.   The following diagram describes the proof of the first isomorphism theorem, where $\pi : G \to G/K$ is the natural map $\pi : a \mapsto aK$.

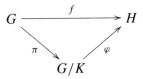

Given any homomorphism $f : G \to H$, one should salivate, like Pavlov's dog, and ask for its kernel and image; the first isomorphism theorem will then provide an isomorphism $G/\ker f \cong \operatorname{im} f$. Since there is no significant difference between isomorphic groups, the first isomorphism theorem also says that there is no significant difference between quotient groups and homomorphic images.

Example 2.33.
Let us revisit Example 2.25, which showed that any two cyclic groups of order m are isomorphic. Let $G = \langle a \rangle$ be a cyclic group of order m. Define a homomorphism $f : \mathbb{Z} \to G$ by $f(n) = a^n$ for all $n \in \mathbb{Z}$. Now f is surjective (because a is a generator of G), while $\ker f = \{n \in \mathbb{Z} : a^n = 1\} = \langle m \rangle$, by Lemma 2.24. The first isomorphism theorem gives an isomorphism $\mathbb{Z}/\langle m \rangle \cong G$. We have shown that every cyclic group of order m is isomorphic to $\mathbb{Z}/\langle m \rangle$, and hence that any two cyclic groups of order m are isomorphic to each other. Of course, Example 2.32 shows that $\mathbb{Z}/\langle m \rangle = \mathbb{Z}_m$, so that every cyclic group of order m is isomorphic to $\mathbb{Z}_m$. ◄

Example 2.34.
What is the quotient group $\mathbb{R}/\mathbb{Z}$? Define $f : \mathbb{R} \to S^1$, where S^1 is the circle group, by

$$f : x \mapsto e^{2\pi i x}.$$

Now f is a homomorphism; that is, $f(x + y) = f(x)f(y)$, by the addition formulas for sine and cosine. The map f is surjective, and $\ker f$ consists of all $x \in \mathbb{R}$ for which $e^{2\pi i x} = \cos 2\pi x + i \sin 2\pi x = 1$. But $\cos 2\pi x = 0 = \sin 2\pi x$ forces x to be an integer; since $1 \in \ker f$, we have $\ker f = \mathbb{Z}$. The first isomorphism theorem now gives

$$\mathbb{R}/\mathbb{Z} \cong S^1. \quad ◄$$

Here is a useful counting result.

Proposition 2.54 (Product Formula).   *If H and K are subgroups of a finite group G, then*

$$|HK||H \cap K| = |H||K|,$$

where $HK = \{hk : h \in H \text{ and } k \in K\}$.

Remark.   The subset HK need not be a subgroup of G; however, Proposition 2.50 shows that if either $H \lhd G$ or $K \lhd G$, then HK is a subgroup (see also Exercise 2.86).   ◄

Proof.   Define a function $f : H \times K \to HK$ by $f : (h, k) \mapsto hk$. Clearly, f is a surjection. It suffices to show, for every $x \in HK$, that $f^{-1}(x) = |H \cap K|$ [because $H \times K$ is the disjoint union $\bigcup_{x \in HK} f^{-1}(x)$].
 We claim that if $x = hk$, then

$$f^{-1}(x) = \{(hd, d^{-1}k) : d \in H \cap K\}.$$

Each $(hd, d^{-1}k) \in f^{-1}(x)$, for $f(hd, d^{-1}k) = hdd^{-1}k = hk = x$. For the reverse inclusion, let $(h', k') \in f^{-1}(x)$, so that $h'k' = hk$. Then $h^{-1}h' = kk'^{-1} \in H \cap K$; call this element d. Then $h' = hd$ and $k' = d^{-1}k$, and so (h', k') lies in the right side. Therefore,

$$|f^{-1}(x)| = |\{(hd, d^{-1}k) : d \in H \cap K\}| = |H \cap K|,$$

because $d \mapsto (hd, d^{-1}h)$ is a bijection.   •

The next two results are variants of the first isomorphism theorem.

Theorem 2.55 (Second Isomorphism Theorem).   *If H and K are subgroups of a group G with H $\lhd$ G, then HK is a subgroup, H $\cap$ K $\lhd$ K, and*

$$K/(H \cap K) \cong HK/H.$$

Proof.   Since $H \lhd G$, Proposition 2.50 shows that HK is a subgroup. Normality of H in HK follows from a more general fact: if $H \le S \le G$ and if H is normal in G, then H is normal in S (if $ghg^{-1} \in H$ for every $g \in G$, then, in particular, $ghg^{-1} \in H$ for every $g \in S$).

We now show that every coset $xH \in HK/H$ has the form kH for some $k \in K$. Of course, $xH = hkH$, where $h \in H$ and $k \in K$. But $hk = kk^{-1}hk = kh'$ for some $h' \in H$, so that $hkH = kh'H = kH$. It follows that the function $f : K \to HK/H$, given by $f : k \mapsto kH$, is surjective. Moreover, f is a homomorphism, for it is the restriction of the natural map $\pi : G \to G/H$. Since $\ker \pi = H$, it follows that $\ker f = H \cap K$, and so $H \cap K$ is a normal subgroup of K. The first isomorphism theorem now gives $K/(H \cap K) \cong HK/H$. •

The second isomorphism theorem gives the product formula in the special case when one of the subgroups is normal: if $K/(H \cap K) \cong HK/H$, then $|K/(H \cap K)| = |HK/H|$, and so $|HK||H \cap K| = |H||K|$.

Theorem 2.56 (Third Isomorphism Theorem). *If H and K are normal subgroups of a group G with $K \leq H$, then $H/K \lhd G/K$ and*

$$(G/K)/(H/K) \cong G/H.$$

Proof. Define $f : G/K \to G/H$ by $f : aK \mapsto aH$. Note that f is a (single-valued) function, for if $a' \in G$ and $a'K = aK$, then $a^{-1}a' \in K \leq H$, and so $aH = a'H$. It is easy to see that f is a surjective homomorphism.

Now $\ker f = H/K$, for $aK = H$ if and only if $a \in H$, and so H/K is a normal subgroup of G/K. Since f is surjective, the first isomorphism theorem gives $(G/K)/(H/K) \cong G/H$. •

The third isomorphism theorem is easy to remember: the K's in the fraction $(G/K)/(H/K)$ can be canceled. One can better appreciate the first isomorphism theorem after having proved the third one. The quotient group $(G/K)/(H/K)$ consists of cosets (of H/K) whose representatives are themselves cosets (of G/K). A direct proof of the third isomorphism theorem could be nasty.

If X and Y are sets, then a function $f : X \to Y$ defines a "forward motion" carrying subsets of X into subsets of Y, namely, if $S \subset X$, then

$$f(S) = \{y \in Y : y = f(x) \text{ for some } x \in X\},$$

and a "backward motion" carrying subsets of Y into subsets of X, namely, if $W \subset Y$, then

$$f^{-1}(W) = \{x \in X : f(x) \in W\}.$$

One calls $f^{-1}(W)$ the ***inverse image*** of W. When f is a surjection, then these motions set up a bijection between all the subsets of Y and some of the subsets of X.

a homomorphism, let $g' = h'k'$, so that $gg' = hkh'k' = hh'kk'$. Hence, $\varphi(gg') = \varphi(hkh'k')$, which is not in the proper form for evaluation. If we knew that if $h \in H$ and $k \in K$, then $hk = kh$, then we could continue:

$$\varphi(hkh'k') = \varphi(hh'kk')$$
$$= (hh', kk')$$
$$= (h, k)(h', k')$$
$$= \varphi(g)\varphi(g').$$

Let $h \in H$ and $k \in K$. Since K is a normal subgroup, $(hkh^{-1})k^{-1} \in K$; since H is a normal subgroup, $h(kh^{-1}k^{-1}) \in H$. But $H \cap K = \{1\}$, so that $hkh^{-1}k^{-1} = 1$ and $hk = kh$. Finally, we show that the homomorphism φ is an isomorphism. If $(h, k) \in H \times K$, then the element $g \in G$ defined by $g = hk$ satisfies $\varphi(g) = (h, k)$; hence φ is surjective. If $\varphi(g) = (1, 1)$, then $g = 1$, so that $\ker \varphi = 1$ and φ is injective. Therefore, φ is an isomorphism.   •

Theorem 2.62.   *If m and n are relatively prime, then*

$$\mathbb{Z}_{mn} \cong \mathbb{Z}_m \times \mathbb{Z}_n.$$

Proof.   If $a \in \mathbb{Z}$, denote its congruence class in $\mathbb{Z}_m$ by $[a]_m$. The reader can show that the function $f : \mathbb{Z} \to \mathbb{Z}_m \times \mathbb{Z}_n$, given by $a \mapsto ([a]_m, [a]_n)$, is a homomorphism. We claim that $\ker f = \langle mn \rangle$. Clearly, $\langle mn \rangle \leq \ker f$. For the reverse inclusion, if $a \in \ker f$, then $[a]_m = [0]_m$ and $[a]_n = [0]_n$; that is, $a \equiv 0 \bmod m$ and $a \equiv 0 \bmod n$; that is, $m \mid a$ and $n \mid a$. Since m and n are relatively prime, Exercise 1.51 gives $mn \mid a$, and so $a \in \langle mn \rangle$, that is, $\ker f \leq \langle mm \rangle$ and $\ker f = \langle mn \rangle$.

We now show that f is surjective. If $([a]_m, [b]_n) \in \mathbb{Z}_m \times \mathbb{Z}_n$, is there $x \in \mathbb{Z}$ with $f(x) = ([x]_m, [x]_n) = ([a]_m, [b]_n)$; that is, is there $x \in \mathbb{Z}$ with $x \equiv a \bmod m$ and $x \equiv b \bmod n$? Since m and n are relatively prime, the Chinese Remainder Theorem provides a solution x. The first isomorphism theorem now gives $\mathbb{Z}_{mn} = \mathbb{Z}/\langle mn \rangle \cong \mathbb{Z}_m \times \mathbb{Z}_n$.   •

For example, it follows that $\mathbb{Z}_6 \cong \mathbb{Z}_2 \times \mathbb{Z}_3$. Note that there is no isomorphism if m and n are not relatively prime. For example, $\mathbb{Z}_4 \not\cong \mathbb{Z}_2 \times \mathbb{Z}_2$, for $\mathbb{Z}_4$ has an element of order 4 and the direct product (which is isomorphic to the four-group **V**) has no such element.

In light of Proposition 2.28, we may say that an element $a \in G$ has order n if $\langle a \rangle \cong \mathbb{Z}_n$. Theorem 2.62 can now be interpreted as saying that if a and b are commuting elements having relatively prime orders m and n, then ab has order mn. Let us give a direct proof of this result.

Proposition 2.63.   *Let G be a group, and let $a, b \in G$ be commuting elements of orders m and n, respectively. If $(m, n) = 1$, then ab has order mn.*

Proof.   Since a and b commute, we have $(ab)^r = a^r b^r$ for all r, so that $(ab)^{mn} = a^{mn} b^{mn} = 1$. It suffices to prove that if $(ab)^k = 1$, then $mn \mid k$. If $1 = (ab)^k = a^k b^k$, then $a^k = b^{-k}$. Since a has order m, we have $1 = a^{mk} = b^{-mk}$. Since b has order n, Lemma 2.24 gives $n \mid mk$. As $(m, n) = 1$, however, Corollary 1.34 gives $n \mid k$; a similar argument gives $m \mid k$. Finally, Exercise 1.51 shows that $mn \mid k$. Therefore, $mn \leq k$, and mn is the order of ab.   •

Corollary 2.64.   *If $(m, n) = 1$, then $\phi(mn) = \phi(m)\phi(n)$, where ϕ is the Euler ϕ-function.*

Proof. [21]   By the theorem, the function $f : \mathbb{Z}_{mn} \to \mathbb{Z}_m \times \mathbb{Z}_n$, given by $[a] \mapsto ([a]_m, [a]_n)$, is an isomorphism. The result will follow once we prove that $f(U(\mathbb{Z}_{mn})) = U(\mathbb{Z}_m) \times U(\mathbb{Z}_n)$, for then

$$\phi(mn) = |U(\mathbb{Z}_{mn})| = |f(U(\mathbb{Z}_{mn}))|$$
$$= |U(\mathbb{Z}_m) \times U(\mathbb{Z}_n)| = |U(\mathbb{Z}_m)| \cdot |U(\mathbb{Z}_n)| = \phi(m)\phi(n).$$

If $[a] \in U(\mathbb{Z}_{mn})$, then $[a][b] = [1]$ for some $[b] \in \mathbb{Z}_{mn}$, and

$$f([ab]) = ([ab]_m, [ab]_n) = ([a]_m[b]_m, [a]_n[b]_n)$$
$$= ([a]_m, [a]_n)([b]_m, [b]_n) = ([1]_m, [1]_n).$$

Hence, $[1]_m = [a]_m[b]_m$ and $[1]_n = [a]_n[b]_n$, so that $f([a]) = ([a]_m, [a]_n) \in U(\mathbb{Z}_m) \times U(\mathbb{Z}_n)$, and $f(U(\mathbb{Z}_{mn})) \leq U(\mathbb{Z}_m) \times U(\mathbb{Z}_n)$.

For the reverse inclusion, if $f([c]) = ([c]_m, [c]_n) \in U(\mathbb{Z}_m) \times U(\mathbb{Z}_n)$, then we must show that $[c] \in U(\mathbb{Z}_{mn})$. There is $[d]_m \in \mathbb{Z}_m$ with $[c]_m[d]_m = [1]_m$, and there is $[e]_n \in \mathbb{Z}_n$ with $[c]_n[e]_n = [1]_n$. Since f is surjective, there is $b \in \mathbb{Z}$ with $([b]_m, [b]_n) = ([d]_m, [e]_n)$, so that

$$f([1]) = ([1]_m, [1]_n) = ([c]_m[b]_m, [c]_n[b]_n) = f([c][b]).$$

Since f if an injection, $[1] = [c][b]$ and $[c] \in U(\mathbb{Z}_{mn})$.   •

The following curious theorem will be used in Chapter 4 to prove that certain specific groups are cyclic.

---

[21] See Exercise 3.50(iii) for a less computational proof.

Proposition 2.65. *If G is a finite abelian group having a unique subgroup of order p for every prime divisor p of $|G|$, then G is cyclic.*

Proof. Choose $a \in G$ of largest order, say, n. If p is a prime divisor of $|G|$, let $C = C_p$ be the unique subgroup of G having order p; the subgroup C must be cyclic, say $C = \langle c \rangle$. We show that $p \mid n$ by showing that $c \in \langle a \rangle$ (and hence $C \leq \langle a \rangle$). If $(p, n) = 1$, then ca has order $pn > n$, by Proposition 2.63, contradicting a being an element of largest order. If $p \mid n$, say, $n = pq$, then a^q has order p, and hence it lies in the unique subgroup $\langle c \rangle$ of order p. Thus, $a^q = c^i$ for some i. Now $(i, p) = 1$, so there are integers u and v with $1 = ui + vp$; hence, $c = c^{ui+vp} = c^{ui}c^{vp} = c^{ui}$. Therefore, $a^{qu} = c^{ui} = c$, so that $c \in \langle a \rangle$, as desired. It follows that $\langle a \rangle$ contains every element $x \in G$ with $x^p = 1$ for some prime p.

If $\langle a \rangle = G$, we are finished. Therefore, we may assume there is $b \in G$ with $b \notin \langle a \rangle$. Now $b^{|G|} = 1 \in \langle a \rangle$; let k be the smallest positive integer with $b^k \in \langle a \rangle$:

$$b^k = a^q.$$

Note that $k \mid |G|$ because k is the order of $b \langle a \rangle$ in $G/\langle a \rangle$. Of course, $k \neq 1$, and so there is a factorization $k = pm$, where p is prime. There are now two possibilities. If $p \mid q$, then $q = pu$ and

$$b^{pm} = b^k = a^q = a^{pu}.$$

Hence, $(b^m a^{-u})^p = 1$, and so $b^m a^{-u} \in \langle a \rangle$. Thus, $b^m \in \langle a \rangle$, and this contradicts k being the smallest exponent with this property. The second possibility is that $p \nmid q$, in which case $(p, q) = 1$. There are integers s and t with $1 = sp + tq$, and so

$$a = a^{sp+tq} = a^{sp}a^{tq} = a^{sp}b^{pmt} = (a^s b^{mt})^p.$$

Therefore, $a = x^p$, where $x = a^s b^{mt}$, and Exercise 2.38, which applies because $p \mid n$, says that the order of x is greater than that of a, a contradiction. We conclude that $G = \langle a \rangle$. •

The proposition is false for nonabelian groups, for the group of quaternions is a counterexample.

EXERCISES

2.81 Prove that $U(\mathbb{Z}_9) \cong \mathbb{Z}_6$ and $U(\mathbb{Z}_{15}) \cong \mathbb{Z}_4 \times \mathbb{Z}_2$.

2.82   (i) If H and K are groups, use the definitions of subgroup and normal subgroup to show that $H^* = \{(h, 1) : h \in H\}$ and $K^* = \{(1, k) : k \in K\}$ are normal subgroups of $H \times K$ with $H \cong H^*$ and $K \cong K^*$.

   (ii) Prove that $f : H \to (H \times K)/K^*$, defined by $f(h) = (h, 1)K^*$, is an isomorphism.

   (iii) Use the first isomorphism theorem to prove that $H^* \lhd H \times K$ and that $(H \times K)/K^* \cong H$.

2.83 If G is a group that is not abelian, prove that $G/Z(G)$ is not cyclic, where $Z(G)$ denotes the center of G.

2.84   (i) Prove that $\mathbf{Q}/Z(\mathbf{Q}) \cong \mathbf{V}$, where $\mathbf{Q}$ is the group of quaternions and $\mathbf{V}$ is the four-group; conclude that the quotient of a group by its center can be abelian.

   (ii) Prove that $\mathbf{Q}$ has no subgroup isomorphic to $\mathbf{V}$. Conclude that the quotient $\mathbf{Q}/Z(\mathbf{Q})$ is not isomorphic to a subgroup of $\mathbf{Q}$.

2.85 Let G be a finite group with $K \lhd G$. If $(|K|, [G : K]) = 1$, prove that K is the unique subgroup of G having order $|K|$.

2.86 If H and K are subgroups of a group G, prove that HK is a subgroup of G if and only if $HK = KH$.

2.87 Let G be a group and regard $G \times G$ as the direct product of G with itself. If the multiplication $\mu : G \times G \to G$ is a group homomorphism, prove that G must be abelian.

2.88 Generalize Theorem 2.62 as follows. Let G be a finite (additive) abelian group of order mn, where $(m, n) = 1$. Define

$$G_m = \{g \in G : \text{order }(g) \mid m\} \text{ and } G_n = \{h \in G : \text{order }(h) \mid n\}.$$

   (i) Prove that G_m and G_n are subgroups with $G_m \cap G_n = \{0\}$.
   (ii) Prove that $G = G_m + G_n = \{g + h : g \in G_m \text{ and } h \in G_n\}$.
   (iii) Prove that $G \cong G_m \times G_n$.

2.89 Let G be a finite group, let p be a prime, and let H be a normal subgroup of G. Prove that if both H and G/H are powers of p, then $|G|$ is a power of p.

Definition.   If $H_1, \ldots, H_n$ are groups, then their ***direct product***

$$H_1 \times \cdots \times H_n$$

is the set of all n-tuples $(h_1, \ldots, h_n)$, where $h_i \in H_i$ for all i, with coordinatewise multiplication:

$$(h_1, \ldots, h_n)(h'_1, \ldots, h'_n) = (h_1 h'_1, \ldots, h_n h'_n).$$

2.90   (i) Generalize Theorem 2.62 by proving that if the prime factorization of an integer m is $m = p_1^{e_1} \cdots p_n^{e_n}$, then

$$\mathbb{Z}_m \cong \mathbb{Z}_{p_1^{e_1}} \times \cdots \times \mathbb{Z}_{p_n^{e_n}}.$$

   (ii) Generalize Corollary 2.64 by proving that if the prime factorization of an integer m is $m = p_1^{e_1} \cdots p_n^{e_n}$, then

$$U(\mathbb{Z}_m) \cong U(\mathbb{Z}_{p_1^{e_1}}) \times \cdots \times U(\mathbb{Z}_{p_n^{e_n}}).$$

2.91 Let p be an odd prime, and assume $a_i \equiv i \bmod p$ for $1 \leq i \leq p - 1$. Prove that there exist $i \neq j$ with $i a_i \equiv j a_j \bmod p$.

2.92   (i) If p is a prime, prove that $\phi(p^k) = p^k(1 - 1/p)$.
   (ii) If the distinct prime divisors of a positive integer h are $p_1, p_2, \ldots, p_n$, prove that

$$\phi(h) = h(1 - 1/p_1)(1 - 1/p_2) \cdots (1 - 1/p_n).$$

2.7 GROUP ACTIONS

Groups of permutations led us to abstract groups; the next result, due to A. Cayley (1821–1895), shows that abstract groups are not so far removed from permutations.

Theorem 2.66 (Cayley).   *Every group G is (isomorphic to) a subgroup of the symmetric group S_G. In particular, if $|G| = n$, then G is isomorphic to a subgroup of S_n.*

Proof.   For each $a \in G$, define "translation" $\tau_a : G \to G$ by $\tau_a(x) = ax$ for every $x \in G$ (if $a \neq 1$, then τ_a is not a homomorphism). For $a, b \in G$, $(\tau_a \circ \tau_b)(x) = \tau_a(\tau_b(x)) = \tau_a(bx) = a(bx) = (ab)x$, by associativity, so that

$$\tau_a \tau_b = \tau_{ab}.$$

It follows that each τ_a is a bijection, for its inverse is $\tau_{a^{-1}}$:

$$\tau_a \tau_{a^{-1}} = \tau_{aa^{-1}} = \tau_1 = 1_G = \tau_{a^{-1}a},$$

and so $\tau_a \in S_G$.

Define $\varphi : G \to S_G$ by $\varphi(a) = \tau_a$. Rewriting,

$$\varphi(a)\varphi(b) = \tau_a \tau_b = \tau_{ab} = \varphi(ab),$$

so that φ is a homomorphism. Finally, φ is an injection. If $\varphi(a) = \varphi(b)$, then $\tau_a = \tau_b$, and hence $\tau_a(x) = \tau_b(x)$ for all $x \in G$; in particular, when $x = 1$, this gives $a = b$, as desired.

The last statement follows from Exercise 2.55, which says that if X is a set with $|X| = n$, then $S_X \cong S_n$.   •

The reader may note, in the proof of Cayley's theorem, that the permutation τ_a is just the ath row of the multiplication table of G.

To tell the truth, Cayley's theorem itself is only mildly interesting. However, the identical proof works in a larger setting that is more interesting.

Theorem 2.67 (Representation on Cosets).   *Let G be a group, and let H be a subgroup of G having finite index n. Then there exists a homomorphism* $\varphi : G \to S_n$ *with* $\ker \varphi \le H$.

Proof.   Even though H may not be a normal subgroup, we still denote the family of all the cosets of H in G by G/H.

For each $a \in G$, define "translation" $\tau_a : G/H \to G/H$ by $\tau_a(xH) = axH$ for every $x \in G$. For $a, b \in G$,

$$(\tau_a \circ \tau_b)(xH) = \tau_a(\tau_b(xH)) = \tau_a(bxH) = a(bxH) = (ab)xH,$$

by associativity, so that

$$\tau_a \tau_b = \tau_{ab}.$$

It follows that each τ_a is a bijection, for its inverse is $\tau_{a^{-1}}$:

$$\tau_a \tau_{a^{-1}} = \tau_{aa^{-1}} = \tau_1 = 1_G = \tau_{a^{-1}} \tau_a,$$

and so $\tau_a \in S_{G/H}$. Define $\varphi : G \to S_{G/H}$ by $\varphi(a) = \tau_a$. Rewriting,

$$\varphi(a)\varphi(b) = \tau_a \tau_b = \tau_{ab} = \varphi(ab),$$

so that φ is a homomorphism. Finally, if $a \in \ker \varphi$, then $\varphi(a) = 1_{G/H}$, so that $\tau_a(xH) = xH$ for all $x \in G$; in particular, when $x = 1$, this gives $aH = H$, and $a \in H$, by Lemma 2.31(i). The result follows from Exercise 2.55, for $|G/H| = n$, and so $S_{G/H} \cong S_n$.   •

When $H = \{1\}$, this is the Cayley theorem.

We are now going to classify all groups of order up to 7. By Example 2.25, every group of prime order p is isomorphic to $\mathbb{Z}_p$, and so, to isomorphism, there is just one group of order p. Of the possible orders through 7, four of them, 2, 3, 5, and 7, are primes, and so we need look only at orders 4 and 6.

Proposition 2.68. *Every group G of order 4 is isomorphic to either $\mathbb{Z}_4$ or the four-group* **V**. *Moreover, $\mathbb{Z}_4$ and* **V** *are not isomorphic.*

Proof. By Lagrange's theorem, every element in G, other than 1, has order either 2 or 4. If there is an element of order 4, then G is cyclic. Otherwise, $x^2 = 1$ for all $x \in G$, so that Exercise 2.41 shows that G is abelian.

If distinct elements x and y in G are chosen, neither being 1, then one quickly checks that $xy \notin \{1, x, y\}$; hence,

$$G = \{1, x, y, xy\}.$$

It is easy to see that the bijection $f : G \to \mathbf{V}$, defined by $f(1) = 1$, $f(x) = (1\ 2)(3\ 4)$, $f(y) = (1\ 3)(2\ 4)$, and $f(xy) = (1\ 4)(2\ 3)$, is an isomorphism. We have already seen, in Example 2.26, that $\mathbb{Z}_4 \not\cong \mathbf{V}$.   •

Proposition 2.69. *If G is a group of order 6, then G is isomorphic to either $\mathbb{Z}_6$ or S_3. Moreover, $\mathbb{Z}_6$ and S_3 are not isomorphic.*[22]

Proof. By Lagrange's theorem, the only possible orders of nonidentity elements are 2, 3, and 6. Of course, $G \cong \mathbb{Z}_6$ if G has an element of order 6. Now Exercise 2.42 shows that G must contain an element of order 2, say, t. We now consider the cases G abelian and G nonabelian separately.
Case 1. G is abelian.

If there is a second element of order 2, say, a, then it is easy to see, using $at = ta$, that $H = \{1, a, t, at\}$ is a subgroup of G. This contradicts

---

[22]Cayley states this proposition in an article he wrote in 1854. However, in 1878, in the *American Journal of Mathematics*, he wrote, "The general problem is to find all groups of a given order n; ... if $n = 6$, there are three groups; a group

$$1, \alpha, \alpha^2, \alpha^3, \alpha^4, \alpha^5 \quad (\alpha^6 = 1),$$

and two more groups

$$1, \beta, \beta^2, \alpha, \alpha\beta, \alpha\beta^2 \quad (\alpha^2 = 1, \beta^3 = 1),$$

viz., in the first of these $\alpha\beta = \beta\alpha$ while in the other of them, we have $\alpha\beta = \beta^2\alpha$, $\alpha\beta^2 = \beta\alpha$." Cayley's list is $\mathbb{Z}_6$, $\mathbb{Z}_2 \times \mathbb{Z}_3$, and S_3. Of course, $\mathbb{Z}_2 \times \mathbb{Z}_3 \cong \mathbb{Z}_6$; even Homer nods.

Lagrange's theorem, because 4 is not a divisor of 6. It follows that G must contain an element b of order 3. But tb has order 6, for its powers are

$$tb, (tb)^2 = b^2, (tb)^3 = t, (tb)^4 = b, (tb)^5 = tb^2, (tb)^6 = 1.$$

Therefore, G is cyclic if it is abelian.

Case 2. G is not abelian.

If G has no elements of order 3, then $x^2 = 1$ for all $x \in G$, and G is abelian, by Exercise 2.41. Therefore, G contains an element s of order 3 as well as the element t of order 2.

Now $|\langle s \rangle| = 3$, so that $[G : \langle s \rangle] = |G|/|\langle s \rangle| = 6/3 = 2$, and so $\langle s \rangle$ is a normal subgroup of G, by Proposition 2.39(ii). Since $t = t^{-1}$, we have $tst \in \langle s \rangle$; hence, $tst = s^i$ for $i = 0, 1$ or 2. Now $i \neq 0$, for $tst = s^0 = 1$ implies $s = 1$. If $i = 1$, then s and t commute, and this gives st of order 6, as in Case 1 (which forces G to be cyclic, hence abelian, contrary to our present hypothesis). Therefore, $tst = s^2 = s^{-1}$.

We now use Theorem 2.67 to construct an isomorphism $G \rightarrow S_3$. Let $H = \langle t \rangle$, and consider the homomorphism $\varphi : G \rightarrow S_{G/\langle t \rangle}$ given by

$$\varphi(g) : x \langle t \rangle \mapsto gx \langle t \rangle.$$

By the theorem, $\ker \varphi \leq \langle t \rangle$, so that either $\ker \varphi = \{1\}$ (and φ is injective), or $\ker \varphi = \langle t \rangle$. Now $G/\langle t \rangle = \{\langle t \rangle, s \langle t \rangle, s^2 \langle t \rangle\}$, and, in two-rowed notation,

$$\varphi(t) = \begin{pmatrix} \langle t \rangle & s \langle t \rangle & s^2 \langle t \rangle \\ t \langle t \rangle & ts \langle t \rangle & ts^2 \langle t \rangle \end{pmatrix}.$$

If $\varphi(t)$ is the identity permutation, then $ts \langle t \rangle = s \langle t \rangle$, so that $s^{-1}ts \in \langle t \rangle = \{1, t\}$, by Lemma 2.31. But now $s^{-1}ts = t$ (it cannot be 1), hence $ts = st$, contradicting t and s not commuting. Therefore, $t \notin \ker \varphi$, and $\varphi : G \rightarrow S_{G/\langle t \rangle} \cong S_3$ is an injective homomorphism. Since both G and S_3 have order 6, φ must be a bijection, and so $G \cong S_3$.

It is clear that $\mathbb{Z}_6$ and S_3 are not isomorphic, for one is abelian and the other is not. •

One consequence of this result is another proof that $\mathbb{Z}_6 \cong \mathbb{Z}_2 \times \mathbb{Z}_3$ (see Theorem 2.62). We also have a new proof of Exercise 2.97 as well as a new way of establishing the existence of an isomorphism in Example 2.24: each of these involves identifying a nonabelian group of order 6.

Example 2.35.
The group of 6 permutations of the extended reals $\hat{\mathbb{R}}$ that is described in Example 2.24 is isomorphic to S_3, for it is a nonabelian group of order 6. One can also see that it acts as a group of permutations on the set $\{0, 1, \infty\}$. ◄

Classifying groups of order 8 is more difficult, for we have not yet developed enough theory (see Theorem 5.49). It turns out that there are 5 nonisomorphic groups of order 8: three are abelian: $\mathbb{Z}_8$; $\mathbb{Z}_4 \times \mathbb{Z}_2$; $\mathbb{Z}_2 \times \mathbb{Z}_2 \times \mathbb{Z}_2$; two are nonabelian: D_8; $\mathbf{Q}$.

One can continue this discussion for larger orders, but things soon get out of hand, as Table 2.4 shows. Making a telephone directory of groups is not the way to study them.

Order of Group	Number of Groups
16	14
32	51
64	267
128	2, 328
256	56, 092
512	10, 494, 213

Table 2.4.

Groups arose by abstracting the fundamental properties enjoyed by permutations. But there is an important feature of permutations that the axioms do not mention: permutations are functions. We shall see that there are interesting consequences when this feature is restored.

Definition. If X is a set and G is a group, then G **acts** on X if, for each $g \in G$, there is a function $\alpha_g : X \to X$, such that

(i) for $g, h \in G$, $\alpha_g \circ \alpha_h = \alpha_{gh}$;

(ii) $\alpha_1 = 1_X$, the identity function.

If G acts on X, we shall usually write gx instead of $\alpha_g(x)$.

Of course, if G is a subgroup of S_X, then G acts on X. As we saw in the proofs of Theorems 2.66 and 2.67 and Proposition 2.38, when G acts on a set X, each function α_g is a bijection; indeed, its inverse is $\alpha_{g^{-1}}$ because $\alpha_g \alpha_{g^{-1}} = \alpha_{gg^{-1}} = \alpha_1 = 1_X$; similarly, $\alpha_{g^{-1}} \alpha_g = 1_X$. Cayley's theorem says that a group G acts on itself by (left) translation, and its generalization,

Theorem 2.67, shows that G also acts on the family of cosets of a subgroup H by (left) translation.

Example 2.36.
We show that G *acts on itself by conjugation*: that is, for each $g \in G$, define $\alpha_g : G \to G$ by

$$\alpha_g(x) = gxg^{-1}.$$

To verify axiom (i), note that for each $x \in G$,

$$
\begin{aligned}
(\alpha_g \circ \alpha_h)(x) &= \alpha_g(\alpha_h(x)) \\
&= \alpha_g(hxh^{-1}) \\
&= g(hxh^{-1})g^{-1} \\
&= (gh)x(gh)^{-1} \\
&= \alpha_{gh}(x).
\end{aligned}
$$

Therefore, $\alpha_g \circ \alpha_h = \alpha_{gh}$.

To prove axiom (ii), note that for each $x \in G$,

$$\alpha_1(x) = 1x1^{-1} = x,$$

and so $\alpha_1 = 1_G$. ◄

The following two definitions are fundamental.

Definition.   If G acts on X and $x \in X$, then the *orbit* of x, denoted by $\mathcal{O}(x)$, is the subset of X

$$\mathcal{O}(x) = \{gx : g \in G\} \subset X;$$

the *stabilizer* of x, denoted by G_x, is

$$G_x = \{g \in G : gx = x\} \leq G.$$

It is easy to check that the stabilizer G_x of a point x is a subgroup of G. Let us find orbits and stabilizers in the examples above.

Example 2.37.
(i) Cayley's theorem says that G acts on itself by translations: $\tau_a : x \mapsto ax$. If $x \in G$, then the orbit $\mathcal{O}(x) = G$, for if $g \in G$, then $g = (gx^{-1})x$. The stabilizer G_x of x is $\{1\}$, for if $x = \tau_a(x) = ax$, then $a = 1$. One says that G acts *transitively* on X when there is some $x \in X$ with $\mathcal{O}(x) = X$.

(ii) When G acts on G/H (the family of cosets of a subgroup H) by translations $\tau_a : xH \mapsto axH$, then the orbit $\mathcal{O}(xH) = G/H$, for if $g \in G$ and
$a = gx^{-1}$, then $\tau_a : xH \mapsto gH$. Thus, G acts transitively on G/H. The
stabilizer G_{xH} of xH is xHx^{-1}, for $axH = xH$ if and only if $x^{-1}ax \in H$ if
and only if $a \in xHx^{-1}$.   ◀

Example 2.38.
Let $X = $ the vertices $\{v_1, v_2, v_3, v_4\}$ of a square, and let G be the dihedral
group D_8 acting on X, as in Figure 2.13 (for clarity, the vertices in the figure
are labeled 1, 2, 3, 4 instead of v_1, v_2, v_3, v_4).

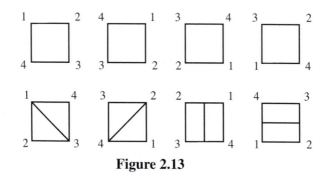

Figure 2.13

$$G = \{\text{rotations} : (1), (v_1\ v_2\ v_3\ v_4), (v_1\ v_3)(v_2\ v_4), (v_1\ v_4\ v_3\ v_2);$$
$$\text{reflections} : (v_2\ v_4), (v_1\ v_3), (v_1\ v_2)(v_3\ v_4), (v_1\ v_4)(v_2\ v_3)\}.$$

For each vertex $v_i \in X$, there is some $g \in G$ with $gv_1 = v_i$; therefore,
$\mathcal{O}(v_1) = X$ and D_8 acts transitively.

What is the stabilizer G_{v_1} of v_1? Aside from the identity, there is only one
$g \in D_8$ fixing v_1, namely, $g = (v_2\ v_4)$; therefore G_{v_1} is a subgroup of order
2. (This example can be generalized to the dihedral group D_{2n} acting on a
regular n–gon.)   ◀

Example 2.39.
When a group G acts on itself by conjugation, then an orbit

$$\mathcal{O}(x) = \{y \in G : y = axa^{-1} \text{ for some } a \in G\};$$

in this case, $\mathcal{O}(x)$ is called the **conjugacy class** of x, commonly denoted by
x^G. For example, Proposition 2.12 shows that if $\alpha \in S_n$, then the conjugacy

class of α consists of all the permutations in S_n having the same cycle structure as α. As a second example, an element z lies in the center $Z(G)$ if and
only if $z^G = \{z\}$; that is, no other elements in G are conjugate to z.

If $x \in G$, then the stabilizer G_x of x is

$$C_G(x) = \{g \in G : gxg^{-1} = x\}.$$

This subgroup of G, consisting of all $g \in G$ that commute with x, is called
the **centralizer** of x in G. ◄

Example 2.40.
Let $X = \{1, 2, \ldots, n\}$, let $\alpha \in S_n$, and regard the cyclic group $G = \langle \alpha \rangle$ as
acting on X. If $i \in X$, then

$$\mathcal{O}(i) = \{\alpha^k(i) : k \in \mathbb{Z}\}.$$

Let the complete factorization of α be $\alpha = \beta_1 \cdots \beta_{t(\alpha)}$, and let $i = i_1$ be
moved by α. If the cycle involving i_1 is $\beta_j = (i_1 \ i_2 \ \ldots \ i_r)$, then the proof of
Theorem 2.10 shows that $i_{k+1} = \alpha^k(i_1)$ for all $k < r$. Therefore,

$$\mathcal{O}(i) = \{i_1, i_2, \ldots, i_r\},$$

where $i = i_1$. It follows that $|\mathcal{O}(i)| = r$. The stabilizer G_ℓ of a symbol ℓ is
G if α fixes ℓ, and it is $\{1\}$ if α moves ℓ. ◄

Proposition 2.70. *If G acts on a set X, then X is the disjoint union of the
orbits. If X is finite, then*

$$|X| = \sum_i |\mathcal{O}(x_i)|,$$

where one x_i is chosen from each orbit.

Proof. If $x \in X$, then $x = 1x \in \mathcal{O}(x)$, and so $X = \bigcup_{x \in G} \mathcal{O}(x)$. If $z \in \mathcal{O}(x) \cap \mathcal{O}(y)$, then there are $g, h \in G$ with $gx = z = hy$; hence, $x = g^{-1}hy$
and $y = h^{-1}gx$. We claim that $\mathcal{O}(x) = \mathcal{O}(y)$. If $u \in \mathcal{O}(x)$, then $u = g'x$
for some $g' \in G$, and so $u = g'g^{-1}hy \in \mathcal{O}(y)$; thus, $\mathcal{O}(x) \subset \mathcal{O}(y)$. For the
reverse inclusion, if $v \in \mathcal{O}(y)$, then $v = h'y = h'h^{-1}gx \in \mathcal{O}(x)$. Therefore,
$\mathcal{O}(x) = \mathcal{O}(y)$, and so distinct orbits are disjoint.

The count given in the second statement is correct: since the orbits are
disjoint, no element in X is counted twice. •

Here is the connection between orbits and stabilizers.

Theorem 2.71.   *If G acts on a set X and $x \in X$, then*

$$|\mathcal{O}(x)| = [G : G_x]$$

the index of the stabilizer G_x in G.

Proof.   Let G/G_x denote the family of all the cosets of G_x in G. We will exhibit a bijection $\varphi : \mathcal{O}(x) \to G/G_x$, and this will give the result, since $|G/G_x| = [G : G_x]$, by Lagrange's theorem. If $y \in \mathcal{O}(x)$, then $y = gx$ for some $g \in G$; define $\varphi(y) = gG_x$. Now φ is single-valued: if $y = hx$ for some $h \in G$, then $h^{-1}gx = x$ and $h^{-1}g \in G_x$; hence $hG_x = gG_x$. To see that φ is injective, suppose that $\varphi(y) = \varphi(z)$; then there are $g, h \in G$ with $y = gx$, $z = hx$, and $gG_x = hG_x$; that is, $h^{-1}g \in G_x$. It follows that $h^{-1}gx = x$, and so $y = gx = hx = z$. Finally, φ is a surjection: if $gG_x \in G/G_x$, then let $y = gx \in \mathcal{O}(x)$, and note that $\varphi(y) = gG_x$.   •

In Example 2.38, D_8 acting on the four corners of a square, we saw that $|\mathcal{O}(v_1)| = 4$, $|G_{v_1}| = 2$, and $[G : G_{v_1}] = 8/2 = 4$. In Example 2.40, $G = \langle \alpha \rangle \le S_n$ acting on $X = \{1, 2, \ldots, n\}$, we saw that if, in the complete factorization of α into disjoint cycles $\alpha = \beta_1 \cdots \beta_{t(\alpha)}$, the r-cycle β_j moves ℓ, then $r = |\mathcal{O}(\ell)|$ for any ℓ occurring in β_j. Theorem 2.71 says that r is a divisor of the order k of α. (But Theorem 2.25 tells us more: k is the lcm of the lengths of the cycles occurring in the factorization.)

Corollary 2.72.   *If a finite group G acts on a set X, then the number of elements in any orbit is a divisor of $|G|$.*

Proof.   This follows at once from Lagrange's theorem.   •

In Example 2.8, there is a table displaying the number of permutations in S_4 of each cycle structure; these numbers are 1, 6, 8, 6, 3. Note that each of these numbers is a divisor of $|S_4| = 24$, In Example 2.9, we saw that the corresponding numbers are 1, 10, 20, 30, 24, 20, and 15, and these are all divisors of $|S_5| = 120$. We now recognize these subsets as being conjugacy classes, and the next corollary explains why these numbers divide the group order.

Corollary 2.73.   *If x lies in a finite group G, then the number of conjugates of x is the index of its centralizer:*

$$|x^G| = [G : C_G(x)],$$

and hence it is a divisor of $|G|$.

Proof.   As in Example 2.39, the orbit of x is its conjugacy class x^G, and the stabilizer G_x is the centralizer $C_G(x)$.   •

When we began classifying groups of order 6, it would have been helpful to be able to assert that any such group has an element of order 3 (we were able to use an earlier exercise to assert the existence of an element of order 2). We now prove that every finite group G contains an element of prime order p, where $p \mid |G|$.

If the conjugacy class x^G of an element x in a group G consists of x alone, then x commutes with every $g \in G$, for $gxg^{-1} = x$; that is, $x \in Z(G)$. Conversely, if $x \in Z(G)$, then $x^G = \{x\}$. Thus, the center $Z(G)$ consists of all those elements in G whose conjugacy class has exactly one element.

Theorem 2.74 (*Cauchy*).   *If G is a finite group whose order is divisible by a prime p, then G contains an element of order p.*

Proof.   We prove the theorem by induction on $|G|$; the base step $|G| = 1$ is vacuously true, for there are no prime divisors of 1. If $x \in G$, then the number of conjugates of x is $|x^G| = [G : C_G(x)]$, where $C_G(x)$ is the centralizer of x in G. As noted above, if $x \notin Z(G)$, then x^G has more than one element, and so $|C_G(x)| < |G|$. If $p \mid |C_G(x)|$ for some noncentral x, then the inductive hypothesis says there is an element of order p in $C_G(x) \leq G$, and we are done. Therefore, we may assume that $p \nmid |C_G(x)|$ for all noncentral $x \in G$. Better, since $|G| = [G : C_G(x)]|C_G(x)|$, Euclid's lemma gives

$$p \mid [G : C_G(x)].$$

After recalling that $Z(G)$ consists of all those elements $x \in G$ with $|x^G| = 1$, we may use Proposition 2.70 to see

$$|G| = |Z(G)| + \sum_i [G : C_G(x_i)],$$

where one x_i is selected from each conjugacy class having more than one element. Since $|G|$ and all $[G : C_G(x_i)]$ are divisible by p, it follows that $|Z(G)|$ is divisible by p. But $Z(G)$ is abelian, and so Proposition 2.59 says that $Z(G)$, and hence G, contains an element of order p.   •

Definition.   The *class equation* of a finite group G is

$$|G| = |Z(G)| + \sum_i [G : C_G(x_i)],$$

where one x_i is selected from each conjugacy class having more than one element.

Definition.   If p is a prime, then a **p-group** is a group of order p^n for some $n \geq 1$.

We have seen examples of groups whose center is trivial; for example, $Z(S_3) = 1$. For p-groups, however, this is never true.

Theorem 2.75.   *If p is a prime and G is a p-group, then $Z(G) \neq \{1\}$.*

Proof.   Consider the class equation

$$|G| = |Z(G)| + \sum_i [G : C_G(x_i)].$$

Each $C_G(x_i)$ is a proper subgroup of G, for $x_i \notin Z(G)$. Since G is a p-group, $[G : C_G(x_i)]$ is a divisor of $|G|$, hence is itself a power of p. Thus, p divides each of the terms in the class equation other than $|Z(G)|$, and so $p \mid |Z(G)|$ as well. Therefore, $Z(G) \neq \{1\}$.   •

Corollary 2.76.   *If p is a prime, then every group G of order p^2 is abelian.*

Proof.   If G is not abelian, then its center $Z(G)$ is a proper subgroup, so that $|Z(G)| = 1$ or p, by Lagrange's theorem. But Theorem 2.75 says that $Z(G) \neq \{1\}$, and so $|Z(G)| = p$. The center is always a normal subgroup, so that the quotient $G/Z(G)$ is defined; it has order p, and hence $G/Z(G)$ is cyclic. This contradicts Exercise 2.83.   •

Example 2.41.
Who would have guessed that Cauchy's theorem and Fermat's theorem are special cases of some common theorem? The elementary yet ingenious proof of this is due to J. H. McKay (as A. Mann has shown me). If G is a finite group and p is a prime, denote the cartesian product of p copies of G by G^p, and define

$$X = \{(a_1, a_2, \ldots, a_p) \in G^p : a_1 a_2 \ldots a_p = 1\}.$$

Note that $|X| = |G|^{p-1}$, for having chosen the first $p-1$ entries arbitrarily, the pth entry must equal $(a_1 a_2 \cdots a_{p-1})^{-1}$. Now make X into a $\mathbb{Z}_p$-set by defining, for $0 \leq i \leq p - 1$,

$$[i](a_1, a_2, \ldots, a_p) = (a_{i+1}, a_{i+2}, \ldots, a_p, a_1, a_2, \ldots, a_i).$$

The product of the entries in the new p-tuple is a conjugate of $a_1 a_2 \cdots a_p$:

$$a_{i+1} a_{i+2} \cdots a_p a_1 a_2 \cdots a_i = (a_1 a_2 \cdots a_i)^{-1} (a_1 a_2 \cdots a_p)(a_1 a_2 \cdots a_i).$$

This conjugate is 1 (for $g^{-1} 1 g = 1$), and so $[i](a_1, a_2, \ldots, a_p) \in X$. By Corollary 2.72, the size of every orbit of X is a divisor of $|\mathbb{Z}_p| = p$; since p is prime, these sizes are either 1 or p. Now orbits with just one element consist of a p-tuple all of whose entries a_i are equal, for all cyclic permutations of the p-tuple are the same. In other words, such an orbit corresponds to an element $a \in G$ with $a^p = 1$. Clearly, $(1, 1, \ldots, 1)$ is such an orbit; if it were the only such, then we would have

$$|G|^{p-1} = |X| = 1 + kp$$

for some $k \geq 0$; that is, $|G|^{p-1} \equiv 1 \bmod p$. If p is a divisor of $|G|$, then we have a contradiction, for $|G|^{p-1} \equiv 0 \bmod p$. We have thus proved Cauchy's theorem: If a prime p is a divisor of $|G|$, then G has an element of order p.

Suppose now that G is a group of order n and that p is not a divisor of n. By Lagrange's theorem, G has no elements of order p, so that if $a^p = 1$, then $a = 1$. Therefore, the only orbit in G^p of size 1 is $(1, 1, \ldots, 1)$, and so

$$n^{p-1} = |G|^{p-1} = |X| = 1 + kp;$$

that is, if p is not a divisor of n, then $n^{p-1} \equiv 1 \bmod p$. Multiplying both sides by n, we have $n^p \equiv n \bmod p$, a congruence also holding when p is a divisor of n; this is Fermat's theorem. ◀

We have seen, in Proposition 2.40, that A_4 is a group of order 12 having no subgroup of order 6. Thus, the assertion that if d is a divisor of $|G|$, then G must have a subgroup of order d, is false. However, this assertion is true when G is a p-group.

Proposition 2.77. *If G is a group of order $|G| = p^e$, then G has a subgroup of order p^k for every $k \leq e$.*

Proof. We prove the result by induction on $e \geq 0$. The base step is obviously true, and so we proceed to the inductive step. By Theorem 2.75, the center of G is nontrivial: $Z(G) \neq \{1\}$. If $Z(G) = G$, then G is abelian, and we have already proved the result in Proposition 2.59. Therefore, we may assume that $Z(G)$ is a proper subgroup of G. Since $Z(G) \lhd G$, we have $G/Z(G)$ a p-group of smaller order than $|G|$. Assume that $|Z(G)| = p^z$. If

$k \leq z$, then $Z(G)$ and, hence G, contains a subgroup of order p^k, because $Z(G)$ is abelian. If $k > z$, then $G/Z(G)$ contains a subgroup S^* of order p^{k-z}, by induction. The correspondence theorem gives a subgroup S of G with

$$Z(G) \leq S \leq G$$

such that $S/Z(G) \cong S^*$. By Lagrange's theorem,

$$|S| = |S^*||Z(G)| = p^k. \quad \bullet$$

Abelian groups (and the quaternions) have the property that every subgroup is normal. At the opposite pole are groups having no normal subgroups other than the two obvious ones: $\{1\}$ and G.

Definition. A group $G \neq \{1\}$ is called *simple* if G has no normal subgroups other than $\{1\}$ and G itself.

Proposition 2.78. *An abelian group G is simple if and only if it is finite and of prime order.*

Proof. If G is finite of prime order p, then G has no subgroups H other than $\{1\}$ and G, otherwise Lagrange's theorem would show that $|H|$ is a divisor of p. Therefore, G is simple.

Conversely, assume that G is simple. Since G is abelian, every subgroup is normal, and so G has no subgroups other than $\{1\}$ and G. Choose $x \in G$ with $x \neq 1$. Since $\langle x \rangle$ is a subgroup, we have $\langle x \rangle = G$. If x has infinite order, then all the powers of x are distinct, and so $\langle x^2 \rangle < \langle x \rangle$ is a forbidden subgroup of $\langle x \rangle$, a contradiction. Therefore, every $x \in G$ has finite order. If x has (finite) order m and if m is composite, say $m = k\ell$, then $\langle x^k \rangle$ is a proper nontrivial subgroup of $\langle x \rangle$, a contradiction. Therefore, $G = \langle x \rangle$ has prime order. $\bullet$

We are now going to show that A_5 is a nonabelian simple group (indeed, it is the smallest such; there is no nonabelian simple group of order less than 60).

Suppose that an element $x \in G$ has k conjugates; that is

$$|x^G| = |\{gxg^{-1} : g \in G\}| = k.$$

If there is a subgroup $H \leq G$ with $x \in H \leq G$, how many conjugates does x have in H? Since

$$x^H = \{hxh^{-1} : h \in H\} \subset \{gxg^{-1} : g \in G\} = x^G,$$

we have $|x^H| \leq |x^G|$. It is possible that there is strict inequality $|x^H| < |x^G|$. For example, take $G = S_3$, $x = (1\ 2)$, and $H = \langle x \rangle$. We know that $|x^G| = 3$ (because all transpositions are conjugate), whereas $|x^H| = 1$ (because H is abelian).

Now let us consider this question, in particular, for $G = S_5$, $x = (1\ 2\ 3)$, and $H = A_5$.

Lemma 2.79.   *All 3-cycles are conjugate in A_5.*

Proof.   Let $G = S_5$, $\alpha = (1\ 2\ 3)$, and $H = A_5$. We know that $|\alpha^{S_5}| = 20$, for there are 20 3-cycles in S_5, as we saw in Example 2.9. Therefore, $20 = |S_5|/|C_{S_5}(\alpha)| = 120/|C_{S_5}(\alpha)|$, by Corollary 2.73, so that $|C_{S_5}(\alpha)| = 6$; that is, there are exactly six permutations in S_5 that commute with α. Here they are:

$$(1),\ (1\ 2\ 3),\ (1\ 3\ 2),\ (4\ 5),\ (4\ 5)(1\ 2\ 3),\ (4\ 5)(1\ 3\ 2).$$

The last three of these are odd permutations, so that $|C_{A_5}(\alpha)| = 3$. We conclude that

$$|\alpha^{A_5}| = |A_5|/|C_{A_5}(\alpha)| = 60/3 = 20;$$

that is, all 3-cycles are conjugate to $\alpha = (1\ 2\ 3)$ in A_5.   •

This lemma can be generalized from A_5 to all A_n for $n \geq 5$; see Exercise 2.101.

Lemma 2.80.   *Every element in A_5 is a 3-cycle or a product of 3-cycles.*

Proof.   If $\alpha \in A_5$, then α is a product of an even number of transpositions: $\alpha = \tau_1 \tau_2 \cdots \tau_{2n-1} \tau_{2n}$. As the transpositions may be grouped in pairs $\tau_{2i} \tau_{2i+1}$, it suffices to consider products $\tau \tau'$, where τ and τ' are transpositions. If τ and τ' are not disjoint, then $\tau = (i\ j)$, $\tau' = (i\ k)$, and $\tau \tau' = (i\ k\ j)$; if τ and τ' are disjoint, then $\tau \tau' = (i\ j)(k\ \ell) = (i\ j)(j\ k)(j\ k)(k\ \ell) = (i\ j\ k)(j\ k\ \ell)$.   •

Theorem 2.81.   *A_5 is a simple group.*

Proof.   We shall show that if H is a normal subgroup of A_5 and $H \neq \{(1)\}$, then $H = A_5$. Now if H contains a 3-cycle, then normality forces H to contain all its conjugates. By Lemma 2.79, H contains every 3-cycle, and by Lemma 2.80, $H = A_5$. Therefore, it suffices to prove that H contains a 3-cycle.

As $H \neq \{(1)\}$, it contains some $\sigma \neq (1)$. We may assume, after a harmless relabeling, that either $\sigma = (1\ 2\ 3)$, $\sigma = (1\ 2)(3\ 4)$, or $\sigma = (1\ 2\ 3\ 4\ 5)$. As we have just remarked, we are done if σ is a 3-cycle.

If $\sigma = (1\ 2)(3\ 4)$, define $\tau = (1\ 2)(3\ 5)$. Now H contains $(\tau \sigma \tau^{-1})\sigma^{-1}$, because it is a normal subgroup, and $\tau \sigma \tau^{-1} \sigma^{-1} = (3\ 5\ 4)$, as the reader should check. If $\sigma = (1\ 2\ 3\ 4\ 5)$, define $\rho = (1\ 3\ 2)$; now H contains $\rho \sigma \rho^{-1} \sigma^{-1} = (1\ 3\ 4)$, as the reader should also check.

We have shown, in all cases, that H contains a 3-cycle. Therefore, the only normal subgroups in A_5 are $\{(1)\}$ and A_5 itself, and so A_5 is simple.   •

As we shall see in Chapter 4, Theorem 2.81 turns out to be the basic reason why the quadratic formula has no generalization giving the roots of polynomials of degree 5 or higher.

Without much more effort, we can prove that the alternating groups A_n are simple for all $n \geq 5$. Observe that A_4 is not simple, for the four-group **V** is a normal subgroup of A_4.

Lemma 2.82.   *A_6 is a simple group.*

Proof.   Let $H \neq \{(1)\}$ be a normal subgroup of A_6; we must show that $H = A_6$. Assume that there is some $\alpha \in H$ with $\alpha \neq (1)$ which fixes some i, where $1 \leq i \leq 6$. Define

$$F = \{\sigma \in A_6 : \sigma(i) = i\}.$$

Note that $\alpha \in H \cap F$, so that $H \cap F \neq \{(1)\}$. The second isomorphism theorem gives $H \cap F \lhd F$. But F is simple, for $F \cong A_5$, and so the only normal subgroups in F are $\{(1)\}$ and F. Since $H \cap F \neq \{(1)\}$, we have $H \cap F = F$; that is, $F \leq H$. It follows that H contains a 3-cycle, and so $H = A_6$, by Exercise 2.101.

We may now assume that there is no $\alpha \in H$ with $\alpha \neq (1)$ which fixes some i with $1 \leq i \leq 6$. If one considers the cycle structures of permutations in A_6, however, any such α must have cycle structure $(1\ 2)(3\ 4\ 5\ 6)$ or $(1\ 2)(3\ 4)(5\ 6)$. In the first case, $\alpha^2 \in H$ is a nontrivial permutation which fixes 1 (and also 2), a contradiction. In the second case, H contains $\alpha(\beta\alpha^{-1}\beta^{-1})$, where $\beta = (2\ 3\ 4)$, and it is easily checked that this is a nontrivial element in H which fixes 1, another contradiction. Therefore, no such normal subgroup H can exist, and so A_6 is a simple group.   •

Theorem 2.83. A_n is a simple group for all $n \geq 5$.

Proof.   If H is a nontrivial normal subgroup of A_n [that is, $H \neq (1)$], then we must show that $H = A_n$; by Exercise 2.101, it suffices to prove that H contains a 3-cycle. If $\beta \in H$ is nontrivial, then there exists some i that β moves; say, $\beta(i) = j \neq i$. Choose a 3-cycle α which fixes i and moves j. The permutations α and β do not commute: $\beta\alpha(i) = \beta(i) = j$, while $\alpha\beta(i) = \alpha(j) \neq j$. It follows that $\gamma = (\alpha\beta\alpha^{-1})\beta^{-1}$ is a nontrivial element of H. But $\beta\alpha^{-1}\beta^{-1}$ is a 3-cycle, by Proposition 2.12, and so $\gamma = \alpha(\beta\alpha^{-1}\beta^{-1})$ is a product of two 3-cycles. Hence, γ moves at most 6 symbols, say, $i_1, \ldots, i_6$ (if γ moves fewer than 6 symbols, just adjoin others so we have a list of 6). Define

$$F = \{\sigma \in A_n : \sigma \text{ fixes all } i \neq i_1, \ldots, i_6\}.$$

Now $F \cong A_6$ and $\gamma \in H \cap F$. Hence, $H \cap F$ is a nontrivial normal subgroup of F. But F is simple, being isomorphic to A_6, and so $H \cap F = F$; that is, $F \leq H$. Therefore, H contains a 3-cycle, and so $H = A_n$; the proof is complete.   •

EXERCISES

2.93 If a and b are elements in a group G, prove that ab and ba have the same order.

2.94 Prove that no pair of the following groups of order 8,

$$\mathbb{Z}_8; \quad \mathbb{Z}_4 \times \mathbb{Z}_2; \quad \mathbb{Z}_2 \times \mathbb{Z}_2 \times \mathbb{Z}_2; \quad D_8; \quad \mathbf{Q},$$

are isomorphic.

2.95 If p is a prime and G is a finite group in which every element has order a power of p, then G is a p-group.

2.96 Show that S_4 has a subgroup isomorphic to D_8.

2.97 Prove that $S_4/\mathbf{V} \cong S_3$.

2.98   (i) Prove that $A_4 \ncong D_{12}$.

  (ii) Prove that $D_{12} \cong S_3 \times \mathbb{Z}_2$.

2.99   (i) If H is a subgroup of G and if $x \in H$, prove that

$$C_H(x) = H \cap C_G(x).$$

  (ii) If H is a subgroup of index 2 in a finite group G and if $x \in H$, prove that $|x^H| = |x^G|$ or $|x^H| = \frac{1}{2}|x^G|$, where x^H is the conjugacy class of x in H.

  (iii) Show that there are two conjugacy classes of 5-cycles in A_5, each of which has 12 elements.

  (iv) Prove that the conjugacy classes in A_5 have sizes 1, 12, 12, 15, and 20.

2.100   (i) Prove that every normal subgroup H of a group G is a union of conjugacy classes of G, one of which is $\{1\}$.

  (ii) Use part (i) and Exercise 2.99 to give a second proof of the simplicity of A_5.

2.101   (i) For all $n \geq 3$, prove that every $\alpha \in A_n$ is a product of 3- cycles.

  (ii) Prove that if a normal subgroup $H \triangleleft A_n$ contains a 3-cycle, where $n \geq 5$, then $H = A_n$. (*Remark.* We have proved this in Lemma 2.80 when $n = 5$.)

2.102 Prove that the only normal subgroups of S_4 are $\{(1)\}$, **V**, A_4, and S_4.

2.103 Prove that A_5 is a group of order 60 that has no subgroup of order 30.

2.104   (i) Prove that if a simple group G has a subgroup of index n, then G is isomorphic to a subgroup of S_n.

  (ii) Prove that an infinite simple group (such do exist) has no subgroups of finite index $n > 1$.

2.105 Let G be a group with $|G| = mp$, where p is a prime and $1 < m < p$. Prove that G is not simple.

   Remark. Of all the numbers smaller than 60, we can now show that all but 11 are not orders of nonabelian simple groups (namely, 12, 18, 24, 30, 36, 40, 45, 48, 50, 54, 56). Theorem 2.75 eliminates all prime powers (for the center is always a normal subgroup), and Exercise 2.105 eliminates all numbers of the form mp, where p is a prime and $m < p$. (One can complete the proof that there are no nonabelian simple groups of order less than 60 using Sylow's theorem, which we present in Chapter 5.)

2.106 Prove that if $n \geq 3$, then A_n is the only subgroup of S_n of order $\frac{1}{2}n!$.

2.8 COUNTING WITH GROUPS

We are now going to use group theory to do some fancy counting.

Lemma 2.84.

(i) *If a group G acts on a set X and if x and y lie in the same orbit, say,
$y = \sigma x$, then $G_y = \sigma G_x \sigma^{-1}$.*

(ii) *If a finite group G acts on a finite set X and if x and y lie in the same
orbit, then $|G_y| = |G_x|$.*

Proof.   (i) If $\tau \in G_x$, then $\tau x = x$. Since $y = \sigma x$, we have

$$\sigma \tau \sigma^{-1} y = \sigma \tau \sigma^{-1} \sigma x = \sigma \tau x = \sigma x = y.$$

Therefore, $\sigma \tau \sigma^{-1}$ fixes y, and so $\sigma G_x \sigma^{-1} \leq G_y$. The reverse inclusion is
proved in the same way, for $x = \sigma^{-1} y$.

(ii) In light of part (i), there is an isomorphism $\gamma_\sigma : G \to G$, namely, conju-
gation by σ, with $\gamma_\sigma(G_x) = G_y$. It follows that $|G_x| = |G_y|$.   •

The next theorem is usually called Burnside's lemma, although Burnside
himself attributed it to Frobenius. To avoid the confusion that would be
caused by changing a popular name, we accept a suggestion of P. M. Neu-
mann and call it "not-Burnside's lemma" (Burnside was a fine mathematician, and there do exist theorems properly attributed to him. For example,
Burnside proved that if p and q are primes, then there are no simple groups
of order $p^m q^n$.)

Theorem 2.85 (not-Burnside's Lemma).   *Let G act on a finite set X. If N
is the number of orbits, then*

$$N = \frac{1}{|G|} \sum_{\tau \in G} F(\tau),$$

where $F(\tau)$ is the number of $x \in X$ fixed by τ.

Proof.   List the elements of X as follows: choose $x_1 \in X$, and then list all
the elements $x_1, x_2, \ldots, x_r$ in the orbit $\mathcal{O}(x_1)$; then choose $x_{r+1} \notin \mathcal{O}(x_1)$, and
list the elements $x_{r+1}, x_{r+2}, \ldots$ in $\mathcal{O}(x_{r+1})$; continue this procedure until all

the elements of X are listed. Now list the elements $\tau_1, \tau_2, \ldots, \tau_n$ of G, and form the following array, where

$$f_{i,j} = \begin{cases} 1 & \text{if } \tau_i \text{ fixes } x_j \\ 0 & \text{if } \tau_i \text{ moves } x_j. \end{cases}$$

	x_1	x_2	$\cdots$	x_{r+1}	x_{r+2}	$\cdots$
τ_1	$f_{1,1}$	$f_{1,2}$	$\cdots$	$f_{1,r+1}$	$f_{1,r+2}$	$\cdots$
τ_2	$f_{2,1}$	$f_{2,2}$	$\cdots$	$f_{2,r+1}$	$f_{2,r+2}$	$\cdots$
τ_i	$f_{i,1}$	$f_{i,2}$	$\cdots$	$f_{i,r+1}$	$f_{i,r+2}$	$\cdots$
τ_n	$f_{n,1}$	$f_{n,2}$	$\cdots$	$f_{n,r+1}$	$f_{n,r+2}$	$\cdots$

Now $F(\tau_i)$, the number of x fixed by τ_i, is the number of 1's in the ith row of the array; therefore, $\sum_{\tau \in G} F(\tau)$ is the total number of 1's in the array. Let us now look at the columns. The number of 1's in the first column is the number of τ_i that fix x_1; by definition, these τ_i comprise G_{x_1}. Thus, the number of 1's in column 1 is $|G_{x_1}|$. Similarly, the number of 1's in column 2 is $|G_{x_2}|$. By Lemma 2.84(ii), $|G_{x_1}| = |G_{x_2}|$. By Theorem 2.71, the number of 1's in the r columns labeled by the $x_i \in \mathcal{O}(x_1)$ is thus

$$r|G_{x_1}| = |x_1^G| \cdot |G_{x_1}| = \left(|G|/|G_{x_1}|\right)|G_{x_1}| = |G|.$$

The same is true for any other orbit: its columns contain exactly $|G|$ 1's. Therefore, if there are N orbits, there are $N|G|$ 1's in the array. We conclude that

$$\sum_{\tau \in G} F(\tau) = N|G|. \quad \bullet$$

We are going to use not-Burnside's lemma to solve problems of the following sort. How many striped flags are there having six stripes (of equal width) each of which can be colored red, white, or blue? Clearly, the two flags in Figure 2.14 on page 197 are the same: the bottom flag is just the top one turned over.

r	w	b	r	w	b

b	w	r	b	w	r

Figure 2.14

Let X be the set of all 6-tuples of colors; if $x \in X$, then

$$x = (c_1, c_2, c_3, c_4, c_5, c_6),$$

where each c_i denotes either red, white, or blue. Let τ be the permutation that reverses all the indices:

$$\tau = \begin{pmatrix} 1 & 2 & 3 & 4 & 5 & 6 \\ 6 & 5 & 4 & 3 & 2 & 1 \end{pmatrix} = (1\ 6)(2\ 5)(3\ 4)$$

(thus, τ "turns over" each 6-tuple x of colored stripes). The cyclic group $G = \langle \tau \rangle$ acts on X; since $|G| = 2$, the orbit of any 6-tuple x consists of either 1 or 2 elements: either τ fixes x or it does not. Since a flag is unchanged by turning it over, it is reasonable to identify a flag with an orbit of a 6-tuple. For example, the orbit consisting of the 6-tuples

$$(r, w, b, r, w, b) \quad \text{and} \quad (b, w, r, b, w, r)$$

describes the flag in Figure 2.14. The number of flags is thus the number N of orbits; by not-Burnside's lemma, $N = \frac{1}{2}[F((1)) + F(\tau)]$. The identity permutation (1) fixes every $x \in X$, and so $F((1)) = 3^6$ (there are 3 colors). Now τ fixes a 6-tuple x if it is a "palindrome," that is, if the colors in x read the same forward as backward. For example,

$$x = (r, r, w, w, r, r)$$

is fixed by τ. Conversely, if

$$x = (c_1, c_2, c_3, c_4, c_5, c_6)$$

is fixed by $\tau = (1\ 6)(2\ 5)(3\ 4)$, then $c_1 = c_6$, $c_2 = c_5$, and $c_3 = c_4$; that is, x is a palindrome. It follows that $F(\tau) = 3^3$, for there are 3 choices for each of c_1, c_2, and c_3. The number of flags is thus

$$N = \frac{1}{2}(3^6 + 3^3) = 378.$$

Let us make the notion of coloring more precise.

Definition.   If a group G acts on $X = \{1, \ldots, n\}$, and if C is a set of q *colors*, then G acts on the set C^n of all n-tuples of colors by

$$\tau(c_1, \ldots, c_n) = (c_{\tau 1}, \ldots, c_{\tau n}) \quad \text{for all } \tau \in G.$$

An orbit of $(c_1, \ldots, c_n) \in C^n$ is called a **(q, G)-coloring** of X.

Example 2.42.
Color each square in a 4×4 grid red or black (adjacent squares may have the same color; indeed, one possibility is that all the squares have the same color).

1	2	3	4
5	6	7	8
9	10	11	12
13	14	15	16

13	9	5	1
14	10	6	2
15	11	7	3
16	12	8	4

Figure 2.15

If X consists of the 16 squares in the grid and if C consists of the two colors red and black, then the cyclic group $G = \langle R \rangle$ of order 4 acts on X, where R is clockwise rotation by $90°$; Figure 2.15 shows how R acts: the right square is R's action on the left square. In cycle notation,

$$R = (1,\ 4,\ 16,\ 13)(2,\ 8,\ 15,\ 9)(3,\ 12,\ 14,\ 5)(6,\ 7,\ 11,\ 10),$$
$$R^2 = (1,\ 16)(4,\ 13)(2,\ 15)(8,\ 9)(3,\ 14)(12,\ 5)(6,\ 11)(7,\ 10),$$
$$R^3 = (1,\ 13,\ 16,\ 4)(2,\ 9,\ 15,\ 8)(3,\ 5,\ 14,\ 12)(6,\ 10,\ 11,\ 7).$$

A red-and-black chessboard does not change when it is rotated; it is merely viewed from a different position. Thus, we may regard a chessboard as a 2-coloring of X; the orbit of a 16-tuple corresponds to the four ways of viewing the board.

By not-Burnside's lemma, the number of chessboards is

$$\tfrac{1}{4}\Big[F((1)) + F(R) + F(R^2) + F(R^3)\Big].$$

Now $F((1)) = 2^{16}$, for every 16-tuple is fixed by the identity. To compute $F(R)$, note that squares 1, 4, 16, 13 must all have the same color in a 16-tuple fixed by R. Similarly, squares 2, 8, 15, 9 must have the same color, squares 3, 12, 14, 5 must have the same color, and squares 6, 7, 11, 10 must have the same color. We conclude that $F(R) = 2^4$; note that the exponent 4 is the number of cycles in the complete factorization of R. A similar analysis shows that $F(R^2) = 2^8$, for the complete factorization of R^2 has 8 cycles, and $F(R^3) = 2^4$, because the cycle structure of R^3 is the same as that of R. Therefore, the number N of chessboards is

$$N = \tfrac{1}{4}\left[2^{16} + 2^4 + 2^8 + 2^4\right] = 16{,}456. \quad \blacktriangleleft$$

We now show, as in Example 2.42, that the cycle structure of a permutation τ allows one to calculate $F(\tau)$.

Lemma 2.86.   *Let C be a set of q colors, and let G be a subgroup of S_n. If $\tau \in G$, then*

$$F(\tau) = q^{t(\tau)},$$

where $t(\tau)$ is the number of cycles in the complete factorization of τ.

Proof.   Since $\tau(c_1, \ldots, c_n) = (c_{\tau 1}, \ldots, c_{\tau n}) = (c_1, \ldots, c_n)$, we see that $c_{\tau i} = c_i$ for all i, and so τi has the same color as i. It follows, for all k, that $\tau^k i$ has the same color as i, that is, all points in the orbit of i acted on by $\langle \tau \rangle$ have the same color. If the complete factorization of τ is $\tau = \beta_1 \cdots \beta_{t(\tau)}$, and if i occurs in β_j, then Example 2.40 shows that the orbit containing i is the set of symbols occurring in β_j. Thus, for an n-tuple to be fixed by τ, all the symbols involved in each of the $t(\tau)$ cycles must have the same color; as there are q colors, there are thus $q^{t(\tau)}$ n-tuples fixed by τ.   •

Corollary 2.87.   *Let G act on a finite set X. If N is the number of orbits, then*

$$N = \frac{1}{|G|} \sum_{\tau \in G} q^{t(\tau)},$$

where $t(\tau)$ is the number of cycles in the complete factorization of τ.

We introduce a polynomial in several variables to allow us to state a more delicate counting result due to Pólya.

Definition. If the complete factorization of $\tau \in S_n$ has $e_r(\tau) \geq 0$ r-cycles, then the *index* of τ is the monomial

$$\text{ind}(\tau) = x_1^{e_1(\tau)} x_2^{e_2(\tau)} \cdots x_n^{e_n(\tau)}.$$

If G is a subgroup of S_n, then the *cycle index* of G is the polynomial in n variables with coefficients in $\mathbb{Q}$:

$$P_G(x_1, \ldots, x_n) = \frac{1}{|G|} \sum_{\tau \in G} \text{ind}(\tau).$$

In our earlier discussion of the striped flags, the group G was a cyclic group of order 2 with generator $\tau = (1\ 6)(2\ 5)(3\ 4)$. Thus, $\text{ind}((1)) = x_1^6$, $\text{ind}(\tau) = x_2^3$, and

$$P_G(x_1, \ldots, x_6) = \tfrac{1}{2}(x_1^6 + x_2^3).$$

As a second example, consider all possible blue-and-white flags having 9 stripes. Here $|X| = 9$ and $G = \langle \tau \rangle \leq S_9$, where

$$\tau = (1\ 9)(2\ 8)(3\ 7)(4\ 6)(5).$$

Now, $\text{ind}((1)) = x_1^9$, $\text{ind}(\tau) = x_1 x_2^4$, and the cycle index of $G = \langle \tau \rangle$ is thus

$$P_G(x_1, \ldots, x_9) = \tfrac{1}{2}(x_1^9 + x_1 x_2^4).$$

In Example 2.42, we saw that the cyclic group $G = \langle R \rangle$ of order 4 acts on a grid with 16 squares, and:

$$\text{ind}((1)) = x_1^{16}; \quad \text{ind}(R) = x_4^4; \quad \text{ind}(R^2) = x_2^8; \quad \text{ind}(R^3) = x_4^4.$$

The cycle index is thus

$$P_G(x_1, \ldots, x_{16}) = \tfrac{1}{4}(x_1^{16} + x_2^8 + 2x_4^4).$$

Proposition 2.88. *If $|X| = n$ and G is a subgroup of S_n, then the number of (q, G)-colorings of X is $P_G(q, \ldots, q)$.*

Proof. By Corollary 2.87, the number of (q, G)-colorings of X is

$$\frac{1}{|G|} \sum_{\tau \in G} q^{t(\tau)},$$

where $t(\tau)$ is the number of cycles in the complete factorization of τ. On the other hand,

$$P_G(x_1, \ldots, x_n) = \frac{1}{|G|} \sum_{\tau \in G} \operatorname{ind}(\tau)$$

$$= \frac{1}{|G|} \sum_{\tau \in G} x_1^{e_1(\tau)} x_2^{e_2(\tau)} \cdots x_n^{e_n(\tau)},$$

and so

$$P_G(q, \ldots, q) = \frac{1}{|G|} \sum_{\tau \in G} q^{e_1(\tau) + e_2(\tau) + \cdots + e_n(\tau)}$$

$$= \frac{1}{|G|} \sum_{\tau \in G} q^{t(\tau)}. \quad \bullet$$

Let us count again the number of red-and-black chessboards with sixteen squares in Example 2.42. Here,

$$P_G(x_1, \ldots, x_{16}) = \tfrac{1}{4}(x_1^{16} + x_2^8 + 2x_4^4).$$

and so the number of chessboards is

$$P_G(2, \ldots, 2) = \tfrac{1}{4}(2^{16} + 2^8 + 2 \cdot 2^4).$$

Doing this count without group theory is more difficult because of the danger of counting the same chessboard more than once. The reason we have introduced the cycle index is that it allows us to state Pólya's generalization of not-Burnside's lemma which solves the following sort of problem. How many blue-and-white flags with 9 stripes have 4 blue stripes and 5 white stripes? More generally, we want to count the number of orbits in which we prescribe the number of "stripes" of any given color.

Theorem 2.89 (Pólya). *Let $G \leq S_X$, where $|X| = n$, let $|\mathcal{C}| = q$, and, for each $i \geq 1$, define $\sigma_i = c_1^i + \cdots + c_q^i$. Then the number of (q, G)-colorings of X having f_r elements of color c_r, for every r, is the coefficient of $c_1^{f_1} c_2^{f_2} \cdots c_q^{f_q}$ in $P_G(\sigma_1, \ldots, \sigma_n)$.*

Proofs of Pólya's theorem can be found in combinatorics books (for example, see Biggs, *Discrete Mathematics*). To solve the flag problem posed

above, first note that the cycle index for blue-and-white flags having 9 stripes
is

$$P_G(x_1, \ldots, x_9) = \tfrac{1}{2}(x_1^9 + x_1 x_2^4).$$

and so the number of flags is $P_G(2, \ldots, 2) = \tfrac{1}{2}(2^9 + 2^5) = 272$. Using
Pólya's theorem, the number of flags with 4 blue stripes and 5 white ones is
the coefficient of $b^4 w^5$ in

$$P_G(\sigma_1, \ldots, \sigma_9) = \tfrac{1}{2}\left[(b+w)^9 + (b+w)(b^2+w^2)^4\right].$$

A short exercise with the binomial theorem shows that the coefficient of $b^4 w^5$
is 66.

EXERCISES

2.107 How many flags are there with n stripes each of which can be colored any one
of q given colors?

2.108 Let X be the squares in an $n \times n$ grid, and let ρ be a rotation by $90°$. Define a
chessboard to be a (q, G)-coloring, where the cyclic group $G = \langle \rho \rangle$ of order
4 is acting. Show that the number of chessboards is

$$\tfrac{1}{4}\left(q^{n^2} + q^{\lfloor (n^2+1)/2 \rfloor} + 2q^{\lfloor (n^2+3)/4 \rfloor}\right),$$

where $\lfloor x \rfloor$ is the greatest integer in the number x.

2.109 Let X be a disk divided into n congruent circular sectors, and let ρ be a rotation
by $(360/n)°$. Define a ***roulette wheel*** to be a (q, G)-coloring, where the cyclic
group $G = \langle \rho \rangle$ of order n is acting. Prove that if $n = 6$, then there are
$\tfrac{1}{6}(2q + 2q^2 + q^3 + q^6)$ roulette wheels having 6 sectors.
        [The formula for the number of roulette wheels with n sectors is

$$\tfrac{1}{n}\sum_{d|n}\phi(n/d)q^d,$$

where ϕ is the Euler ϕ-function.]

2.110 Let X be the vertices of a regular n-gon, and let the dihedral group $G = D_{2n}$
act (as the usual group of symmetries [see Example 2.18]). Define a ***bracelet***
to be a (q, G)-coloring of a regular n-gon, and call each of its vertices a ***bead***.
(Not only can one rotate a bracelet; one can also flip it.)

   (i) How many bracelets are there having 5 beads, each of which can be
       colored any one of q available colors?

   (ii) How many bracelets are there having 6 beads, each of which can be
       colored any one of q available colors?

3

Commutative Rings I

3.1 FIRST PROPERTIES

In high school algebra, one is usually presented with a list of "rules" for ordinary addition and multiplication of real numbers and polynomials; these lists[1] are often quite long, having perhaps 20 or more items. For example, one rule is the *additive cancellation law*:

$$\text{if } a + c = b + c, \text{ then } a = b.$$

Some rules, as this one, follow from $\mathbb{R}$ being an additive group, but there are also rules involving both addition and multiplication. One such is the *distributive law*:

$$(a + b)c = ac + bc;$$

when read from left to right, it says that c can be "multiplied through" $a + b$; when read from right to left, it says that c can be "factored out" of $ac + bc$. There is also the "mysterious" rule:

$$(-1) \times (-1) = 1, \tag{M}$$

which has no group analog because it involves multiplying additive inverses. Lists of rules can be shrunk by deleting redundant items, but there is a good reason for so shrinking them aside from the obvious economy provided by

---

[1]For example, see H. S. Hall and S. R. Knight, *Algebra for Colleges and* ͡ lan, 1923, or J. C. Stone and V. S. Mallory, *A Second Course in Algebra*, San

shorter list: a short list makes it easier to see analogies between numbers and other realms in which one can both add and multiply. Before exploring such other realms, let us dispel the mystery of (M).

Lemma 3.1. $0 \cdot a = 0$ *for every number a.*

Proof.   Since $0 = 0 + 0$, the distributive law gives

$$0 \cdot a = (0 + 0) \cdot a = 0 \cdot a + 0 \cdot a.$$

Since adding 0 does not change a number, we have

$$0 \cdot a = 0 \cdot a + 0.$$

Therefore,

$$0 \cdot a + 0 \cdot a = 0 \cdot a + 0,$$

and subtracting $0 \cdot a$ from both sides gives $0 \cdot a = 0$.   •

Incidentally, we can now see why dividing by 0 is forbidden. By definition, dividing by a number b is the opposite of multiplying by b [see Example 2.6]. Thus, $b(1/b)x = x$ for all x; in particular, for $x = 1$, we have $b(1/b) = 1$. If we could set $b = 0$, then $1/0$ would be a number satisfying $0 \cdot (1/0) = 1$. But Lemma 3.1 gives $0 \cdot (1/0) = 0$, contradicting $1 \neq 0$.

Lemma 3.2.   *If $-a$ is that number which, when added to a, gives 0, then* $(-1)(-a) = a$.

Proof.   The distributive law and Lemma 3.1 give

$$0 = 0 \cdot (-a) = (-1 + 1)(-a) = (-1)(-a) + (-a);$$

now add a to both sides to get $a = (-1)(-a)$.   •

Setting $a = 1$ gives the (no longer) mysterious (M).
While we are proving elementary properties, let us show that, fortunately, the product $(-1)a$ is the same as $-a$.

Corollary 3.3. $(-1)a = -a$ *for every number a.*

Proof.   By Lemma 3.2, $(-1)(-a) = a$. Multiplying both sides by -1 gives

$$(-1)(-1)(-a) = (-1)a.$$

ut Lemma 3.2 gives $(-1)(-1) = 1$, so that $-a = (-1)a$.   •

Mathematical objects other than numbers can be added and multiplied. For example, derivatives obey the ***product rule***:

$$[f(x)g(x)]' = f'(x)g(x) + f(x)g'(x)$$

in which differentiable functions are multiplied and added. If $f(x)$ is a differentiable function, then so is $-f(x)$, and the constant function $c(x) \equiv 1$, which is also differentiable, behaves just like the number 1 under multiplication. Is the analog of Proposition 3.2 true; is $(-c(x))(-f(x)) = f(x)$? The answer is yes, and the proof of this fact is exactly the same as the proof given for numbers: just replace every occurrence of the letter a by $f(x)$.

We focus on certain simple properties enjoyed by ordinary addition and multiplication and elevate them to the status of axioms. In essence, we are describing more general realms in which we shall be working.

Definition.   A *commutative ring*[2] R is a set with two operations, addition and multiplication, such that:

(i) R is an abelian group under addition;

(ii) $ab = ba$ for all $a, b \in R$;

(iii) $a(bc) = (ab)c$ for every $a, b, c \in R$;

(iv) there is an element $1 \in R$ with $1 \neq 0$ and with $1a = a$ for every $a \in R$;[3]

(v) $a(b + c) = ab + ac$ for every $a, b, c \in R$.

Addition and multiplication in a commutative ring R are operations, so there are functions

$$\alpha : R \times R \to R \qquad \text{with} \qquad \alpha(r, r') = r + r' \in R$$

and

$$\mu : R \times R \to R \qquad \text{with} \qquad \mu(r, r') = rr' \in R$$

---

[2]This term was probably coined by D. Hilbert, in 1897, when he wrote *Zahlring*. One of the meanings of the word *ring*, in German as in English, is collection, as in the phrase "a ring of thieves." (It has also been suggested that Hilbert used this term because, for a ring of algebraic integers, an appropriate power of each element "cycles back" to being a linear combination of lower powers.)

[3]Some authors do not demand that commutative rings have 1. For them, the set of all even integers is a commutative ring, but we do not recognize it as such.

for all $r, r' \in R$. The law of substitution holds here, as it does for any operation: if $r = r'$ and $s = s'$, then $r + s = r' + s'$ and $rs = r's'$. For example,
the proof of Lemma 3.1 begins with $\mu(0, a) = \mu(0 + 0, a)$, and the proof of
Lemma 3.2 begins with $\alpha(0, -a) = \alpha(-1 + 1, -a)$.

Example 3.1.

(i) The reader may assume that $\mathbb{Z}$, $\mathbb{Q}$, $\mathbb{R}$, and $\mathbb{C}$ are commutative rings with
the usual addition and multiplication (the ring axioms are verified in courses
in the foundations of mathematics).

(ii) Let $\mathbb{Z}[i]$ be the set of all complex numbers of the form $a + bi$, where
$a, b \in \mathbb{Z}$ and $i^2 = -1$. It is a boring exercise to check that $\mathbb{Z}[i]$ is, in fact,
a commutative ring (this exercise will be significantly shortened once the
notion of *subring* has been introduced). $\mathbb{Z}[i]$ is called the ring of **Gaussian
integers**.

(iii) Consider the set R of all real numbers x of the form

$$x = a + b\omega,$$

where $a, b \in \mathbb{Q}$ and $\omega = \sqrt[3]{2}$. It is easy to see that R is closed under ordinary
addition. However, if R is closed under multiplication, then $\omega^2 \in R$, and
there are rationals a and b with

$$\omega^2 = a + b\omega.$$

Multiplying both sides by ω gives the equations:

$$\begin{aligned}
2 &= a\omega + b\omega^2 \\
&= a\omega + b(a + b\omega) \\
&= a\omega + ab + b^2\omega \\
&= ab + (a + b^2)\omega.
\end{aligned}$$

If $a + b^2 = 0$, then $a = -b^2$, and the last equation gives $2 = ab$; hence,
$2 = (-b^2)b = -b^3$. But this says that the cube root of 2 is rational, contradicting Exercise 1.45(ii). Therefore, $a + b^2 \neq 0$ and $\omega = (2 - ab)/(a + b^2)$.
Since a and b are rational, we have ω rational, again contradicting Exercise 1.45(ii). Therefore, R is not closed under multiplication, and so R is not
a commutative ring. ◄

Remark.   There are noncommutative rings; that is, there are sets with addition and multiplication satisfying all the axioms of a commutative ring except the axiom: $ab = ba$. [Actually, the definition replaces the axiom $1a = a$ by $1a = a = a1$, and it replaces the distributive law by two distributive laws, one on either side: $a(b + c) = ab + ac$ and $(b + c)a = ba + ca$.] For example, let M denote the set of all 2×2 matrices. Example 2.15(i) defines multiplication of matrices; we now define addition by

$$\begin{bmatrix} a & b \\ c & d \end{bmatrix} + \begin{bmatrix} a' & b' \\ c' & d' \end{bmatrix} = \begin{bmatrix} a + a' & b + b' \\ c + c' & d + d' \end{bmatrix}.$$

It is easy to see that M, equipped with this addition and multiplication, satisfies all the new ring axioms except the commutativity of multiplication.

Even though there are interesting examples of noncommutative rings, we shall consider only commutative rings in this chapter.   ◄

Proposition 3.4.   *Lemma 3.1, Lemma 3.2, and Corollary 3.3 hold for every commutative ring.*

Proof.   Each of these results has been proved using only the defining axioms of a commutative ring. To illustrate, we now prove Corollary 3.3, $(-1)a = -a$, in slow motion, assuming the reader has already proved that Lemmas 3.1 and 3.2 hold for commutative rings.

By Lemma 3.2, $(-1)(-a) = a$. Multiply both sides by -1. The law of substitution gives

$$(-1)\big[(-1)(-a)\big] = (-1)a.$$

By associativity of multiplication, $(-1)\big[(-1)(-a)\big] = \big[(-1)(-1)\big](-a) = 1(-a)$, by Lemma 3.2. But $1(-a) = -a$, so that $-a = (-1)a$.   •

What have we shown? Formulas such as $(-1)(-a) = a$ hold, not because of the nature of the numbers a and 1, but as consequences of the properties of addition and multiplication stated in the definition of a commutative ring. For example, these results also hold for differentiable functions (once we see that differentiable functions form a commutative ring). Thus, a theorem about commutative rings applies not only to numbers but to other realms as well, thereby giving many theorems all at once instead of one at a time. The abstract approach allows us to be more efficient; the same result need not be proved over and over again.

There is another advantage of abstraction. As we have just seen, the fact that $-1 \times -1 = +1$ has nothing to do with the "nature" of the numbers 1 and -1; it is a consequence of the properties of addition and multiplication. The things one adds and multiplies may be very complicated, but many of their properties may be consequences of the rules of manipulating them and not of their intrinsic structure. For example, we have just proved that if $f(x)$ is a differentiable function, then $(-c(x))(-f(x)) = f(x)$, and nowhere in the proof did we have to use the definition of differentiability. Thus, as we have seen when we studied groups, the abstract approach allows us to focus on the essential parts of a problem; we need not be distracted by any features irrelevant to a specific problem.

Definition.   If R is a commutative ring and $a, b \in R$, then **subtraction** is defined by
$$a - b = a + (-b).$$

In light of Corollary 3.3,

$$a - b = a + (-1)b.$$

Here is an ultrafussy proof that a distributive law $ca - cb = c(a - b)$ always holds (we shall not be so fussy again!):

$$\begin{aligned}
a(b - c) = a[b + (-1)c] \;\; &= ab + a[(-1)c] \\
&= ab + [a(-1)]c = ab + [(-1)a]c \\
&= ab + (-1)(ac) = ab - ac.
\end{aligned}$$

Definition.   A **domain** (often called an *integral domain*) is a commutative ring R that satisfies an extra axiom, the **cancellation law** for multiplication:

$$\text{if } ca = cb \text{ and } c \neq 0, \text{ then } a = b.$$

The familiar examples of commutative rings: $\mathbb{Z}$, $\mathbb{Q}$, $\mathbb{R}$, $\mathbb{C}$, are domains, but we shall soon exhibit honest examples of commutative rings that are not domains.

Proposition 3.5.   *A commutative ring R is a domain if and only if the product of any two nonzero elements of R is nonzero.*

Proof.   Assume that R is a domain, so that the cancellation law holds. Suppose, by way of contradiction, that there are nonzero elements $a, b \in R$ with $ab = 0$. Proposition 3.4 gives $0 \cdot b = 0$, so that $ab = 0 \cdot b$. The cancellation law now gives $a = 0$ (for $b \neq 0$), and this is a contradiction.

Conversely, assume that the product of nonzero elements in R is always nonzero. If $ca = cb$ with $c \neq 0$, then $0 = ca - cb = c(a - b)$. Since $c \neq 0$, the hypothesis that the product of nonzero elements is nonzero forces $a - b = 0$. Therefore, $a = b$, as desired. •

Definition.   A subset S of a commutative ring R is a *subring* of R if:

(i) $1 \in S$;[4]

(ii) if $a, b \in S$, then $a - b \in S$;

(iii) if $a, b \in S$, then $ab \in S$.

Just as a subgroup is a group in its own right, so is a subring of a commutative ring a commutative ring in its own right.

Proposition 3.6.   *A subring S of a commutative ring R is itself a commutative ring.*

Proof.   We show first that S is a subgroup of the additive group of R. Now $S \neq \varnothing$ because $1 \in S$; by axiom (ii) in the definition of subring, $0 = 1 - 1 \in S$. Another application of (ii) shows that if $b \in S$, then $0 - b = -b \in S$; finally, if $a, b \in S$, then Lemma 3.3 shows that S contains

$$a - (-b) = a + (-1)(-b)$$
$$= a + (-1)(-1)b$$
$$= a + b.$$

Therefore, S is an additive subgroup[5] of R, and hence it is an abelian group.

By axiom (iii), (the restriction of) multiplication is an operation on S. Associativity, commutativity, and distributivity are all inherited by S from their holding in the commutative ring R. It follows that S is a commutative ring. •

Of course, one advantage of the notion of subring is that fewer ring axioms need to be checked to determine whether a subset of a commutative ring is

---

[4]The even integers do *not* form a subring of $\mathbb{Z}$ because 1 is not even. Their special structure will be recognized when *ideals* are introduced.

[5]We have actually proved a group-theoretic result: Proposition 2.26, which states, in multiplicative notation: If S is a nonempty subset of a (not necessarily abelian) group G with the property that $a, b \in S$ implies $ab^{-1} \in S$, then S is a subgroup of G.

itself a commutative ring. For example, it is simpler to show that the ring of Gaussian integers

$$\mathbb{Z}[i] = \{z \in \mathbb{C} : z = a + ib : a, b \in \mathbb{Z}\}$$

is a subring of $\mathbb{C}$ than to verify all the axioms in the definition of a commutative ring.

Example 3.2.
If $n \geq 3$ is an integer, let $\zeta_n = e^{2\pi i/n}$ be a primitive nth root of unity, and define

$$\mathbb{Z}[\zeta_n] = \{z \in \mathbb{C} : z = a_0 + a_1\zeta_n + a_2\zeta_n^2 + \cdots + a_{n-1}\zeta_n^{n-1}, \text{ all } a_i \in \mathbb{Z}\}.$$

(When $n = 4$, then $\mathbb{Z}[\zeta_4]$ is the Gaussian integers $\mathbb{Z}[i]$.) It is easy to check that $\mathbb{Z}[\zeta_n]$ is a subring of $\mathbb{C}$ (to prove that $\mathbb{Z}[\zeta_n]$ is closed under multiplication, note that if $m \geq n$, then $m = qn + r$, where $0 \leq r < n$, and $\zeta_n^m = \zeta_n^r$). ◄

Here is an example of a commutative ring that is not a domain.

Example 3.3.
(i) Let $\mathcal{F}(\mathbb{R})$ be the set of all the functions $\mathbb{R} \to \mathbb{R}$ equipped with the operations of *pointwise addition* and *pointwise multiplication*: given $f, g \in \mathcal{F}(\mathbb{R})$, define functions $f + g$ and fg by

$$f + g : a \mapsto f(a) + g(a) \qquad \text{and} \qquad fg : a \mapsto f(a)g(a)$$

(notice that fg is *not* their composite).

Pointwise addition and pointwise multiplication are precisely those operations on functions that occur in calculus. For example, recall the formulas

$$\int f(x) + g(x)\, dx = \int f(x)\, dx + \int g(x)\, dx$$

and

$$D(fg) = D(f)g + fD(g),$$

where D denotes derivative. The sum $f + g$ in the first integrand is pointwise addition, and the product fg in the derivative formula is pointwise multiplication.

We claim that $\mathcal{F}(\mathbb{R})$ with these operations is a commutative ring. Verification of the axioms is left to the reader with the following hint: the "zero" in

$\mathcal{F}(\mathbb{R})$ is the constant function z with value 0 [that is, $z(a) = 0$ for all $a \in \mathbb{R}$] and the "one" is the constant function ε with $\varepsilon(a) = 1$ for all $a \in \mathbb{R}$.

We now show that $\mathcal{F}(\mathbb{R})$ is not a domain. Define f and g by:

$$f(a) = \begin{cases} a & \text{if } a \le 0 \\ 0 & \text{if } a \ge 0; \end{cases} \qquad g(a) = \begin{cases} 0 & \text{if } a \le 0 \\ a & \text{if } a \ge 0. \end{cases}$$

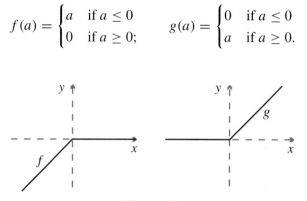

Figure 3.1

Clearly, neither f nor g is zero (i.e., $f \ne z$ and $g \ne z$). On the other hand, for each $a \in \mathbb{R}$, $fg : a \mapsto f(a)g(a) = 0$, because at least one of the factors $f(a)$ or $g(a)$ is the number zero. Therefore, $fg = z$, by Proposition 2.1, and $\mathcal{F}(\mathbb{R})$ is not a domain.

(ii) All differentiable functions $f : \mathbb{R} \to \mathbb{R}$ form a subring of $\mathcal{F}(\mathbb{R})$, and so the differentiable functions form a ring.   ◄

Proposition 3.7.

   (i) $\mathbb{Z}_m$, *the integers* mod m, *is a commutative ring.*

   (ii) *The commutative ring* $\mathbb{Z}_m$ *is a domain if and only if m is a prime.*

Proof.   (i) We have already seen, in Theorems 2.43 and 2.45, that $\mathbb{Z}_m$ has an addition and a multiplication: $[a] + [b] = [a + b]$ and $[a][b] = [ab]$. Indeed, the only ring axiom that was not checked is the distributive law, and we check it now. Since distributivity does hold in $\mathbb{Z}$, we have

$$\begin{aligned} [a]([b] + [c]) &= [a][b + c] \\ &= [a(b + c)] \\ &= [ab + ac] \\ &= [ab] + [ac] \\ &= [a][b] + [a][c]. \end{aligned}$$

Therefore, $\mathbb{Z}_m$ is a commutative ring.

(ii) If m is not a prime, then $m = ab$, where $0 < a, b < m$. Now both $[a]$ and $[b]$ are not $[0]$ in $\mathbb{Z}_m$, because m divides neither a nor b, but $[a][b] = [m] = [0]$. Thus, $\mathbb{Z}_m$ is not a domain.

Conversely, suppose that m is prime. If $[a][b] = [0]$, then $ab \equiv 0 \bmod m$, that is, $m \mid ab$. Since m is a prime, Euclid's lemma gives $m \mid a$ or $m \mid b$, that is, $a \equiv 0 \bmod m$ or $b \equiv 0 \bmod m$, that is, $[a] = [0]$ or $[b] = [0]$. Therefore, $\mathbb{Z}_m$ is a domain.   •

For example, $\mathbb{Z}_6$ is not a domain because $[2] \neq 0$ and $[3] \neq 0$, yet $[2][3] = [6] = [0]$.

Many theorems of ordinary arithmetic, that is, properties of the commutative ring $\mathbb{Z}$, hold in more generality. We now generalize some familiar definitions from $\mathbb{Z}$ to arbitrary commutative rings.

Definition.   Let a and b be elements of a commutative ring R. Then a *divides* b *in* R (or a is a *divisor* of b or b is a *multiple* of a), denoted[6] by $a \mid b$, if there exists an element $c \in R$ with $b = ca$.

As an extreme example, if $0 \mid a$, then $a = 0 \cdot b$ for some $b \in R$. Since $0 \cdot b = 0$, however, we must have $a = 0$. Thus, $0 \mid a$ if and only if $a = 0$.

Notice that whether $a \mid b$ depends not only on the elements a and b but on the ambient ring R as well. For example, 3 does divide 2 in $\mathbb{Q}$, for $2 = 3 \times \frac{2}{3}$, and $\frac{2}{3} \in \mathbb{Q}$; on the other hand, 3 does not divide 2 in $\mathbb{Z}$, because there is no *integer* c with $3c = 2$.

The reader can quickly check each of the following facts. For every $a \in R$, we have $a \mid a$, $1 \mid a$, $-a \mid a$, and $-1 \mid a$; moreover, $a \mid 0$, but $0 \mid a$ if and only if $a = 0$.

Lemma 3.8.   *Let R be a commutative ring, and let a, b, c be elements of R.*

(i) *If $a \mid b$ and $b \mid c$, then $a \mid c$.*

(ii) *If $a \mid b$ and $a \mid c$, then a divides every* **linear combination** *$sb + tc$, where $s, t \in R$.*

Proof.   Exercises for the reader.   •

---

[6]Do not confuse the notations $a \mid b$ and a/b. The first one denotes the *statement* "a is a divisor of b," whereas the second one denotes an *element* $c \in R$ with $bc = a$.

Definition.   An element u in a commutative ring R is called a *unit* if $u \mid 1$ in R, that is, if there exists $v \in R$ with $uv = 1$; the element v is called the *inverse* of u, and v is often denoted by u^{-1}.

Units are of interest because one can always divide by them: if u is a unit in R, so that there is $v \in R$ with $uv = 1$, and if $a \in R$, then

$$a = u(va)$$

is a factorization of a in R, for $va \in R$; thus, it is reasonable to define the quotient a/u as $va = u^{-1}a$. (Recall that this last equation is the reason why zero is never a unit, that is, why dividing by zero is forbidden.)

Just as divisibility depends on the ambient ring R, so does the question whether an element $u \in R$ is a unit depend on R (for it is a question whether $u \mid 1$ in R). For example, the number 2 is a unit in $\mathbb{Q}$, for $\frac{1}{2}$ lies in $\mathbb{Q}$ and $2 \times \frac{1}{2} = 1$, but 2 is not a unit in $\mathbb{Z}$, because there is no *integer* v with $2v = 1$. In fact, the only units in $\mathbb{Z}$ are 1 and -1, as the reader can check.

The following theorem generalizes Exercise 1.42.

Proposition 3.9.   *Let R be a domain, and let $a, b \in R$ be nonzero. Then $a \mid b$ and $b \mid a$ if and only if $b = ua$ for some unit $u \in R$.*

Proof.   If $a \mid b$ and $b \mid a$, there are elements $u, v \in R$ with $b = ua$ and $a = vb$. Substituting, $b = ua = uvb$. Since $b = 1b$ and $b \neq 0$, the cancellation law in the domain R gives $1 = uv$, and so u is a unit.

Conversely, assume that $b = ua$, where u is a unit in R. Plainly, $a \mid b$. If $v \in R$ satisfies $uv = 1$, then $vb = vua = a$, and so $b \mid a$.   •

There exist examples of commutative rings in which Proposition 3.9 is false, and so the hypothesis in this theorem that R be a domain is needed.

What are the units in $\mathbb{Z}_m$?

Proposition 3.10.   *If a is an integer, then $[a]$ is a unit in $\mathbb{Z}_m$ if and only if a and m are relatively prime. In fact, if $sa + tm = 1$, then $[a]^{-1} = [s]$.*

Proof.   If $[a]$ is a unit in $\mathbb{Z}_m$, Theorem 1.53 shows there is $[s] \in \mathbb{Z}_m$ with $[s][a] = [1]$. Therefore, $sa \equiv 1 \bmod m$, and so there is an integer t with $sa - 1 = tm$; hence, $1 = sa - tm$. By Exercise 1.48, a and m are relatively prime.

Conversely, if a and m are relatively prime, there are integers s and t with $1 = sa + tm$. Hence, $sa - 1 = -tm$ and so $sa \equiv 1 \bmod m$. Thus, $[s][a] = [1]$, and $[a]$ is a unit in $\mathbb{Z}_m$.   •

Corollary 3.11. *If p is a prime, then every nonzero $[a]$ in $\mathbb{Z}_p$ is a unit.*

Proof. If $[a] \neq [0]$, then $a \not\equiv 0 \bmod p$, and hence $p \nmid a$. Therefore, a and p are relatively prime because p is prime. ●

Definition. If R is a commutative ring, then the **group of units** of R is

$$U(R) = \{\text{all units in } R\}.$$

It is easy to check that $U(R)$ is a multiplicative group (we have already met $U(\mathbb{Z}_m)$ in the proof of Theorem 2.64). It follows that a unit u in R has exactly one inverse in R, for each element in a group has a unique inverse.

The introduction of the commutative ring $\mathbb{Z}_m$ makes the solution of congruence problems much more natural. A congruence $ax \equiv b \bmod m$ in $\mathbb{Z}$ becomes an equation $[a][x] = [b]$ in $\mathbb{Z}_m$. If $[a]$ is a unit in $\mathbb{Z}_m$, that is, if $(a, m) = 1$, then it has an inverse $[a]^{-1} = [s]$, and we can divide by it; the solution is $[x] = [a]^{-1}[b] = [s][b] = [sb]$. In other words, congruences are solved just as ordinary linear equations $ax = b$ are solved over $\mathbb{R}$.

EXERCISES

3.1 Suppose that R satisfies all the axioms in the definition of a commutative ring except the condition $1 \neq 0$, that is, assume that $1 = 0$. Show that R consists of the single element 0.

3.2 Prove that a commutative ring R has a unique 1.

3.3 Prove that the binomial theorem holds in any commutative ring R: If $n \geq 1$ and $a, b \in R$, then

$$(b + a)^n = \sum_{k=0}^{n} \binom{n}{k} b^{n-k} a^k.$$

Note: Let us agree to define $a^0 = 1$ for every $a \in R$, even for $a = 0$.

3.4 (i) Prove that subtraction in $\mathbb{Z}$ is not an associative operation.

(ii) Give an example of a commutative ring R in which subtraction is associative.

3.5 Assume that S is a subset of a commutative ring R such that

(i) $1 \in S$;

(ii) if $a, b \in S$, then $a + b \in S$;

(iii) if $a, b \in S$, then $ab \in S$.

Prove that S is a subring of R. (In the definition of subring, one assumes that $a - b \in S$ instead of $a + b \in S$.)

3.6 Give an example to show, in axiom (ii) of the definition of subring, that one cannot replace "$a - b \in S$" by "$a + b \in S$."

3.7 (i) If R is a domain and $a \in R$ satisfies $a^2 = a$, prove that either $a = 0$ or $a = 1$.

(ii) Show that the commutative ring $\mathcal{F}(\mathbb{R})$ in Example 3.3(i) contains elements $f \neq 0, 1$ with $f^2 = f$.

3.8 (i) If X is a set, prove that the Boolean group $\mathcal{B}(X)$ in Example 2.14 with elements the subsets of X and with addition given by

$$U + V = (U - V) \cup (V - U)$$

is a commutative ring if one defines multiplication

$$UV = U \cap V.$$

One calls $\mathcal{B}(X)$ a **Boolean ring**.

(ii) Prove that $\mathcal{B}(X)$ contains exactly one unit.

(iii) If Y is a proper subset of X (that is, $Y \subset X$ but $Y \neq X$), show that the unit in $\mathcal{B}(Y)$ is distinct from the unit in $\mathcal{B}(X)$. Conclude that $\mathcal{B}(Y)$ is *not* a subring of $\mathcal{B}(X)$.

3.9 Find all the units in the commutative ring $\mathcal{F}(\mathbb{R})$ defined in Example 3.3(i).

3.10 Generalize the construction of $\mathcal{F}(\mathbb{R})$ to arbitrary commutative rings R: let $\mathcal{F}(R)$ be the set of all functions from R to R, with pointwise addition $f + g : r \mapsto f(r)+g(r)$ and pointwise multiplication $fg : r \mapsto f(r)g(r)$ for $r \in R$.

(i) Show that $\mathcal{F}(R)$ is a commutative ring.

(ii) Show that $\mathcal{F}(R)$ is not a domain.

(iii) Show that $\mathcal{F}(\mathbb{Z}_2)$ has exactly four elements, and that $f + f = 0$ for every $f \in \mathcal{F}(\mathbb{Z}_2)$.

3.11 (i) If R is a domain and S is a subring of R, then S is a domain.

(ii) Prove that $\mathbb{C}$ is a domain, and conclude that $\mathbb{Z}$, $\mathbb{Q}$, and $\mathbb{R}$ are domains.

(iii) Prove that the ring of Gaussian integers is a domain.

3.12 Prove that the only subring of $\mathbb{Z}$ is $\mathbb{Z}$ itself.

3.13 Let a and b be relatively prime integers. Prove that if $sa+tm = 1 = s'a+t'm$, then $s \equiv s' \bmod m$. See Exercise 1.48.

3.14 (i) Is $R = \{a + b\sqrt{2} : a, b \in \mathbb{Z}\}$ a domain?

(ii) Is $R = \{\frac{1}{2}(a + b\sqrt{2}) : a, b \in \mathbb{Z}\}$ a domain?

(iii) Using the fact that $\alpha = \frac{1}{2}(1 + \sqrt{-19})$ is a root of $x^2 - x + 5$, prove that $R = \{a + b\alpha : a, b \in \mathbb{Z}\}$ is a domain.

3.15 A function $f : \mathbb{R} \to \mathbb{R}$ is *differentiable* if its derivative $(Df)(a)$ exists for all $a \in \mathbb{R}$. Show that the set $\mathcal{D}(\mathbb{R})$ of all differentiable functions is a subring of $\mathcal{F}(\mathbb{R})$.

3.16 Prove that the set of all C^∞-functions is a subring of $\mathcal{F}(\mathbb{R})$. (See Exercise 1.33.)

3.2 FIELDS

There is an obvious difference between $\mathbb{Q}$ and $\mathbb{Z}$: every nonzero element of $\mathbb{Q}$ is a unit.

Definition.   A *field*[7] F is a commutative ring in which every nonzero element a is a unit; that is, there is $a^{-1} \in F$ with $a^{-1}a = 1$.

The first examples of fields are $\mathbb{Q}$, $\mathbb{R}$, and $\mathbb{C}$.

The definition of field can be restated in terms of the group of units; a commutative ring R is a field if and only if $U(R) = R^\times$, the nonzero elements of R. To say this another way, R is a field if and only if $R^\times$ is a multiplicative group.

Proposition 3.12.   *Every field F is a domain.*

Proof.   Assume that $ab = ac$, where $a \neq 0$. Multiplying both sides by a^{-1} gives $a^{-1}ab = a^{-1}ac$, and so $b = c$.   •

Of course, the converse of this proposition is false, for $\mathbb{Z}$ is a domain that is not a field.

Proposition 3.13.   *The commutative ring $\mathbb{Z}_m$ is a field if and only if m is prime.*

Proof.   If m is prime, then Corollary 3.11 shows that $\mathbb{Z}_m$ is a field.

Conversely, if m is composite, then Proposition 3.7(ii) shows that $\mathbb{Z}_m$ is not a domain. By Proposition 3.12, $\mathbb{Z}_m$ is not a field.   •

---

[7]The derivation of the mathematical usage of the English term *field* (first used by E. H. Moore in 1893 in his article classifying the finite fields) as well as the German term *Körper* and the French term *corps* is probably similar to the derivation of the words *group* and *ring*: each word denotes a "realm" or a "collection of things." The word *domain* abbreviates the usual English translation *integral domain* of the German word *Integretätsbereich*, a collection of integers.

Later we shall see that there are finite fields other than $\mathbb{Z}_p$ for p prime.

When I was a graduate student, one of my fellow students was hired to tutor a mathematically gifted 10-year-old boy. To illustrate how gifted the boy was, the tutor described the session in which he introduced 2×2 matrices and matrix multiplication to the boy. The boy's eyes lit up when he was shown multiplication by the identity matrix, and he immediately went off in a corner by himself. In a few minutes, he told his tutor that a matrix

$$\begin{bmatrix} a & b \\ c & d \end{bmatrix}$$

has a multiplicative inverse if and only if $ad - bc \neq 0$!

In another session, the boy was shown the definition of a field. He was quite content as the familiar examples of the rationals, reals, and complex numbers were displayed. But when he was shown a field with 2 elements, he became very agitated. After carefully checking that every axiom really does hold, he exploded in a rage. I tell this story to illustrate how truly surprising and unexpected are the finite fields.

In Chapter 2 we introduced GL$(2, \mathbb{R})$, the group of nonsingular matrices. Afterward, we observed that $\mathbb{R}$ could be replaced by $\mathbb{Q}$ or by $\mathbb{C}$. We now observe that $\mathbb{R}$ can be replaced by any field k: GL$(2, k)$ is a group for every field k. In particular, GL$(2, \mathbb{Z}_p)$ is a finite group for every prime p.

It was shown in Exercise 3.11 that every subring of a domain is itself a domain. Since fields are domains, it follows that every subring of a field is a domain. The converse of this exercise is true, and it is much more interesting: Every domain is a subring of a field. In order to prove this result, we pause for an interlude of set theory.

Definition. A relation $x \equiv y$ on a set X is

> *reflexive* : if $x \equiv x$ for all $x \in X$;
> *symmetric* : if $x \equiv y$ implies $y \equiv x$ for all $x, y \in X$;
> *transitive* : if $x \equiv y$ and $y \equiv z$ imply $x \equiv z$ for all $x, y, z \in X$.

A relation that has all three properties: reflexivity, symmetry, and transitivity, is called an *equivalence relation*.

Example 3.4.

(i) Ordinary equality is an equivalence relation on any set.

(ii) If $m \geq 0$, then Proposition 1.45 says that $x \equiv y \bmod m$ is an equivalence relation on $X = \mathbb{Z}$.

(iii) We now show that if H is a subgroup of a group G, then the relation on G, defined by

$$a \equiv b \quad \text{if} \quad a^{-1}b \in H,$$

is an equivalence relation on G. If $a \in G$, then $a^{-1}a = 1 \in H$, and $a \equiv a$; hence, $\equiv$ is reflexive. If $a \equiv b$, then $a^{-1}b \in H$; since subgroups are closed under inverses, $(a^{-1}b)^{-1} = b^{-1}a \in H$ and $b \equiv a$; hence $\equiv$ is symmetric. If $a \equiv b$ and $b \equiv c$, then $a^{-1}b, b^{-1}c \in H$; since subgroups are closed under multiplication, $(a^{-1}b)(b^{-1}c) = a^{-1}c \in H$, and $a \equiv c$. Therefore, $\equiv$ is transitive, and hence it is an equivalence relation.

(iv) A group G acting on a set X gives an equivalence relation on X. Define

$$x \equiv y \text{ if there exists } g \in G \text{ with } y = gx.$$

If $x \in X$, then $1x = x$, where $1 \in G$, and so $x \equiv x$; hence, $\equiv$ is reflexive. If $x \equiv y$, so that $y = gx$, then

$$g^{-1}y = g^{-1}(gx) = (g^{-1}g)x = 1x = x,$$

so that $x = g^{-1}y$ and $y \equiv x$; hence, $\equiv$ is symmetric. If $x \equiv y$ and $y \equiv z$, there are $g, h \in G$ with $y = gx$ and $z = hy$, so that $z = hy = h(gx) = (hg)x$, and $x \equiv z$. Therefore, $\equiv$ is transitive, and hence it is an equivalence relation.

(v) Let $X = \{(a, b) \in \mathbb{Z} \times \mathbb{Z} : b \neq 0\}$, and define a relation $\equiv$ on X by cross-multiplication:

$$(a, b) \equiv (c, d) \quad \text{if} \quad ad = bc.$$

We claim that $\equiv$ is an equivalence relation. Verification of reflexivity and symmetry is easy. For transitivity, assume that $(a, b) \equiv (c, d)$ and $(c, d) \equiv (e, f)$. Now $ad = bc$ gives $adf = bcf$, and $cf = de$ gives $bcf = bde$; thus, $adf = bde$. We may cancel the nonzero integer d to get $af = be$; that is, $(a, b) \equiv (e, f)$.

(vi) If R is a commutative ring, define $r \equiv s$ to mean there is a unit $u \in R$ with $r = us$; in this case, one says that r and s are **associates**. It is easy to check that this is an equivalence relation on R. For example, if $R = \mathbb{Z}$, then m and n are associates if and only if $m = \pm n$. ◄

An equivalence relation on a set X yields a family of subsets of X.

Definition.   Let $\equiv$ be an equivalence relation on a set X. If $a \in X$, the *equivalence class* of a, denoted by $[a]$, is defined by

$$[a] = \{x \in X : x \equiv a\} \subset X.$$

We now display the equivalence classes arising from each of the equivalence relations given above.

Example 3.5.
(i) Let $\equiv$ be equality on a set X. If $a \in X$, then $[a] = \{a\}$, the subset having only one element, namely, a. After all, if $x = a$, then x and a are equal!

(ii) Consider the relation congruence mod m on $\mathbb{Z}$, and let $a \in \mathbb{Z}$. By definition, the congruence class of a is

$$\{x \in \mathbb{Z} : x = a + km \text{ where } k \in \mathbb{Z}\};$$

we have been denoting this class by $[a]$. On the other hand, the equivalence class of a is, by definition,

$$\{x \in \mathbb{Z} : x \equiv a \bmod m\}.$$

Since $x \equiv a \bmod m$ if and only if there is some $k \in \mathbb{Z}$ with $x = a + km$, these two subsets coincide; that is, the equivalence class is the congruence class $[a]$.

(iii) Let G be a group, and let $H \leq G$ be a subgroup; consider the equivalence relation on G given by $a \equiv b$ if $a^{-1}b \in H$. We claim that the equivalence class of $a \in H$ is the coset aH. If $x \equiv a$, then $x^{-1}a \in H$, and Lemma 2.31 gives $x \in xH = aH$. Thus, $[a] \subset aH$. For the reverse inclusion, it is easy to see that if $x = ah \in aH$, then $x^{-1}a = (ah)^{-1}a = h^{-1}a^{-1}a = h \in H$, so that $x \equiv a$ and $x \in [a]$. Hence, $aH \subset [a]$, and so $[a] = aH$.

(iv) If a group G acts on a set X, then the equivalence class of $a \in X$ is its orbit, for

$$[a] = \{x \in X : x \equiv a\} = \{ga : g \in G\} = \mathcal{O}(a).$$

In particular, if G acts on itself by conjugation (see Example 2.36), then the equivalence class of $x \in G$ is its conjugacy class x^{G}.

(v) The equivalence class of (a, b) under cross-multiplication, where $a, b \in \mathbb{Z}$ and $b \neq 0$, is

$$[(a, b)] = \{(c, d) : ad = bc\}.$$

If we denote $[(a, b)]$ by a/b, then this equivalence class is precisely the fraction having the same notation. After all, it is plain that $(1, 2) \neq (2, 4)$, but $[(1, 2)] = [(2, 4)]$; that is, $1/2 = 2/4$.

(vi) There is no special name for the equivalence classes of the relation on a ring R that elements are associates; however, Exercise 3.27 observes that the equivalence class of 1 is $U(R)$.   ◄

Here is a set-theoretic idea that we shall show is intimately involved with equivalence relations.

Definition.   A family of nonempty subsets A_i of a set X is called *pairwise disjoint* if $A_i \cap A_j = \varnothing$ for all $i \neq j$.
    A *partition* of a set X is a family of nonempty pairwise disjoint subsets whose union is all of X.

Notice that if X is a finite set and $A_1, A_2, \ldots, A_n$ is a partition of X, then

$$|X| = |A_1| + |A_2| + \cdots + |A_n|.$$

The next result generalizes Proposition 2.41.

Lemma 3.14.   *If $\equiv$ is an equivalence relation on a set X, then $x \equiv y$ if and only if $[x] = [y]$.*

Proof.   Assume that $x \equiv y$. If $z \in [x]$, then $z \equiv x$, and so transitivity gives $z \equiv y$; hence $[x] \subset [y]$. By symmetry, $y \equiv x$, and this gives the reverse inclusion $[y] \subset [x]$. Thus, $[x] = [y]$.
    Conversely, if $[x] = [y]$, then $x \in [x]$, by reflexivity, and so $x \in [x] = [y]$. Therefore, $x \equiv y$.   •

In words, this lemma says that one can replace equivalence by honest equality at the cost of replacing elements by their equivalence classes.
    There is only one theorem in this generality: equivalence relations and partitions are different views of the same thing. Proposition 2.70 is a special case of the next theorem.

Proposition 3.15.   *If $\equiv$ is an equivalence relation on a set X, then the equivalence classes form a partition of X. Conversely, given a partition $\{A_i : i \in I\}$ of X, there is an equivalence relation on X whose equivalence classes are the A_i.*

Proof.   Assume that an equivalence relation $\equiv$ on X is given. Each $x \in X$ lies in the equivalence class $[x]$ because $\equiv$ is reflexive; it follows that the equivalence classes are nonempty subsets whose union is X. To prove pairwise disjointness, assume that $a \in [x] \cap [y]$, so that $a \equiv x$ and $a \equiv y$. By symmetry, $x \equiv a$, and so transitivity gives $x \equiv y$. Therefore, $[x] = [y]$, by the lemma, and the equivalence classes form a partition of X.

Conversely, let $\{A_i : i \in I\}$ be a partition of X. If $x, y \in X$, define $x \equiv y$ if there is $i \in I$ with both $x \in A_i$ and $y \in A_i$. It is plain that $\equiv$ is reflexive and symmetric. To see that $\equiv$ is transitive, assume that $x \equiv y$ and $y \equiv z$; that is, there are $i, j \in I$ with $x, y \in A_i$ and $y, z \in A_j$. Since $y \in A_i \cap A_j$, pairwise disjointness gives $A_i = A_j$, so that $i = j$ and $x, z \in A_i$; that is, $x \equiv z$. We have shown that $\equiv$ is an equivalence relation.

It remains to show that the equivalence classes are the A_i's. If $x \in X$, then $x \in A_i$, for some i. By definition of $\equiv$, if $y \in A_i$, then $y \equiv x$ and $y \in [x]$; hence, $A_i \subset [x]$. For the reverse inclusion, let $z \in [x]$, so that $z \equiv x$. There is some j with $x \in A_j$ and $z \in A_j$; thus, $x \in A_i \cap A_j$. By pairwise disjointness, $i = j$, so that $z \in A_i$, and $[x] \subset A_i$. Hence, $[x] = A_i$.   •

Example 3.6.
(i) If $\equiv$ is the identity relation on a set X, then X is the union of its one-point subsets.

(ii) If H is a subgroup of a group G, then G is the disjoint union of the cosets of H (we have seen, in Example 3.5(iii), that the cosets are the equivalence classes of an equivalence relation).

The proof of Lagrange's theorem uses the fact that the cosets of a subgroup H of a group G partition G [together with the fact, not necessarily true for other partitions, that all the parts (cosets) have the same size].

(iii) Example 3.6(ii). If G is a group acting on a set X, then X is the disjoint union of the orbits. In particular, regarding a group G as acting on itself by conjugation, one sees that G is the disjoint union of its conjugacy classes (see Proposition 2.70).   ◄

Let us return to domains and fields. Given four elements a, b, c, and d in a field F with $b \neq 0$ and $d \neq 0$, assume that $ab^{-1} = cd^{-1}$. Multiply both sides by bd to obtain $ad = bc$. In other words, were ab^{-1} written as a/b, then we have just shown that $a/b = c/d$ implies $ad = bc$; that is, "cross-multiplication" is valid. Conversely, if $ad = bc$ and both b and d are nonzero, then multiplication by $b^{-1}d^{-1}$ gives $ab^{-1} = cd^{-1}$, that is, $a/b = c/d$.

The proof of the next theorem is a straightforward generalization of the usual construction of the field of rational numbers $\mathbb{Q}$ from the domain of integers $\mathbb{Z}$.

Theorem 3.16. *If R is a domain, then there is a field F containing R as a subring. Moreover, F can be chosen so that, for each $f \in F$, there are a, $b \in R$ with $b \neq 0$ and $f = ab^{-1}$.*

Proof.   We proceed as in Examples 3.4(v) and 3.5(v). Let $X = \{(a, b) \in R \times R : b \neq 0\}$, and define a relation $\equiv$ on X by $(a, b) \equiv (c, d)$ if $ad = bc$.

We claim that $\equiv$ is an equivalence relation. Verification of reflexivity and symmetry is easy. For transitivity, assume that $(a, b) \equiv (c, d)$ and $(c, d) \equiv (e, f)$. Now $ad = bc$ gives $adf = bcf$, and $cf = de$ gives $bcf = bde$; thus, $adf = bde$. Since R is a domain, we may cancel the nonzero element d to get $af = be$; that is, $(a, b) \equiv (e, f)$.

Denote the equivalence class of (a, b) by $[a, b]$. Define F as the set of all $[a, b]$ with $(a, b) \in X$, and equip F with the following addition and multiplication (if we pretend that $[a, b]$ is the fraction a/b, then these are just the usual formulas):

$$[a, b] + [c, d] = [ad + bc, bd]$$

and

$$[a, b][c, d] = [ac, bd].$$

Notice that the symbols on the right make sense, for $b \neq 0$ and $d \neq 0$ imply $bd \neq 0$ because R is a domain. The proof that F is a field is now a series of routine steps.

Addition $F \times F \to F$ is single-valued: if $[a, b] = [a', b']$ and $[c, d] = [c', d']$, then $[ad + bc, bd] = [a'd' + b'c', b'd']$. We are told that $ab' = a'b$ and $cd' = c'd$. Hence,

$$(ad + bc)b'd' = adb'd' + bcb'd' = (ab')dd' + bb'(cd')$$
$$= a'bdd' + bb'c'd = (a'd' + b'c')bd;$$

that is, $(ad + bc, bd) \equiv (a'd' + b'c', b'd')$, as desired. A similar computation shows that multiplication $F \times F \to F$ is single-valued.

The verification that F is a commutative ring is also routine, and it is left to the reader, with the remark that the zero element is $[0, 1]$, the one is $[1, 1]$, and the negative of $[a, b]$ is $[-a, b]$. If we identify $a \in R$ with $[a, 1] \in F$, then it is easy to see that the family R' of all such elements is a subring of F:

$$[1, 1] \in R';$$
$$[a, 1] - [c, 1] = [a, 1] + [-c, 1] = [a - c, 1] \in R';$$
$$[a, 1][c, 1] = [ac, 1] \in R'.$$

To see that F is a field, observe first that if $[a, b] \neq 0$, then $a \neq 0$ (for the zero element of F is $[0, 1] = [0, b]$). The inverse of $[a, b]$ is $[b, a]$, for $[a, b][b, a] = [ab, ab] = [1, 1]$.

Finally, if $b \neq 0$, then $[1, b] = [b, 1]^{-1}$ (as we have just seen). Therefore, if $[a, b] \in F$, then $[a, b] = [a, 1][1, b] = [a, 1][b, 1]^{-1}$. This completes the proof, for $[a, 1]$ and $[b, 1]$ are in R'.   •

Definition.   The field F just constructed from R in Theorem 3.16 is called the *fraction field* of R; we denote it by $\text{Frac}(R)$, and we denote the element $[a, b] \in \text{Frac}(R)$ by a/b; in particular, the elements $[a, 1]$ of R' are denoted by $a/1$ or, more simply, by a.

Notice that the fraction field of $\mathbb{Z}$ is $\mathbb{Q}$; that is, $\text{Frac}(\mathbb{Z}) = \mathbb{Q}$.

Remark.   In calculus, equivalence relations are implicit in the discussion of vectors. An *arrow* from a point P to a point Q can be denoted by the ordered pair (P, Q); call P its *foot* and Q its *head*. An equivalence relation on arrows can be defined by saying that $(P, Q) \equiv (P', Q')$ if these arrows have the same length and the same direction. More precisely, if the quadrilateral obtained by joining P to P' and Q to Q' is a parallelogram [this definition is incomplete, for one must also relate collinear arrows as well as "degenerate" arrows (P, P)]. Define a *vector* as an equivalence class of arrows $[(P, Q)]$, and denote it by $\overrightarrow{PQ}$.

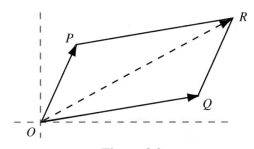

Figure 3.2

It is instructive to compare rational numbers and vectors, for both are defined as equivalence classes. Every rational a/b has a "favorite" name: its expression in lowest terms; every vector has a favorite name: an arrow (O, Q) with its foot at the origin. Working with fractions in lowest terms is not always convenient; for example, even if both a/b and c/d are in lowest terms, their sum $(ad+bc)/bd$ may not be in lowest terms. Vector addition is defined by the parallelogram law: $\overrightarrow{OP} + \overrightarrow{OQ} = \overrightarrow{OR}$, where O, P, Q, and R are the vertices of a parallelogram. But $\overrightarrow{OQ} = \overrightarrow{PR}$, because $(O, Q) \equiv (P, R)$, and it is more natural to write $\overrightarrow{OP} + \overrightarrow{OQ} = \overrightarrow{OP} + \overrightarrow{PR} = \overrightarrow{OR}$. ◄

EXERCISES

3.17   (i) If R is a commutative ring, define the *circle operation* $a \circ b$ by

$$a \circ b = a + b - ab.$$

Prove that the circle operation is associative and that $0 \circ a = a$ for all $a \in R$.

(ii) Prove that a commutative ring R is a field if and only if the set

$$R^{\#} = \{r \in R : r \neq 1\}$$

is an abelian group under the circle operation.

3.18 A *subfield* of a field K is a subring k of K which is also a field. Show that a subset k of a field K is a subfield if and only if it is a subring that is closed under inverses; that is, if $a \in k$ and $a \neq 0$, then $a^{-1} \in k$.

3.19 Show that any intersection of subfields of F is itself a subfield of F. (Why is the intersection not equal to $\{0\}$?)

Definition.   If k is a field, the intersection of all the subfields of k is called the *prime field* of k.

3.20   (i) Show that every subfield of $\mathbb{C}$ contains $\mathbb{Q}$.
(ii) Show that the prime field of $\mathbb{R}$ is $\mathbb{Q}$.
(iii) Show that the prime field of $\mathbb{C}$ is $\mathbb{Q}$.

3.21 Find the inverses of the nonzero elements of $\mathbb{Z}_{11}$.

3.22 Write the four-group additively and call it $\mathbb{F}_4$:

$$\mathbb{F}_4 = \{0, \ e, \ a, \ a + e\}.$$

Define a multiplication $\mathbb{F}_4 \times \mathbb{F}_4 \to \mathbb{F}_4$ by the following table:

Prove that $\mathbb{F}_4$ is a field with exactly four elements.

$\mathbb{F}_4$	0	e	a	$a+e$
0	0	0	0	0
e	0	e	a	$a+e$
a	0	a	$a+e$	e
$a+e$	0	$a+e$	e	a

3.23 Prove that every domain R with a finite number of elements must be a field. Using Proposition 3.7(ii), this gives a new proof of sufficiency in Proposition 3.13.

3.24 Find all the units in the ring $\mathbb{Z}[i]$ of Gaussian integers.

3.25 Show that $F = \{a + b\sqrt{2} : a, b \in \mathbb{Q}\}$ is a field.

3.26   (i) Show that $F = \{a + bi : a, b \in \mathbb{Q}\}$ is a field.

    (ii) Show that F is the fraction field of the Gaussian integers.

3.27 If R is a commutative ring, define a relation $\equiv$ on R, as in Example 3.4(vi), by $a \equiv b$ if there is a unit $u \in R$ with $b = ua$. Prove that if $a \equiv b$, then $(a) = (b)$, where $(a) = \{ra : r \in R\}$. Conversely, prove that if R is a domain, then $(a) = (b)$ implies $a \equiv b$.

3.28   (i) For any field F, prove that $\Sigma(2, F) \cong \text{Aff}(1, F)$, where $\Sigma(2, F)$ denotes the stochastic group. P134

    (ii) If F is a finite field with q elements, prove that $|\Sigma(2, F)| = q(q-1)$.

    (iii) Prove that $\Sigma(2, \mathbb{Z}_3) \cong S_3$.

3.3 POLYNOMIALS

Even though the reader is familiar with polynomials, we now introduce them carefully. One modest consequence is that the mystery surrounding the "unknown" x will vanish.

The key observation is that one should pay attention to where the coefficients of polynomials live.

Definition.   If R is a commutative ring, then a **sequence** σ in R is

$$\sigma = (s_0, s_1, s_2, \ldots, s_i \ \ldots);$$

the entries $s_i \in R$, for all $i \geq 0$, are called the **coefficients**[8] of the sequence.

---

[8]Coefficient means "acting together to some single end." Here, coefficients combine to determine the sequence.

To determine when two sequences are equal, let us recognize that a sequence σ is really a function $\sigma : \mathbb{N} \to R$, where $\mathbb{N}$ is the set of natural numbers, with $\sigma(i) = s_i$ for all $i \geq 0$. Thus, if $\tau = (t_0, t_1, t_2, \ldots, t_i, \ldots)$ is a sequence, then $\sigma = \tau$ if and only if $\sigma(i) = \tau(i)$ for all $i \geq 0$; that is, $\sigma = \tau$ if and only if $s_i = t_i$ for all $i \geq 0$.

Definition. A sequence $\sigma = (s_0, s_1, \ldots, s_i, \ldots)$ in a commutative ring R is called a ***polynomial*** if there is some integer $m \geq 0$ with $s_i = 0$ for all $i > m$; that is,

$$\sigma = (s_0, s_1, \ldots, s_m, 0, 0, \ldots).$$

A polynomial has only finitely many nonzero coefficients.

The sequence $\sigma = (0, 0, 0, \ldots)$ is a polynomial, called the ***zero polynomial***; it is denoted by $\sigma = 0$.

Definition. If $\sigma = (s_0, s_1, \ldots, s_n, 0, 0, \ldots) \neq 0$ is a polynomial, then there is $s_n \neq 0$ with $s_i = 0$ for all $i > n$. One calls s_n the ***leading coefficient*** of σ, one calls n the ***degree***[9] of σ, and one denotes the degree n by $\deg(\sigma)$.

The zero polynomial 0 does not have a degree because it has no nonzero coefficients.

Notation. If R is a commutative ring, then the set of all polynomials with coefficients in R is denoted by $R[x]$.

We will soon prove that a polynomial $(s_0, s_1, \ldots, s_n, 0, 0, \ldots)$ of degree n can be written as $s_0 + s_1 x + s_2 x^2 + \cdots + s_n x^n$, but, until then, we proceed formally. Equip $R[x]$ with the following operations. Define

$$\sigma + \tau = (s_0 + t_0, s_1 + t_1, \ldots, s_i + t_i, \ldots)$$

and

$$\sigma\tau = (a_0, a_1, \ldots, a_k, \ldots),$$

where $a_k = \sum_{i+j=k} s_i t_j = \sum_{i=0}^{k} s_i t_{k-i}$; thus,

$$\sigma\tau = (s_0 t_0, s_0 t_1 + s_1 t_0, s_0 t_2 + s_1 t_1 + s_2 t_0, \ldots).$$

This formula for multiplication was forced upon us.

---

[9] The word *degree* comes from the Latin word meaning "step."

Proposition 3.17.   *If R is a commutative ring and r, s_i, $t_j \in R$ for $i \geq 0$ and $j \geq 0$, then*

$$(s_0 + s_1 r + \cdots + s_m r^m)(t_0 + t_1 r + \cdots + t_n r^n)$$
$$= a_0 + a_1 r + \cdots + a_{m+n} r^{m+n},$$

where $a_k = \sum_{i+j=k} s_i t_j$ for all $k \geq 0$.

Proof.   Write $\sum_i s_i r^i = f(r)$ and $\sum_j t_j r^j = g(r)$. Then

$$f(r)g(r) = (s_0 + s_1 r + s_2 r^2 + \cdots)g(r)$$
$$= s_0 g(r) + s_1 r g(r) + s_2 r^2 g(r) + \cdots$$
$$= s_0(t_0 + t_1 r + \cdots) + s_1 r (t_0 + t_1 r + \cdots)$$
$$\quad + s_2 r^2 (t_0 + t_1 r + \cdots) + \cdots$$
$$= s_0 t_0 + (s_1 t_0 + s_0 t_1)r + (s_2 t_0 + s_1 t_1 + s_0 t_2)r^2 + \cdots . \quad \bullet$$

Lemma 3.18.   *Let R be a commutative ring and let σ, $\tau \in R[x]$ be nonzero polynomials.*

(i) *Either $\sigma \tau = 0$ or $\deg(\sigma \tau) \leq \deg(\sigma) + \deg(\tau)$.*

(ii) *If R is a domain, then $\sigma \tau \neq 0$ and*

$$\deg(\sigma \tau) = \deg(\sigma) + \deg(\tau).$$

Proof.   (i) Let $\sigma = (s_0, s_1, \dots)$ have degree m, let $\tau = (t_0, t_1, \dots)$ have degree n, and let $\sigma \tau = (a_0, a_1, \dots)$. It suffices to prove that $a_k = 0$ for all $k > m + n$. By definition,

$$a_k = \sum_{i+j=k} s_i t_j.$$

If $i \leq m$, then $j = k - i \geq k - m > n$ (because $k > m + n$), and so $t_j = 0$ because τ has degree n; if $i > m$, then $s_i = 0$ because σ has degree m. In either case, each term $s_i t_j = 0$, and so $a_k = \sum_{i+j=k} s_i t_j = 0$.

(ii) Now let $k = m + n$. The same calculation as in part (i) shows, with the possible exception of $s_m t_n$ (the product of the leading coefficients of σ and

τ), that each term $s_i t_j$ in

$$a_{m+n} = \sum_{i+j=m+n} s_i t_j$$

$$= s_0 t_{m+n} + \cdots + s_{n-1} t_{m+1} + s_n t_m + s_{n+1} t_{m-1} + \cdots$$

is 0. If $i < m$, then $m - i > 0$, hence $j = m - i + n > n$, and so $t_j = 0$; if $i > m$, then $s_i = 0$. Hence

$$a_{m+n} = s_m t_n.$$

Since R is a domain, $s_m \neq 0$ and $t_n \neq 0$ imply $s_m t_n \neq 0$; hence, $\sigma \tau \neq 0$ and $\deg(\sigma \tau) = m + n$.   •

Proposition 3.19.   *If R is a commutative ring, then $R[x]$ is a commutative ring that contains R as a subring. Moreover, if R is a domain, then $R[x]$ is a domain.*

Proof.   Addition and multiplication are operations on $R[x]$: the sum of two polynomials σ and τ is a sequence which is also a polynomial (indeed, either $\sigma + \tau = 0$ or $\deg(\sigma + \tau) \leq \max\{\deg(\sigma), \deg(\tau)\}$), while the lemma shows that the sequence which is the product of two polynomials is a polynomial as well. Verifications of the axioms for a commutative ring are again routine, and they are left to the reader. Note that *zero* is the zero polynomial, *one* is the polynomial $(1, 0, 0, \ldots)$, and the negative of $(s_0, s_1, \ldots, s_t, \ldots)$ is $(-s_0, -s_1, \ldots, -s_i, \ldots)$. The only possible problem is proving associativity of multiplication; we give the hint that if $\rho = (r_0, r_1, \ldots, r_i, \ldots)$, then the ℓth coordinate of the polynomial $\rho(\sigma \tau)$ turns out to be $\sum_{i+j+k=\ell} r_i(s_j t_k)$, while the ℓth coordinate of the polynomial $(\rho \sigma)\tau$ turns out to be $\sum_{i+j+k=\ell} (r_i s_j) t_k$; these are equal because of associativity of the multiplication in R.

It is easy to check that $R' = \{(r, 0, 0, \ldots) : r \in R\}$ is a subring of $R[x]$, and we identify R' with R by identifying $r \in R$ with $(r, 0, 0, \ldots)$.

If R is a domain and if σ and τ are nonzero polynomials, then the lemma shows that $\sigma \tau \neq 0$. Therefore, $R[x]$ is a domain.   •

Here is the link between this discussion and the usual notation.

Definition.   The element $x \in R[x]$ is defined by

$$x = (0, 1, 0, 0, \ldots).$$

Thus, x is neither "the unknown" nor is it a variable; it is a specific element in the ring $R[x]$, namely, the polynomial $(t_0, t_1, t_2, \ldots)$ with $t_1 = 1$ and all other $t_i = 0$.

Lemma 3.20.

(i) *If* $\sigma = (s_0, s_1, \ldots, s_j, \ldots)$, *then*

$$x\sigma = (0, s_0, s_1, \ldots, s_j, \ldots);$$

that is, multiplying by x shifts each coefficient one step to the right.

(ii) *If* $n \geq 1$, *then* x^n *is the polynomial having* 0 *everywhere except for* 1 *in the nth coordinate.*

(iii) *If* $r \in R$, *then*

$$(r, 0, 0, \ldots)(s_0, s_1, \ldots, s_j, \ldots) = (rs_0, rs_1, \ldots, rs_j, \ldots).$$

Proof.   (i) Write $x = (t_0, t_1, \ldots, t_i, \ldots)$, where $t_1 = 1$ and all other $t_i = 0$, and let $x\sigma = (a_0, a_1, \ldots, a_k, \ldots)$. Now $a_0 = t_0 s_0 = 0$ because $t_0 = 0$. If $k \geq 1$, then the only nonzero term in the sum $a_k = \sum_{i+j=k} s_i t_j$ is $s_{k-1} t_1 = s_{k-1}$, because $t_1 = 1$ and $t_i = 0$ for $i \neq 1$; thus, for $k \geq 1$, the kth coordinate a_k of $x\sigma$ is s_{k-1}, and $x\sigma = (0, s_0, s_1, \ldots, s_i, \ldots)$.

(ii) An easy induction, using (i).

(iii) This follows easily from the definition of multiplication.   •

If we identify $(r, 0, 0, \ldots)$ with r, then Lemma 3.20(iii) reads

$$r(s_0, s_1, \ldots, s_i, \ldots) = (rs_0, rs_1, \ldots, rs_i, \ldots).$$

We can now recapture the usual notation.

Proposition 3.21.   *If* $\sigma = (s_0, s_1, \ldots, s_n, 0, 0, \ldots)$, *then*

$$\sigma = s_0 + s_1 x + s_2 x^2 + \cdots + s_n x^n,$$

where each element $s \in R$ *is identified with the polynomial* $(s, 0, 0, \ldots)$.

Proof.

$$
\begin{aligned}
\sigma &= (s_0, s_1, \ldots, s_n, 0, 0, \ldots) \\
&= (s_0, 0, 0, \ldots) + (0, s_1, 0, \ldots) + \cdots + (0, 0, \ldots, s_n, 0, \ldots) \\
&= s_0(1, 0, 0, \ldots) + s_1(0, 1, 0, \ldots) + \cdots + s_n(0, 0, \ldots, 1, 0, \ldots) \\
&= s_0 + s_1 x + s_2 x^2 + \cdots + s_n x^n. \quad •
\end{aligned}
$$

We shall use this familiar (and standard) notation from now on. As is customary, we shall write

$$f(x) = s_0 + s_1 x + s_2 x^2 + \cdots + s_n x^n$$

instead of $\sigma = (s_0, s_1, \ldots, s_n, 0, 0, \ldots)$.

Definition. If R is a commutative ring, then $R[x]$ is called the ***ring of polynomials over*** R.

Here is some standard vocabulary associated with polynomials. If $f(x) = s_0 + s_1 x + s_2 x^2 + \cdots + s_n x^n$, where $s_n \neq 0$, then s_0 is called its ***constant term*** and, as we have already said, s_n is called its leading coefficient. If its leading coefficient $s_n = 1$, then $f(x)$ is called ***monic***. Every polynomial other than the zero polynomial 0 (having all coefficients 0) has a degree. A ***constant polynomial*** is either the zero polynomial or a polynomial of degree 0. Polynomials of degree 1, namely, $a + bx$ with $b \neq 0$, are called ***linear***, polynomials of degree 2 are ***quadratic***,[10] degree 3's are ***cubic***, then ***quartics***, ***quintics***, etc.

Corollary 3.22. *Polynomials* $f(x) = s_0 + s_1 x + s_2 x^2 + \cdots + s_n x^n$ *and* $g(x) = t_0 + t_1 x + t_2 x^2 + \cdots + t_m x^m$ *of degrees n and m, respectively, are equal if and only if $n = m$ and $s_i = t_i$ for all i.*

Proof. We have merely restated the definition of equality of polynomials in terms of the familiar notation. •

We can now describe the usual role of x in $f(x)$ as a variable. If R is a commutative ring, each polynomial $f(x) = s_0 + s_1 x + s_2 x^2 + \cdots + s_n x^n \in R[x]$ defines a ***polynomial function*** $f : R \to R$ by evaluation: if $a \in R$, define $f(a) = s_0 + s_1 a + s_2 a^2 + \cdots + s_n a^n \in R$. The reader should realize that polynomials and polynomial functions are distinct objects. For example, if R is a finite ring, e.g., $\mathbb{Z}_m$, then there are only finitely many functions from R to itself, and so there are only finitely many polynomial functions. On the other hand, there are infinitely many polynomials: for example, all the powers $1, x, x^2, \ldots, x^n, \ldots$ are distinct, by Corollary 3.22.

---

[10]Quadratic polynomials are so called because the particular quadratic x^2 gives the area of a square (*quadratic* comes from the Latin word meaning "four," which is to remind one of the 4-sided figure); similarly, cubic polynomials are so called because x^3 gives the volume of a cube. Linear polynomials are so called because the graph of a linear polynomial in $\mathbb{R}[x]$ is a line.

Definition.   Let F be a field. The fraction field of $F[x]$, denoted by $F(x)$, is called the *field of rational functions* over F.

Proposition 3.23.   *The elements of $F(x)$ have the form $f(x)/g(x)$, where $f(x), g(x) \in F[x]$ and $g(x) \neq 0$.*

Proof.   By Theorem 3.16, every element in the fraction field $F(x)$ has the form $f(x)g(x)^{-1}$.   •

Proposition 3.24.   *If p is a prime, then the field of rational functions $\mathbb{Z}_p(x)$ is an infinite field containing $\mathbb{Z}_p$ as a subfield.*

Proof.   By Proposition 3.19, $\mathbb{Z}_p[x]$ is a domain. Its fraction field $\mathbb{Z}_p(x)$ is a field containing $\mathbb{Z}_p[x]$ as a subring, while $\mathbb{Z}_p[x]$ contains $\mathbb{Z}_p$ as a subring, by Proposition 3.19.   •

In spite of the difference between polynomials and polynomial functions (we shall see in Corollary 3.35 that these objects coincide when the coefficient ring R is an infinite field), one often calls $R[x]$ the ring of all *polynomials over R in one variable*. If we write $A = R[x]$, then the polynomial ring $A[y]$ is called the ring of all *polynomials over R in two variables x and y*, and it is denoted by $R[x, y]$. For example, the quadratic polynomial $ax^2 + bxy + cy^2 + dx + ey + f$ can be written $cy^2 + (bx + e)y + (ax^2 + dx + f)$, a polynomial in y with coefficients in $R[x]$. By induction, one can form the commutative ring $R[x_1, x_2, \dots, x_n]$ of all *polynomials in n variables* with coefficients in R. Proposition 3.19 can now be generalized, by induction on n, to say that if R is a domain, then so is $R[x_1, x_2, \dots, x_n]$. Moreover, when F is a field, we can describe $\mathrm{Frac}(F[x_1, x_2, \dots, x_n])$ as all rational functions in n variables; its elements have the form $f(x_1, x_2, \dots, x_n)/g(x_1, x_2, \dots, x_n)$, where f and g lie in $F[x_1, x_2, \dots, x_n]$.

EXERCISES

3.29 Show that if R is a commutative ring, then $R[x]$ is never a field.

3.30   (i) If R is a domain, show that a polynomial in $R[x]$ is a unit is a nonzero constant (the converse is true if R is a field).

  (ii) Show that $(2x + 1)^2 = 1$ in $\mathbb{Z}_4[x]$. Conclude that the hypothesis in part (i) that R be a domain is necessary.

3.31 Prove that $\mathbb{Z}_p$ is the prime field of $\mathbb{Z}_p(x)$.

3.32 Show that the polynomial function defined by $f(x) = x^p - x \in \mathbb{Z}_p[x]$ is identically zero.

3.33 If R is a commutative ring and $f(x) = \sum_{i=0}^{n} s_i x^i \in R[x]$ has degree $n \geq 1$, define its **derivative** $Df \in R[x]$ by

$$Df(x) = s_1 + 2s_2 x + 3s_3 x^2 + \cdots + n s_n x^{n-1};$$

if $f(x)$ is a constant polynomial, define its derivative to be the zero polynomial. Prove that the usual rules of calculus hold for this definition of derivative:

$$D(f + g) = Df + Dg;$$
$$D(rf) = r Df \quad \text{if } r \in R;$$
$$D(fg) = f Dg + (Df)g;$$
$$D(f^n) = n f^{n-1} Df \quad \text{for all } n \geq 1.$$

3.34 Assume that $x - a \mid f(x)$ in $R[x]$. Prove that $(x - a)^2 \mid f(x)$ if and only if $x - a \mid Df$ in $R[x]$.

3.35   (i) If $f(x) = ax^{2p} + bx^p + c \in \mathbb{Z}_p[x]$, prove that $Df = 0$.

   (ii) State and prove a necessary and sufficient condition that a polynomial $f(x) \in \mathbb{Z}_p[x]$ have $Df = 0$.

3.36 If R is a commutative ring, define $R[[x]]$, the ring of **formal power series over** R, as the set of all sequences in R.

   (i) Show that the formulas defining addition and multiplication on $R[x]$ make sense for $R[[x]]$, and prove that $R[[x]]$ is a commutative ring under these operations.

   (ii) Prove that $R[x]$ is a subring of $R[[x]]$.

   (iii) Prove that if R is a domain, then $R[[x]]$ is a domain.

3.37   (i) Denote a formal power series $\sigma = (s_0, s_1, s_2, \ldots, s_n, \ldots)$ by

$$\sigma = s_0 + s_1 x + s_2 x^2 + \cdots .$$

Prove that if $\sigma = 1 + x + x^2 + \cdots$, then $\sigma = 1/(1 - x)$ in $R[[x]]$.

   (ii) Prove that if k is a field, then a formal power series $\sigma \in k[[x]]$ is a unit if and only if its constant term is nonzero; that is, $\operatorname{ord}(\sigma) = 0$.

   (iii) Prove that if $\sigma \in k[[x]]$ and $\operatorname{ord}(\sigma) = n$, then

$$\sigma = x^n u,$$

where u is a unit in $k[[x]]$.

3.4 HOMOMORPHISMS

Just as one can use homomorphisms to compare groups, so one can use homomorphisms to compare commutative rings.

Definition.   If A and R are (commutative) rings, a (*ring*) *homomorphism* is a function $f : A \to R$ such that

(i) $f(1) = 1$;

(ii) $f(a + a') = f(a) + f(a')$ for all $a, a' \in A$;

(iii) $f(aa') = f(a)f(a')$ for all $a, a' \in A$.

A homomorphism that is also a bijection is called an *isomorphism*. Commutative rings A and R are called *isomorphic*, denoted by $A \cong R$, if there is an isomorphism $f : A \to R$.

Example 3.7.
(i) Let R be a domain and let $F = \text{Frac}(R)$ denote its fraction field. In Theorem 3.16 we said that R is a subring of F, but that is not the truth; R is not even a subset of F. We did find a subring R' of F, however, that has a very strong resemblance to R, namely, $R' = \{[a, 1] : a \in R\} \subset F$. The function $f : R \to R'$, given by $f(a) = [a, 1]$, is easily seen to be an isomorphism.

(ii) If R is a commutative ring, we implied that R is a subring of $R[x]$ when we "identified" an element $r \in R$ with a constant polynomial [see Lemma 3.20(iii)]. Using the sequence notation, we set $R' = \{(r, 0, 0, \dots) : r \in R\}$, which is a subring of $R[x]$. It is easy to see that the function $f : R \to R'$, defined by $f(r) = (r, 0, 0, \dots)$, is an isomorphism.   ◄

Example 3.8.
(i) Complex conjugation $z = a + ib \mapsto a - ib$ is an isomorphism $\mathbb{C} \to \mathbb{C}$ because $\overline{1} = 1, \overline{z + w} = \overline{z} + \overline{w}$, and $\overline{zw} = \overline{z}\,\overline{w}$

(ii) Here is an example of a homomorphism of rings that is not an isomorphism. Choose $m \geq 2$ and define $f : \mathbb{Z} \to \mathbb{Z}_m$ by $f(n) = [n]$. Notice that f is surjective (but not injective).

(iii) The preceding example can be generalized. If R is a commutative ring with its "one" denoted by ε, then the function $\chi : \mathbb{Z} \to R$, defined by $\chi(n) = n\varepsilon$, is a ring homomorphism.

(iv) Let R be a commutative ring, and let $a \in R$. Define the **evaluation homomorphism** $e_a : R[x] \to R$ by $e_a(f(x)) = f(a)$; that is, if $f(x) = \sum r_i x^i$, then $f(a) = \sum r_i a^i$. We let the reader check that e_a is a ring homomorphism. ◄

Certain properties of a ring homomorphism follow from its being a homomorphism between the additive groups A and R. For example: $f(0) = 0$, $f(-a) = -f(a)$, and $f(na) = nf(a)$ for all $n \in \mathbb{Z}$.

Lemma 3.25.   *If $f : A \to R$ is a ring homomorphism, then, for all $a \in A$,*

   (i) $f(a^n) = f(a)^n$ *for all $n \geq 0$;*

   (ii) *if a is a unit, then $f(a)$ is a unit and $f(a^{-1}) = f(a)^{-1}$;*

   (iii) *if a is a unit, then $f(a^{-n}) = f(a)^{-n}$ for all $n \geq 1$.*

Proof.   (i) If $n = 0$, then $f(a^0) = 1 = (f(a))^0$; this follows from our convention that $r^0 = 1$ for any ring element r together with the property $f(1) = 1$ satisfied by every ring homomorphism. The statement for positive n is proved by induction on $n \geq 1$.

(ii) Applying f to the equation $a^{-1}a = 1$ shows that $f(a)$ is a unit with inverse $f(a^{-1})$.

(iii) Recall that $a^{-n} = (a^{-1})^n$, and invoke (i) and (ii).   •

Corollary 3.26.   *If $f : A \to R$ is a ring homomorphism, then*

$$f(U(A)) \leq U(R),$$

where $U(A)$ is the group of units of A; if f is an isomorphism, then

$$U(A) \cong U(R).$$

Proof.   The first statement is just a rephrasing of part (ii) of the lemma: if a is a unit in A, then $f(a)$ is a unit in R.

   If f is an isomorphism, then its inverse $f^{-1} : R \to A$ is also a ring homomorphism, by Exercise 3.38(i); hence, if r is a unit in R, then $f^{-1}(r)$ is a unit in A. It is now easy to check that $\varphi : U(A) \to U(R)$, defined by $a \mapsto f(a)$, is a (group) isomorphism, for its inverse $\psi : U(R) \to U(A)$ is given by $r \mapsto f^{-1}(r)$.   •

   The proof of Corollary 2.64: If $(m, n) = 1$, then $\phi(mn) = \phi(m)\phi(n)$, where ϕ is Euler's function, should now seem more natural.

Definition.   If $f : A \to R$ is a ring homomorphism, then its *kernel* is

$$\ker f = \{a \in A \text{ with } f(a) = 0\},$$

and its *image* is

$$\operatorname{im} f = \{r \in R : r = f(a) \text{ for some } a \in R\}.$$

Notice that if we forget the multiplications, then the rings A and R are additive abelian groups and these definitions coincide with the group-theoretic ones.

Let k be a field, let $a \in k$ and, as in Example 3.8(iv), consider the evaluation homomorphism $e_a : k[x] \to k$ sending $f(x) \mapsto f(a)$. Now e_a is always surjective, for if $b \in k$, then $b = e_a(f)$, where $f(x) = x - a + b$. By definition, $\ker e_a$ consists of all those polynomials $g(x)$ for which $g(a) = 0$.

The kernel of a group homomorphism is not merely a subgroup; it is a *normal* subgroup; that is, it is also closed under conjugation by any element in the ambient group. Similarly, the kernel of a ring homomorphism f is almost a subring ($\ker f$ is not a subring because it never contains 1: $f(1) = 1 \neq 0$), and we shall see that it is closed under multiplication by any element in the ambient ring.

Definition.   An *ideal* in a commutative ring R is a subset I of R such that

(i) $0 \in I$;

(ii) if $a, b \in I$, then $a - b \in I$;

(iii) if $a \in I$ and $r \in R$, then $ra \in I$.

Remark.   Condition (ii) in the definition of ideal can be replaced by

(ii′). if $a, b \in I$, then $a + b \in I$.

Condition (iii), with $r = -1$, says $b \in I$ if and only if $-b \in I$. If (ii) holds, then $a - (-b) = a + b \in I$, and so (ii′) holds. Conversely, if (ii′) holds, then $a + (-b) = a - b \in I$, and so (ii) holds. ◄

There are always two ideals in a commutative ring R: the ring R itself and the subset consisting of 0 alone, which we denote by $\{0\}$. An ideal $I \neq R$ is called a *proper ideal*.

Example 3.9.
If $b_1, b_2, \ldots, b_n$ lie in R, then the set of all linear combinations

$$I = \{r_1 b_1 + r_2 b_2 + \cdots + r_n b_n : r_i \in R \text{ for all } i\}$$

is an ideal in R. One writes $I = (b_1, b_2, \ldots, b_n)$ in this case.
  In particular, if $n = 1$, then

$$I = (b) = \{rb : r \in R\}$$

is an ideal in R; (b) consists of all the multiples of b, and it is called the **principal ideal** generated by b.
  Notice that R and $\{0\}$ are always principal ideals: $R = (1)$ and $\{0\} = (0)$. In $\mathbb{Z}$, the even integers form the principal ideal (2). ◀

Proposition 3.27. *If $f : A \to R$ is a ring homomorphism, then* $\ker f$ *is a proper ideal in A and* $\operatorname{im} f$ *is a subring of R.*

Proof. Since f is a homomorphism between the additive groups, $0 \in \ker f$ and $x + y \in \ker f$ if $x, y \in \ker f$. If $x \in \ker f$ and $a \in A$, then $f(ax) = f(a)f(x) = f(a)0 = 0$, and so $ax \in \ker f$. Therefore, $\ker f$ is an ideal in A. Note that $\ker f$ is a proper ideal, for $f(1) = 1 \neq 0$, and so $1 \notin \ker f$.
  Since f is a homomorphism between the additive groups, $0 \in \operatorname{im} f$ and $r - r' \in \operatorname{im} f$ if $r, r' \in \operatorname{im} f$. If $r, r' \in \operatorname{im} f$, then $rr' = f(a)f(a') = f(aa') \in \operatorname{im} f$. Since $f(1) = 1$ (by the definition of a ring homomorphism), $\operatorname{im} f$ is a subring of R. •

Example 3.10.
(i) If an ideal I in a commutative ring R contains 1, then $I = R$, for now I contains $r = r1$ for every $r \in R$. Indeed, if I contains a unit u, then $I = R$, for then I contains $u^{-1}u = 1$.

(ii) It follows that if R is a field, then the only ideals I in R are $\{0\}$ and R itself: if $I \neq \{0\}$, it contains some nonzero element, and every nonzero element in a field is a unit.
  Conversely, assume that R is a commutative ring whose only ideals are R itself and $\{0\}$. If $a \in R$ and $a \neq 0$, then the principal ideal $(a) = R$, for $(a) \neq 0$, and so $1 \in R = (a)$. There is thus $r \in R$ with $1 = ra$; that is, a has an inverse in R, and so R is a field. ◀

Proposition 3.28.   *A ring homomorphism* $f : A \to R$ *is an injection if and only if* $\ker f = \{0\}$.

Proof.   If one forgets the multiplication, then A and R are additive groups and f is a group homomorphism. The result now follows from Proposition 2.37(iii).   •

Corollary 3.29.   *If* $f : k \to R$ *is a ring homomorphism and* k *is a field, then* f *is an injection.*

Proof.   By the theorem, it suffices to prove that $\ker f = \{0\}$. But $\ker f$ is a proper ideal in k, by Proposition 3.27, and Example 3.10(ii) shows that k has only two ideals: k and $\{0\}$. Therefore, $\ker f = \{0\}$ and f is an injection.   •

EXERCISES

3.38   (i) Let $\varphi : A \to R$ be an isomorphism, and let $\psi : R \to A$ be its inverse. Show that ψ is an isomorphism.

(ii) Show that the composite of two homomorphisms (isomorphisms) is again a homomorphism (isomorphism).

(iii) Show that $A \cong R$ defines an equivalence relation on the class of all commutative rings.

3.39 Let R be a commutative ring and let $\mathcal{F}(R)$ be the commutative ring of all functions $f : R \to R$ (see Exercise 3.10).

(i) Show that R is isomorphic to the subring of $\mathcal{F}(R)$ consisting of all the constant functions.

(ii) If $f(x) \in R[x]$, let $\varphi_f : R \to R$ be defined by $r \mapsto f(r)$ [thus, φ_f is the polynomial function associated to $f(x)$]. Show that the function $\varphi : R[x] \to \mathcal{F}(R)$, defined by $\varphi[f(x)] = \varphi_f$, is a ring homomorphism. (It will be shown, in Theorem 3.33, that φ is injective if R is an infinite field.)

3.40 Let R be a commutative ring. Show that the function $\varepsilon : R[x] \to R$, defined by

$$\varepsilon : a_0 + a_1 x + a_2 x + \cdots + a_n x^n \mapsto a_0,$$

is a homomorphism. Describe $\ker \varepsilon$ in terms of roots of polynomials.

3.41 Let R and S be commutative rings and let $\varphi : R \to S$ be a homomorphism. Show that $\widetilde{\varphi} : R[x] \to S[x]$, defined by

$$r_0 + r_1 x + r_2 x^2 + \cdots \mapsto \varphi(r_0) + \varphi(r_1)x + \varphi(r_2)x^2 + \cdots ,$$

is a homomorphism.

3.42 If R is a commutative ring and $c \in R$, prove that the function $\varphi : R[x] \to R[x]$, defined by $f(x) \mapsto f(x + c)$, is an isomorphism. In more detail, $\varphi(\sum_i s_i x^i) = \sum_i s_i (x + c)^i$.

3.43 Prove that any two fields having exactly four elements are isomorphic.

3.44   (i) Show that the function $\varphi : \mathbb{Z}_p \to \mathbb{Z}_p$, given by $\varphi(a) = a^p$, is an isomorphism.

   (ii) Show that every element $a \in \mathbb{Z}_p$ has a pth root, i.e., there is $b \in \mathbb{Z}_p$ with $a = b^p$.

   (iii) Let k be a field that contains $\mathbb{Z}_p$ as a subfield [e.g., $k = \mathbb{Z}_p(x)$]. For every positive integer n, show that the function $\varphi_n : k \to k$, given by $\varphi(a) = a^{p^n}$, is a ring homomorphism.

3.45 If R is a field, show that $R \cong \operatorname{Frac}(R)$. More precisely, show that the homomorphism $f : R \to \operatorname{Frac}(R)$ in Example 3.15, namely, $r \mapsto [r, 1]$, is an isomorphism.

3.46   (i) If A and R are domains and $\varphi : A \to R$ is a ring isomorphism, then $[a, b] \mapsto [\varphi(a), \varphi(b)]$ is a ring isomorphism $\operatorname{Frac}(A) \to \operatorname{Frac}(R)$.

   (ii) Show that a field k containing $\mathbb{Z}$ as a subring must contain an isomorphic copy of $\mathbb{Q}$.

3.47 Let R be a domain with fraction field $F = \operatorname{Frac}(R)$.

   (i) Prove that $\operatorname{Frac}(R[x]) \cong F(x)$.

   (ii) Prove that $\operatorname{Frac}(R[x_1, x_2, \ldots, x_n]) \cong F(x_1, x_2, \ldots, x_n)$.

3.48 Prove that F, the field with 4 elements (see Exercise 3.22), and $\mathbb{Z}_4$ are not isomorphic commutative rings.

3.49   (i) If R and S are commutative rings, show that their ***direct product*** $R \times S$ is also a commutative ring, where addition and multiplication in $R \times S$ are defined "coordinatewise":

$$(r, s) + (r', s') = (r + r', s + s') \quad \text{and} \quad (r, s)(r', s') = (rs, r's').$$

   (ii) Show that if m and n are relatively prime, then $\mathbb{Z}_{mn} \cong \mathbb{Z}_m \times \mathbb{Z}_n$ as rings.

   (iii) Show that $R \times S$ is never a domain.

   (iv) Show that $R \times \{0\}$ is an ideal in $R \times S$.

3.50   (i) If R and S are commutative rings, prove that

$$U(R \times S) = U(R) \times U(S),$$

   where $U(R)$ is the group of units of R.

   (ii) Redo Exercise 2.81(iii) using part (i).

(iii) Give a new proof of Corollary 2.64 using Exercise 3.49(ii).

3.51 Let F be the set of all 2×2 real matrices of the form

$$A = \begin{bmatrix} a & b \\ -b & a \end{bmatrix}.$$

Prove that F is a field (with operations matrix addition and matrix multiplication) isomorphic to $\mathbb{C}$.

3.5 GREATEST COMMON DIVISORS

We are now going to see, when k is a field, that virtually all the theorems proved for $\mathbb{Z}$ have polynomial analogs in $k[x]$; moreover, we shall see that the proofs there can be translated into proofs here.

The division algorithm for polynomials with coefficients in a field says that long division is possible. We introduce the following term to simplify the notation in its proof.

Definition.   If $f(x) = s_n x^n + \cdots + s_1 x + s_0$ is a polynomial of degree n, then its **leading term** is

$$\mathrm{LT}(f) = s_n x^n.$$

Let k be a field and let $f(x) = s_n x^n + \cdots + s_1 x + s_0$ and $g(x) = t_m x^m + \cdots + t_1 x + t_0$ be polynomials in $k[x]$ with $\deg(f) \le \deg(g)$; that is, $n \le m$. Then $s_n^{-1} \in k$, because k is a field, and

$$\frac{\mathrm{LT}(g)}{\mathrm{LT}(f)} = s_n^{-1} t_m x^{m-n};$$

thus, $\mathrm{LT}(f) \mid \mathrm{LT}(g)$.

Theorem 3.30 (Division Algorithm).   *Assume that k is a field and that $f(x), g(x) \in k[x]$ with $f(x) \neq 0$. Then there are unique polynomials $q(x)$, $r(x) \in k[x]$ with*

$$g(x) = q(x)f(x) + r(x)$$

and either $r(x) = 0$ or $\deg(r) < \deg(f)$.

Proof.   We first prove the existence of such q and r. If $f \mid g$, then $g = qf$ for some q; define the remainder $r = 0$, and we are done. If $f \nmid g$, then consider all (necessarily nonzero) polynomials of the form $g - qf$ as q varies over $k[x]$. The least integer axiom provides a polynomial $r = g - qf$ having least degree among all such polynomials. Since $g = qf + r$, it suffices to show that $\deg(r) < \deg(f)$. Write $f(x) = s_n x^n + \cdots + s_1 x + s_0$ and $r(x) = t_m x^m + \cdots + t_1 x + t_0$. Now $s_n \neq 0$ implies that s_n is a unit, because k is a field, and so s_n^{-1} exists in k. If $\deg(r) \geq \deg(f)$, define

$$h(x) = r(x) - t_m s_{n-m}^{-1} x^{n-m} f(x);$$

that is,

$$h = r - \frac{\mathrm{LT}(r)}{\mathrm{LT}(f)} f;$$

note that $h = 0$ or $\deg(h) < \deg(r)$. If $h = 0$, then $r = [\mathrm{LT}(r)/\mathrm{LT}(f)] f$ and

$$g = qf + r$$
$$= qf + \frac{\mathrm{LT}(r)}{\mathrm{LT}(f)} f$$
$$= \left[q + \frac{\mathrm{LT}(r)}{\mathrm{LT}(f)} \right] f,$$

contradicting $f \nmid g$. If $h \neq 0$, then $\deg(h) < \deg(r)$ and

$$g - qf = r = h + \frac{\mathrm{LT}(r)}{\mathrm{LT}(f)} f.$$

Thus, $g - \left[q + \mathrm{LT}(r)/\mathrm{LT}(f) \right] f = h$, contradicting r being a polynomial of least degree having this form. Therefore, $\deg(r) < \deg(f)$.

To prove uniqueness of $q(x)$ and $r(x)$, assume that $g = q'f + r'$, where $\deg(r') < \deg(f)$. Then

$$(q - q')f = r' - r.$$

If $r' \neq r$, then each side has a degree. But $\deg((q - q')f) = \deg(q - q') + \deg(f) \geq \deg(f)$, while $\deg(r' - r) \leq \max\{\deg(r'), \deg(r)\} < \deg(f)$, a contradiction. Hence, $r' = r$ and $(q - q')f = 0$. As $k[x]$ is a domain and $f \neq 0$, it follows that $q - q' = 0$ and $q = q'$.   •

Definition.   If $f(x)$ and $g(x)$ are polynomials in $k[x]$, where k is a field, then the polynomials $q(x)$ and $r(x)$ occurring in the division algorithm are called the **quotient** and the **remainder** after dividing $g(x)$ by $f(x)$.

The hypothesis that k is a field is much too strong; long division can be carried out in $R[x]$ for every commutative ring R as long as the leading coefficient of $f(x)$ is a unit in R [for then $\mathrm{LT}(f)$ is defined and $\mathrm{LF}(f) \mid \mathrm{LT}(g)$]; in particular, long division is always possible when $f(x)$ is a monic polynomial.

Our proof of the division algorithm for polynomials is written as an indirect proof, but the proof can be recast so that it is a true algorithm in the sense that the division algorithm for integers is. Here is a pseudocode implementing it.

$$\text{Input: } g, f$$
$$\text{Output: } q, r$$
$$q := 0; \quad r := g$$
$$\text{WHILE } r \neq 0 \text{ AND } \mathrm{LT}(f) \mid \mathrm{LT}(r) \text{ DO}$$
$$q := q + \mathrm{LT}(r)/\mathrm{LT}(f)$$
$$r := r - \big[\mathrm{LT}(r)/\mathrm{LT}(f)\big] f$$
$$\text{END WHILE}$$

We now turn our attention to roots of polynomials.

Definition.   If $f(x) \in k[x]$, where k is a field, then a **root** of $f(x)$ **in** k is an element $a \in k$ with $f(a) = 0$.

Remark.   The polynomial $f(x) = x^2 - 2$ has its coefficients in $\mathbb{Q}$, but we usually say that $\sqrt{2}$ is a root of $f(x)$ even though $\sqrt{2}$ is irrational; that is, $\sqrt{2} \notin \mathbb{Q}$. We shall see later, in Theorem 3.77, that for every polynomial $f(x) \in k[x]$, where k is any field, there is a larger field E which contains k as a subfield and which contains all the roots of $f(x)$.   ◄

Lemma 3.31.   *Let $f(x) \in k[x]$, where k is a field, and let $a \in k$. Then there is $q(x) \in k[x]$ with*

$$f(x) = q(x)(x - a) + f(a).$$

Proof.   Use the division algorithm to obtain

$$f(x) = q(x)(x - a) + r;$$

the remainder r is a constant because $x - a$ has degree 1. Now evaluate: $f(a) = q(a)(a - a) + r$, and so $r = f(a)$.   •

Proposition 3.32. *If $f(x) \in k[x]$, where k is a field, then $a \in k$ is a root of $f(x)$ in k if and only if $x - a$ divides $f(x)$ in $k[x]$.*

Proof. If a is a root of $f(x)$ in k, then $f(a) = 0$ and the lemma gives $f(x) = q(x)(x - a)$. Conversely, if $f(x) = q(x)(x - a)$, then evaluating at a gives $f(a) = q(a)(a - a) = 0$. •

Theorem 3.33. *Let k be a field and let $f(x) \in k[x]$.*

(i) *If $f(x)$ has degree n, then $f(x)$ has at most n roots in k.*

(ii) *If $f(x)$ has degree n and $a_1, a_2, \ldots, a_n \in k$ are distinct roots of $f(x)$ in k, then there is $c \in k$ and a factorization*

$$f(x) = c(x - a_1)(x - a_2) \cdots (x - a_n) \text{ in } k[x].$$

Proof. (i) We prove the statement true by induction on $n \geq 0$. If $n = 0$, then $f(x)$ is a nonzero constant, and so the number of its roots in k is zero. Now let $n > 0$. If $f(x)$ has no roots in k, then we are done, for $0 \leq n$. Otherwise, we may assume that there is $a \in k$ with a a root of $f(x)$; hence, by Proposition 3.32,

$$f(x) = q(x)(x - a),$$

and $q(x) \in k[x]$ has degree $n - 1$. If there is a root $b \in k$ with $b \neq a$, then

$$0 = f(b) = q(b)(b - a).$$

Since $b - a \neq 0$, we have $q(b) = 0$, so that b is a root of $q(x)$. Now $\deg(q) = n - 1$, so that the inductive hypothesis says that $q(x)$ has at most $n - 1$ roots in k. Therefore, $f(x)$ has at most n roots in k.

(ii) A modification of the induction just given is left as an exercise for the reader. •

We can now complete the proof of Corollary 1.25 describing the complex nth root of unity.

Corollary 3.34. *Every nth root of unity is equal to*

$$e^{2\pi i k/n} = \cos(2\pi k/n) + i \sin(2\pi k/n),$$

where $k = 0, 1, 2, \ldots, n - 1$.

Proof.   In the proof of Corollary 1.25 we showed that each of these n complex numbers is an nth root of unity; that is, each is a root of $x^n - 1$. By Theorem 3.33, there can be no other complex roots.   •

Recall that every polynomial $f(x) \in k[x]$ determines the polynomial function $k \to k$ that sends a into $f(a)$ for all $a \in k$. In Exercise 3.32, however, we saw that a nonzero polynomial in $\mathbb{Z}_p[x]$ (e.g., $x^p - x$) can determine the constant function zero. This pathology vanishes when the field k is infinite.

Corollary 3.35.   *Let k be an infinite field and let $f(x)$ and $g(x)$ be distinct polynomials in $k[x]$. If $f(x)$ and $g(x)$ determine the same polynomial function, i.e., if $f(a) = g(a)$ for all $a \in k$, then $f(x) = g(x)$.*

Proof.   If $f(x) \neq g(x)$, then the polynomial $h(x) = f(x) - g(x)$ is nonzero, so that it has some degree, say, n. Now every element of k is a root of $h(x)$; since k is infinite, $h(x)$ has more than n roots, and this contradicts the theorem.   •

In fact, the last proof gives a bit more.

Corollary 3.36.   *Let k be any (possibly finite) field, and let $f(x)$, $g(x) \in k[x]$. If $\deg(f) \leq \deg(g) = n$, and if $f(a) = g(a)$ for $n + 1$ values $a \in k$, then $f(x) = g(x)$.*

Proof.   If $f(x) \neq g(x)$, then $h(x) = f(x) - g(x) \neq 0$, and

$$\deg(h) \leq \max\{\deg(f), \deg(g)\} = n.$$

By hypothesis, there are $n + 1$ elements $a \in k$ with $h(a) = f(a) - g(a) = 0$, contradicting Theorem 3.33(i). Therefore, $h(x) = 0$ and $f(x) = g(x)$.   •

The definition of a greatest common divisor of polynomials is essentially the same as the corresponding definition for integers.

Definition.   If $f(x)$ and $g(x)$ are polynomials in $k[x]$, where k is a field, then a ***common divisor*** is a polynomial $c(x) \in k[x]$ with $c(x) \mid f(x)$ and $c(x) \mid g(x)$.

Given polynomials $f(x)$ and $g(x)$ in $k[x]$, not both 0, define their ***greatest common divisor***, abbreviated gcd, as the monic common divisor of them having largest degree. If $f(x) = 0 = g(x)$, define their gcd = 0. The gcd of $f(x)$ and $g(x)$ is often denoted by $(f(x), g(x))$.

Suppose that $f(x) \neq 0$; for convenience, suppose further that $f(x)$ is monic. If $g(x) = 0$, then the $\gcd(f(x), g(x)) = \gcd(f(x), 0) = f(x)$, for every polynomial divides 0; hence, if at least one of f or g is not 0, then their gcd is not 0.

We now proceed just as we did in $\mathbb{Z}$, beginning with the analog of Theorem 1.29.

Theorem 3.37.   *If k is a field and $f(x)$, $g(x) \in k[x]$, then their gcd $d(x)$ is a linear combination of $f(x)$ and $g(x)$.*

Remark.   By *linear combination*, we now mean $d = sf + tg$, where both $s = s(x)$ and $t = t(x)$ are polynomials in $k[x]$.   ◄

Proof.   We may assume that at least one of f and g is not zero (the gcd is 0 otherwise). Consider the set I of all the linear combinations:

$$I = \{s(x)f(x) + t(x)g(x) : s(x), t(x) \in k[x]\}.$$

Now f and g are in I (take $s = 1$ and $t = 0$ or vice versa). It follows that if N is the set of all those nonnegative integers occurring as degrees of polynomials in I, then N is nonempty. By the least integer axiom, N contains a smallest integer, say, n and there is some $d(x) \in I$ with $\deg(d) = n$; indeed, because k is a field, there is such a polynomial $d(x)$ that is monic [if the leading coefficient of $d(x)$ is s_n, then $s_n^{-1} \in k$, and $s_n^{-1}d(x)$ is a monic polynomial in I having the same degree as $d(x)$]. We claim that $d(x)$ is the $\gcd(f(x), g(x))$.

Since $d \in I$, it is a linear combination of f and g:

$$d = sf + tg.$$

Let us show that d is a common divisor by trying to divide each of f and g by d. The division algorithm gives $f = qd + r$, where $r = 0$ or $\deg(r) < \deg(d)$. If $r \neq 0$, then

$$r = f - qd = f - q(sf + tg) = (1 - qs)f - qtg \in I,$$

contradicting d having smallest degree among all linear combinations of f and g. Hence, $r = 0$ and $d \mid f$; a similar argument shows that $d \mid g$.

Finally, if c is a common divisor of f and g, then c divides $d = sf + tg$. But $c \mid d$ implies $\deg(c) \leq \deg(d)$. Therefore, d is a gcd of f and g.   •

Corollary 3.38.   *Let k be a field and let $f(x)$, $g(x) \in k[x]$. A monic common divisor $d(x)$ is the gcd if and only if $d(x)$ is divisible by every common divisor; that is, if $c(x)$ is a common divisor, then $c(x) \mid d(x)$.*

Moreover, $f(x)$ and $g(x)$ have a unique gcd.

Proof.   The last paragraph of the proof of Theorem 3.37 shows that every common divisor c of f and g is a divisor of $d = sf + tg$.

Conversely, let d denote a gcd of f and g, and let d' be a common divisor divisible by every common divisor c. By definition, $d \mid d'$. On the other hand, d is divisible by every common divisor, as we have just seen in the first paragraph, so that $d' \mid d$. By Proposition 3.9, there is a unit $u(x) \in k[x]$ with $d'(x) = u(x)d(x)$; by Exercise 3.30(i), $u(x)$ is a nonzero constant. Since both $d(x)$ and $d'(x)$ are monic, it follows that $u(x) = 1$ and $d(x) = d'(x)$.

To prove uniqueness of the gcd, note that if $d(x)$ and $d_1(x)$ are both gcd's of $f(x)$ and $g(x)$, then each divides the other, and so the proof in the previous paragraph gives $d(x) = d_1(x)$.   •

Analysis of the proof of Theorem 3.37, as an analysis of the corresponding proof of Theorem 1.29, the existence of gcd's in $\mathbb{Z}$, yields an ideal in $k[x]$: the set of all linear combinations of $f(x)$ and $g(x)$.

Corollary 1.31 says that every ideal in $\mathbb{Z}$ is a principal ideal; its analog below has essentially the same proof.

Theorem 3.39.   *If k is a field, then every ideal I in $k[x]$ is a principal ideal. Moreover, if $I \neq \{0\}$, there is a monic polynomial that generates I.*

Proof.   If I consists of 0 alone, take $d = 0$. If there are nonzero polynomials in I, the least integer axiom allows us to choose a monic polynomial $d(x) \in I$ of least degree.

We claim that every f in I is a multiple of d. The division algorithm gives polynomials q and r with $f = qd + r$, where either $r = 0$ or $\deg(r) < \deg(d)$. Now $d \in I$ gives $qd \in I$, by part (iii) in the definition of ideal; hence part (ii) gives $r = f - qd \in I$. If $r \neq 0$, then it has a degree and $\deg(r) < \deg(d)$, contradicting d having least degree among all the polynomials in I. Therefore, $r = 0$ and f is a multiple of d.   •

The proof of Theorem 3.37 identifies the gcd of $f(x)$ and $g(x)$ (when at least one of them is not the zero polynomial) as the monic generator of the ideal $I = (f(x), g(x))$ consisting of all the linear combinations of $f(x)$ and $g(x)$. This explains the notation $d = (f, g)$ for the gcd. Notice that this notation makes sense even when both $f(x)$ and $g(x)$ are the zero polynomial:

the gcd is 0, which is the generator of the ideal $\{0\}$.

Definition.   A domain R is a ***principal ideal domain*** if every ideal in R is a principal ideal. One often abbreviates this name to PID.

Example 3.11.
(i) The ring of integers is a PID, by Corollary 1.31.

(ii) Every field is a PID, by Example 3.10(ii).

(iii) If k is a field, then the polynomial ring $k[x]$ is a PID, by Theorem 3.39.

(iv) There are rings other than $\mathbb{Z}$ and $k[x]$, where k is a field, that have a division algorithm; they are called *euclidean rings*, and they, too, are PID's. We shall consider them at the end of this section.   ◄

It is not true that ideals in arbitrary commutative rings are always principal ideals.

Example 3.12.
Let $R = \mathbb{Z}[x]$, the commutative ring of all polynomials over $\mathbb{Z}$. It is easy to see that the set I of all polynomials with even constant term is an ideal in $\mathbb{Z}[x]$. We show that I is *not* a principal ideal.

Suppose there is $d(x) \in \mathbb{Z}[x]$ with $I = (d(x))$. The constant $2 \in I$, so that there is $f(x) \in \mathbb{Z}[x]$ with $2 = d(x)f(x)$. Since the degree of a product is the sum of the degrees of the factors, $0 = \deg(2) = \deg(d) + \deg(f)$. Since degrees are nonnegative, it follows that $\deg(d) = 0$, i.e., $d(x)$ is a nonzero constant. As constants here are integers, the candidates for $d(x)$ are ± 1 and ± 2. Suppose $d(x) = \pm 2$; since $x \in I$, there is $g(x) \in \mathbb{Z}[x]$ with $x = d(x)g(x) = \pm 2g(x)$. But every coefficient on the right side is even, while the coefficient of x on the left side is 1. This contradiction gives $d(x) = \pm 1$. By Example 3.10(ii), $I = \mathbb{Z}[x]$, another contradiction. Therefore, no such $d(x)$ exists, that is, the ideal I is not a principal ideal.   ◄

Example 3.13.
If I and J are ideals in a commutative ring R, we now show that $I \cap J$ is also an ideal in R. Since $0 \in I$ and $0 \in J$, we have $0 \in I \cap J$. If $a, b \in I \cap J$, then $a - b \in I$ and $a - b \in J$, for each is an ideal, and so $a - b \in I \cap J$. If $a \in I \cap J$ and $r \in R$, then $ra \in I$ and $ra \in J$, hence $ra \in I \cap J$. Therefore, $I \cap J$ is an ideal. With minor alterations, this argument also proves that the intersection of any family of ideals in R is also an ideal in R.   ◄

Definition. If $f(x)$ and $g(x)$ are polynomials in $k[x]$, where k is a field, then a ***common multiple*** is a polynomial $m(x) \in k[x]$ with $f(x) \mid m(x)$ and $g(x) \mid m(x)$.

Given polynomials $f(x)$ and $g(x)$ in $k[x]$, not both 0, define their ***least common multiple***, abbreviated lcm, as the monic common multiple of them having smallest degree. If $f(x) = 0 = g(x)$, define their lcm $= 0$. The lcm of $f(x)$ and $g(x)$ is often denoted by $[f(x), g(x)]$.

Proposition 3.40. *Assume that k is a field and that $f(x), g(x) \in k[x]$ are nonzero.*

(i) *$[f(x), g(x)]$ is the monic generator of $(f(x)) \cap (g(x))$.*

(ii) *Let $m(x)$ be a monic common multiple of $f(x)$ and $g(x)$. Then $m(x) = [f(x), g(x)]$ if and only if $m(x)$ divides every common multiple of $f(x)$ and $g(x)$.*

Proof. (i) Since $f(x) \neq 0$ and $g(x) \neq 0$, we have $(f) \cap (g) \neq 0$, because $0 \neq fg \in (f) \cap (g)$. By Theorem 3.39, $(f) \cap (g) = (m)$, where m is the monic polynomial of least degree in $(f) \cap (g)$. As $m \in (f)$, $m = qf$ for some $q(x) \in k[x]$, and so $f \mid m$; similarly, $g \mid m$, so that m is a common multiple of f and g. If M is another common multiple, then $M \in (f)$ and $M \in (g)$, hence $M \in (f) \cap (g) = (m)$, and so $m \mid M$. Therefore, $\deg(m) \leq \deg(M)$, and $m = [f, g]$.

(ii) We have just seen that $[f, g]$ divides every common multiple M of f and g. Conversely, assume that m' is a monic common multiple that divides every other common multiple. Now $m' \mid [f, g]$, because $[f, g]$ is a common multiple, while $[f, g] \mid m'$, by part (i). Proposition 3.9 provides a unit $u(x) \in k[x]$ with $m'(x) = u(x)m(x)$; by Exercise 3.30, $u(x)$ is a nonzero constant. Since both $m(x)$ and $m'(x)$ are monic, it follows that $m(x) = m'(x)$. •

Every polynomial $f(x)$ is divisible by units u and by $uf(x)$. The analog of a prime number is a polynomial having only divisors of these trivial sorts.

Definition. Let k be a field. A polynomial $p(x) \in k[x]$ is ***irreducible*** if $\deg(p) = n \geq 1$ and there is no factorization in $k[x]$ of the form $p(x) = f(x)g(x)$ in which both factors have degree smaller than n.

Corollary 3.41.   *Let k be a field and let $f(x) \in k[x]$ be a quadratic or cubic polynomial. Then $f(x)$ is irreducible in $k[x]$ if and only if $f(x)$ does not have a root in k.*

Proof.   If $f(x)$ has a root a in k, then the theorem shows that $f(x)$ has an honest factorization, and so it is not irreducible.

Conversely, assume that $f(x)$ is not irreducible, i.e., there is a factorization $f(x) = g(x)h(x)$ in $k[x]$ with $\deg(g) < \deg(f)$ and $\deg(h) < \deg(f)$. By Lemma 3.18, $\deg(f) = \deg(g) + \deg(h)$. Since $\deg(f) = 2$ or 3, one of $\deg(g), \deg(h)$ must be 1, and Proposition 3.32 says that $f(x)$ has a root in k.   •

This corollary is false for larger degrees. For example,

$$x^4 + 2x^2 + 1 = (x^2 + 1)^2$$

obviously factors in $\mathbb{R}[x]$, but it has no real roots.

As the definition of divisibility depends on the ambient ring, so irreducibility of a polynomial $p(x) \in k[x]$ also depends on the commutative ring $k[x]$ and hence on the field k. For example, $p(x) = x^2 + 1$ is irreducible in $\mathbb{R}[x]$, but it factors as $(x+i)(x-i)$ in $\mathbb{C}[x]$. On the other hand, a linear polynomial $f(x)$ is always irreducible [if $f = gh$, then $1 = \deg(f) = \deg(g) + \deg(h)$, and so one of g or h must have degree 0 while the other has degree $1 = \deg(f)$].

It is easy to see that if $p(x)$ and $q(x)$ are irreducibles, then $p(x) \mid q(x)$ if and only if there is a unit u with $q(x) = up(x)$. If, in addition, both $p(x)$ and $q(x)$ are monic, then $p(x) \mid q(x)$ implies $p(x) = q(x)$.

Lemma 3.42.   *Let k be a field, let $p(x), f(x) \in k[x]$, and let $d(x) = (p(x), f(x))$ be their gcd. If $p(x)$ is a monic irreducible, then*

$$d(x) = \begin{cases} 1 & \text{if } p(x) \nmid f(x) \\ p(x) & \text{if } p(x) \mid f(x). \end{cases}$$

Proof.   The only monic divisors of $p(x)$ are 1 and $p(x)$. If $p(x) \mid f(x)$, then $d(x) = p(x)$, for $p(x)$ is monic. If $p(x) \nmid f(x)$, then the only monic common divisor is 1, and so $d(x) = 1$.   •

Theorem 3.43 (Euclid's Lemma).   *Let k be a field and let $f(x), g(x) \in k[x]$. If $p(x)$ is an irreducible polynomial in $k[x]$, and $p(x) \mid f(x)g(x)$, then either $p(x) \mid f(x)$ or $p(x) \mid g(x)$.*
More generally, if $p(x) \mid f_1(x) \cdots f_n(x)$, then $p(x) \mid f_i(x)$ for some i.

Proof. If $p \mid f$, we are done. If $p \nmid f$, then the lemma says that $\gcd(p, f) = 1$. There are thus polynomials $s(x)$ and $t(x)$ with $1 = sp + tf$, and so

$$g = spg + tfg.$$

Since $p \mid fg$, it follows that $p \mid g$, as desired. The second statement follows by induction on $n \geq 2$.   •

Definition.   Two polynomials $f(x), g(x) \in k[x]$, where k is a field, are called *relatively prime* if their gcd is 1.

Corollary 3.44.   *Let $f(x), g(x), h(x) \in k[x]$, where k is a field, and let $h(x)$ and $f(x)$ be relatively prime. If $h(x) \mid f(x)g(x)$, then $h(x) \mid g(x)$.*

Proof.   By hypothesis, $fg = hq$ for some $q(x) \in k[x]$. There are polynomials s and t with $1 = sf + th$, and so $g = sfg + thg = shq + thg = h(sq + tg)$; that is, $h \mid g$.   •

Definition.   If k is a field, then a rational function $f(x)/g(x) \in k(x)$ is in *lowest terms* if $f(x)$ and $g(x)$ are relatively prime.

Proposition 3.45.   *If k is a field, every nonzero $f(x)/g(x) \in k(x)$ can be put in lowest terms.*

Proof.   If $d = (f, g)$, then $f = df'$ and $g = dg'$ in $k[x]$. Moreover, f' and g' are relatively prime, for if h were a nonconstant common divisor of f' and g', then hd would be a common divisor of f and g of degree greater than that of d. Now $f/g = df'/dg' = f'/g'$, and the latter is in lowest terms.   •

The same complaint about computing gcd's that arose in $\mathbb{Z}$ arises here, and it has the same resolution.

Theorem 3.46 (Euclidean Algorithm).   *If k is a field and $f(x), g(x) \in k[x]$, then there are algorithms for computing the gcd $d(x) = (f(x), g(x))$ and for finding a pair of polynomials $s(x)$ and $t(x)$ with $d(x) = s(x)f(x) + t(x)g(x)$.*

Proof.   The proof is just a repetition of the euclidean algorithm in $\mathbb{Z}$: iterated application of the division algorithm.

$$
\begin{aligned}
g &= q_0 f + r_1 &\qquad \deg(r_1) &< \deg(f) \\
f &= q_1 r_1 + r_2 &\qquad \deg(r_2) &< \deg(r_1) \\
r_1 &= q_2 r_2 + r_3 &\qquad \deg(r_3) &< \deg(r_2) \\
r_2 &= q_3 r_3 + r_4 &\qquad \deg(r_4) &< \deg(r_3) \\
&\;\;\vdots & &\;\;\vdots
\end{aligned}
$$

We refer the reader to the proof of Theorem 1.37 for further details. $\bullet$

Example 3.14.
Use the euclidean algorithm to find the gcd $(x^5 + 1, x^3 + 1)$.

$$
\begin{aligned}
x^5 + 1 &= x^2(x^3 + 1) + (-x^2 + 1) \\
x^3 + 1 &= (-x)(-x^2 + 1) + (x + 1) \\
-x^2 + 1 &= (-x + 1)(x + 1).
\end{aligned}
$$

We conclude that $x + 1$ is the gcd. $\blacktriangleleft$

Example 3.15.
Find the gcd in $\mathbb{Q}[x]$ of

$$ f(x) = x^3 - x^2 - x + 1 \qquad \text{and} \qquad g(x) = x^3 + 4x^2 + x - 6. $$

Note that $f(x), g(x) \in \mathbb{Z}[x]$, and $\mathbb{Z}$ is not a field. As we proceed, rational numbers will enter, for $\mathbb{Q}$ is the smallest field containing $\mathbb{Z}$. Here are the equations.

$$
\begin{aligned}
g &= 1 \cdot f + (5x^2 + 2x - 7) \\
f &= (\tfrac{1}{5}x - \tfrac{7}{25})(5x^2 + 2x - 7) + (\tfrac{24}{25}x - \tfrac{24}{25}) \\
5x^2 + 2x - 7 &= (\tfrac{25}{24}5x + \tfrac{25}{24})(\tfrac{24}{25}x - \tfrac{24}{25}).
\end{aligned}
$$

We conclude that the gcd is $x - 1$ [which is $\tfrac{24}{25}x - \tfrac{24}{25}$ made monic]. The reader should find $s(x), t(x)$ expressing $x - 1$ as a linear combination (as in arithmetic, work from the bottom up).

As a computational aid, one can clear denominators at any stage. For example, one can replace the second equation above by

$$(5x - 7)(5x^2 + 2x - 7) + (24x - 24);$$

after all, we ultimately multiply by a unit to obtain a monic gcd. However, clearing denominators at some intermediate stage may interfere with finding $s(x)$ and $t(x)$, so do not clear denominators if a linear combination is being sought. ◄

Example 3.16.
Find the gcd in $\mathbb{Z}_5[x]$ of

$$f(x) = x^3 - x^2 - x + 1 \quad \text{and} \quad g(x) = x^3 + 4x^2 + x - 6.$$

The euclidean algorithm simplifies considerably.

$$g = 1 \cdot f + 2x + 2$$
$$f = (3x^2 - x + 2)(2x + 2).$$

The gcd is $x + 1$ (which is $2x + 2$ made monic). ◄

Here are factorizations of the polynomials $f(x)$ and $g(x)$ in Example 3.15:

$$f(x) = x^3 - x^2 - x + 1 = (x - 1)^2(x + 1)$$

and

$$g(x) = x^3 + 4x^2 + x - 6 = (x - 1)(x + 2)(x + 3);$$

one could have seen that $x - 1$ is the gcd, had these factorizations been known at the outset. This suggests that an analog of the fundamental theorem of arithmetic could be useful. Such an analog does exist, and it will be proved in the next section. As a practical matter, however, factoring polynomials is a very difficult task, and the euclidean algorithm is the best way to compute gcd's.

Here is an unexpected bonus from the euclidean algorithm.

Corollary 3.47. *Let k be a subfield of a field K, so that $k[x]$ is a subring of $K[x]$. If $f(x), g(x) \in k[x]$, then their gcd in $k[x]$ is equal to their gcd in $K[x]$.*

Proof.   The division algorithm in $K[x]$ gives

$$g(x) = Q(x)f(x) + R(x),$$

where $Q(x), R(x) \in K[x]$; since $f(x), g(x) \in k[x]$, the division algorithm in $k[x]$ gives

$$g(x) = q(x)f(x) + r(x),$$

where $q(x), r(x) \in k[x]$. But the equation $g(x) = q(x)f(x) + r(x)$ also holds in $K[x]$ because $k[x] \subset K[x]$, so that the uniqueness of quotient and remainder in the division algorithm in $K[x]$ gives $Q(x) = q(x) \in k[x]$ and $R(x) = r(x) \in k[x]$. Therefore, the list of equations occurring in the euclidean algorithm in $K[x]$ is exactly the same list occurring in the euclidean algorithm in the smaller ring $k[x]$, and so the same gcd is obtained in both polynomial rings.   •

For example, the gcd of $x^3 - x^2 + x - 1$ and $x^4 - 1$ is $x^2 + 1$, whether computed in $\mathbb{R}[x]$ or in $\mathbb{C}[x]$, in spite of the fact that there are more divisors with complex coefficients.

We have seen, when k is a field, that there are many analogs for $k[x]$ of theorems proved for $\mathbb{Z}$. The essential reason for this is that both rings are PID's.

Euclidean Rings

There are rings other than $\mathbb{Z}$ and $k[x]$, where k is a field, that have a division algorithm. In particular, we present an example of such a ring in which the quotient and remainder are not unique. We begin by generalizing a property shared by both $\mathbb{Z}$ and $k[x]$.

Definition.   A domain R is a ***euclidean ring*** if R is not a field and there is a function

$$\partial : R - \{0\} \to \mathbb{N},$$

called a ***degree function***, such that

(i) $\partial(f) \leq \partial(fg)$ for all $f, g \in R$ with $f, g \neq 0$;

(ii) for all $f, g \in R$ with $f \neq 0$, there exist $q, r \in R$ with

$$g = qf + r,$$

where either $r = 0$ or $\partial(r) < \partial(f)$.

Example 3.17.
(i) The domain $\mathbb{Z}$ is a euclidean ring with degree function $\partial(m) = |m|$. In $\mathbb{Z}$, we have

$$\partial(mn) = |mn| = |m||n| = \partial(m)\partial(n).$$

(ii) When k is a field, the domain $k[x]$ is a euclidean ring with degree function the usual degree of a nonzero polynomial. In $k[x]$, we have

$$\begin{aligned}
\partial(fg) &= \deg(fg) \\
&= \deg(f) + \deg(g) \\
&= \partial(f) + \partial(g).
\end{aligned}$$

Since $\partial(mn) = \partial(m)\partial(n)$ in $\mathbb{Z}$, the behavior of the degree of a product is not determined by the axioms in the definition of a degree function. If a degree function ∂ is multiplicative, that is, if

$$\partial(fg) = \partial(f)\partial(g),$$

then ∂ is called a **norm**.

(iii) The Gaussian[11] integers $\mathbb{Z}[i]$ form a euclidean ring whose degree function

$$\partial(a + bi) = a^2 + b^2$$

is a norm.

To see that ∂ is a multiplicative degree function, note first that if $\alpha = a+bi$, then

$$\partial(\alpha) = \alpha\overline{\alpha},$$

where $\overline{\alpha} = a - bi$ is the complex conjugate of α. It follows that $\partial(\alpha\beta) = \partial(\alpha)\partial(\beta)$ for all $\alpha, \beta \in \mathbb{Z}[i]$, because

$$\partial(\alpha\beta) = \alpha\beta\overline{\alpha\beta} = \alpha\beta\overline{\alpha}\,\overline{\beta} = \alpha\overline{\alpha}\beta\overline{\beta} = \partial(\alpha)\partial(\beta);$$

indeed, this is even true for all $\alpha, \beta \in \mathbb{Q}[i] = \{x + yi : x, y \in \mathbb{Q}\}$, by Corollary 1.21.

---

[11] The Gaussian integers are so called because Gauss tacitly used $\mathbb{Z}[i]$ and its norm ∂ to investigate biquadratic residues.

We now show that ∂ satisfies the first property of a degree function. If $\beta = c + id \in \mathbb{Z}[i]$ and $\beta \neq 0$, then

$$1 \leq \partial(\beta),$$

for $\partial(\beta) = c^2 + d^2$ is a positive integer; it follows that if $\alpha, \beta \in \mathbb{Z}[i]$ and $\beta \neq 0$, then

$$\partial(\alpha) \leq \partial(\alpha)\partial(\beta) = \partial(\alpha\beta).$$

Let us show that ∂ also satisfies the second desired property. Given α, $\beta \in \mathbb{Z}[i]$ with $\beta \neq 0$, regard α/β as an element of $\mathbb{C}$. Rationalizing the denominator gives $\alpha/\beta = \alpha\overline{\beta}/\beta\overline{\beta} = \alpha\overline{\beta}/\partial(\beta)$, so that

$$\alpha/\beta = x + yi,$$

where $x, y \in \mathbb{Q}$. Write $x = a + u$ and $y = b + v$, where $a, b \in \mathbb{Z}$ are integers closest to x and y, respectively; thus, $|u|, |v| \leq \frac{1}{2}$. (If x or y has the form $m + \frac{1}{2}$, where m is an integer, then there is a choice of nearest integer: $x = m + \frac{1}{2}$ or $x = (m+1) - \frac{1}{2}$; a similar choice arises if x or y has the form $m - \frac{1}{2}$.) It follows that

$$\alpha = \beta(a + bi) + \beta(u + vi).$$

Notice that $\beta(u + vi) \in \mathbb{Z}[i]$, for it is equal to $\alpha - \beta(a + bi)$. Finally, we have $\partial\big(\beta(u + vi)\big) = \partial(\beta)\partial(u + vi)$, and so ∂ will be a degree function if $\partial(u + vi) < 1$. And this is so, for the inequalities $|u| \leq \frac{1}{2}$ and $|v| \leq \frac{1}{2}$ give $u^2 \leq \frac{1}{4}$ and $v^2 \leq \frac{1}{4}$, and hence $\partial(u + vi) = u^2 + v^2 \leq \frac{1}{4} + \frac{1}{4} = \frac{1}{2} < 1$. Therefore, $\partial(\beta(u+vi)) < \partial(\beta)$, and so $\mathbb{Z}[i]$ is a euclidean ring whose degree function is a norm.

One reason for showing that $\mathbb{Z}[i]$ is a euclidean ring is that quotients and remainders may not be unique (because of the choices noted above). For example, let $\alpha = 3 + 5i$ and $\beta = 2$. Then $\alpha/\beta = \frac{3}{2} + \frac{5}{2}i$; the choices are:

$$a = 1 \text{ and } u = \tfrac{1}{2} \quad \text{or} \quad a = 2 \text{ and } u = -\tfrac{1}{2};$$
$$b = 2 \text{ and } v = \tfrac{1}{2} \quad \text{or} \quad b = 3 \text{ and } v = -\tfrac{1}{2}.$$

There are four quotients and remainders after dividing $3 + 5i$ by 2 in $\mathbb{Z}[i]$, for each of the remainders, e.g., $1 + i$, has degree $2 < 4 = \partial(2)$:

$$3 + 5i = 2(1 + 2i) + (1 + i);$$
$$= 2(1 + 3i) + (1 - i);$$
$$= 2(2 + 2i) + (-1 + i);$$
$$= 2(2 + 3i) + (-1 - i). \quad \blacktriangleleft$$

Example 3.18.
The reader can adapt the proof of Proposition 3.39 to prove that every euclidean ring is a PID. In particular, the ring of Gaussian integers $\mathbb{Z}[i]$ is a principal ideal domain. The converse is false: there are PID's which are not euclidean rings, as we see in the next example. ◄

Example 3.19.
It is shown in algebraic number theory that the ring

$$R = \{a + b\alpha : a, b \in \mathbb{Z}\},$$

where $\alpha = \frac{1}{2}(1 + \sqrt{-19})$, is a PID [$R$ is the ring of algebraic integers in the quadratic number field $\mathbb{Q}(\sqrt{-19})$]. In 1949, T. S. Motzkin showed that R is not a euclidean ring. He found the following property of euclidean rings that does not mention its degree function.

Definition.   An element u in a domain R is a ***universal side divisor*** if u is not a unit and, for every $x \in R$, either $u \mid x$ or there is a unit $z \in R$ with $u \mid x + z$.

Proposition 3.48.   *If R is a euclidean ring, then R has a universal side divisor.*

Proof.   Define

$$S = \{\partial(v) : v \neq 0 \text{ or } v \text{ is not a unit}\},$$

where ∂ is the degree function on R. Since R is not a field (by definition, euclidean rings are not fields), S is a nonempty subset of the natural numbers. By the Least Integer Axiom, S has a smallest element, say, $\partial(u)$. We claim that u is a universal side divisor. If $x \in R$, there are elements q and r with $x = qu + r$, where either $r = 0$ or $\partial(r) < \partial(u)$. If $r = 0$, then $u \mid x$; if $r \neq 0$, then r must be a unit, otherwise its existence contradicts $\partial(u)$ being the smallest number in S. We have shown that u is a universal side divisor. •

Motzkin then showed that the ring $\{a + b\alpha : a, b \in \mathbb{Z}\}$, where $\alpha = \frac{1}{2}(1 + \sqrt{-19})$, has no universal side divisors, concluding that this PID is not a euclidean ring. For details, we refer the reader to K. S. Williams, "Note on Non-euclidean Principal Ideal Domains," *Math. Mag.* 48 (1975), 176–177. ◄

Certain theorems holding in $\mathbb{Z}$ carry over to PID's, and hence to euclidean rings, once one generalizes the standard definitions. For example, let us describe gcd's and primes. We have already defined *divisors* in any commutative ring, and so common divisors of $\alpha, \beta \in \mathbb{Z}[i]$ are defined.

Definition. An element p in a PID R is **irreducible** if p is neither 0 nor a unit and, in any factorization $p = uv$ in R, either u or v is a unit.

For example, a prime in $\mathbb{Z}$ is an irreducible element, as is an irreducible polynomial in $k[x]$, where k is a field.

Definition. An element $\delta \in \mathbb{Z}[i]$ is a gcd of elements $\alpha, \beta \in \mathbb{Z}[i]$ if

(i) δ is a common divisor of α and β;

(ii) if γ is any common divisor of α and β, then $\gamma \mid \delta$.

Notice that we say nothing about the uniqueness of a gcd should it exist. For an example of a domain in which a pair of elements does not have a gcd, see Exercise 3.71.

Remark. If π is an irreducible element in a PID R, then every divisor of π is either a unit or an associate $u\pi$, where u is a unit in R. It follows that if $\beta \in R$ and $\pi \nmid \beta$, then 1 is a gcd of π and β. ◂

We now pass from general PID's back to the special euclidean ring $\mathbb{Z}[i]$.

Proposition 3.49.

(i) *Every* $\alpha = a + ib, \beta = c + id \in \mathbb{Z}[i]$ *has a gcd,* δ, *which is a linear combination of* α *and* β:

$$\delta = \sigma\alpha + \tau\beta,$$

where $\sigma, \tau \in \mathbb{Z}[i]$.

(ii) *If an irreducible element* $\pi \in \mathbb{Z}[i]$ *divides a product* $\alpha\beta$, *then either* $\pi \mid \alpha$ *or* $\pi \mid \beta$.

Proof. (i) We may assume that at least one of α and β is not zero (otherwise, the gcd is 0 and the result is obvious). Consider the set I of all the linear combinations:

$$I = \{\sigma\alpha + \tau\beta : \sigma, \tau \text{ in } \mathbb{Z}[i]\}.$$

Now α and β are in I (take $\sigma = 1$ and $\tau = 0$ or vice versa). It is easy to check that I is an ideal in $\mathbb{Z}[i]$, and so there is $\delta \in I$ with $I = (\delta)$, because $\mathbb{Z}[i]$ is a PID; we claim that δ is a gcd of α and β.

Since $\alpha \in I = (\delta)$, we have $\alpha = \rho\delta$ for some $\rho \in \mathbb{Z}[i]$; that is, δ is a divisor of α; similarly, δ is a divisor of β, and so δ is a common divisor of α and β.

Since $\delta \in I$, it is a linear combination of α and β: there are $\sigma, \tau \in \mathbb{Z}[i]$ with

$$\delta = \sigma\alpha + \tau\beta.$$

Finally, if γ is any common divisor of α and β, then $\alpha = \gamma\alpha'$ and $\beta = \gamma\beta'$, so that γ divides δ, for $\delta = \sigma\alpha + \tau\beta = \gamma(\sigma\alpha' + \tau\beta')$. We conclude that δ is a gcd.

(ii) If $\pi \mid \alpha$, we are done. If $\pi \nmid \alpha$, then the remark says that 1 is a gcd of π and α. There are thus Gaussian integers σ and τ with $1 = \sigma\pi + \tau\alpha$, and so

$$\beta = \sigma\pi\beta + \tau\alpha\beta.$$

Since $\pi \mid \alpha\beta$, it follows that $\pi \mid \beta$, as desired.   •

If n is an odd number, then either $n \equiv 1 \bmod 4$ or $n \equiv 3 \bmod 4$. In particular, the odd prime numbers are divided into two classes. Thus, 5, 13, 17 are congruent to 1 mod 4, for example, while 3, 7, 11 are congruent to 3 mod 4.

Lemma 3.50.   *If p is a prime and $p \equiv 1 \bmod 4$, then there is an integer m with*

$$m^2 \equiv -1 \bmod p.$$

Proof.   If $G = (\mathbb{Z}_p)^\times$ is the multiplicative group of nonzero elements in $\mathbb{Z}_p$, then $|G| = p - 1 \equiv 0 \bmod 4$; that is, 4 is a divisor of $|G|$. By Proposition 2.59, G contains a subgroup S of order 4. By Exercise 2.52, either S is cyclic or $a^2 = 1$ for all $a \in S$. Since $\mathbb{Z}_p$ is a field, however, it cannot contain 4 roots of the quadratic $x^2 - 1$. Therefore, S is cyclic, say, $S = \langle [m] \rangle$, where $[m]$ is the congruence class of m mod 4. Since $[m]$ has order 4, we have $[m^4] = [1]$. Moreover, $[m^2] \neq [1]$ (lest $[m]$ have order $\leq 2 < 4$), and so $[m^2] = [-1]$, for $[-1]$ is the unique element in S of order 2. Therefore, $m^2 \equiv -1 \bmod p$.   •

Proposition 3.51 (Fermat's Two-Squares Theorem[12]).   *An odd prime p*
is a sum of two squares,

$$p = a^2 + b^2,$$

where a and b are integers, if and only if $p \equiv 1$ mod 4.

Proof.   For any integer a we have $a \equiv r$ mod 4, where $r = 0, 1, 2$ or 3, and
so $a^2 \equiv r^2$ mod 4. But, mod 4,

$$0^2 \equiv 0, \ 1^2 \equiv 1, \ 2^2 = 4 \equiv 0, \ \text{and} \ 3^2 = 9 \equiv 1,$$

so that $a^2 \equiv 0$ or 1 mod 4. It follows, for any integers a and b, that $a^2 + b^2 \not\equiv 3$ mod 4. Therefore, if $p = a^2 + b^2$, where a and b are integers, then
$p \not\equiv 3$ mod 4. Since p is odd, either $p \equiv 1$ mod 4 or $p \equiv 3$ mod 4. We have
just ruled out the latter possibility, and so $p \equiv 1$ mod 4.

   Conversely, assume that $p \equiv 1$ mod 4. By the lemma, there is an integer
m such that

$$p \mid m^2 + 1.$$

In $\mathbb{Z}[i]$, there is a factorization $m^2 + 1 = (m + i)(m - i)$, and so

$$p \mid (m + i)(m - i) \ \text{in} \ \mathbb{Z}[i].$$

If $p \mid m + i$ in $\mathbb{Z}[i]$, then there are integers u and v with $m + i = p(u + iv)$.
Taking complex conjugates, we have $m - i = p(u - iv)$, so that $p \mid m - i$
as well. Therefore, $p \mid (m + i) - (m - i) = 2i$, which is a contradiction,
for $\partial(p) = p^2 > \partial(2i) = 4$. We conclude that p is not an irreducible element, for it does not satisfy the analog of Euclid's lemma in Proposition 3.49.
Hence, there is a factorization

$$p = \alpha\beta \ \text{in} \ \mathbb{Z}[i]$$

in which neither $\alpha = a + ib$ nor $\beta = c + id$ is a unit. Therefore, taking norms
gives an equation in $\mathbb{Z}$:

$$\begin{aligned}
p^2 &= \partial(p) \\
&= \partial(\alpha\beta) \\
&= \partial(\alpha)\partial(\beta) \\
&= (a^2 + b^2)(c^2 + d^2).
\end{aligned}$$

---

[12]Fermat was the first to state this theorem, but the first published proof is due to Euler.
Gauss proved that there is only one pair of numbers a and b with $p = a^2 + b^2$.

Since $a^2 + b^2 \neq 1$ and $c^2 + d^2 \neq 1$, Euclid's lemma gives $p \mid a^2 + b^2$ and $p \mid c^2 + d^2$, and these equations give $p = a^2 + b^2$ (and $p = c^2 + d^2$), as desired.   •

EXERCISES

3.52 Find the gcd of $x^2 - x - 2$ and $x^3 - 7x + 6$ in $\mathbb{Z}_5[x]$, and express it as a linear combination of them.

3.53   (i) If R is a domain and $f(x) \in R[x]$ has degree n, show that $f(x)$ has at most n roots in R.

(ii) Show that $x^2 - 1 \in \mathbb{Z}_8[x]$ has 4 roots in $\mathbb{Z}_8$.

3.54 Show that the following pseudocode implements the euclidean algorithm finding gcd $f(x)$ and $g(x)$ in $\mathbb{Z}_3[x]$, where $f(x) = x^2 + 1$ and $g(x) = x^3 + x + 1$.

```
Input: g, f
Output: d
d := f;   s := g
WHILE s ≠ 0 DO
    rem := remainder(h, s)
    h := s
    s := rem
END WHILE
```

3.55 Prove the converse of Euclid's lemma. Let k be a field and let $f(x) \in k[x]$ be a polynomial of degree ≥ 1; if, whenever $f(x)$ divides a product of two polynomials, it necessarily divides one of the factors, then $f(x)$ is irreducible.

3.56 Let $f(x), g(x) \in R[x]$, where R is a domain. If the leading coefficient of $f(x)$ is a unit in R, then the division algorithm gives a quotient $q(x)$ and a remainder $r(x)$ after dividing $g(x)$ by $f(x)$. Prove that $q(x)$ and $r(x)$ are uniquely determined by $g(x)$ and $f(x)$.

3.57 Let k be a field, and let $f(x), g(x) \in k[x]$ be relatively prime. If $h(x) \in k[x]$, prove that $f(x) \mid h(x)$ and $g(x) \mid h(x)$ imply $f(x)g(x) \mid h(x)$.

3.58 If k is a field, prove that $\sqrt{1 - x^2} \notin k(x)$, where $k(x)$ is the field of rational functions.

3.59 Prove that every euclidean ring R is a PID.

3.60 Let ∂ be the degree function of a euclidean ring R. If $m, n \in \mathbb{N}$ and $m \geq 1$, prove that ∂' is also a degree function on R, where

$$\partial'(x) = m\partial(x) + n$$

for all $x \in R$. Conclude that a euclidean ring may have no elements of degree 0 or degree 1.

3.61 Let R be a euclidean ring with degree function ∂.

  (i) Prove that $\partial(1) \leq \partial(a)$ for all nonzero $a \in R$.

  (ii) Prove that a nonzero $u \in R$ is a unit if and only if $\partial(u) = \partial(1)$.

3.62 Let R be a euclidean ring, and assume that $b \in R$ is neither zero nor a unit. Prove, for every $i \geq 0$, that $\deg(b^i) < \deg(b^{i+1})$.

3.63 (i) If k is a field, prove that the ring of formal power series $k[[x]]$ is a PID.

  (ii) Prove that every nonzero ideal in $k[[x]]$ is equal to (x^n) for some $n \geq 0$.

Definition. Let k be a field. A ***common divisor*** of $a_1(x), a_2(x), \ldots, a_n(x)$ in $k[x]$ is a polynomial $c(x) \in k[x]$ with $c(x) \mid a_i(x)$ for all i; the ***greatest common divisor*** is the monic common divisor of largest degree.

3.64 Let k be a field, and let polynomials $a_1(x), a_2(x), \ldots, a_n(x)$ in $k[x]$ be given.

  (i) Show that the greatest common divisor $d(x)$ of these polynomials has the form $\sum t_i(x)a_i(x)$, where $t_i(x) \in k[x]$ for $1 \leq i \leq n$.

  (ii) Prove that if $c(x)$ is a monic common divisor of these polynomials, then $c(x) \mid d(x)$.

3.65 If $[f(x), g(x)]$ denotes the lcm of $f(x), g(x) \in k[x]$, where k is a field, show that

$$[f, g](f, g) = fg.$$

3.66 If k is a field, show that the ideal (x, y) in $k[x, y]$ is not a principal ideal.

3.67 For every $m \geq 1$, prove that every ideal in $\mathbb{Z}_m$ is a principal ideal. (If m is composite, then $\mathbb{Z}_m$ is not a PID because it is not a domain.)

3.68 (i) Show that $x, y \in k[x, y]$ are relatively prime, but that 1 is not a linear combination of them, i.e., there do not exist $s(x, y), t(x, y) \in k[x, y]$ with $1 = xs(x, y) + yt(x, y)$.

  (ii) Show that 2 and x are relatively prime in $\mathbb{Z}[x]$, but that 1 is not a linear combination of them; that is, there do not exist $s(x), t(x) \in \mathbb{Z}[x]$ with $1 = 2s(x) + xt(x)$.

3.69 A student claims that $x - 1$ is not irreducible because it factors as $x - 1 = (\sqrt{x} + 1)(\sqrt{x} - 1)$. Explain the error of his ways.

3.70 If R is a PID and $a, b \in R$, prove that a gcd of a and b exists.

3.71 Prove that there are domains R containing a pair of elements having no gcd.

3.6 UNIQUE FACTORIZATION

Here is the analog for polynomials of the fundamental theorem of arithmetic; it shows that irreducible polynomials are "building blocks" of arbitrary polynomials in the same sense that primes are building blocks of arbitrary integers. To avoid long sentences, let us agree that a "product" may have only one factor. Thus, when we say that a polynomial $f(x)$ is a product of irreducibles, we allow the possibility that the product has only one factor, that is, that $f(x)$ is itself irreducible.

Theorem 3.52 (Unique Factorization).   *If k is a field, then every polynomial $f(x) \in k[x]$ of degree ≥ 1 is a product of a nonzero constant and monic irreducibles. Moreover, if $f(x)$ has two such factorizations*

$$f(x) = ap_1(x) \cdots p_m(x) \qquad and \qquad f(x) = bq_1(x) \cdots q_n(x),$$

where a and b are nonzero constants and the p's and q's are monic irreducibles, then $a = b$, $m = n$, and the q's may be reindexed so that $q_i = p_i$ for all i.

Proof.   We prove the existence of a factorization for a polynomial $f(x)$ by (the second form of) induction on $\deg(f) \geq 1$. If $\deg(f) = 1$, then $f(x) = ax + c = a(x + a^{-1}c)$; as every linear polynomial, $x + a^{-1}c$ is irreducible, and so it is a product of irreducibles (in our present usage of the word). Assume now that $\deg(f) \geq 1$. If $f(x)$ is irreducible and its leading coefficient is a, write $f(x) = a(a^{-1}f(x))$; we are done, for $a^{-1}f(x)$ is monic. If $f(x)$ is not irreducible, then $f(x) = g(x)h(x)$, where $\deg(g) < \deg(f)$ and $\deg(h) < \deg(f)$. By the inductive hypothesis, there are factorizations $g(x) = bp_1(x) \cdots p_m(x)$ and $h(x) = cq_1(x) \cdots q_n(x)$, where the p's and q's are monic irreducibles. It follows that

$$f(x) = (bc)p_1(x) \cdots p_m(x)q_1(x) \cdots q_n(x),$$

as desired.

   We now prove, by induction on $M = \max\{m, n\} \geq 1$, that if there is an equation

$$ap_1(x) \cdots p_m(x) = bq_1(x) \cdots q_n(x)$$

in which a and b are nonzero constants and the p's and q's are monic irreducibles, then $a = b$, $m = n$, and the q's may be reindexed so that $q_i = p_i$ for all i. For the base step $M = 1$, the hypothesis gives a polynomial, call it $g(x)$, with $g(x) = ap_1(x) = bq_1(x)$. Now a is the leading coefficient of $g(x)$, because $p_1(x)$ is monic; similarly, b is the leading coefficient of $g(x)$ because $q_1(x)$ is monic. Therefore, $a = b$, and canceling gives $p_1(x) = q_1(x)$. For the inductive step, the given equation shows that $p_m(x) \mid q_1(x) \cdots q_n(x)$. By Euclid's lemma for polynomials, there is some i with $p_m(x) \mid q_i(x)$. But $q_i(x)$, being monic irreducible, has no monic divisors other than 1 and itself, so that $q_i(x) = p_m(x)$. Reindexing, we may assume that $q_n(x) = p_m(x)$. Canceling this factor, we have $ap_1(x) \cdots p_{m-1}(x) = bq_1(x) \cdots q_{n-1}(x)$. By the inductive hypothesis, $a = b$, $m - 1 = n - 1$ (hence $m = n$), and after possible reindexing, $q_i = p_i$ for all i. •

Let k be a field, and assume that all the roots of $f(x) \in k[x]$ lie in k; that is, there are $a, r_1, \ldots, r_n \in k$ with

$$f(x) = a \prod_{i=1}^{n} (x - r_i).$$

If $r_1, \ldots, r_s$, where $s \leq n$, are the distinct roots of $f(x)$, then collecting terms gives

$$f(x) = a(x - r_1)^{e_1} (x - r_2)^{e_2} \cdots (x - r_s)^{e_s},$$

where $e_j \geq 1$ for all j. We call e_j the **multiplicity** of the root r_j. As linear polynomials are always irreducible, unique factorization shows that multiplicities of roots are well-defined.

Remark.   A domain R is called a UFD, *unique factorization domain*, if every nonzero nonunit $r \in R$ is a product of irreducibles and, moreover, such a factorization of r is essentially unique. The domains $\mathbb{Z}[\zeta_p]$ in Example 3.2, where $\zeta = e^{2\pi i/p}$ for p an odd prime, are quite interesting in this regard. Positive integers a, b, c for which

$$a^2 + b^2 = c^2,$$

for example, 3, 4, 5 and 5, 12, 13, are called *Pythagorean triples*, and they have been recognized for four thousand years (a Babylonian tablet of roughly this age has been found containing a dozen of them), and they were classified by Diophantus about two thousand years ago. Around 1637, Fermat wrote in the margin of his copy of a book by Diophantus what is nowadays called

Fermat's Last Theorem: For all integers $n \geq 3$, there do not exist positive integers a, b, c for which

$$a^n + b^n = c^n.$$

Fermat claimed that he had a wonderful proof of this result, but that the margin was too small to contain it. Elsewhere, he did prove this result for $n = 4$ and, later, others proved it for small values of n. However, the general statement challenged mathematicians for centuries.

Call a positive integer $n \geq 2$ a *Fermat integer* if there are no positive integers a, b, c with $a^n + b^n = c^n$. If n is a Fermat integer, then so is any multiple nk of it: otherwise, there are positive integers r, s, t with $r^{nk} + s^{nk} = t^{nk}$; this gives the contradiction $a^n + b^n = c^n$, where $a = r^k, b = s^k$, and $c = t^k$. For example, any integer of the form $4k$ is a Fermat integer. Since every positive integer is a product of primes, Fermat's Last Theorem would follow if every odd prime is a Fermat integer.

As in Exercise 3.74, a solution $a^p + b^p = c^p$, for p an odd prime, gives a factorization

$$c^p = (a + b)(a + \zeta b)(a + \zeta^2 b) \cdots (a + \zeta^{p-1} b),$$

where $\zeta = \zeta_p = e^{2\pi i/p}$. In the 1840s, E. Kummer considered this factorization in the domain $\mathbb{Z}[\zeta_p]$. He proved that if unique factorization into irreducibles holds in $\mathbb{Z}[\zeta_p]$, then there do not exist positive integers a, b, c (none of which is divisible by p) with $a^p + b^p = c^p$. Kummer realized, however, that even though unique factorization does hold in $\mathbb{Z}[\zeta_p]$ for some primes p, it does not hold in all $\mathbb{Z}[\zeta_p]$. To extend his proof, he invented what he called "ideal numbers," and he proved that there is unique factorization of ideal numbers as products of "prime ideal numbers." These ideal numbers motivated R. Dedekind to define ideals in arbitrary commutative rings (our definition of ideal is that of Dedekind), and he proved that ideals in the special rings $\mathbb{Z}[\zeta_p]$ correspond to Kummer's ideal numbers. Over the years these investigations have been vastly developed, so that, in 1995, A. Wiles, with the assistance of R. Taylor, proved Fermat's Last Theorem. ◄

The next result is an analog of Proposition 1.39: For $b \geq 2$, every positive integer has an expression in base b.

Lemma 3.53.   *Let k be a field, and let $b(x) \in k[x]$ have $\deg(b) \geq 1$. Each nonzero $f(x) \in k[x]$ has an expression*

$$f(x) = d_m(x)b(x)^m + d_{m-1}(x)b(x)^{m-1} + \cdots + d_0(x),$$

where, for every j, *either* $d_j(x) = 0$ *or* $\deg(d_j) < \deg(b)$.

Proof.   Let $f(x) \in k[x]$. Since $\deg(b) \geq 1$, there is $m \geq 0$ with $\deg(f) < (m+1)\deg(b) = \deg(b^{m+1})$. We prove, by induction on $m \geq 0$, that if

$$\deg(b^m) \leq \deg(f) < \deg(b^{m+1}),$$

then f has an expression $d_m b^m + d_{m-1} b^{m-1} + \cdots + d_0$ of the desired form.

If $m = 0$, then $0 = \deg(b^0) \leq \deg(f) < \deg(b)$, and we may define $d_0 = f$. If $m > 0$, then $\deg(b^m) \leq \deg(f) < \deg(b^{m+1})$, and the division algorithm gives polynomials d_m and r with $f = d_m b^m + r$, where either $r = 0$ or $\deg(r) < \deg(b^m)$. Notice that $d_m \neq 0$ [otherwise $\deg(f) = \deg(r) < \deg(b^m)$] and $\deg(d_m) < \deg(b)$ [otherwise $\deg(f) \geq \deg(b^{m+1})$]. If $r = 0$, define $d_{m-1} = \cdots = d_0 = 0$, and $f = d_m b^m$ is an expression of the desired form. If $r \neq 0$, then the inductive hypothesis shows that r, and hence f, has an expression of the desired form.   •

Just as for integers, it can be proved that the "digits" $d_i(x)$ are unique.

Definition.   Polynomials $q_1(x), \ldots, q_n(x)$ are ***pairwise relatively prime*** if $(q_i(x), q_j(x)) = 1$ for all $i \neq j$.

It is easy to see that if $q_1(x), \ldots, q_m(x)$ are pairwise relatively prime, then $q_1(x)$ and the product $q_2(x) \cdots q_m(x)$ are relatively prime.

Lemma 3.54.   *Let* k *be a field, and let* $f(x)/g(x) \in k(x)$. *If* $g(x) = q_1(x) \cdots q_m(x)$, *where* $q_1(x), \ldots, q_m(x)$ *are pairwise relatively prime, then there are* $a_i(x) \in k[x]$ *with*

$$\frac{f(x)}{g(x)} = \sum_{i=1}^{m} \frac{a_i(x)}{q_i(x)}.$$

Proof.   The proof is by induction on $m \geq 1$. The base step $m = 1$ is clearly true. Since q_1 and $q_2 \ldots q_m$ are relatively prime, there are polynomials s and

t with $1 = sq_1 + tq_2 \cdots q_m$. Therefore,

$$
\begin{aligned}
\frac{f}{g} &= (sq_1 + tq_2 \cdots q_m)\frac{f}{g} \\
&= \frac{sq_1 f}{g} + \frac{tq_2 \cdots q_m f}{g} \\
&= \frac{sq_1 f}{q_1 q_2 \cdots q_m} + \frac{tq_2 \cdots q_m f}{q_1 q_2 \cdots q_m} \\
&= \frac{sf}{q_2 \cdots q_m} + \frac{tf}{q_1}.
\end{aligned}
$$

The polynomials $q_2(x), \ldots, q_m(x)$ are pairwise relatively prime, and the inductive hypothesis now rewrites the first summand. $\quad\bullet$

We now prove the algebraic portion of the method of partial fractions used in calculus to integrate rational functions.

Theorem 3.55 (Partial Fractions).   *Let k be a field, and let the factorization into irreducibles of a polynomial $g(x) \in k[x]$ be*

$$ g(x) = p_1(x)^{e_1} \cdots p_m(x)^{e_m}. $$

If $f(x)/g(x) \in k(x)$, then

$$
\frac{f(x)}{g(x)} = h(x) + \sum_{i=1}^{m} \left(\frac{d_1(x)}{p_i(x)} + \frac{d_2(x)}{p_i(x)^2} + \cdots + \frac{d_{e_i}(x)}{p_i(x)^{e_i}} \right),
$$

where $h(x) \in k[x]$ and either $d_j(x) = 0$ or $\deg(d_j) < \deg(p_i)$.

Proof.   Clearly, the polynomials $p_1(x)^{e_1}, p_2(x)^{e_2}, \ldots, p_m(x)^{e_m}$ are pairwise relatively prime. By Lemma 3.54, there are $a_i(x) \in k[x]$ with

$$
\frac{f(x)}{g(x)} = \sum_{i=1}^{m} \frac{a_i(x)}{p_i(x)^{e_i}}.
$$

For each i, the division algorithm gives polynomials $Q_i(x)$ and $R_i(x)$ with $a_i(x) = Q_i(x)p_i(x)^{e_i} + R_i(x)$, where either $R_i(x) = 0$ or $\deg(R_i) < \deg(p_i(x)^{e_i})$. Hence,

$$
\frac{a_i(x)}{p_i(x)^{e_i}} = Q_i(x) + \frac{R_i(x)}{p_i(x)^{e_i}}.
$$

By Lemma 3.53,

$$R_i(x) = d_m(x)p_i(x)^m + d_{m-1}(x)p_i(x)^{m-1} + \cdots + d_0(x),$$

where, for all j, either $d_j(x) = 0$ or $\deg(d_j) < \deg(p_i)$; moreover, since $\deg(R_i) < \deg(p_i^{e_i})$, we have $m \leq e_i$. Therefore,

$$\frac{a_i(x)}{p_i(x)^{e_i}} = Q_i(x) + \frac{d_m(x)p_i(x)^m + d_{m-1}(x)p_i(x)^{m-1} + \cdots + d_0(x)}{p_i(x)^{e_i}}$$

$$= Q_i(x) + \frac{d_m(x)p_i(x)^m}{p_i(x)^{e_i}} + \frac{d_{m-1}(x)p_i(x)^{m-1}}{p_i(x)^{e_i}} + \cdots + \frac{d_0(x)}{p_i(x)^{e_i}}.$$

After cancellation, each of the summands $d_j(x)p_i(x)^j/p_i(x)^{e_i}$ is either a polynomial or a rational function of the form $d_j(x)/p_i(x)^s$, where $1 \leq s \leq e_i$. If we call $h(x)$ the sum of all those polynomials which are not rational functions, then this is the desired expression.   ●

It is known that the only irreducible polynomials in $\mathbb{R}[x]$ are linear or quadratic, so that all the numerators in the partial fraction decomposition in $\mathbb{R}[x]$ are either constant or linear. Theorem 3.55 is used in proving that all rational functions in $\mathbb{R}(x)$ can be integrated in closed form using logs and arctans.

Here is the integer version of partial fractions. If a/b is a positive rational, where the prime factorization of b is $b = p_1^{e_1} \cdots p_m^{e_m}$, then

$$\frac{a}{b} = h + \sum_{i=1}^{m}\left(\frac{c_1}{p_i} + \frac{c_2}{p_i^2} + \cdots + \frac{c_{e_i}}{p_i^{e_i}}\right),$$

where $h \in \mathbb{Z}$ and $0 \leq c_j < p_i$ for all j.

EXERCISES

3.72 (i) In $R[x]$, where R is a field, let $f = p_1^{e_1} \cdots p_m^{e_m}$ and $g = p_1^{\varepsilon_1} \cdots p_m^{\varepsilon_m}$, where the p_i's are distinct monic irreducibles and $e_i, \varepsilon_i \geq 0$ for all i (as with integers, the device of allowing zero exponents allows us to have the same irreducible factors in the two factorizations). Prove that $f \mid g$ if and only if $e_i \leq \varepsilon_i$ for all i.

(ii) Use the (unique) factorization into irreducibles to give formulas for the gcd and lcm of two polynomials analogous to the formulas in Proposition 1.44.

3.73   (i) Let $f(x) = (x - a_1) \cdots (x - a_n) \in k[x]$, where k is a field. Show that $f(x)$ has **no repeated roots** (that is, all the a_i are distinct) if and only if the gcd $(f, Df) = 1$, where Df is the derivative of f.

(ii) If $f(x) \in \mathbb{R}[x]$, show that $f(x)$ has no repeated roots in $\mathbb{C}$ if and only if $(f, Df) = 1$.

(iii) Prove that if $p(x) \in \mathbb{Q}[x]$ is an irreducible polynomial, then $p(x)$ has no repeated roots.

3.74 Let $\zeta = e^{2\pi i/n}$.

(i) Prove that

$$x^n - 1 = (x - 1)(x - \zeta)(x - \zeta^2) \cdots (x - \zeta^{n-1})$$

and, if n is odd, that

$$x^n + 1 = (x + 1)(x + \zeta)(x + \zeta^2) \cdots (x + \zeta^{n-1}).$$

(ii) For numbers a and b, prove that

$$a^n - b^n = (a - b)(a - \zeta b)(a - \zeta^2 b) \cdots (a - \zeta^{n-1} b)$$

and, if n is odd, that

$$a^n + b^n = (a + b)(a + \zeta b)(a + \zeta^2 b) \cdots (a + \zeta^{n-1} b).$$

3.7 IRREDUCIBILITY

Although there are some techniques to help decide whether an integer is prime, the general problem is a very difficult one. It is also very difficult to determine whether a polynomial is irreducible, but we now present some useful techniques that frequently work.

We know that if $f(x) \in k[x]$ and r is a root of $f(x)$ in a field k, then there is a factorization $f(x) = (x - r)g(x)$ in $k[x]$, so that $f(x)$ is not irreducible. In Corollary 3.41, we saw that this decides the matter for quadratic and cubic polynomials in $k[x]$: such polynomials are irreducible in $k[x]$ if and only if they have no roots in k.

Theorem 3.56.   *Let $f(x) = a_0 + a_1 x + \cdots + a_n x^n \in \mathbb{Z}[x] \subset \mathbb{Q}[x]$. Every rational root r of $f(x)$ has the form b/c, where $b \mid a_0$ and $c \mid a_n$.*

Proof. We may assume that $r = b/c$ is in lowest terms, that is, $(b, c) = 1$. Substituting r into $f(x)$ gives

$$0 = f(b/c) = a_0 + a_1 b/c + \cdots + a_n b^n/c^n,$$

and multiplying through by c^n gives

$$0 = a_0 c^n + a_1 b c^{n-1} + \cdots + a_n b^n.$$

Hence, $a_0 c^n = b(-a_1 c^{n-1} - \cdots - a_n b^{n-1})$, that is, $b \mid a_0 c^n$. Since b and c are relatively prime, it follows that b and c^n are relatively prime, and so Euclid's lemma in $\mathbb{Z}$ gives $b \mid a_0$. Similarly, $a_n b^n = c(-a_{n-1} b^{n-1} - \cdots - a_0 c^{n-1})$, $c \mid a_n b^n$, and $c \mid a_n$. •

Definition.   A complex number α is called an ***algebraic integer*** if α is a root of a monic $f(x) \in \mathbb{Z}[x]$.

We note that it is crucial, in the definition of algebraic integer, that $f(x) \in \mathbb{Q}[x]$ be monic. Every algebraic number z, that is, every complex number z that is a root of some polynomial $g(x) \in \mathbb{Q}[x]$, is necessarily a root of some polynomial $h(x) \in \mathbb{Z}[x]$; just clear the denominators of the coefficients of $g(x)$.

Corollary 3.57.   *A rational number z that is an algebraic integer must lie in $\mathbb{Z}$. More precisely, if $f(x) \in \mathbb{Z}[x] \subset \mathbb{Q}[x]$ is a monic polynomial, then every rational root of $f(x)$ is an integer that divides the constant term.*

Proof.   If $f(x) = a_0 + a_1 x + \cdots + a_n x^n$ is monic, then $a_n = 1$, and Theorem 3.56 applies at once. •

For example, consider $f(x) = x^3 + 4x^2 - 2x - 1 \in \mathbb{Q}[x]$. By Corollary 3.41, this cubic is irreducible if and only if it has no rational root. As $f(x)$ is monic, the candidates for rational roots are ± 1, for these are the only divisors of -1 in $\mathbb{Z}$. But $f(1) = 2$ and $f(-1) = 4$, so that neither 1 nor -1 is a root. Thus, $f(x)$ has no roots in $\mathbb{Q}$, and hence $f(x)$ is irreducible in $\mathbb{Q}[x]$.

This corollary gives a new solution of Exercise 1.61(ii). If m is an integer that is not a perfect square, then the polynomial $x^2 - m$ has no integer roots, and so $\sqrt{m}$ is irrational. Indeed, the reader can now generalize to nth roots: If m is not an nth power of an integer, then $\sqrt[n]{m}$ is irrational, for any rational root of $x^n - m$ must be an integer.

The next proposition gives a way of proving that a complex number is an algebraic integer; moreover, it shows that the sum and product of algebraic integers is another such (if α and β are algebraic integers, then it is not too difficult to give monic polynomials having $\alpha + \beta$ and $\alpha\beta$ as roots, but it takes a bit of work to find such polynomials having all coefficients in $\mathbb{Z}$).

Definition.   An abelian group G is *finitely generated* if there is a finite subset $x_1, \ldots, x_n$ that generates it; that is, every element in G is a linear combination $\sum_i m_i x_i$, where $m_i \in \mathbb{Z}$ for all i.

Proposition 3.58.

(i) *A complex number α is an algebraic integer if and only if the additive group of $\mathbb{Z}[\alpha]$ is finitely generated, where $\mathbb{Z}[\alpha]$ is the smallest subring containing α.*

(ii) *The set of all algebraic integers is a subring of $\mathbb{C}$.*

Proof.   (i) If α is an algebraic integer, then there is a monic $f(x) = x^n + b_{n-1}x^{n-1} + \cdots + b_0 \in \mathbb{Z}[x]$ having α as a root. Let $G = \langle \alpha^{n-1}, \ldots, \alpha^2, \alpha, 1 \rangle$; clearly, $\alpha^k \in G$ for all $k < n$. We now show that $\alpha^k \in G$ for all $k \geq n$ by induction on k. If $k = n$, then $\alpha^n = -(b_{n-1}\alpha^{n-1} + \cdots + b_1\alpha + b_0) \in G$ (notice how we have used the fact that $f(x)$ is monic). For the inductive step, assume that there are integers c_i so that $\alpha^k = c_{n-1}\alpha^{n-1} + \cdots + c_1\alpha + c_0$. Then

$$
\begin{aligned}
\alpha^{k+1} &= \alpha\alpha^k \\
&= c_{n-1}\alpha^n + c_{n-2}\alpha^{n-1} + \cdots + c_1\alpha^2 + c_0\alpha \\
&= c_{n-1}[-(b_{n-1}\alpha^{n-1} + \cdots + b_2\alpha^2 + b_1\alpha + b_0)] \\
&\quad + c_{n-2}\alpha^{n-1} + \cdots + c_1\alpha^2 + c_0\alpha,
\end{aligned}
$$

which lies in G. Thus, the additive group of $\mathbb{Z}[\alpha]$ is finitely generated.

Conversely, if the additive group of the commutative ring $\mathbb{Z}[\alpha]$ is finitely generated, that is, $\mathbb{Z}[\alpha] = \langle g_1, \ldots, g_m \rangle$ as an abelian group, then each g_j is a $\mathbb{Z}$-linear combination of powers of α. Let m be the largest power of α occurring in any of these g's. Now $\alpha^{m+1} \in \mathbb{Z}[\alpha]$, so that it can be expressed as a $\mathbb{Z}$-linear combination of smaller powers of α; say, $\alpha^{m+1} = \sum_{i=0}^m b_i\alpha^i$, where $b_i \in \mathbb{Z}$. Therefore, α is a root of $f(x) = x^{m+1} - \sum_{i=0}^m b_i x^i$, which is a monic polynomial in $\mathbb{Z}[x]$, and so α is an algebraic integer.

(ii) Suppose that α and β are algebraic integers; let α be a root of a monic $f(x) \in \mathbb{Z}[x]$ of degree n, and let β be a root of a monic $g(x) \in \mathbb{Z}[x]$ of degree m. We let the reader prove that $\mathbb{Z}[\alpha\beta]$ is the abelian group generated by all $\alpha^i \beta^j$, where $0 \le i < n$ and $0 \le j < m$; hence, $\alpha\beta$ is algebraic, by part (i). Finally, $\mathbb{Z}[\alpha + \beta]$ is the abelian group generated by all $\alpha^i \beta^j$, where $i + j \le n + m - 1$, and so $\alpha + \beta$ is also an algebraic integer.   •

This last theorem gives a technique for proving that an integer a is a divisor of an integer b. If one can prove that b/a is an algebraic integer, then it must be an integer, for it is obviously rational.

We are going to find several conditions that imply that a polynomial $f(x) \in \mathbb{Z}[x]$ does not factor in $\mathbb{Z}[x]$. However, $\mathbb{Z}$ is not a field, and we are more interested in whether $f(x)$ factors in $\mathbb{Q}[x]$. C. F. Gauss (1777–1855) found the relation between the factorizations of $f(x)$ in $\mathbb{Z}[x]$ and in $\mathbb{Q}[x]$; we prove this result after first proving several lemmas.

Definition.   A polynomial $f(x) = a_0 + a_1 x + a_2 x^2 + \cdots + x^n \in \mathbb{Z}[x]$ is called **primitive** if the gcd of its coefficients is 1.

Of course, every monic polynomial is primitive. It is easy to see that if d is the gcd of the coefficients of $f(x)$, then $(1/d)f(x)$ is a primitive polynomial in $\mathbb{Z}[x]$.

Observe that if $f(x)$ is not primitive, then there exists a prime p that divides each of its coefficients: if the gcd is $d > 1$, then take for p any prime divisor of d.

Lemma 3.59.   *If $f(x)$, $g(x) \in \mathbb{Z}[x]$ are both primitive, then their product $f(x)g(x)$ is also primitive.*

Proof.   Let $f(x) = \sum a_i x^i$, $g(x) = \sum b_j x^j$, and $f(x)g(x) = \sum c_k x^k$. If $f(x)g(x)$ is not primitive, then there is a prime p that divides every c_k. Since $f(x)$ is primitive, at least one of its coefficients is not divisible by p; let a_i be the first such; similarly, let b_j be the first coefficient of $g(x)$ that is not divisible by p. The definition of multiplication of polynomials gives

$$a_i b_j = c_{i+j} - (a_0 b_{i+j} + \cdots + a_{i-1} b_{j+1} + a_{i+1} b_{j-1} + \cdots + a_{i+j} b_0).$$

Each term on the right side is divisible by p, and so p divides $a_i b_j$. As p divides neither a_i nor b_j, however, this contradicts Euclid's lemma in $\mathbb{Z}$.   •

Remark.   We can give a more elegant proof of this last lemma; after all, the hypothesis that a polynomial $h(x) \in \mathbb{Z}[x]$ is not primitive says that all its coefficients are 0 in $\mathbb{Z}_p$ for some prime p. Assume that the product $f(x)g(x)$ is not primitive, so there is some prime p dividing each of its coefficients. If $\varphi : \mathbb{Z} \to \mathbb{Z}_p$ is the natural map $a \mapsto [a]$, then Exercise 3.41 shows that the function $\widetilde{\varphi} : \mathbb{Z}[x] \to \mathbb{Z}_p[x]$, which reduces all the coefficients of a polynomial mod p, is a ring homomorphism. In particular, since p divides every coefficient of $f(x)g(x)$, we have $0 = \widetilde{\varphi}(fg) = \widetilde{\varphi}(g)\widetilde{\varphi}(h)$ in $\mathbb{Z}_p[x]$. On the other hand, neither $\widetilde{\varphi}(f)$ nor $\widetilde{\varphi}(g)$ is 0, because they are primitive. We have contradicted the fact that $\mathbb{Z}_p[x]$ is a domain.   ◄

Lemma 3.60.   *Every nonzero $f(x) \in \mathbb{Q}[x]$ has a unique factorization*

$$f(x) = c(f)f^*(x),$$

where $c(f) \in \mathbb{Q}$ is positive and $f^*(x) \in \mathbb{Z}[x]$ is primitive.

Proof.   There are integers a_i and b_i with

$$f(x) = (a_0/b_0) + (a_1/b_1)x + \cdots + (a_n/b_n)x^n \in \mathbb{Q}[x].$$

Define $B = b_0 b_1 \ldots b_n$, so that $g(x) = Bf(x) \in \mathbb{Z}[x]$. Now define $D = \pm d$, where d is the gcd of the coefficients of $g(x)$; the sign is chosen to make the rational number D/B positive. Now $(B/D)f(x) = (1/D)g(x)$ lies in $\mathbb{Z}[x]$, and it is a primitive polynomial. If we define $c(f) = D/B$ and $f^*(x) = (B/D)f(x)$, then $f(x) = c(f)f^*(x)$ is a desired factorization.

Suppose that $f(x) = eh(x)$ is a second such factorization, so that e is a positive rational and $h(x) \in \mathbb{Z}[x]$ is primitive. Now $c(f)f^*(x) = f(x) = eh(x)$, so that $f^*(x) = [e/c(f)]h(x)$. Write $e/c(f)$ in lowest terms: $e/c(f) = u/v$, where u and v are relatively prime positive integers. The equation $vf^*(x) = uh(x)$ holds in $\mathbb{Z}[x]$; equating like coefficients, v is a common divisor of each coefficient of $uh(x)$. Since $(u, v) = 1$, Euclid's lemma in $\mathbb{Z}$ shows that v is a (positive) common divisor of the coefficients of $h(x)$. Since $h(x)$ is primitive, it follows that $v = 1$. A similar argument shows that $u = 1$. Since $e/c(f) = u/v = 1$, we have $e = c(f)$ and hence $h(x) = f^*(x)$.   •

Definition.   The rational $c(f)$ in Lemma 3.60 is called the ***content*** of $f(x)$.

Corollary 3.61.   *If $f(x) \in \mathbb{Z}[x]$, then $c(f) \in \mathbb{Z}$.*

Proof.   If d is the gcd of the coefficients of $f(x)$, then $(1/d)f(x) \in \mathbb{Z}[x]$ is primitive. Since $d[(1/d)f(x)]$ is a factorization of $f(x)$ as the product of a positive rational d (which is even a positive integer) and a primitive polynomial, the uniqueness in the lemma gives $c(f) = d \in \mathbb{Z}$.   •

Corollary 3.62.   *If $f(x) \in \mathbb{Q}[x]$ factors as $f(x) = g(x)h(x)$, then*

$$c(f) = c(g)c(h) \qquad and \qquad f^*(x) = g^*(x)h^*(x).$$

Proof.   We have

$$f(x) = g(x)h(x)$$
$$f(c) = [c(g)g^*(x)][c(h)h^*(x)]$$
$$= c(g)c(h)g^*(x)h^*(x).$$

Since $g^*(x)h^*(x)$ is primitive, by Lemma 3.59, and $c(g)c(h)$ is a positive rational, the uniqueness of the factorization in Lemma 3.60 gives $c(f) = c(g)c(h)$ and $f^*(x) = g^*(x)h^*(x)$.   •

Theorem 3.63 (*Gauss*).   *Let $f(x) \in \mathbb{Z}[x]$. If*

$$f(x) = G(x)H(x) \text{ in } \mathbb{Q}[x],$$

then there is a factorization

$$f(x) = g(x)h(x) \text{ in } \mathbb{Z}[x],$$

where $\deg(g) = \deg(G)$ and $\deg(h) = \deg(H)$. Therefore, if $f(x)$ does not factor into polynomials of smaller degree in $\mathbb{Z}[x]$, then $f(x)$ is irreducible in $\mathbb{Q}[x]$.

Proof.   By Corollary 3.62, there is a factorization

$$f(x) = c(G)c(H)G^*(x)H^*(x) \text{ in } \mathbb{Q}[x],$$

where $G^*(x)$, $H^*(x) \in \mathbb{Z}[x]$ are primitive polynomials. But $c(G)c(H) = c(f)$, by Corollary 3.62, and $c(f) \in \mathbb{Z}$, by Corollary 3.61 (which applies because $f(x) \in \mathbb{Z}[x]$). Therefore, $f(x) = g(x)h(x)$ is a factorization in $\mathbb{Z}[x]$, where $g(x) = c(f)G^*(x)$ and $h(x) = H^*(x)$.   •

Remark.   Gauss used these ideas to prove that the ring $k[x_1, \ldots , x_n]$ of all polynomials in n variables with coefficients in a field k is a unique factorization domain. We shall give this proof in Chapter 6.   ◀

Each algebraic integer has an irreducible polynomial associated with it.

Corollary 3.64.   *If α is an algebraic integer and $p(x) \in \mathbb{Z}[x]$ is the monic polynomial of least degree having α as a root, then $p(x)$ is an irreducible polynomial in $\mathbb{Q}[x]$.*

Proof.   Assume that $p(x) = G(x)H(x)$ in $\mathbb{Q}[x]$, where $\deg(G) < \deg(p)$ and $\deg(H) < \deg(p)$, then α is a root of either $G(x)$ or $H(x)$. By Gauss's theorem, there is a factorization $p(x) = g(x)h(x)$ in $\mathbb{Z}[x]$ with $\deg(g) = \deg(G)$ and $\deg(h) = \deg(H)$; in fact, there are rationals c and d with $g(x) = cG(x)$ and $h(x) = dH(x)$. If a is the leading coefficient of $g(x)$ and b is the leading coefficient of $h(x)$, then $ab = 1$, for $p(x)$ is monic. Therefore, we may assume that $a = 1 = b$, for $a, b \in \mathbb{Z}$; that is, we may assume that both $g(x)$ and $h(x)$ are monic. Since α is a root of $g(x)$ or $h(x)$, we have contradicted $p(x)$ being a monic polynomial in $\mathbb{Z}[x]$ of least degree having α as a root.   •

Definition.   If α is an algebraic integer, then its ***minimum polynomial*** is the monic polynomial in $\mathbb{Z}[x]$ of least degree having α as a root.

Corollary 3.64 shows that every algebraic integer α has a unique minimum polynomial $m(x) \in \mathbb{Z}[x]$, and $m(x)$ is irreducible in $\mathbb{Q}[x]$. One defines the (algebraic) ***conjugates*** of α to be the roots of $m(x)$, and one defines the ***norm*** of α to be the absolute value of the product of the conjugates of α. Of course, the norm of α is just the absolute value of the constant term of $m(x)$, and so it is an (ordinary) integer. Norms are of great importance in algebraic number theory.

The next criterion uses the integers mod p.

Theorem 3.65.   *Let $f(x) = a_0 + a_1 x + a_2 x^2 + \cdots + x^n \in \mathbb{Z}[x]$ be monic, and let p be a prime. If $\overline{f}(x) = [a_0] + [a_1]x + [a_2]x^2 + \cdots + x^n \in \mathbb{Z}_p[x]$ is irreducible, then $f(x)$ is irreducible in $\mathbb{Q}[x]$.*

Proof.   By Exercise 3.41, the natural map $\varphi : \mathbb{Z} \to \mathbb{Z}_p$ defines a homomorphism $\widetilde{\varphi} : \mathbb{Z}[x] \to \mathbb{Z}_p[x]$ by

$$\widetilde{\varphi}(b_0 + b_1 x + b_2 x^2 + \cdots) = [b_0] + [b_1]x + [b_2]x^2 + \cdots,$$

that is, just reduce all the coefficients mod p. If $g(x) \in \mathbb{Z}[x]$, denote its image $\widetilde{\varphi}(g(x)) \in \mathbb{Z}_p[x]$ by $\widetilde{g}(x)$. Suppose that $f(x)$ factors in $\mathbb{Z}[x]$; say, $f(x) = g(x)h(x)$, where $\deg(g) < \deg(f)$ and $\deg(h) < \deg(f)$; of course, $\deg(f) = \deg(g) + \deg(h)$. Now $\widetilde{f}(x) = \widetilde{g}(x)\widetilde{h}(x)$, because $\widetilde{\varphi}$ is a ring homomorphism, so that $\deg(\widetilde{f}) = \deg(\widetilde{g}) + \deg(\widetilde{h})$. Since $f(x)$ is monic,

$\widetilde{f}(x)$ is also monic, and so $\deg(\widetilde{f}) = \deg(f)$. Thus, both $\widetilde{g}(x)$ and $\widetilde{h}(x)$ have degrees less than $\deg(\widetilde{f})$, contradicting the irreducibility of $\widetilde{f}(x)$. Therefore, $f(x)$ is irreducible in $\mathbb{Z}[x]$, and, by Gauss's theorem, $f(x)$ is irreducible in $\mathbb{Q}[x]$. •

The converse of Theorem 3.65 is false; its criterion does not always work. It is not difficult to find an irreducible $f(x) \in \mathbb{Z}[x] \leq \mathbb{Q}[x]$ with $f(x)$ factoring mod p for some prime p, and Exercise 3.86 gives an example of an irreducible polynomial in $\mathbb{Q}[x]$ that factors in $\mathbb{Z}_p[x]$ for every prime p.

Theorem 3.65 says that if one can find a prime p with $\widetilde{f}(x)$ irreducible in $\mathbb{Z}_p[x]$, then $f(x)$ is irreducible in $\mathbb{Q}[x]$. Until now, the finite fields $\mathbb{Z}_p$ have been oddities; $\mathbb{Z}_p$ has appeared only as a curious artificial construct. Now the finiteness of $\mathbb{Z}_p$ is a genuine advantage, for there are only a finite number of polynomials in $\mathbb{Z}_p[x]$ of any given degree. In principle, then, one can test whether a polynomial of degree n in $\mathbb{Z}_p[x]$ is irreducible by just looking at *all* the possible factorizations of it. Since it becomes tiresome not to do so, we are now going to write the elements of $\mathbb{Z}_p$ without brackets.

Example 3.20.
We determine the irreducible polynomials in $\mathbb{Z}_2[x]$ of small degree.

As always, the linear polynomials x and $x + 1$ are irreducible.

There are four quadratics: x^2; $x^2 + x$; $x^2 + 1$; $x^2 + x + 1$ (more generally, there are p^n monic polynomials of degree n in $\mathbb{Z}_p[x]$, for there are p choices for each of the n coefficients $a_0, \ldots, a_{n-1}$). Since each of the first three has a root in $\mathbb{Z}_2$, there is only one irreducible quadratic.

There are eight cubics, of which four are reducible because their constant term is 0. The remaining polynomials are

$$x^3 + 1; \qquad x^3 + x + 1; \qquad x^3 + x^2 + 1; \qquad x^3 + x^2 + x + 1.$$

Since 1 is a root of the first and fourth, the middle two are the only irreducible cubics.

There are 16 quartics, of which eight are reducible because their constant term is 0. Of the eight with nonzero constant term, those having an even number of nonzero coefficients have 1 as a root. There are now only four surviving polynomials $f(x)$, and each of them has no roots in $\mathbb{Z}_2$, i.e., they have no linear factors. If $f(x) = g(x)h(x)$, then both $g(x)$ and $h(x)$ must be irreducible quadratics. But there is only one irreducible quadratic, namely, $x^2 + x + 1$, and so $(x^2 + x + 1)^2 = x^4 + x^2 + 1$ is reducible while the other three quartics are irreducible. The following list summarizes these observations.

Irreducible Polynomials of Low Degree over $\mathbb{Z}_2$

degree 2: $x^2 + x + 1$.

degree 3: $x^3 + x + 1$; $x^3 + x^2 + 1$.

degree 4: $x^4 + x^3 + 1$; $x^4 + x + 1$; $x^4 + x^3 + x^2 + x + 1$.   ◀

Example 3.21.

Here is a list of the monic irreducible quadratics and cubics in $\mathbb{Z}_3[x]$. The reader can verify that the list is correct by first enumerating all such polynomials; there are 6 monic quadratics having nonzero constant term, and there are 18 monic cubics having nonzero constant term. It must then be checked which of these have 1 or -1 as a root (it is more convenient to write -1 instead of 2).

Monic Irreducible Quadratics and Cubics over $\mathbb{Z}_3$

degree 2: $x^2 + 1$; $x^2 + x - 1$; $x^2 - x - 1$.

degree 3: $x^3 - x + 1$; $x^3 + x^2 + x + 1$; $x^3 + x^2 - x + 1$;

$x^3 - x^2 + 1$; $x^3 - x^2 + x + 1$;

$x^3 - x - 1$; $x^3 + x^2 - 1$; $x^3 + x^2 + x - 1$;

$x^3 - x^2 - 1$; $x^3 - x^2 - x - 1$.   ◀

Example 3.22.

(i) We show that $f(x) = x^4 - 5x^3 + 2x + 3$ is an irreducible polynomial in $\mathbb{Q}[x]$.

By Corollary 3.57, the only candidates for rational roots of $f(x)$ are $1, -1, 3, -3$, and the reader may check that none of these is a root. Since $f(x)$ is a quartic, one cannot yet conclude that $f(x)$ is irreducible, for it might be a product of (irreducible) quadratics.

Let us try the criterion of Theorem 3.65. Since $\widetilde{f}(x) = x^4 + x^3 + 1$ in $\mathbb{Z}_2[x]$ is irreducible, by Example 3.20, it follows that $f(x)$ is irreducible in $\mathbb{Q}[x]$. [It was not necessary to check that $f(x)$ has no rational roots; irreducibility of $\widetilde{f}(x)$ is enough to conclude irreducibility of $f(x)$.]

(ii) Let $\Phi_5(x) = x^4 + x^3 + x^2 + x + 1 \in \mathbb{Q}[x]$.

In Example 3.20, we saw that $\widetilde{\Phi}_5(x) = x^4 + x^3 + x^2 + x + 1$ is irreducible in $\mathbb{Z}_2[x]$, and so $\Phi_5(x)$ is irreducible in $\mathbb{Q}[x]$.   ◀

Definition.   If p is a prime, then the pth *cyclotomic*[13] *polynomial* is

$$\Phi_p(x) = (x^p - 1)/(x - 1) = x^{p-1} + x^{p-2} + \cdots + x + 1.$$

---

[13]The roots of $x^n - 1$ are the nth roots of unity: $1, \zeta, \zeta^2, \ldots, \zeta^{n-1}$, where $\zeta = e^{2\pi i/n} =$

As any linear polynomial, $\Phi_2(x) = x + 1$ is irreducible in $\mathbb{Q}[x]$; $\Phi_3(x) = x^2 + x + 1$ is irreducible in $\mathbb{Q}[x]$ because it has no rational roots; we have just seen that $\Phi_5(x)$ is irreducible in $\mathbb{Q}[x]$. Let us introduce another irreducibility criterion in order to prove that $\Phi_p(x)$ is irreducible in $\mathbb{Q}[x]$ for all primes p.

Lemma 3.66.   *Let $g(x) \in \mathbb{Z}[x]$. If there is $c \in \mathbb{Z}$ with $g(x + c)$ irreducible in $\mathbb{Z}[x]$, then $g(x)$ is irreducible in $\mathbb{Q}[x]$.*

Proof.   By Exercise 3.42, the function $\varphi : \mathbb{Z}[x] \to \mathbb{Z}[x]$, given by $f(x) \mapsto f(x + c)$, is an isomorphism. If $g(x) = s(x)t(x)$, then $g(x + c) = \varphi(g(x)) = \varphi(st) = \varphi(s)\varphi(t)$ is a forbidden factorization of $g(x + c)$. Therefore, $g(x)$ is irreducible in $\mathbb{Z}[x]$ and hence, by Gauss's theorem, $g(x)$ is irreducible in $\mathbb{Q}[x]$.   •

Theorem 3.67 (Eisenstein Criterion).   *Let $f(x) = a_0 + a_1 x + \cdots + a_n x^n \in \mathbb{Z}[x]$. If there is a prime p dividing a_i for all $i < n$ but with $p \nmid a_n$ and $p^2 \nmid a_0$, then $f(x)$ is irreducible in $\mathbb{Q}[x]$.*

Proof.   Assume, on the contrary, that

$$f(x) = (b_0 + b_1 x + \cdots + b_m x^m)(c_0 + c_1 x + \cdots + c_k x^k),$$

where $m < n$ and $k < n$; by Theorem 3.63, we may assume that both factors lie in $\mathbb{Z}[x]$. Now $p \mid a_0 = b_0 c_0$, so that Euclid's lemma in $\mathbb{Z}$ gives $p \mid b_0$ or $p \mid c_0$; since $p^2 \nmid a_0$, only one of them is divisible by p, say, $p \mid c_0$ but $p \nmid b_0$. By hypothesis, the leading coefficient $a_n = b_m c_k$ is not divisible by p, so that p does not divide c_k (or b_m). Let c_r be the first coefficient not divisible by p (so that p does divide $c_0, \ldots, c_{r-1}$). If $r < n$, then $p \mid a_r$, and so $b_0 c_r = a_r - (b_1 c_{r-1} + \cdots + b_r c_0)$ is also divisible by p. This contradicts Euclid's lemma, for $p \mid b_0 c_r$, but p divides neither factor. It follows that $r = n$; hence $n \geq k \geq r = n$, and so $k = n$, contradicting $k < n$. Therefore, $f(x)$ is irreducible in $\mathbb{Q}[x]$.   •

Remark.   The following elegant proof of Eisenstein's criterion is due to Peter Cameron.

Let $\widetilde{\varphi} : \mathbb{Z}[x] \to \mathbb{Z}_p[x]$ be the ring homomorphism that reduces coefficients mod p, and let $\overline{f}(x)$ denote $\widetilde{\varphi}(f(x))$. If $f(x)$ is not irreducible in $\mathbb{Q}[x]$,

---

$\cos(2\pi/n) + i \sin(2\pi/n)$. Now these roots divide the unit circle $\{\zeta \in \mathbb{C} : |z| = 1\}$ into n equal arcs (see Figure 1.7). When p is prime, then $x^p - 1 = (x - 1)\Phi_p(x)$. This explains the term cyclotomic, for its Greek origin means *circle splitting*.

then Gauss's theorem gives polynomials $g(x), h(x) \in \mathbb{Z}[x]$ with $f(x) = g(x)h(x)$, where $g(x) = b_0 + b_1 x + \cdots + b_m x^m$ and $h(x) = c_0 + c_1 x + \cdots + c_k x^k$. There is thus an equation $\tilde{f}(x) = \tilde{g}(x)\tilde{h}(x)$ in $\mathbb{Z}_p[x]$.

Since $p \nmid a_n$, we have $\tilde{f}(x) \neq 0$; in fact, $\tilde{f}(x) = ux^n$ for some unit $u \in \mathbb{Z}_p$, because all its coefficients aside from its leading coefficient are 0. By Theorem 3.52, unique factorization in $\mathbb{Z}_p[x]$, we must have $\tilde{g}(x) = vx^m$, where v is a unit in $\mathbb{Z}_p$, for any irreducible factor of $\tilde{g}(x)$ is an irreducible factor of $\tilde{f}(x)$; similarly, $\tilde{h}(x) = wx^k$, where w is a unit in $\mathbb{Z}_p$. It follows that each of $\tilde{g}(x)$ and $\tilde{h}(x)$ has constant term 0; that is, $[b_0] = 0 = [c_0]$ in $\mathbb{Z}_p$; equivalently, $p \mid b_0$ and $p \mid c_0$. But $a_0 = b_0 c_0$, and so $p^2 \mid a_0$, a contradiction. Therefore, $f(x)$ is irreducible in $\mathbb{Q}[x]$. ◀

Corollary 3.68 (Gauss). *For every prime p, the pth cyclotomic polynomial $\Phi_p(x)$ is irreducible in $\mathbb{Q}[x]$.*

Proof. Since $\Phi_p(x) = (x^p - 1)/(x - 1)$, we have

$$\Phi_p(x + 1) = [(x + 1)^p - 1]/x$$

$$= x^{p-1} + \binom{p}{1}x^{p-2} + \binom{p}{2}x^{p-3} + \cdots + p.$$

Since p is prime, Proposition 1.33 shows that Eisenstein's criterion applies; we conclude that $\Phi_p(x + 1)$ is irreducible in $\mathbb{Q}[x]$. By Lemma 3.66, $\Phi_p(x)$ is irreducible in $\mathbb{Q}[x]$. •

We do not say that $x^{n-1} + x^{n-2} + \cdots + x + 1$ is irreducible when n is not prime. For example, when $n = 4$, $x^3 + x^2 + x + 1 = (x + 1)(x^2 + 1)$.

EXERCISES

3.75 Determine whether the following polynomials are irreducible in $\mathbb{Q}[x]$.
   (i) $f(x) = 3x^2 - 7x - 5$.
   (ii) $f(x) = 2x^3 - x - 6$.
   (iii) $f(x) = 8x^3 - 6x - 1$.
   (iv) $f(x) = x^3 + 6x^2 + 5x + 25$.
   (v) $f(x) = x^5 - 4x + 2$.
   (vi) $f(x) = x^4 + x^2 + x + 1$.
   (vii) $f(x) = x^4 - 10x^2 + 1$.

3.76 Prove that there are 6 irreducible quintics in $\mathbb{Z}_2[x]$.

3.77 If α is an algebraic integer, prove that there is a unique monic polynomial $p(x) \in \mathbb{Z}[x]$ of least degree having α as a root.

3.78 (i) If $a \neq \pm 1$ is a squarefree integer, show that $x^n - a$ is irreducible in $\mathbb{Q}[x]$ for every $n \geq 1$. Conclude that there are irreducible polynomials in $\mathbb{Q}[x]$ of every degree $n \geq 1$.

    (ii) If $a \neq \pm 1$ is a squarefree integer, prove that $\sqrt[n]{a}$ is irrational.

3.79 Let k be a field, and let $f(x) = a_0 + a_1 x + \cdots + a_n x^n \in k[x]$ have degree n. If $f(x)$ is irreducible, then so is $a_n + a_{n-1}x + \cdots x a_0 x^n$.

3.8 QUOTIENT RINGS AND FINITE FIELDS

The fundamental theorem of algebra states that every nonconstant polynomial in $\mathbb{C}[x]$ is a product of linear polynomials in $\mathbb{C}[x]$, that is, $\mathbb{C}$ contains all the roots of every polynomial in $\mathbb{C}[x]$. We now return to the study of ideals and homomorphisms in order to prove a "local" analog of the Fundamental Theorem of Algebra for polynomials over an arbitrary field k: given a polynomial $f(x) \in k[x]$, then there is some field K containing k that also contains all the roots of $f(x)$ (we call this a local analog for even though the larger field K contains all the roots of the polynomial $f(x)$, it may not contain roots of other polynomials in $k[x]$). The main idea behind the construction of K involves quotient rings, a construction akin to quotient groups.

Let I be an ideal in a commutative ring R. If we forget the multiplication, then I is a subgroup of the additive group R; since R is an abelian group, the subgroup I is necessarily normal, and so the quotient group R/I is defined, as is the natural map $\pi : R \to R/I$ given by $\pi(a) = a + I$. Recall Lemma 2.31(i), which we now write in additive notation: $a + I = b + I$ in R/I if and only if $a - b \in I$.

Theorem 3.69.   *If I is a proper ideal in a commutative ring R, then the additive abelian group R/I can be made into a commutative ring in such a way that the natural map $\pi : R \to R/I$ is a ring homomorphism.*

Proof.   Define multiplication on the additive abelian group R/I by

$$(a + I)(b + I) = ab + I.$$

To see that this is a single-valued function $R/I \times R/I \to R/I$, assume that $a + I = a' + I$ and $b + I = b' + I$, that is, $a - a' \in I$ and $b - b' \in I$. We must show that $(a' + I)(b' + I) = a'b' + I = ab + I$, that is, $ab - a'b' \in I$. But

$$ab - a'b' = ab - a'b + a'b - a'b'$$
$$= (a - a')b + a'(b - b') \in I,$$

as desired.

We know that R/I is an additive abelian group; to verify that R/I is a commutative ring, it suffices to show associativity and commutativity of multiplication, distributivity, and that one is $1 + I$. Proofs of these properties can be transcribed from their proofs for $\mathbb{Z}_m$ in Propositions 2.43, 2.45, and 3.7(i). Note that we need I to be a proper ideal in order to guarantee that $1 \neq 0$.

Rewriting the equation $(a + I)(b + I) = ab + I$ using the definition of π, namely, $a + I = \pi(a)$, gives $\pi(a)\pi(b) = \pi(ab)$. Since $\pi(1) = 1 + I$, it follows that π is a ring homomorphism.   •

Definition.   The commutative ring R/I constructed above is called the *quotient ring*[14] of R modulo I (briefly, R mod I).

We saw in Example 2.32 that the additive abelian group $\mathbb{Z}/(m)$ is identical to $\mathbb{Z}_m$. They have the same elements: the coset $a + (m)$ and the congruence class $[a]$ are the same subset of $\mathbb{Z}$; they have the same addition: $a + (m) + b + (m) = a + b + (m) = [a + b] = [a] + [b]$. We can now see that the quotient ring $\mathbb{Z}/(m)$ coincides with the commutative ring $\mathbb{Z}_m$, for the two multiplications coincide as well:

$$(a + (m))(b + (m)) = ab + (m) = [ab] = [a][b].$$

Example 3.23.
Let k be a field, let $f(x) \in k[x]$, and let $I = (f(x))$. We show that if $k[x]/I$ is a domain, then $f(x)$ is an irreducible polynomial in $k[x]$. Otherwise, there is a factorization $f(x) = g(x)h(x)$ in $k[x]$ with $\deg(g) < \deg(f)$ and $\deg(h) < \deg(f)$. It follows that neither $g(x) + I$ nor $h(x) + I$ is zero in $k[x]/I$. After all, the zero in $k[x]/I$ is $0 + I = I$, and $g(x) + I = I$ if and only if $g(x) \in I = (f(x))$; but if this were so, then $f(x) \mid g(x)$, giving the contradiction $\deg(f) \leq \deg(g)$. The product

$$(g(x) + I)(h(x) + I) = f(x) + I = I$$

---

[14]Presumably, *quotient rings* are so called in analogy with quotient groups.

is zero in the quotient ring, and this contradicts $k[x]/I$ being a domain. Therefore, $f(x)$ must be an irreducible polynomial.

The converse is also true, and we prove it in Proposition 3.74(i). In fact, we will see that if k is a field and $p(x) \in k[x]$ is irreducible, then $k[x]/(p(x))$ is a field. ◄

We can now prove a converse to Proposition 3.27.

Corollary 3.70.   *If I is a proper ideal in a commutative ring R, then there are a commutative ring A and a ring homomorphism $\pi : R \to A$ with $I = \ker \pi$.*

Proof.   If we forget the multiplication, then the natural map $\pi : R \to R/I$ is a homomorphism between the additive groups and, by Corollary 2.51,

$$I = \ker \pi = \{r \in R : \pi(a) = \{0\}\}.$$

Now remember the multiplication. We have just seen that π is a ring homomorphism, for $(a + I)(b + I) = ab + I$ and $1 + I$ is the one in R/I; that is, $\pi(ab) = \pi(a)\pi(b)$ and $\pi(1) = 1$. Moreover, $\ker \pi$ is equal to I whether the function π is regarded as a ring homomorphism or as a homomorphism of additive groups. •

Theorem 3.71 (First Isomorphism Theorem).   *If $f : R \to A$ is a homomorphism of rings, then $\ker f$ is a proper ideal in R, $\operatorname{im} f$ is a subring of A, and*

$$R/\ker f \cong \operatorname{im} f.$$

Proof.   Let $I = \ker f$. We have already seen, in Proposition 3.27, that I is a proper ideal in R and that $\operatorname{im} f$ is a subring of A.

If we forget the multiplication, then the proof of Theorem 2.53 shows that the function $\varphi : R/I \to A$, given by $\varphi(r + I) = f(r)$, is an isomorphism of the additive groups. Since $\varphi(1 + I) = f(1) = 1$, it now suffices to prove that φ preserves multiplication. But $\varphi\big((r + I)(s + I)\big) = \varphi(rs + I) = f(rs) = f(r)f(s) = \varphi(r + I)\varphi(s + I)$. Therefore, φ is a ring isomorphism. •

As for groups, the first isomorphism theorem creates an isomorphism from a homomorphism once one knows its kernel and image. It also says that there is no significant difference between a quotient ring and the image of a homomorphism. There are analogs of the second and third isomorphism theorems, but they are less useful for rings than they are for groups.

Recall that the prime field of a field k is the intersection of all the subfields of K.

Proposition 3.72.   *If k is a field, then its prime field is isomorphic to $\mathbb{Q}$ or to $\mathbb{Z}_p$ for some prime p.*

Proof.   Consider the ring homomorphism $\chi : \mathbb{Z} \to k$, defined by $\chi(n) = n\varepsilon$, where we denote the *one* in k by ε. Since every ideal in $\mathbb{Z}$ is principal, there is an integer m with ker $\chi = (m)$. If $m = 0$, then χ is an injection, and so there is an isomorphic copy of $\mathbb{Z}$ which is a subring of k. By Exercise 3.46(ii), we have the prime field of k isomorphic to $\mathbb{Q}$ in this case. If $m \neq 0$, the first isomorphism theorem gives $\mathbb{Z}_m = \mathbb{Z}/(m) \cong$ im $\chi \subset k$. Since k is a field, im χ is a domain, and so Proposition 3.7(ii) gives m prime. If we now write p instead of m, then im $\chi = \{\{0\}, \varepsilon, 2\varepsilon, \ldots, (p-1)\varepsilon\}$ is a subfield of k isomorphic to $\mathbb{Z}_p$. Clearly, im χ is the prime field of k, for every subfield contains ε, hence contains im χ.   •

This last result is the first step in classifying different types of fields.

Definition.   A field k has ***characteristic 0*** if its prime field is isomorphic to $\mathbb{Q}$; if its prime field is isomorphic to $\mathbb{Z}_p$ for some prime p, then one says that k has ***characteristic p***.

The fields $\mathbb{Q}$, $\mathbb{R}$, $\mathbb{C}$ have characteristic 0, as does any subfield of them; every finite field has characteristic p for some prime p, as does $\mathbb{Z}_p(x)$, the ring of all rational functions over $\mathbb{Z}_p$.

The structure of R/I can be rather complicated; for special choices of R and I, however, R/I can be nicely described.

Proposition 3.73.   *If k is a field and $I = (p(x))$, where $p(x)$ is a polynomial in $k[x]$, then $k[x]/I$ is a field if and only if $p(x)$ is irreducible.*

Proof.   Assume that $p(x)$ is irreducible. If $f(x) + I \in k[x]/I$ is nonzero, then $f(x) \notin I$, that is, $f(x)$ is not a multiple of $p(x)$ or, to say it another way, $p \nmid f$. By Lemma 3.42, p and f are relatively prime, and so there are polynomials s and t with $sf + tp = 1$. Thus, $sf - 1 \in I$, and so $1 + I = sf + I = (s + I)(f + I)$. Therefore, every nonzero element of $k[x]/I$ has an inverse, and so $k[x]/I$ is a field.

Conversely, if $k[x]/(p(x))$ is a field, then it is a domain, and the result follows from Example 3.23.   •

Note the resemblance to $\mathbb{Z}_m$, which is a quotient ring of $\mathbb{Z}$: Proposition 3.13 shows that if m is a prime, then $\mathbb{Z}_m$ is a field.

Example 3.24.

Consider the homomorphism $\varphi : \mathbb{R}[x] \to \mathbb{C}$, defined by $f(x) \mapsto f(i)$, where $i^2 = -1$; that is, $\varphi : \sum_k a_k x^k \mapsto \sum_k a_k i^k$. The first isomorphism theorem teaches us to seek $\operatorname{im} \varphi$ and $\ker \varphi$.

First, φ is surjective: if $a + bi \in \mathbb{C}$, then $a + bi = \varphi(a + bx) \in \operatorname{im} \varphi$.

Second,

$$\ker \varphi = \{ f(x) \in \mathbb{R}[x] : f(i) = 0 \},$$

the set of all polynomials having i as a root. Of course, $x^2 + 1 \in \ker \varphi$, and we claim that $\ker \varphi = (x^2 + 1)$. Clearly, $(x^2 + 1) \subset \ker \varphi$, for $x^2 + 1$ has i as a root. For the reverse inclusion, if $f(x) \in \ker \varphi$, then $f(i) = 0$, and so $x - i \mid f(x)$ in $\mathbb{C}[x]$, by Proposition 3.32. Thus, the $\gcd(f(x), x^2 + 1) \neq 1$ in $\mathbb{C}[x]$, and hence this gcd is distinct from 1 in $\mathbb{R}[x]$, by Corollary 3.47. Since $x^2 + 1$ is irreducible in $\mathbb{R}[x]$, we have $x^2 + 1 \mid f(x)$ for all $f(x) \in \ker \varphi$. The first isomorphism theorem now gives $\mathbb{R}[x]/(x^2 + 1) \cong \mathbb{C}$.

Thus, the quotient ring construction builds the complex numbers from the reals. One advantage of constructing $\mathbb{C}$ in this way is that it is not necessary to check all the field axioms, for Proposition 3.73 shows that they hold automatically. ◀

The reasoning in the previous example works in more generality, and it gives the converse of the result in Example 3.23.

Proposition 3.74.

(i) *If k is a field and $p(x) \in k[x]$ is an irreducible polynomial, then $K = k[x]/(p(x))$ is a field containing (an isomorphic copy of) k and a root z of $p(x)$.*

(ii) *If $g(x) \in k[x]$ and z is a root of $g(x)$, then $p(x) \mid g(x)$.*

Proof. (i) Let us denote the ideal $(p(x))$ in $k[x]$ by I. The quotient ring $K = k[x]/I$ is a field, by Proposition 3.73, because $p(x)$ is irreducible. It is easy to see, using Corollary 3.29, that the restriction of the natural map, $\varphi : k \to K$, defined by $\varphi(a) = a + I$, is an isomorphism from k to the subfield $\{ a + I : a \in k \}$ of K; let us identify k with this subfield of K.

Remember that x is a particular element of $k[x]$; we claim that $z = x + I \in K$ is a root of $p(x)$. Now

$$p(x) = a_0 + a_1 x + \cdots + a_n x^n,$$

where $a_i \in k$ for all i. In $K = k[x]/I$, we have

$$p(z) = (a_0 + I) + (a_1 + I)z + \cdots + (a_n + I)z^n$$
$$= (a_0 + I) + (a_1 + I)(x + I) + \cdots + (a_n + I)(x + I)^n$$
$$= (a_0 + I) + (a_1 x + I) + \cdots + (a_n x^n + I)$$
$$= a_0 + a_1 x + \cdots + a_n x^n + I$$
$$= p(x) + I = I,$$

because $p(x) \in I = (p(x))$. But I is the zero element of $K = k[x]/I$, and so z is a root of $p(x)$.

(ii) We may regard both $p(x)$ and $g(x)$ as lying in $k[x]$, and each is a multiple of $x - z$ in $k[x]$; it follows that their $\gcd(g, p)$ in $k[x]$ is not 1. By Corollary 3.47, however, this gcd is the same whether computed in $k[x]$ or in $k[x]$. But $p(x)$ is irreducible, so that $(g, p) \neq 1$ in $k[x]$ gives $p \mid g$ in $k[x]$. •

In Example 3.24, $z = x + (x^2 + 1)$ corresponds to i, and we may denote the quotient ring by $\mathbb{R}(i)$.

The next theorem, giving a compact description of $k[x]/I$ when I is generated by an irreducible polynomial, is similar to Corollary 1.47, the description of $\mathbb{Z}_p$ as $\{[0], [1], \ldots, [p-1]\}$.

Theorem 3.75.   *Let $p(x) \in k[x]$ be irreducible of degree n, let $I = (p(x))$, and let $K = k[x]/I$. Then every element in $K = k[z]$ has a unique expression of the form*

$$b_0 + b_1 z + \cdots + b_{n-1} z^{n-1},$$

where $z = x + I$ is a root of $p(x)$ and all $b_i \in k$.

Proof.   Every element of K has the form $f(x) + I$, where $f(x) \in k[x]$. By the division algorithm, there are polynomials $q(x), r(x) \in k[x]$ with $f(x) = q(x)p(x) + r(x)$ and either $r(x) = 0$ or $\deg(r) < n = \deg(p)$. Since $f - r = qp \in I$, it follows that $f(x) + I = r(x) + I$. As in the proof of Proposition 3.74(i), we may rewrite $r(x) + I$ as $r(z) = b_0 + b_1 z + \cdots + b_{n-1} z^{n-1}$ with all $b_i \in k$.

To prove uniqueness, suppose that

$$r(z) = b_0 + b_1 z + \cdots + b_{n-1} z^{n-1} = c_0 + c_1 z + \cdots + c_{n-1} z^{n-1}.$$

Define $g(x) \in k[x]$ by $g(x) = \sum (b_i - c_i)x^i$. If $g(x)$ is not the zero polynomial, i.e., if some c_i differs from b_i, then $\deg(g) < n = \deg(p)$; but z is a root of $g(x)$, contradicting Proposition 3.74(i). •

Applying this theorem to Example 3.24 (in which $p(x) = x^2 + 1 \in \mathbb{R}[x]$, $n = 2$, and the coset $x + (p(x))$ is denoted by i), we see that every complex number has a unique expression of the form $a + bi$, where $a, b \in \mathbb{R}$, and that $i^2 + 1 = 0$, that is, $i^2 = -1$.

The easiest way to multiply in $\mathbb{C}$ is to first treat i as a variable and then to impose the condition $i^2 = -1$. For example, to compute $(a + bi)(c + di)$, first write $ac + (ad + bc)i + bdi^2$, and then observe that $i^2 = -1$. The proper way to multiply $(b_0 + b_1 z + \cdots + b_{n-1} z^{n-1})(c_0 + c_1 z + \cdots + c_{n-1} z^{n-1})$ in the quotient ring $k[x]/(p(x))$ is to first regard the factors as polynomials in z and then to impose the condition that $p(z) = 0$.

We are now going to generalize Example 3.24.

Definition. Let K be a field and let k be a subfield. If $z \in K$, then we define $k(z)$ to be the smallest subfield of K containing k and z; that is, $k(z)$ is the intersection of all the subfields of K containing k and z.

Proposition 3.76.   *Let k be a subfield of a field K and let $z \in K$.*

(i) *If z is a root of some nonzero polynomial $f(x) \in k[x]$, then z is a root of an irreducible polynomial $p(x) \in k[x]$.*

(ii) *There is an isomorphism*

$$\varphi : k[x]/(p(x)) \to k(z)$$

with $\varphi(x + (p(x))) = z$ and $\varphi(a) = a$ for all $a \in k$.

(iii) *If $z' \in K$ is another root of $p(x)$, then there is an isomorphism $\theta : k(z) \to k(z')$ with $\theta(z) = z'$ and $\theta(a) = a$ for all $a \in k$.*

Proof.   Our proof is essentially the same as that of Proposition 3.72. (i) Define a homomorphism $\varphi : k[x] \to K$ by $\varphi(x) = z$; in more detail,

$$\varphi : \sum a_i x^i \mapsto \sum a_i z^i.$$

Notice that $\varphi(a) = a$ for all $a \in k$. Now $f(x) \in \ker \varphi$ because z is a root of $f(x)$, and so $\ker \varphi$ is a nonzero ideal in $k[x]$; indeed, $\ker \varphi$ must have the form $(p(x))$ for some nonzero $p(x) \in k[x]$ because $k[x]$ is a PID. Since im φ is a subring of the field K, it is a domain, and so Example 3.23 says that $p(x)$ is irreducible.

(ii) Since $p(x)$ is irreducible, Proposition 3.73 gives im φ a field; that is, im φ is a subfield of K containing k and z, and so $k(z) \leq \text{im } \varphi$. On the other hand, every element in im φ has the form $g(x) + I$, where $g(x) \in k[x]$, so that any subfield S of K containing k and z must also contain im φ; that is, im $\varphi = k(z)$.

(iii) As in part (ii), there are isomorphisms $\varphi : k[x]/(p(x)) \rightarrow k(z)$ and $\psi : k[x]/(p(x)) \rightarrow k(z')$ with $\varphi(a) = a$ and $\psi(a) = a$ for all $a \in k$; moreover, $\varphi : x + (p(x)) \mapsto z$ and $\psi : x + (p(x)) \mapsto z'$. The composite $\theta = \varphi^{-1}\psi$ is the desired isomorphism.   •

We now prove two main results: the first, due to Kronecker, says that if $f(x) \in k[x]$, where k is any field, then there is some larger field E which contains k and all the roots of $f(x)$; the second, due to Galois, constructs finite fields other than $\mathbb{Z}_p$.

Theorem 3.77 (Kronecker).   *If k is a field and $f(x) \in k[x]$, then there exists a field E, containing k as a subfield, such that $f(x)$ is a product of linear polynomials in $E[x]$.*

Proof.   The proof is by induction on $\deg(f)$. If $\deg(f) = 1$, then $f(x)$ is linear and we can choose $E = k$. If $\deg(f) > 1$, write $f(x) = p(x)g(x)$, where $p(x)$ is irreducible. If $p(x)$ is linear, then $f(x)$ factors as a product of linears whenever $g(x)$ does; moreover, such a factorization of $g(x)$ (over some larger field) does exist, by induction, because $\deg(g) < \deg(f)$. If $\deg(p) > 1$, then Proposition 3.74(i)(i) provides a field F containing k and a root z of $p(x)$. Hence $p(x) = (x - z)h(x)$ in $F[x]$. By induction, there is a field E containing F (and hence k) so that $h(x)g(x)$, and hence $f(x) = (x - z)h(x)g(x)$, is a product of linear factors in $E[x]$.   •

For the familiar fields $\mathbb{Q}$, $\mathbb{R}$, and $\mathbb{C}$, Kronecker's theorem offers nothing new. The *Fundamental Theorem of Algebra*, first proved by Gauss in 1799 (completing earlier attempts of Euler and of Lagrange), says that every non-constant $f(x) \in \mathbb{C}[x]$ has a root in $\mathbb{C}$; it follows, by induction on the degree of $f(x)$, that all the roots of $f(x)$ lie in $\mathbb{C}$; that is, $f(x) = a(x - r_1) \ldots (x - r_n)$, where $a \in \mathbb{C}$ and $r_j \in \mathbb{C}$ for all j. On the other hand, if $k = \mathbb{Z}_p$ or $k = \mathbb{C}(x) = \mathrm{Frac}(\mathbb{C}[x])$, then Kronecker's theorem does apply to tell us, for any given $f(x)$, that there is always some larger field E that contains all the roots of $f(x)$. For example, there is some field containing $\mathbb{C}(x)$ and $\sqrt{x}$. There is a general version of the fundamental theorem: Every field k is a subfield of an **algebraically closed** field K, that is, K is a field containing k such that every $f(x) \in k[x]$ is a product of linear polynomials in $K[x]$. In contrast, Kronecker's theorem gives roots of just one polynomial at a time.

The next group-theoretic result will help us describe finite fields.

Theorem 3.78. *If k is a field and G is a finite subgroup of the multiplicative group $k^{\times}$, then G is cyclic. In particular, if k itself is finite (e.g., $k = \mathbb{Z}_p$), then $k^{\times}$ is cyclic.*

Proof. Let p be a prime divisor of $|G|$. If there are two subgroups of G of order p, say, S and T, then $|S \cup T| > p$. But each $a \in S \cup T$ satisfies $a^p = 1$, and hence it is a root of $x^p - 1$. This contradicts Theorem 3.33(i), for $x^p - 1$ now has too many roots in k. Thus, G is cyclic, by Proposition 2.65.   •

Although the multiplicative groups $\mathbb{Z}_p^{\times}$ are cyclic, no explicit formula giving generators of each of them is known; that is, no algorithm is known that computes $s(p)$ for every prime p, where $[s(p)]$ is a generator of $\mathbb{Z}_p^{\times}$.

Corollary 3.79. *Every finite field k has exactly p^n elements for some prime p and some $n \geq 1$.*

Proof. Since k is finite, it must have characteristic p for some prime p (see Proposition 3.72); let $k \cong \mathbb{Z}_p$ be its prime field. As k is finite, $k^{\times}$ is a cyclic group, say, with generator z (by Theorem 3.78). Let $\varphi : k[x] \to k$ be the homomorphism with $\varphi(x) = z$ and $\varphi(a) = a$ for all $a \in \mathbb{Z}_p$ (see the proof of Proposition 3.76). Now φ is surjective, for the powers of z already give all of $k^{\times}$, while $\ker \varphi = (q(x))$ for some $q(x) \in k[x]$. By the first isomorphism theorem, $k[x]/(q(x)) \cong k$, and by Example 3.23, $q(x)$ is irreducible in $k[x]$. Finally, Theorem 3.75 shows that every element of k has a unique expression of the form $b_0 + b_1 z + \cdots + b_{n-1} z^{n-1}$, where $b_i \in k \cong \mathbb{Z}_p$ and $n = \deg(q)$, which shows that there are exactly p^n elements in k.   •

Remark.   There is a much simpler proof of this last theorem that uses linear algebra. One shows that k is an n-dimensional vector space over $\mathbb{Z}_p$ for some prime p and some $n \geq 1$ and, as such, it consists of all n-tuples with entries in $\mathbb{Z}_p$.   ◄

We are now going to display the finite fields.

Theorem 3.80 (*Galois*).   *If p is a prime and n is a positive integer, then there is a field that has exactly p^n elements.*

Proof.   Write $q = p^n$, and consider the polynomial

$$g(x) = x^q - x \in \mathbb{Z}_p[x].$$

By Kronecker's theorem, there is a field E containing $\mathbb{Z}_p$ such that $g(x)$ is a product of linear factors in $E[x]$. Define

$$F = \{\alpha \in E : g(\alpha) = 0\};$$

thus, F is the set of all the roots of $g(x)$. Since the derivative $Dg(x) = qx^{q-1} - 1 = p^n x^{q-1} - 1 = -1$, it follows that the $\gcd(g, Dg)$ is 1. By Exercise 3.73, all the roots of $g(x)$ are distinct; that is, F has exactly $q = p^n$ elements.

We claim that F is a subfield of E, and this will complete the proof. If $a, b \in F$, then $a^q = a$ and $b^q = b$. Therefore, $(ab)^q = a^q b^q = ab$, and $ab \in F$. By Exercise 3.44(iii), $(a-b)^q = a^q - b^q = a - b$, so that $a - b \in F$. Finally, if $a \neq 0$, then the cancellation law applied to $a^q = a$ gives $a^{q-1} = 1$, and so the inverse of a is a^{q-2} (which lies in F because F is closed under multiplication).   •

We will see in the next chapter that any two finite fields k with the same number of elements are isomorphic.

Example 3.25.
In Exercise 3.22, we constructed a field with 4 elements $k = \{0, e, a, a + e\}$. Under addition, k is isomorphic to the four-group. We now know that $k^\times$ is cyclic of order 3, so that the multiplication is given by $a^2 = a + e$ and $a^3 = e$.

On the other hand, we may construct a field of order 4 as the quotient $F = \mathbb{Z}_2[x]/(q(x))$, where $q(x) = x^2 + x + 1 \in \mathbb{Z}_2[x]$ is irreducible. By Corollary 3.75, F is a field consisting of all $u + vz$, where $z = x + (q(x))$ is a root of $q(x)$ and $u, v \in \mathbb{Z}_2$. Since $z^2 + z + 1 = 0$, we have $z^2 = -z - 1 = z + 1$; moreover, $z^3 = zz^2 = z(z+1) = z^2 + z = 1$. It is now easy to see that there is a ring isomorphism $\varphi : k \to F$ with $\varphi(a) = z$.   ◀

Example 3.26.
According to the table in Example 3.21, there are three monic irreducible quadratics in $\mathbb{Z}_3[x]$, namely,

$$p(x) = x^2 + 1, \quad q(x) = x^2 + x - 1, \quad \text{and } r(x) = x^2 - x - 1;$$

each gives rise to a field with $9 = 3^2$ elements. Let us look at the first two in more detail. Corollary 3.75 says that $E = \mathbb{Z}_3[x]/(q(x))$ is given by

$$E = \{a + b\alpha : \text{ where } \alpha^2 + 1 = 0\}.$$

Similarly, if $F = \mathbb{Z}_3[x]/(p(x))$, then

$$F = \{a + b\beta : \text{ where } \beta^2 + \beta - 1 = 0\}.$$

The reader can show that these two fields are isomorphic by checking that $\varphi : E \to F$, defined by

$$\varphi(a + b\alpha) = a + b(1 - \beta),$$

is an isomorphism.

Now $\mathbb{Z}_3[x]/(x^2 - x - 1)$ is also a field with 9 elements, and one can show that it is isomorphic to both of the two fields E and F given above.

We have exhibited 10 monic irreducible cubics $p(x) \in \mathbb{Z}_3[x]$; each of them gives rise to a field $\mathbb{Z}_3[x]/(p(x))$ having $27 = 3^3$ elements. All ten of these fields are isomorphic. ◄

EXERCISES

3.80 If $I = \{0\}$, then $R/I \cong R$.

3.81 For every commutative ring R, prove that $R[x]/(x) \cong R$.

3.82 Let k be a field and $f(x), g(x) \in k[x]$ be relatively prime. If each divides $h(x)$ in $k[x]$, prove that their product $f(x)g(x)$ also divides $h(x)$.

3.83 *Chinese Remainder Theorem.*

    (i) Prove that if k is a field and $f(x), f'(x) \in k[x]$ are relatively prime, then given $b(x), b'(x) \in k[x]$, there exists $c(x) \in k[x]$ with

$$c - b \in (f) \text{ and } c - b' \in (f');$$

    moreover, if $d(x)$ is another common solution, then $c - d \in (ff')$.

    (ii) Prove that if k is a field and $f(x), g(x) \in k[x]$ are relatively prime, then

$$k[x]/(f(x)g(x)) \cong k[x]/(f(x)) \times k[x]/(g(x)).$$

3.84   (i) Prove that a field K cannot have subfields k' and k'' with $k' \cong \mathbb{Q}$ and $k'' \cong \mathbb{Z}_p$ for some prime p.

    (ii) Prove that a field K cannot have subfields k' and k'' with $k' \cong \mathbb{Z}_p$ and $k'' \cong \mathbb{Z}_q$, where $p \neq q$.

3.85 If p is a prime and $p \equiv 3 \mod 4$, prove that $a^2 \equiv 2 \mod p$ or $a^2 \equiv -2 \mod p$ is solvable.

3.86   (i) Prove that $x^4 + 1$ factors in $\mathbb{Z}_2[x]$.

      (ii) If $x^4 + 1 = (x^2 + ax + b)(x^2 + cx + d)$, prove that $c = -a$ and

$$d + b - a^2 = 0$$
$$a(d - b) = 0$$
$$bd = 1.$$

      (iii) Prove that $x^4 + 1$ factors in $\mathbb{Z}_p[x]$ if any of the following congruences are solvable:

$$b^2 \equiv -1 \bmod p,$$
$$a^2 \equiv 2 \bmod p,$$
$$a^2 \equiv -2 \bmod p.$$

      (iv) Prove that $x^4 + 1$ factors in $\mathbb{Z}_p[x]$ for all primes p.

3.87 If F is a field with four elements, prove that the stochastic group $\Sigma(2, F) \cong A_4$.

3.88 Let $f(x) = s_0 + s_1 x + \cdots + s_{n-1}x^{n-1} + s_n x^n \in k[x]$, where k is a field, and suppose that $f(x) = (x - \alpha_1)(x - \alpha_2) \ldots (x - \alpha_n)$. Prove that $s_{n-1} = -(\alpha_1 + \alpha_2 + \cdots + \alpha_n)$ and that $s_0 = (-1)^n \alpha_1 \alpha_2 \ldots \alpha_n$. Conclude that the sum and the product of all the roots of $f(x)$ lie in k.

3.89 Write addition and multiplication tables for a field with eight elements.

3.90 This exercise gives another proof of Proposition 3.79.

      (i) Let F be a finite field, but consider it only as a group under addition. Show, for each pair x and y of nonzero elements in F, that there is an isomorphism $\varphi : F \to F$ with $y = \varphi(x)$.

      (ii) Prove that $|F| = p^n$ for some prime p and some $n \geq 1$.

3.9 OFFICERS, FERTILIZER, AND A LINE AT INFINITY

OFFICERS

In 1782, L. Euler posed the following problem in an article he was writing about magic squares. Suppose there are 36 officers of 6 ranks and from 6 regiments. If the regiments are numbered 1 through 6 and the ranks are captain, major, lieutenant, ... , then each officer has a double label: e.g., captain 3 or major 4. Euler asked whether there is a 6×6 formation of these officers so that each row and each column contains exactly one officer of each rank and one officer from each regiment. Thus, no row can have two captains in it, nor can any column; no row can have two officers from the same regiment, nor can any column.

The problem is made clearer by the following definitions.

Definition. An $n \times n$ **Latin square** is an $n \times n$ matrix whose entries are taken from a set of n objects so that no object occurs twice in any row or column.

It is easy to see that an $n \times n$ matrix A all of whose entries lie in $X = \{1, 2, \ldots, n\}$ is a Latin square if and only if every row and every column of A is a permutation of X.

As is customary, we label the entries of a matrix A as a_{ij}, the first index i describing the ith row

$$a_{i1} \; a_{i2} \; a_{i3} \; \ldots \; a_{in},$$

and the second index j describing the jth column

$$a_{1j}$$
$$a_{2j}$$
$$a_{3j}$$
$$\vdots$$
$$a_{nj}$$

The matrix A is often denoted by $[a_{ij}]$.

Example 3.27.
Since the cancellation laws hold in groups, the multiplication table of a group $G = \{a_1, \ldots, a_n\}$, namely, $[a_i a_j]$, is a Latin square, for $a_i a_j = a_i a_k$ implies $a_j = a_k$ and $a_i a_\ell = a_j a_\ell$ implies $a_i = a_j$. ◀

Example 3.28.
There are exactly two 2×2 Latin squares having entries 1 and 2:

$$\begin{bmatrix} 1 & 2 \\ 2 & 1 \end{bmatrix} \quad \text{and} \quad \begin{bmatrix} 2 & 1 \\ 1 & 2 \end{bmatrix} \quad ◀$$

Example 3.29.
Here are two 4×4 Latin squares.

$$\begin{bmatrix} 1 & 2 & 3 & 4 \\ 2 & 1 & 4 & 3 \\ 3 & 4 & 1 & 2 \\ 4 & 3 & 2 & 1 \end{bmatrix} \quad \text{and} \quad \begin{bmatrix} 1 & 2 & 3 & 4 \\ 3 & 4 & 1 & 2 \\ 4 & 3 & 2 & 1 \\ 2 & 1 & 4 & 3 \end{bmatrix} \quad ◀$$

Sec. 3.9   (

Thus, $ax +$

Since $a \neq$ (
that is, x'
squares.

Corollary
thogonal p

Proof. If
by Corollai
need $|k^\times| \geq$

Remark.
constructed
cated) way.

We now
ones. First,
then aB is
$k \times k$ matr
product A

Theorem
pair of $n \times$

Proof. W
if A and E
proves that
are orthog
$k\ell \times k\ell$ Là
We have
squares, bu

Example 3.30.

If $A = [a_{ij}]$ is an $n \times n$ Latin square, and if $\alpha \in S_n$, then A^α is defined to be the matrix A after its columns have been permuted by α; that is, the entries of A^α are $a_{i,\alpha j}$.

We claim that if A is a Latin square, then so is A^α. Of course, each column of A^α is a permutation of $\{1, 2, \ldots, n\}$. The entries in the ith row of A are $a_{i,1}, a_{i,2}, \ldots, a_{i,n}$, which is some permutation of the first row: say,

$$a_{i,1} = \sigma(a_{1,1}), \quad \ldots, \quad a_{i,n} = \sigma(a_{1,n}).$$

The ith row of A^α is

$$\alpha\sigma(a_{1,1}), \quad \ldots, \quad \alpha\sigma(a_{1,n}),$$

which is also a permutation of the first row. Thus, the columns and the rows are permutations, and so A^α is a Latin square. ◀

Definition.   Two $n \times n$ Latin squares $[a_{ij}]$ and $[b_{ij}]$ are called ***orthogonal*** if the n^2 ordered pairs (a_{ij}, b_{ij}) are all distinct.

Since there are n^2 ordered pairs in $X \times X$, where $X = \{1, 2, \ldots, n\}$, every possible ordered pair (a_{ij}, b_{ij}) in $X \times X$ arises from an orthogonal pair $A = [a_{ij}]$ and $B = [b_{ij}]$ of Latin squares.

There is no orthogonal pair of 2×2 Latin squares: as we saw in Example 3.28, there are only two 2×2 Latin squares, and if we superimpose one on the other, we get

$$\begin{bmatrix} 12 & 21 \\ 21 & 12 \end{bmatrix}.$$

There are only two distinct ordered pairs, not four as the definition requires. On the other hand, the two 4×4 Latin squares in Example 3.28 are orthogonal, for all 16 ordered pairs are distinct.

$$\begin{bmatrix} 11 & 22 & 33 & 44 \\ 23 & 14 & 41 & 32 \\ 34 & 43 & 12 & 21 \\ 42 & 31 & 24 & 13 \end{bmatrix}$$

Example 3.
If A and B
B^β are orth
  We have
suffices to
would prov
pairs that a

If we call α

Orthogonal
orthogonal

  Euler's
squares (th
regiment).
than he wa
construct s

Propositio

  (i) *If k*

        whe

  (ii) *If a,*

Proof. (i
fixed. The
the yth col
are distinc
cancellatic

(ii) Suppo

any odd prime p and any $e \geq 1$, there is an orthogonal pair of $p^e \times p^e$ Latin squares. Taking Kronecker products thus produces orthogonal pairs of $n \times n$ Latin squares for every positive n that either is odd or is a multiple of 4. Now every number $n \equiv 0, 1, 2,$ or $3 \bmod 4$. If $n \equiv 0 \bmod 4$, then n is a multiple of 4; if $n \equiv 1$ or $3 \bmod 4$, then n is odd. In these cases, therefore, there exists an orthogonal pair of $n \times n$ Latin squares; that is, an orthogonal pair exists if $n \not\equiv 2 \bmod 4$. •

The smallest n not covered by Euler's theorem is $n = 6$, and this is why Euler posed the question of the 36 officers. Indeed, he conjectured that there is no orthogonal pair of $n \times n$ Latin squares if $n \equiv 2 \bmod 4$. In 1901, G. Tarry proved that there does not exist an orthogonal pair of 6×6 Latin squares, thus answering Euler's question posed at the beginning of this section: there is no such formation of 36 officers. However, in 1958, E. T. Parker discovered an orthogonal pair of 10×10 Latin squares, thereby disproving Euler's conjecture (Parker's example is displayed on the front cover of this book; Table 3.1 is a version of it). Indeed, Parker, R. C. Bose, and S. S. Shrikhande went on to prove that there exists a pair of orthogonal $n \times n$ Latin squares for all n except 2 and 6.

00	15	23	32	46	51	64	79	87	98
94	77	10	25	52	49	01	83	68	36
71	34	88	17	20	02	43	65	96	59
45	81	54	66	18	27	72	90	39	03
82	40	61	04	99	16	28	37	53	75
26	62	47	91	74	33	19	58	05	80
13	29	92	48	31	84	55	06	70	67
69	93	35	50	07	78	86	44	12	21
57	08	76	89	63	95	30	11	24	42
38	56	09	73	85	60	97	22	41	14

Table 3.1.

FERTILIZER

Here is a fertilizer story. To maximize his corn production, a farmer has to choose the best type of seed. But he knows that the amount of fertilizer also affects his crop. How can he design an experiment to show him what is the best combination? We give a simple illustration. Suppose there are three types of seed: A, B, and C. To measure the effect of using different amounts of fertilizer, the farmer can divide a plot into 9 subplots, as follows:

Amount of Fertilizer	Seed Type		
High	A	B	C
Medium	A	B	C
Low	A	B	C

In each position, an observation x_{sf} is made, where x_{sf} is the number of ears harvested according to the seed type s and level f of fertilizer.

The farmer now wants to see the effect of differing uses of pesticide. He could have 27 observations x_{sfp} (more generally, if he had n different dosages and n different seed types, there would be n^3 observations). On the other hand, suppose he arranges his experiment as follows (again, we illustrate with $n = 3$).

Amount of Fertilizer	Amount of Pesticide		
	High	Medium	Low
High	A	B	C
Medium	C	A	B
Low	B	C	A

The seed types are now arranged in a Latin square. For example, the observation from the northwest subplot is the number of ears from seed type A, with a high level of fertilizer, and high level of pesticide. There are only 9 observations instead of 27 (more generally, there are n^2 observations instead of n^3). Obviously we do not have all possible observations. To infer properties about a large collection from measuring a small sample is what statistics is all about. And it turns out that the Latin square organization of data gives essentially the same statistical information as that given by the complete set of all n^3 observations. A discussion of the analysis of variance for such designs can be found, for example, in Li, *An Introduction to Experimental Statistics*.

The farmer now wants to consider water amounts. Again we illustrate with $n = 3$. In addition to the seed types A, B, C, and the various levels of fertilizer and pesticide, let there be three water levels: $\alpha > \beta > \gamma$.

The observation from the northwest subplot, for example, is the number of ears of seed type A, high level of fertilizer, high level of pesticide, and high level of water. Again, the statistical data arising from this small number, namely 9, of observations are essentially the same as one would get from 81 observations x_{sfpw} (more generally, n^2 observations instead of n^4). Euler called such matrices *Graeco-Latin squares*, because he described them, as

Amount of Fertilizer		Amount of Pesticide	
	High	Medium	Low
High	$A\alpha$	$B\beta$	$C\gamma$
Medium	$C\beta$	$A\gamma$	$B\alpha$
Low	$B\gamma$	$C\alpha$	$A\beta$

above, using Latin and Greek fonts; he invented the term *Latin square* for this same notational reason. We recognize an orthogonal pair of Latin squares. One could test more variables if one could find an *orthogonal set* of Latin squares as defined below.

Definition. A set $A_1, A_2, \ldots, A_t$ of $n \times n$ Latin squares is an **orthogonal set** if each pair of them is orthogonal.

Lemma 3.84. *If $A_1, A_2, \ldots, A_t$ is an orthogonal set of $n \times n$ Latin squares, then $t \leq n - 1$.*

Proof. We may assume that the first row of A_1 has entries $1, 2, \ldots, n$ in this order, for after permuting the columns of A_1 so that its first row is as above, then Example 3.31 shows that the rearranged A_1 is still orthogonal to all the other A_i. Now rearrange A_2 so that its first row has entries $1, 2, \ldots, n$ in this order; the argument just given shows that the rearranged Latin squares are still orthogonal; now rearrange $A_3, \ldots, A_t$ in the same way. There is thus no loss in generality in assuming that all the A_i have first row with entries $1, 2, \ldots, n$ in this order. Thus, the first rows from A_i and A_j give the ordered pairs

$$(1, 1), (2, 2), \ldots, (n, n).$$

In any A_i, there are only $n - 1$ possible entries for its $2, 1$ position because 1 already occurs in the $1, 1$ position. We claim that if $A_i \neq A_j$, then they do not have the same $2, 1$ entry, say, k. Otherwise, $a_{21}^i = k = a_{21}^j$, and

$$(a_{21}^i, a_{21}^j) = (k, k) = (a_{1k}^i, a_{1k}^j).$$

This contradicts the orthogonality of A_i and A_j, for the ordered pair (k, k) has already arisen from their first rows. Therefore, distinct A_i have distinct entries in the $2, 1$ position, and so there are at most $n - 1$ distinct A_i's. •

Definition. A *complete orthogonal set* of $n \times n$ Latin squares is an orthogonal set of $n - 1$ Latin squares.

Theorem 3.85. *If* $n = p^e$, *then there exists a complete orthogonal set of* $n - 1$ *Latin squares.*

Proof. If k is a finite field with n elements, then there are $n - 1$ elements $a \in k^{\times}$, and so there are $n - 1$ Latin squares L_a, each pair of which is orthogonal, by Theorem 3.81.   ●

One Latin square can test two variables (e.g., levels of fertilizer and pesticide) on different varieties (e.g., of corn). A Graeco-Latin square, i.e., a pair of orthogonal Latin squares, allows testing for a third variable (e.g., levels of water). More generally, a set of t orthogonal Latin squares allows one to test levels of $t + 1$ different variables on different varieties.

A LINE AT INFINITY

And now a third stream enters the story. By the early 1800s, mathematicians were studying the problems of perspective arising from artists painting pictures of three-dimensional scenes on two-dimensional canvases. To the eye, parallel lines seem to meet at the horizon, and this suggests adjoining a "line at infinity" to the ordinary plane. Every line is parallel to a line ℓ passing through the origin O. For each such line, define a new point ω_ℓ, and we "lengthen" every line parallel to ℓ by adjoining this new point to it. Finally, we decree that all the new points ω_ℓ, for all lines ℓ through the origin, comprise a new line, the ***line at infinity***, or the horizon. If ℓ_1 and ℓ_2 are (lengthened) parallel lines, and if ℓ is the line through O parallel to each of them, then ℓ_1 and ℓ_2 intersect in the point ω_ℓ. The reader may check that the familiar property that every two points determine a unique line is now accompanied by the property that every two lines determine a unique point: it is the usual point of intersection if the lines are not parallel, and it is a point ω_ℓ on the horizon if the lines are parallel.

Since we are now interested in finite structures, let us replace the plane $\mathbb{R} \times \mathbb{R}$ by a finite "plane" $k \times k$, where k is a finite field with q elements. We regard this finite plane as the additive group $k \times k$. Define a ***line*** ℓ ***through the origin*** O to be a subgroup of the form

$$\ell = \{(ax, ay) : a \in k \quad \text{and} \quad (x, y) \neq 0\},$$

and, more generally, define a ***line*** to be a coset $(u, v) + \ell$, where ℓ is a line through the origin. Since k is finite, we can do some counting. There are q^2 points in the plane, and there are q points on every line. As usual, two points determine a line. Call two lines ***parallel*** if they do not intersect, and say that two lines have the same **direction** if they are parallel. How many directions are there? Every line, being a coset of a line ℓ through the origin, has the same direction as ℓ, whereas distinct lines through the origin have different directions, for they intersect. Thus, the number of directions is the same as the number of lines through the origin. There are $q^2 - 1$ points $V \neq O$, each of which determines a line $\ell = OV$ through the origin. Since there are q points on ℓ, there are $q - 1$ points on ℓ other than O, and each of them determines ℓ. There are thus

$$(q^2 - 1)/(q - 1) = q + 1$$

directions. We adjoin $q + 1$ new points ω_ℓ to $k \times k$, one for each direction, that is, one for each line ℓ through the origin. Define ℓ_∞, the **line at infinity**, by

$$\ell_\infty = \{\omega_\ell : \ell \text{ is a line through the origin}\},$$

and define the **projective plane** over k

$$P(k) = (k \times k) \cup \ell_\infty.$$

Define a (projective) **line** in $P(k)$ to be either ℓ_∞ or an old line $(u, v) + \ell$ in $k \times k$ with the point ω_ℓ adjoined, where ℓ is a line through the origin. It follows that $|P(k)| = q^2 + q + 1$, every line has $q + 1$ points, and any two points determine a unique line.

Example 3.32.
If $k = \mathbb{Z}_2$, then $k \times k$ has 4 points:

$$O = (0, 0), \qquad a = (1, 0), \qquad b = (0, 1), \qquad \text{and} \qquad c = (1, 1)$$

and 6 lines, each with two points, as in Figure 3.3.

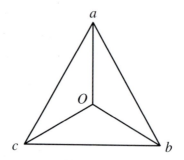

Figure 3.3

There are three sets of parallel lines:

(i) Oa and bc

(ii) Ob and ac

(iii) Oc and ab.

The projective plane $P(\mathbb{Z}_2)$ is obtained by adding new points $\omega_1, \omega_2, \omega_3$ and forcing parallel lines to meet.

There are now 7 lines: the 6 original lines (each lengthened) and the line at infinity $\{\omega_1, \omega_2, \omega_3\}$. This figure is often called a ***Fano plane*** (see Figure 3.4).

◀

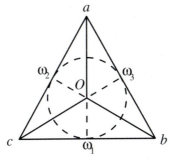

Figure 3.4

We now abstract the features we need.

Definition.   A ***projective plane of order n*** is a set X with $|X| = n^2 + n + 1$, a family of subsets called ***lines***, each having $n + 1$ points, such that every two points determine a unique line.

We have seen above that if k is a finite field with q elements, then $P(k)$ is a projective plane of order q. It is possible to construct projective planes without using finite fields. For example, it is known that there are four projective planes of order 9, only one of which arises from a finite field.

The following theorem is the reason we have introduced projective planes.

Theorem.   *If $n \geq 3$, then there exists a projective plane of order n if and only if there exists a complete orthogonal set of $n \times n$ Latin squares.*

Proof.   See Ryser, *Combinatorial Mathematics*, p. 92.   •

A natural question is to find those n for which there exists a projective plane of order n. Notice that this is harder than Euler's original question;

instead of asking whether there is an orthogonal pair of $n \times n$ Latin squares, we are now asking whether there is an orthogonal set of $n - 1$ $n \times n$ Latin squares. If $n = p^e$, then we have constructed a projective plane of order n above. Since Tarry proved that there is no orthogonal pair of 6×6 Latin squares, there is no set of 5 pairwise orthogonal 6×6 Latin squares, and so there is no projective plane of order 6. The following theorem was proved in 1949.

Theorem (Bruck–Ryser).   *If either $n \equiv 1 \bmod 4$ or $n \equiv 2 \bmod 4$ and, further, if n is not a sum of two squares, then there does not exist a projective plane of order n.*

Proof.   See Ryser, *Combinatorial Mathematics*, p. 111.   •

The first few $n \equiv 1$ or $2 \bmod 4$ are

$$5, \ 6, \ 9, \ 10, \ 13, \ 14, \ 17, \ 18, \ 21, \ 22.$$

Some of these are primes or prime powers, and so they must be sums of squares (because projective planes of these orders do exist), and so it is:

$$5 = 1 + 4; \qquad 9 = 0 + 9; \qquad 13 = 4 + 9; \qquad 17 = 1 + 16.$$

Of the remaining numbers, $10 = 1 + 9$ and $18 = 9 + 9$ are sums of squares (and the theorem does not apply), but the others are not sums of squares. It follows that there is no projective plane of order 6, 14, 21, or 22 (thus, Tarry's result follows from the Bruck–Ryser theorem).

The smallest n not covered by the Bruck–Ryser theorem is $n = 10$. The question whether there exists a projective plane of order 10 was the subject of much investigation (after Tarry, 10 was also the first open case of Euler's conjecture). This is a question about a set with 111 points, and so one would expect that a computer could solve it quickly. But it is really a question about 11-point subsets of a set with 111 points, the order of magnitude of which is the binomial coefficient $\binom{111}{11}$, a huge number. In spite of this, C. Lam was able to show, in 1988, that there does not exist a projective plane of order 10. He used a massive amount of calculation: 19,200 hours on VAX 11/780 followed by 3000 hours on CRAY-1S. Thus, two and half years of actual computer running time (not counting the years of human thought and ingenuity involved in instructing the machines) solved the problem. As of this writing, it is unknown whether a projective plane of order 12 exists (of course, $12 \equiv 0 \bmod 4$, and so it is not mentioned in the Bruck–Ryser theorem).

4

Goodies

4.1 LINEAR ALGEBRA

Vector Spaces

Linear algebra is the study of vector spaces and their homomorphisms, with applications to systems of linear equations. From now on, we are going to assume that most readers have had some course involving $n \times n$ matrices, probably with real entries or with complex entries (we have already discussed elementary facts about 2×2 matrices in some detail). Nowadays, such courses deal mainly with computational aspects of the subject, such as Gaussian elimination, and finding inverses, determinants, eigenvalues, and characteristic polynomials of matrices, but here we do not emphasize this important aspect of linear algebra. Instead, we discuss more theoretical properties of vector spaces (with scalars in any field) and linear transformations (which are homomorphisms between vector spaces).

Dimension is a rather subtle idea. One thinks of a curve in the plane, that is, the image of a continuous function $f : \mathbb{R} \to \mathbb{R}^2$, as a one-dimensional subset of a two-dimensional ambient space. Imagine the confusion at the end of the nineteenth century when a "space-filling curve" was discovered: there exists a continuous $f : \mathbb{R} \to \mathbb{R}^2$ with image the whole plane! We are going to describe a way of defining dimension that works for analogs of euclidean space, called vector spaces (there are topological ways of defining dimension of more general spaces).

Definition.   If k is a field, then a ***vector space over k*** is an (additive) abelian group V equipped with a ***scalar multiplication***; that is, there is a function $k \times V \to V$, denoted by $(a, v) \mapsto av$, such that, for all $a, b, 1 \in k$ and all $u, v \in V$,

  (i) $a(u + v) = au + av$;

  (ii) $(a + b)v = av + bv$;

  (iii) $(ab)v = a(bv)$;

  (iv) $1v = v$.

The elements of V are called ***vectors***[1] and the elements of k are called ***scalars***.[2]

Example 4.1.
(i) Euclidean space $V = \mathbb{R}^n$ is a vector space over $\mathbb{R}$. Vectors are n-tuples $v = (a_1, \ldots, a_n)$, where $a_i \in \mathbb{R}$ for all i. One pictures a vector v as an arrow from the origin to the point having coordinates $(a_1, \ldots, a_n)$. Addition is given by

$$(a_1, \ldots, a_n) + (b_1, \ldots, b_n) = (a_1 + b_1, \ldots, a_n + b_n);$$

geometrically, the sum of two vectors is described by the *parallelogram law*. Scalar multiplication is given by

$$av = a(a_1, \ldots, a_n) = (aa_1, \ldots, aa_n).$$

Scalar multiplication $v \mapsto av$ "stretches" v by a factor $|a|$, reversing its direction when a is negative (we put quotes around *stretches* because av is shorter than v when $|a| < 1$).

(ii) The example above can be generalized. If k is any field, define $V = k^n$, the set of all n-tuples $v = (a_1, \ldots, a_n)$, where $a_i \in k$ for all i. Addition is given by

$$(a_1, \ldots, a_n) + (b_1, \ldots, b_n) = (a_1 + b_1, \ldots, a_n + b_n),$$

---

[1]The word *vector* comes from the Latin root meaning "to carry." In calculus, vectors "carry" the data of length and direction.

[2]The word *scalar* arises from regarding scalar multiplication $v \mapsto av$ as a change of scale. The terms *scalar* and *scale* come from the Latin word meaning "ladder," for the rungs of a ladder are equally spaced.

and scalar multiplication is given by

$$av = a(a_1, \ldots, a_n) = (aa_1, \ldots, aa_n).$$

It is easy to check that V so defined is a vector space over k.

(iii) If k is a subfield of a field K, then K is a vector space over k. Regard the elements of K as vectors and the elements of k as scalars; define scalar multiplication av, where $a \in k$ and $v \in K$, to be the given product of two elements in K. Notice that the axioms in the definition of vector space are just particular cases of some of the axioms holding in the field K.

For example, $\mathbb{C}$ is a vector space over $\mathbb{R}$, and both $\mathbb{C}$ and $\mathbb{R}$ are vector spaces over $\mathbb{Q}$. Indeed, every field is a vector space over its prime field. In particular every finite field is a vector space over $\mathbb{Z}_p$ for some prime p.

(iv) One can relax the hypothesis in the example just given: if K is a commutative ring containing a subring k that is a field, then K is a vector space over k. For example, if k is a field, then the polynomial ring $K = k[x]$ is a vector space over k. Vectors are polynomials $f(x)$, scalars are elements $a \in k$, and scalar multiplication gives the polynomial $af(x)$; that is, if

$$f(x) = b_n x^n + \cdots + b_1 x + b_0,$$

then

$$af(x) = ab_n x^n + \cdots + ab_1 x + ab_0. \quad \blacktriangleleft$$

Definition.   If V is a vector space over a field k, then a **subspace** of V is a subset U of V such that

  (i) $0 \in U$;

  (ii) $u, u' \in U$ imply $u + u' \in U$;

  (iii) $u \in U$ and $a \in k$ imply $au \in U$.

When U is a subspace of V, we often write $U \leq V$.

Example 4.2.
(i) The extreme cases $U = V$ and $U = \{0\}$ (where $\{0\}$ denotes the subset consisting of the zero vector alone) are always subspaces of a vector space. A subspace $U \leq V$ with $U \neq V$ is called a **proper subspace** of V. We may write $U < V$ to denote U being a proper subspace of V.

(ii) If $v = (a_1, \ldots, a_n)$ is a nonzero vector in $\mathbb{R}^n$, then

$$\ell = \{av : a \in \mathbb{R}\}$$

is a subspace of $\mathbb{R}^n$; it is a line through the origin.

Similarly, a plane through the origin consists of all vectors of the form $av_1 + bv_2$, where v_1, v_2 is a fixed pair of noncollinear vectors, and a, b vary over $\mathbb{R}$. It is easy to check that planes through the origin are subspaces of $\mathbb{R}^n$.

(iii) If $m \leq n$ and $\mathbb{R}^m$ is regarded as the set of all those vectors in $\mathbb{R}^n$ whose last $n - m$ coordinates are 0, then $\mathbb{R}^m$ is a subspace of $\mathbb{R}^n$. For example, we may regard the plane $\mathbb{R}^2$ as all points $(x, y, 0)$ in $\mathbb{R}^3$.

(iv) A homogeneous linear system of m equations in n unknowns is a set of equations

$$a_{11}x_1 + \cdots + a_{1n}x_n = 0$$
$$a_{21}x_1 + \cdots + a_{2n}x_n = 0$$
$$\vdots \qquad \qquad \vdots$$
$$a_{m1}x_1 + \cdots + a_{mn}x_n = 0,$$

where $a_{ji} \in \mathbb{R}$. A *solution* of this system is a vector $(c_1, \ldots, c_n) \in \mathbb{R}^n$, where $\sum_j a_{ji}c_i = 0$ for all j; a solution $(c_1, \ldots, c_n)$ is *nontrivial* if some $c_i \neq 0$. The set of all solutions forms a subspace of $\mathbb{R}^n$, called the *solution space* of the system.

(v) The previous example can be generalized from vector spaces over $\mathbb{R}$ to vector spaces over any field k. In particular, one can solve systems of linear equations over $\mathbb{Z}_p$, where p is a prime. This says that one can treat a system of congruences mod p just as one treats an ordinary system of equations.

For example, the system of congruences

$$3x - 2y + z \equiv 1 \bmod 7$$
$$x + y - 2z \equiv 0 \bmod 7$$
$$-x + 2y + z \equiv 4 \bmod 7$$

can be regarded as a system of equations over the field $\mathbb{Z}_7$. The reader should solve this system just as one would in high school were inverses mod 7 known then ($[2][4] = [1]$; $[3][5] = [1]$; $[6][6] = [1]$); the solution is

$$(x, y, z) = ([5], [4], [1]). \quad \blacktriangleleft$$

The key observation in getting the "right" definition of dimension is to understand why $\mathbb{R}^3$ is 3-dimensional. Every vector (x, y, z) is a linear combination of the three vectors $e_1 = (1, 0, 0)$, $e_2 = (0, 1, 0)$, and $e_3 = (0, 0, 1)$; that is,

$$(x, y, z) = xe_1 + ye_2 + ze_3.$$

It is not so important that every vector is a linear combination of these specific vectors; what is important is that there are three of them, for it turns out that three is the smallest number of vectors with this property; that is, one cannot find two vectors $u = (a, b, c)$ and $u' = (a', b', c')$ with every vector a linear combination of u and u'.

Definition.   A *list* in a vector space V is an ordered set $v_1, \ldots, v_n$ of vectors in V.

More precisely, we are saying that there is some $n \geq 1$ and some function

$$\varphi : \{1, 2, \ldots, n\} \to V,$$

with $\varphi(i) = v_i$ for all i. Thus, $X = \operatorname{im} \varphi$; note that X is ordered in the sense that there is a first vector v_1, a second vector v_2, and so forth. A vector may appear several times on a list; that is, φ need not be injective.

Definition.   Let V be a vector space over a field k. A *k-linear combination* of a list $v_1, \ldots, v_n$ in V is a vector v of the form

$$v = a_1 v_1 + \cdots + a_n v_n,$$

where $a_i \in k$ for all i.

Lemma 4.1.   *Let V be a vector space over a field k.*

(i) *Every intersection of subspaces of V is itself a subspace.*

(ii) *If $X = v_1, \ldots, v_m$ is a list in V, then there is a **smallest** subspace U of V containing X; that is, U is a subspace containing X and $U \leq S$ for every subspace S containing X.*

(iii) *The subspace U spanned by X consists of all the k-linear combinations of $\{v_1, \ldots, v_m\}$.*

Proof.   (i) Let $\{S_i : i \in I\}$ be a family of subspaces of V, and let $U = \bigcap_{i \in I} S_i$. Now $0 \in S_i$ for all i, so that $0 \in U$. If $u, u' \in U$, then the definition of intersection gives $u, u' \in S_i$ for all i; since each S_i is a subspace, $u + u' \in S_i$, and so $u + u' \in U$. Similarly, if $a \in k$ and $u \in U$, then $au \in S_i$ for every i, and so $au \in U$. Therefore, U is a subspace of V.

(ii) There is at least one subspace of V containing X, namely, V itself. Let $\{S_i : i \in I\}$ be the family of all the subspaces of V containing X, and define $U = \bigcap_{i \in I} S_i$. Clearly, $X \subset U$ and, by (i), U is a subspace of V. If S is any subspace containing X, then $S = S_j$ for some j, and so $U = \bigcap_{i \in I} S_i \subset S_j = S$.

(iii) Let L denote the family of all the k-linear combinations of $\{v_1, \ldots, v_m\}$. Clearly, $L \subset U = \bigcap_{i \in I} S_i$, where S_i varies over all the subspaces of V containing X, so it suffices to prove that the reverse inclusion holds. Each v_j lies in L, for if we choose $a_j = 1$ and all the other a's $= 0$, then $v_j = a_1 v_1 + \cdots + a_m v_m$; thus, $X \subset L$. In light of (ii), it now suffices to prove that L is a subspace, for then $L = S_i$ (i.e., L is a subspace of V containing X) and so $U \leq S_i = L$. Of course, $0 \in L$. If $u, v \in L$, then $u = a_1 v_1 + \cdots + a_m v_m$ and $v = b_1 v_1 + \cdots + b_m v_m$; it follows that $u + v = (a_1 + b_1)v_1 + \cdots + (a_m + b_m)v_m \in L$. Finally, if $a \in k$ and $v = a_1 v_1 + \cdots + a_m v_m \in L$, then $av = aa_1 v_1 + \cdots + aa_m v_m \in L$.   •

Definition.   If $X = v_1, \ldots, v_m$ is a list in a vector space V, one calls the smallest subspace containing X the **subspace spanned by X**; it is denoted by

$$\langle v_1, \ldots, v_m \rangle,$$

and it consists of all the k-linear combinations of $v_1, \ldots, v_m$. One also says that $v_1, \ldots, v_m$ **spans** $\langle v_1, \ldots, v_m \rangle$.

It follows from the third part of the lemma that the subspace spanned by a list $X = v_1, \ldots, v_m$ does not depend on the ordering of the vectors, but only on the set of vectors themselves.

Were all terminology in algebra consistent, we would call $\langle v_1, \ldots, v_m \rangle$ the subspace *generated by X*. The reason for the different terms is that the theories of groups, rings, and vector spaces developed independently of each other.

If $X = \varnothing$, then we define $\langle X \rangle = \langle \varnothing \rangle = \{0\}$. This is consistent with the definition $\langle X \rangle = \bigcap_{i \in I} S_i$, where S_i ranges over all the subspaces of V containing a subset X. As every subspace contains $X = \varnothing$, $\{0\}$ itself is one of the subspaces occurring in the intersection of all the subspaces of V, and so $\{0\} = \bigcap_{S \leq V} S$.

Example 4.3.
(i) Let $V = \mathbb{R}^2$, let $e_1 = (1, 0)$, and let $e_2 = (0, 1)$. Now $V = \langle e_1, e_2 \rangle$, for if $v = (a, b) \in V$, then

$$v = (a, 0) + (0, b)$$
$$= a(1, 0) + b(0, 1)$$
$$= ae_1 + be_2 \in \langle e_1, e_2 \rangle .$$

(ii) If k is a field and $V = k^n$, define e_i as the n-tuple having 1 in the ith coordinate and 0's elsewhere. The reader may adapt the argument in part (i) to show that $e_1, \ldots, e_n$ spans k^n.

(iii) A vector space V need not be spanned by a finite set. For example, let $V = k[x]$, and suppose that $X = \{f_1(x), \ldots, f_m(x)\}$ is a finite subset of V. If d is the largest degree of any of the $f_i(x)$, then every (nonzero) k-linear combination of $f_1(x), \ldots, f_m(x)$ has degree at most d. Thus, x^{d+1} is not a k-linear combination of vectors in X, and so X does not span $k[x]$.   ◀

The following definition makes sense even though we have not yet defined *dimension*.

Definition.   A vector space V is called *finite-dimensional* if it is spanned by a finite set; otherwise, V is called *infinite-dimensional*.

Example 4.3(ii) shows that k^n is finite-dimensional, while Example 4.3(iii) shows that $k[x]$ is infinite-dimensional. By Example 4.1(iii), both $\mathbb{R}$ and $\mathbb{C}$ are vector spaces over $\mathbb{Q}$; one can prove that both are infinite-dimensional.

Given a subspace U of a vector space V, we seek a list X which spans U. Notice that U can have many such lists; for example, if $X = v_1, v_2, \ldots, v_m$ spans U and u is any vector in U, then $v_1, v_2, \ldots, v_m, u$ also spans U. Let us, therefore, seek a *shortest* list that spans U.

Notation.   If $v_1, \ldots, v_m$ is a list, then $v_1, \ldots, \widehat{v_i} \ldots, v_m$ is the shorter list with v_i deleted.

Proposition 4.2. *If V is a vector space, then the following conditions on a list $X = v_1, \ldots, v_m$ spanning V are equivalent:*

(i) *X is not a shortest spanning list;*

(ii) *some v_i is in the subspace spanned by the others; that is,*

$$v_i \in \langle v_1, \ldots, \widehat{v_i}, \ldots, v_m \rangle ;$$

(iii) *there are scalars $a_1, \ldots, a_m$, not all zero, with*

$$\sum_{\ell=1}^{m} a_\ell v_\ell = 0.$$

Proof. (i) $\Rightarrow$ (ii). If X is not a shortest spanning list, then we can throw out some v_i from X and still have a spanning list. Since $V = \langle v_1, \ldots, \widehat{v_i}, \ldots, v_m \rangle$, we have $v_i \in \langle v_1, \ldots, \widehat{v_i}, \ldots, v_m \rangle$.

(ii) $\Rightarrow$ (iii). By Lemma 4.1, there are scalars b_j, for $j \neq i$, with $v_i = \sum_{j \neq i} b_j v_j$. If we define $a_i = -1$ (so that $a_i \neq 0!$) and $a_j = b_j$ for $j \neq i$, then $\sum_{\ell=1}^{m} a_\ell v_\ell = 0$.

(iii) $\Rightarrow$ (i). Assume that $\sum_{\ell=1}^{m} a_\ell v_\ell = 0$ for scalars $a_1, \ldots, a_m$ that are not all zero. If $a_i \neq 0$, then $-a_i v_i = \sum_{j \neq i} a_j v_j$. Since $a_i \neq 0$, we may rewrite this equation as

$$v_i = -\sum_{j \neq i} a_i^{-1} a_j v_j.$$

To show that X is not a shortest spanning list, it suffices to prove that if $u \in \langle X \rangle$, then $u \in \langle v_1, \ldots, \widehat{v_i}, \ldots, v_m \rangle$. There are scalars b_j with

$$
\begin{aligned}
u &= b_1 v_1 + \cdots + b_i v_i + \cdots + b_m v_m \\
&= b_1 v_1 + \cdots + (-b_i \sum_{j \neq i} a_i^{-1} a_j v_j) + \cdots + b_m v_m \\
&= b_1 v_1 + \cdots - \sum_{j \neq i} b_i a_i^{-1} a_j v_j + \cdots + b_m v_m \\
&= \sum_{j \neq i} (b_j - b_i a_i^{-1} a_j) v_j \in \langle v_1, \ldots, \widehat{v_i}, \ldots, v_m \rangle . \quad \bullet
\end{aligned}
$$

Definition. A list $X = v_1, \ldots, v_m$ in a vector space V is **linearly dependent** if there are scalars $a_1, \ldots, a_m$, not all zero, with $\sum_{\ell=1}^{m} a_\ell v_\ell = 0$; otherwise, X is called **linearly independent**.

The empty set $\varnothing$ is defined to be linearly independent (we may interpret $\varnothing$ as a list of length 0).

Example 4.4.

(i) Any list $X = v_1, \ldots, v_m$ containing the zero vector is linearly dependent: if $v_i = 0$, setting $a_i = 1$ and all the other a's $= 0$ gives $\sum_{\ell=1}^{m} a_\ell v_\ell = 0$.

(ii) A list v_1 of length 1 is linearly dependent if and only if $v_1 = 0$; hence, a list v_1 of length 1 is linearly independent if and only if $v_1 \neq 0$

(iii) A list v_1, v_2 is linearly dependent if and only if one of the vectors is a scalar multiple of the other. Suppose that $a_1 v_1 + a_2 v_2 = 0$, where $a_1 \neq 0$. Then $a_1 v_1 = -a_2 v_2$, and so $v_1 = (-a_2/a_1)v_2$.

(iv) If there is a repetition in the list $v_1, \ldots, v_m$; that is, if $v_i = v_j$ for some $i \neq j$, then $v_1, \ldots, v_m$ is linearly dependent: set $a_i = 1$, $a_j = -1$, and all the other a's zero. Therefore, if $v_1, \ldots, v_m$ is linearly independent, then all the vectors v_i are distinct.   ◄

Linear independence has been defined indirectly, as not being linearly dependent. Because of the importance of linear independence, let us define it directly. A list $X = v_1, \ldots, v_m$ is **linearly independent** if, whenever a k-linear combination $\sum_{\ell=1}^{m} a_\ell v_\ell = 0$, then every $a_i = 0$. It follows that every sublist of a linearly independent list is itself linearly independent (this is one reason for decreeing that $\varnothing$ be linearly independent).

The contrapositive of Theorem 4.2 is also interesting.

Corollary 4.3. *If $X = v_1, \ldots, v_m$ is a list spanning a vector space V, then X is a shortest spanning list if and only if X is linearly independent.*

We have arrived at the notion we have been seeking.

Definition. A *basis* of a vector space V is a linearly independent list that spans V.

Thus, bases are shortest spanning lists. Of course, all the vectors in a linearly independent list $v_1, \ldots, v_n$ are distinct.

Example 4.5.
In Example 4.3(ii), we saw that $X = e_1, \ldots, e_n$ spans k^n, where e_i is the n-tuple having 1 in the ith coordinate and 0's elsewhere. We now show that X is linearly independent. Since $a_i e_i = (0, \ldots, 0, a_i, 0, \ldots, 0)$, where a_i is in the ith position, we have

$$\sum_{i=1}^{n} a_i e_i = (a_1, \ldots, a_n).$$

Therefore, $\sum_{i=1}^{n} a_i e_i = 0$ implies $a_i = 0$ for all i. It follows that $e_1, \ldots, e_n$ is a basis of k^n; it is called the **standard basis**. The standard basis of $\mathbb{R}^2$ is often denoted by i, j; the standard basis of $\mathbb{R}^3$ is often denoted by i, j, k. ◄

Proposition 4.4. *Let $X = v_1, \ldots, v_n$ span a vector space V over a field k. Then X is a basis if and only if each vector in V has a unique expression as a k-linear combination of vectors in X.*

Proof. Let $v \in V$. Since X spans V, there is an equation $v = \sum a_i v_i$. If, also, $v = \sum b_i v_i$, then $\sum (a_i - b_i) v_i = 0$, and so linear independence of X gives $a_i = b_i$ for all i.

Conversely, assume that expressions are unique. If $0 = \sum b_i v_i$ is a k-linear combination in which some $b_i \neq 0$, then there are two different expressions for 0 as a k-linear combination of vectors in X, a contradiction. Therefore, X is linearly independent; since X spans V, by hypothesis, X is a basis. •

If $v_1, \ldots, v_n$ is a basis of a vector space V over a field k, then each vector $v \in V$ has a unique expression

$$v = a_1 v_1 + a_2 v_2 + \cdots + a_n v_n,$$

where $a_i \in k$ for all i. Since there is a first vector v_1, a second vector v_2, and so forth, the coefficients in this k-linear combination determine a unique n-tuple $(a_1, a_2, \ldots, a_n)$. Were a basis merely a subset of V and not an ordered subset, then there would be $n!$ n-tuples corresponding to any vector.

Definition. If $X = v_1, \ldots, v_n$ is a basis of a vector space V and $v \in V$, then there are unique scalars $a_1, \ldots, a_n$ with $v = \sum_{i=1}^{n} a_i v_i$. The n-tuple $(a_1, \ldots, a_n)$ consists of the **coordinates**[3] of a vector $v \in V$ relative to the basis X.

---

[3] The term *coordinates* comes from the Latin word meaning "ordered" or "arranged."

Observe that if $v_1, \ldots, v_n$ is the standard basis of $V = k^n$, then these coordinates coincide with the usual coordinates.

We are going to define the *dimension* of a vector space V to be the number of vectors in a basis. Two questions arise at once.

(i) Does every vector space have a basis?

(ii) Do all bases of a vector space have the same number of elements?

The first question is easy to answer; the second needs some thought.

Theorem 4.5.   *Every finite-dimensional vector space V has a basis.*

Proof.   If $V = \{0\}$, then $X = \varnothing$ does span V; it is a basis, by our agreement that $\varnothing$ is linearly independent.

We now assume that $V \neq \{0\}$. Since V is finite-dimensional, it is spanned by some list $X = v_1, \ldots, v_m$. If X is linearly independent, then X is a basis. Otherwise we may apply Proposition 4.2: one of the v's can be thrown out of X, leaving a spanning list X' with $m - 1$ elements. If X' is linearly independent, it is a basis; otherwise, we may throw out an element of X', leaving a spanning list X'' with $m - 2$ elements. This process eventually ends (after at most m steps) with a shortest spanning list, which is linearly independent, by Corollary 4.3, and hence it is a basis of V.   •

One can extend the definitions of spanning and linear independence to infinite lists in a vector space, and one can then prove that infinite-dimensional vector spaces also have bases. For example, it turns out that a basis of $k[x]$ is $1, x, x^2, x^3, \ldots, x^n, \ldots$.

A modification of the proof of Proposition 4.2 gives a lemma that will help prove that all bases of a vector space have the same size.

Lemma 4.6.   *If $X = v_1, \ldots, v_n$ is a linearly dependent list of vectors in a vector space V, then there exists v_r with $r \geq 1$ with $v_r \in \langle v_1, v_2, \ldots, v_{r-1} \rangle$ [we interpret $\langle v_1, \ldots, v_{r-1} \rangle$ to mean $\{0\}$ when $r = 1$].*

Remark.   We compare Proposition 4.2 with this one. The earlier result says that if v_1, v_2, v_3 is linearly dependent, then either $v_1 \in \langle v_2, v_3 \rangle$, $v_2 \in \langle v_1, v_3 \rangle$, or $v_3 \in \langle v_1, v_2 \rangle$. This lemma says that either $v_1 \in \{0\}$, $v_2 \in \langle v_1 \rangle$, or $v_3 \in \langle v_1, v_2 \rangle$.   ◄

Proof.   Let r be the largest integer for which $v_1, \ldots, v_{r-1}$ is linearly independent. If $v_1 = 0$, then $v_1 \in \{0\}$, and we are done. If $v_1 \neq 0$, then $r \geq 2$;

since $v_1, v_2, \ldots, v_n$ is linearly dependent, we have $r - 1 < n$. As $r - 1$ is largest, the list $v_1, v_2, \ldots, v_r$ is linearly dependent. There are thus scalars $a_1, \ldots, a_r$, not all zero, with $a_1 v_1 + \cdots + a_r v_r = 0$. In this expression, we must have $a_r \neq 0$, for otherwise $v_1, \ldots, v_{r-1}$ would be linearly dependent. Therefore,

$$v_r = \sum_{i=1}^{r-1} (-a_r^{-1}) a_i v_i \in \langle v_1, \ldots, v_{r-1} \rangle. \quad \bullet$$

Theorem 4.7 (Invariance of Dimension). *If* $X = x_1, \ldots, x_n$ *and* $Y = y_1, \ldots, y_m$ *are bases of a vector space* V, *then* $m = n$.

Proof. Let us show that one of the x's in X can be replaced by y_m so that the new list still spans V. Now $y_m \in \langle X \rangle = V$, since X spans V, so that the list

$$y_m, x_1, \ldots, x_n$$

is linearly dependent, by Proposition 4.2. By Lemma 4.6, there is some x_i and scalars $a, a_1, \ldots, a_{i-1}$ with $x_i = a y_m + \sum_{j \neq i} a_j x_j$. If we throw out x_i, then

$$X' = y_m, x_1, \ldots, \widehat{x_i}, \ldots, x_n$$

spans V, for if $v = \sum_{j=1}^{n} b_j x_j$, then just collect terms after replacing x_i by its expression as a k-linear combination of y_m and the other x's (as in the proof of Proposition 4.2).

Since X' spans V, this argument can be repeated for the list $y_{m-1}, y_m, x_1, \ldots, \widehat{x_1}, \ldots, x_n$. The options offered by Lemma 4.6 for this linearly dependent list are $y_m \in \langle y_{m-1} \rangle$, $x_1 \in \langle y_{m-1}, y_m \rangle$, $x_2 \in \langle y_{m-1}, y_m, x_1 \rangle$, and so forth. Since Y is linearly independent, so is its sublist y_{m-1}, y_m, and the first option $y_m \in \langle y_{m-1} \rangle$ cannot be correct. It follows that the disposable vector (provided by Lemma 4.6) must be one of the remaining x's, say x_ℓ. After throwing out x_ℓ, we thus have a new list X'' that spans V. Iterate this construction of spanning lists; each time a new y is adjoined as the first vector, an x is thrown out. If $m > n$, that is, if there are more y's than x's, then this procedure ends with a spanning list consisting of n y's (one for each of the n x's thrown out) and no x's, contradicting $Y = y_1, \ldots, y_m$ being a shortest spanning list. Therefore, $m \leq n$.

A similar procedure, beginning by adjoining x_n to Y, yields the reverse inequality $n \leq m$, and so $m = n$, as desired. $\bullet$

There is a more straightforward proof of the last theorem if one uses the fact that if k is a field and if $m > n$, then a homogeneous system of n k-linear equations in m unknowns [see Example 4.2(iv)] always has a nontrivial solution.

Theorem 4.8 (Invariance of Dimension).   *If $X = x_1, \ldots, x_n$ and $Y = y_1, \ldots, y_m$ are bases of a vector space V, then $m = n$.*

Proof.   Suppose that $m > n$, and consider a linear combination

$$z = c_1 y_1 + c_2 y_2 + \cdots + c_m y_m.$$

Since the x's are a basis, there are scalars a_{ji} with $y_j = \sum_i a_{ji} x_i$. Hence,

$$z = c_1(a_{11}x_1 + a_{12}x_2 + \cdots) + c_2(a_{21}x_1 + a_{22}x_2 + \cdots) + \cdots$$
$$= (c_1 a_{11} + c_2 a_{21} + \cdots)x_1 + (c_1 a_{12} + c_2 a_{22} + \cdots)x_2 + \cdots.$$

Thus, $z = \sum_i b_i x_i$, where

$$b_i = \sum_j c_j a_{ji}.$$

Setting $b_i = 0$ for all i gives a homogeneous k-linear system of n equations in the m unknowns c_j. Since $m > n$, the result quoted above about systems says that there is a nontrivial solution; that is, there are $c_1, \ldots, c_m$, not all zero, with $b_i = \sum c_j a_{ji} = 0$ for all i. But if all the $b_i = 0$, then $z = 0$; since not all the c_j are 0, the list $y_1, \ldots, y_m$ is linearly dependent, a contradiction. We conclude that $m \leq n$. A similar argument proves the reverse inequality, and so $m = n$.   •

It is now permissible to make the following definition.

Definition.   If V is a finite-dimensional vector space, then its ***dimension***, denoted by $\dim(V)$, is the number of elements in a basis of V.

Example 4.6.
(i) Example 4.5 shows that k^n has dimension n, which agrees with our intuition when $k = \mathbb{R}$; the plane $\mathbb{R} \times \mathbb{R}$ is two-dimensional!

(ii) If $V = \{0\}$, then $\dim(V) = 0$, for there are no elements in its basis $\varnothing$.

(iii) Let $X = \{x_1, \ldots, x_n\}$ be a finite set. Define

$$k^X = \{\text{functions } f : X \to k\}.$$

Now k^X is a vector space if we define addition $f + f'$ to be

$$f + f' : x \mapsto f(x) + f'(x)$$

and scalar multiplication af, for $a \in k$ and $f : X \to k$, by

$$af : x \mapsto af(x).$$

It is easy to check that the set of n functions of the form f_x, where $x \in X$, defined by

$$f_x(y) = \begin{cases} 1 & \text{if } y = x; \\ 0 & \text{if } y \neq x, \end{cases}$$

form a basis, and so $\dim(k^X) = n = |X|$.

The reader should note that this not a new example. If one recalls that an n-tuple $(a_1, \dots, a_n)$ is really a function $f : \{1, \dots, n\} \to k$ with $f(i) = a_i$ for all i, then the functions f_x comprise the standard basis.   ◀

Definition.   Call a linearly independent list $u_1, \dots, u_m$ a ***longest***[4] such if there is no vector $v \in V$ such that $u_1, \dots, u_m, v$ is linearly independent.

Lemma 4.9.   *If V is a finite-dimensional vector space, then a longest linearly independent list is a basis of V.*

Proof.   Let $X = u_1, \dots, u_m$ be a longest linearly independent list in V. To see that X is a basis, it suffices to prove that it spans V. If $v \in V$, then the list $u_1, \dots, u_m, v$ must be linearly dependent (because X is longest). Thus, there are scalars $\alpha_1, \dots, \alpha_m, \beta$, not all 0, with

$$\beta v + \sum \alpha_i u_i = 0.$$

If $\beta = 0$, then we have a dependency relation of X, contradicting its linear independence; therefore, $\beta \neq 0$, and so

$$v = \sum (-\beta^{-1} \alpha_i) u_i \in \langle u_1, \dots, u_m \rangle.$$

Therefore, X spans V, and hence it is a basis.   •

It is not obvious that there are any longest linearly independent lists; that they do exist follows from the next result, which is quite useful in its own right.

---

[4]Most people call this a ***maximal*** linearly independent list.

Proposition 4.10.   *Let* $Z = u_1, \ldots, u_m$ *be a linearly independent list in an n-dimensional vector space V. Then there are vectors* $v_{m+1}, \ldots, v_n$ *so that* $u_1, \ldots, u_m, v_{m+1}, \ldots, v_n$ *is a basis of V (indeed, it is a longest linearly independent list in V).*

Proof.   Let $Z = u_1, \ldots, u_m$ be a linearly independent list in a vector space V of dimension n. If Z spans V, then Z is a basis; if Z does not span V, then there is some vector $v_{m+1} \in V$ that is not a linear combination of the vectors in Z. As in the proof of Lemma 4.9, the longer list $u_1, \ldots, u_m, v_{m+1}$ is linearly independent. If this new list spans V, then it is a basis and we are done; otherwise, there is $v_{m+2} \in V$ that is not a linear combination of them and, as above, $u_1, \ldots, u_m, v_{m+1}, v_{m+2}$ is linearly independent. This procedure cannot be iterated indefinitely, for suppose there were a linearly independent list Y so obtained with $\ell = n + 1$ elements. Compare Y to a basis X of V (which has n elements). By the proof of Theorem 4.7, we get the contradiction $n + 1 = \ell \leq n$. We conclude from Lemma 4.9 that this longest linearly independent list is a basis of V which extends the list Z.   •

Corollary 4.11.   *If* $\dim(V) = n$, *then any list of* $n + 1$ *vectors is linearly dependent.*

Proof.   If, on the contrary, a list $v_1, \ldots, v_{n+1}$ were linearly independent, then there would be a basis X of V extending it, and this would contradict Theorem 4.7, for X would be a basis of V not having n elements.   •

Corollary 4.12.   *Let V be a vector space with* $\dim(V) = n$.

  (i) *A list of n vectors which spans V must be linearly independent.*

  (ii) *Any linearly independent list of n vectors must span V.*

Proof.   (i) Assume that $\dim(W) = n = \dim(V)$. If $u_1, \ldots, u_n$ is a basis of W, then Proposition 4.10 provides vectors $v_{n+1}, v_{n+2}, \ldots$ such that $u_1, \ldots, u_n, v_{n+1}, \ldots, v_m$ is a basis of V. But since $\dim(V) = n$, no v's can be adjoined to $u_1, \ldots, u_n$ lest we get a basis of V with more than n elements. Hence, $u_1, \ldots, u_n$ must be a basis of V. It follows that $W = V$.

(ii) If $v_1, \ldots, v_n$ is a linearly independent list which does not span V, then it can be extended to be a basis, and this contradicts Corollary 4.11.   •

Corollary 4.13. *Let $U \leq V$ be a subspace of a finite-dimensional vector space.*

  (i) *U is finite-dimensional and $\dim(U) \leq \dim(V)$.*

  (ii) *If $\dim(U) = \dim(V)$, then $U = V$.*

Proof. (i) Suppose that $\dim(V) = n$. By Corollary 4.11, a list of $n + 1$ vectors in V, hence in its subspace U, is linearly dependent. In particular, a longest linearly independent list in U can have at most n elements. Therefore, $\dim(U) \leq \dim(V)$, because a longest linearly independent list must be a basis, by Lemma 4.9.

(ii) Let $X = w_1, \ldots, w_n$ be a basis of U. Of course, X is a linearly independent list in V. Since $\dim(U) = \dim(V)$, Corollary 4.11 shows that X is a longest linearly independent list in V, and so it is a basis of V. In particular, X spans V, so that each $v \in V$, being a linear combination of X, lies in U. Therefore, $U = V$.   •

EXERCISES

4.1 (i) If $f : k \to k$ is a function, where k is a field, and if $\alpha \in k$, define a new function $\alpha f : k \to k$ by $a \mapsto \alpha f(a)$. Prove that with this definition of scalar multiplication, the ring $\mathcal{F}(k)$ of all functions on k is a vector space over k.

  (ii) If $\mathcal{P}(k) \leq \mathcal{F}(k)$ denotes the family of all polynomial functions $a \mapsto a_n a^n + \cdots + \alpha_1 a + \alpha_0$, prove that $\mathcal{P}(k)$ is a subspace of $\mathcal{F}(k)$.

4.2 Prove that if the only subspaces of a vector space V are $\{0\}$ and V itself, then $\dim(V) \leq 1$.

4.3 If $u = (\alpha_1, \ldots, \alpha_n)$, $v = (\beta_1, \ldots, \beta_n) \in \mathbb{R}^n$, define their ***inner product*** (or *dot product*) to be the number

$$u \cdot v = \sum_{i=1}^{n} \alpha_i \beta_i.$$

We say that u and v are ***orthogonal***[5] (or *perpendicular*) if $u \cdot v = 0$.

  (i) Prove, for all $u, v, v' \in \mathbb{R}^n$ and $\alpha \in \mathbb{R}$, that

$$u \cdot v = v \cdot u;$$
$$u \cdot (v + v') = u \cdot v + u \cdot v';$$
$$u \cdot \alpha v = \alpha(u \cdot v).$$

---

[5]In Greek, *ortho* means "right" and *gon* means "angle." Thus, *orthogonal* means right angled or perpendicular.

(ii) If X is a nonempty subset of $\mathbb{R}^n$, define its **orthogonal complement**

$$X^{\perp} = \{v \in \mathbb{R}^n : x \cdot v = 0 \text{ for all } x \in X\}.$$

Prove that $X^{\perp}$ is a subspace of $\mathbb{R}^n$.

4.4 It is shown in analytic geometry that if ℓ_1 and ℓ_2 are lines with slopes m_1 and m_2, respectively, then ℓ_1 and ℓ_2 are perpendicular if and only if $m_1 m_2 = -1$. If

$$\ell_i = \{\alpha v_i + u_i : \alpha \in \mathbb{R}\},$$

for $i = 1, 2$, prove that $m_1 m_2 = -1$ if and only if the dot product $v_1 \cdot v_2 = 0$. (Since both lines have slopes, neither of them is vertical.)

4.5 If V is a vector space over $\mathbb{Z}_2$ and if $v_1 \neq v_2$ are nonzero vectors in V, prove that v_1, v_2 is linearly independent. Is this true for vector spaces over any other field?

4.6 Prove that the list of polynomials $1, x, x^2, x^3, \ldots, x^{100}$ is a linearly independent list in $k[x]$, where k is a field.

4.7   (i) In calculus, a *line in space passing through a point u* is defined as

$$\{u + \alpha w : \alpha \in \mathbb{R}\} \subset \mathbb{R}^3,$$

where w is a fixed nonzero vector. Show that every line through the origin is a one-dimensional subspace of $\mathbb{R}^3$.

(ii) In calculus, a *plane in space passing through a point u* is defined as the subset

$$\{v \in \mathbb{R}^3 : (v - u, n) = 0\} \subset \mathbb{R}^3,$$

where $n \neq 0$ is a fixed *normal vector*. Prove that a plane through the origin is a two-dimensional subspace of $\mathbb{R}^3$.

4.8   (i) If U and U' are subspaces of a vector space V, define

$$U + U' = \{u + u' : u \in U \text{ and } u' \in U'\}.$$

Prove that $U + U'$ is a subspace of V.

(ii) If U and U' are subspaces of a finite-dimensional vector space V, prove that

$$\dim(U) + \dim(U') = \dim(U \cap U') + \dim(U + U').$$

Linear Transformations

Homomorphisms between vector spaces are called *linear transformations*.

Definition. If V and W are vector spaces over a field k, then a function $T : V \rightarrow W$ is a **linear transformation** if, for all vectors u, $v \in V$ and all scalars $a \in k$,

(i) $T(u + v) = T(u) + T(v)$;

(ii) $T(av) = aT(v)$.

We say that a linear transformation T is **nonsingular** (or is an **isomorphism**) if T is a bijection.

If one forgets the scalar multiplication, then a vector space is an (additive) abelian group and a linear transformation T is a group homomorphism. It is easy to see that T preserves all k-linear combinations:

$$T(a_1 v_1 + \cdots + a_m v_m) = a_1 T(v_1) + \cdots + a_m T(v_m).$$

Example 4.7.
(i) The identity function $1_V : V \rightarrow V$ on any vector space V is a nonsingular linear transformation.

(ii) An $m \times n$ matrix A with entries in a field k determines a linear transformation $k^n \rightarrow k^m$, namely,

$$X \mapsto AX,$$

where X is an $n \times 1$ column vector.   ◄

Definition. If V is a vector space over a field k, then the **general linear group**, denoted by $GL(V)$, is the set of all nonsingular linear transformations $V \rightarrow V$.

A composite ST of linear transformations S and T is again a linear transformation, and ST is nonsingular if both S and T are; moreover, the inverse of a nonsingular linear transformation is again nonsingular. It follows that $GL(V)$ is a group with composition as operation.

We now show how to construct linear transformations $T : V \rightarrow W$, where V and W are vector spaces over a field k. The next theorem says that there is a linear transformation that does anything to a basis.

Theorem 4.14.   *Let $v_1, \ldots, v_n$ be a basis of a vector space V over a field k. If W is a vector space over k and $u_1, \ldots, u_n$ is a list in W, then there exists a unique linear transformation $T : V \to W$ with $T(v_i) = u_i$ for all i.*

Proof.   By Theorem 4.4, each $v \in V$ has a unique expression of the form $v = \sum_i a_i v_i$, and so $T : V \to W$, given by $T(v) = \sum a_i u_i$, is a (single-valued!) function. It is now a routine verification to check that T is a linear transformation.

To prove uniqueness of T, assume that $S : V \to W$ is a linear transformation with

$$S(v_i) = u_i = T(v_i)$$

for all i. If $v \in V$, then $v = \sum a_i v_i$ and

$$
\begin{aligned}
S(v) &= S\left(\sum a_i v_i\right) \\
&= \sum S(a_i v_i) \\
&= \sum a_i S(v_i) \\
&= \sum a_i T(v_i) = T(v).
\end{aligned}
$$

Since v is arbitrary, $S = T$.   •

It follows that if two linear transformations $S, T : V \to W$ agree on a basis, then $S = T$.

Theorem 4.14 establishes the connection between linear transformations and matrices, and the definition of matrix multiplication arises from applying this construction to the composite of two linear transformations.

Definition.   Let $v_1, \ldots, v_n$ be a basis of V and let $w_1, \ldots, w_m$ be a basis of W. If $T : V \to W$ is a linear transformation, then the ***matrix of T*** is the $m \times n$ matrix $\mu(T) = [a_{ij}]$ whose ith column $a_{1i}, a_{2i}, \ldots, a_{mi}$ consists of the coordinates of $T(v_i)$ determined by the w's. Thus,

$$T(v_i) = \sum_{j=1}^{m} a_{ji} w_j.$$

In case $V = W$, one usually lets the bases $v_1, \ldots, v_n$ and $w_1, \ldots, w_m$ coincide. For example, if $1_V : v \mapsto v$ is the identity linear transformation, then $\mu(1_V)$ is the identity matrix (having 1's on the diagonal and 0's elsewhere). The identity linear transformation is unusual, for $\mu(T)$ usually depends on the choice of basis, as we see in the next example.

Example 4.8.

If ε_1, ε_2 is the standard basis of $\mathbb{R}^2$ and T is the linear transformation with

$$T(\varepsilon_1) = 2\varepsilon_1 + 3\varepsilon_2$$
$$T(\varepsilon_2) = \varepsilon_1 - \varepsilon_2,$$

then the matrix of T is

$$\mu(T) = \begin{bmatrix} 2 & 1 \\ 3 & -1 \end{bmatrix}.$$

On the other hand, if we write $\mu(T)$ relative to the basis ε_2, ε_1, then

$$T(\varepsilon_2) = 3\varepsilon_2 + 2\varepsilon_1$$
$$T(\varepsilon_1) = -\varepsilon_2 + \varepsilon_1$$

and

$$\mu(T) = \begin{bmatrix} 3 & -1 \\ 2 & 1 \end{bmatrix}.$$

Thus, the notation $\mu(T)$, where $T : V \to V$, is too simple; it should show its dependence on the choice of basis: $\mu(T, v_1, \ldots, v_n)$. ◄

The following definition of matrix multiplication makes sense if the entries lie in any ring R.

Definition.   Let $A = [a_{ij}]$ be an $m \times n$ matrix, and let $B = [b_{jk}]$ be an $n \times p$ matrix. Then their ***product*** $C = AB$ is the $m \times p$ matrix $C = [c_{ik}]$, where

$$c_{ik} = \sum_{j=1}^{n} a_{ij} b_{jk}.$$

The ik entry of AB can be regarded as the dot product of the ith row $(a_{i1}, a_{i2}, \ldots, a_{in})$ of A and the kth column $(b_{1k}, b_{2k}, \ldots, b_{nk})$ of B. Note that this definition agrees with the earlier definition of multiplication of 2×2 matrices (when $m = 2 = n = p$). If the sizes of A and B do not match, i.e., if A is an $m \times n$ matrix, B is a $q \times p$ matrix, and $n \neq q$, then the product AB is not defined.

The obvious question is, for a fixed linear transformation $T : V \to V$, to survey all the matrices of the form $\mu(T, v_1, \ldots, v_n)$ as the basis $v_1, \ldots, v_n$ varies. This is the portion of linear algebra usually called *change of basis*; the main result is that if $v_1, \ldots, v_n$ and $u_1, \ldots, u_n$ are bases of V, then

$\mu(T, v_1, \ldots, v_n)$ and $\mu(T, u_1, \ldots, u_n)$ are **similar**; that is, there is a matrix P (arising from the linear transformation $S : V \to V$ with $S(u_i) = v_i$) with

$$\mu(T, u_1, \ldots, u_n) = P\mu(T, v_1, \ldots, v_n)P^{-1},$$

and P^{-1} is a matrix satisfying $P^{-1}P = E = PP^{-1}$.

   The next proposition shows where the definition of matrix multiplication comes from.

Proposition 4.15.   *Let $T : V \to W$ and $S : W \to Y$ be linear transformations. Choose bases $v_1, v_2, \ldots, v_n$ of V, $w_1, w_2, \ldots, w_m$ of W, and $y_1, y_2, \ldots, y_p$ of Y, so that $\mu(T)$ and $\mu(S)$ are defined. Then $\mu(S \circ T)$ is defined, and*

$$\mu(S \circ T) = \mu(S)\mu(T).$$

Proof.   Now $\mu(T) = [a_{ij}]$, where $T(v_i) = \sum_{j=1}^{m} a_{ji} y_j$, and $\mu(S) = [b_{j\ell}]$, where $S(w_j) = \sum_{\ell=1}^{p} b_{\ell j} y_\ell$ The matrix $\mu(S \circ T) = [c_{i\ell}]$, where

$$(S \circ T)(v_i) = \sum_{\ell=1}^{p} c_{\ell i} y_\ell;$$

that is, we write $(S \circ T)(v_i)$ as a linear combination of the y's, and the ith column of $\mu(S \circ T)$ is comprised of the coordinates of $S(T(v_i))$. Now

$$
\begin{aligned}
(S \circ T)(v_i) &= S\big(a_{1i}w_1 + a_{2i}w_2 + \cdots + a_{mi}w_m\big) \\
&= a_{1i}S(w_1) + a_{2i}S(w_2) + \cdots + a_{mi}S(w_m) \\
&= a_{1i}(b_{11}y_1 + b_{21}y_2 + \cdots) + a_{2i}(b_{12}y_1 + b_{22}y_2 + \cdots) + \cdots \\
&= (a_{1i}b_{11} + a_{2i}b_{12} + \cdots)y_1 + (a_{1i}b_{12} + a_{2i}b_{22} + \cdots)y_2 + \cdots \\
&= (\sum_j a_{ji}b_{1j})y_1 + (\sum_j a_{ji}b_{2j})y_2 + \cdots.
\end{aligned}
$$

Thus, the coefficient of y_ℓ is

$$\sum_j a_{ji}b_{\ell j} = \sum_j b_{\ell j}a_{ji}.$$

That is, the coefficient of y_ℓ is the dot product of row ℓ of B and column i of A, which is, by definition of matrix multiplication, the ℓi entry of the product BA. In symbols,

$$\mu(S \circ T) = BA = \mu(S)\mu(T). \quad \bullet$$

Corollary 4.16. *Matrix multiplication is associative.*

Proof.   Let A be an $m \times n$ matrix, let B be an $n \times p$ matrix, and let C be a $p \times q$ matrix. By Theorem 4.14, there are linear transformations

$$k^q \xrightarrow{T} k^p \xrightarrow{S} k^n \xrightarrow{R} k^m$$

with $C = \mu(T)$, $B = \mu(S)$, and $A = \mu(R)$.
   Then

$$\mu\big(R \circ (S \circ T)\big) = \mu(R)\mu(S \circ T) = \mu(R)[\mu(S)\mu(T)] = A(BC).$$

On the other hand,

$$\mu\big((R \circ S) \circ T\big) = \mu(R \circ S)\mu(T) = [\mu(R)\mu(S)]\mu(T) = (AB)C.$$

Since composition of functions is associative,

$$R \circ (S \circ T) = (R \circ S) \circ T,$$

and so

$$\begin{aligned} A(BC) &= \mu(R \circ (S \circ T)) \\ &= \mu((R \circ S) \circ T) \\ &= (AB)C. \quad \bullet \end{aligned}$$

One can prove Corollary 4.16 directly, although it is rather tedious, but the connection with linear transformations is the "theological" reason why matrix multiplication is associative.

Definition.   An $n \times n$ matrix A with entries in a field k is ***nonsingular*** if there exists a nonsingular linear transformation T with $A = \mu(T)$.
   Proposition 4.15 shows that a matrix A is nonsingular if and only if there is a matrix B, its ***inverse***, denoted by A^{-1}, with entries in k, namely, $B = \mu(T^{-1})$, with $AB = E = BA$.
   The set of all nonsingular $n \times n$ matrices with entries in k is denoted by $\mathrm{GL}(n, k)$.

It is easy to check (now that we have proven associativity) that $\mathrm{GL}(n, k)$ is a group under matrix multiplication.
   A choice of basis gives an isomorphism between the general linear group and the group of nonsingular matrices.

Proposition 4.17.   *Let V be an n-dimensional vector space over a field k, and let $v_1, \ldots, v_n$ be a basis of V. Then $\mu : \mathrm{GL}(V) \to \mathrm{GL}(n, k)$, defined by $T \mapsto \mu(T, v_1, \ldots, v_n)$, is an isomorphism.*

Proof.   By Proposition 4.15, $\mu : \mathrm{GL}(V) \to \mathrm{GL}(n, k)$ is a homomorphism. Theorem 4.14 shows that μ is surjective. The definition of $\mu(T)$ says that its ith column consists of the coordinates of $T(v_i)$. Thus, if $\mu(T) = E$, then $T(v_i) = v_i$ for all i, and so $T = 1_V$. We conclude that μ is injective, and hence that it is an isomorphism.   •

The center of the general linear group is easily identified; we now generalize Exercise 2.72.

Definition.   A linear transformation $T : V \to V$ is a ***scalar transformation*** if there is $c \in k$ with $T(v) = cv$ for all $v \in V$; that is, $T = c1_V$.

A scalar transformation $T = c1_V$ is nonsingular if and only if $c \neq 0$ (its inverse is $c^{-1}1_V$).

Corollary 4.18.

(i) *The center of the group $\mathrm{GL}(V)$ consists of all the nonsingular scalar transformations.*

(ii) *The center of the group $\mathrm{GL}(n, k)$ consists of all the nonsingular scalar matrices.*

Proof.   (i) If $T \in \mathrm{GL}(V)$ is not scalar, then there exists $v \in V$ with $v, T(v)$ linearly independent. By Proposition 4.10, there is a basis $v, T(v), u_3, \ldots, u_n$ of V. It is easy to see that $v, v + T(v), u_3, \ldots, u_n$ is also a basis of V, and so there is a nonsingular linear transformation S with $S(v) = v$, $S(T(v)) = v + T(v)$, and $S(u_i) = u_i$ for all i. Now S and T do not commute, for $ST(v) = v + T(v)$ while $TS(v) = T(v)$. Therefore, T is not in the center of $\mathrm{GL}(V)$.

(ii) If $f : G \to H$ is a group isomorphism, then $f(Z(G)) = Z(H)$. But it is easily seen that if $T = c1_V$ is a scalar transformation, then $\mu(T) = cE$ is a scalar matrix.   •

Linear transformations defined on k^n are easy to describe.

Proposition 4.19. *If $T : k^n \to k^n$ is a linear transformation, then there exists an $n \times n$ matrix P (namely, the matrix $P = \mu(T)$ relative to the standard basis $e_1, \ldots, e_n$ of k^n) so that*

$$T(y) = Py$$

for all $y \in k^n$ (here, y is an $n \times 1$ column matrix and Py is matrix multiplication).

Moreover, T is an isomorphism if and only if P is nonsingular.

Proof.   Define P to be the matrix whose ith column is $\varphi(e_i)$. The definition of matrix multiplication gives $\varphi(e_i) = Pe_i$. By Example 4.7(ii), the function $S : k^n \to k^n$, defined by $S(y) = Py$, is a linear transformation. But $S(e_i) = T(e_i)$ for $i = 1, 2, \ldots, n$, and so $S = T$, by Theorem 4.14.

If T is an isomorphism, then it has an inverse T^{-1}; if P is the matrix of T and B is the matrix of T^{-1} (relative to the standard basis of k^n), then the proposition gives PB to be the matrix of TT^{-1} and BP to be the matrix of $T^{-1}T$. As the matrix of the identity transformation is the identity matrix E, we have P and B inverse matrices. Therefore, P is nonsingular.

Conversely, if P is nonsingular, then P^{-1} exists; define $S : k^n \to k^n$ by $S(y) = P^{-1}y$ for every $y \in k^n$. Now the matrix of ST is $P^{-1}P = E$, so that ST is the identity linear transformation. Similarly, TS is also the identity, so that T is an isomorphism (with inverse S).   •

Just as for group homomorphisms and ring homomorphisms, we can define kernel and image of linear transformations.

Definition.   If $T : V \to W$ is a linear transformation, then **kernel T** (or **null space T**) is

$$\ker T = \{v \in V : T(v) = 0\},$$

and **image T** is

$$\operatorname{im} T = \{w \in W : w = T(v) \text{ for some } v \in V\}.$$

As in Example 4.7(ii), an $m \times n$ matrix A with entries in a field k determines a linear transformation $k^n \to k^m$, namely, $X \mapsto AX$, where X is an $n \times 1$ column vector. The kernel of this linear transformation is usually called the solution space of A [see Example 4.2(iv)].

Proposition 4.20. *Let $T : V \to W$ be a linear transformation.*

(i) $\ker T$ *is a subspace of V and $\operatorname{im} T$ is a subspace of W.*

(ii) *T is injective if and only if $\ker T = \{0\}$.*

Proof. The routine argument, analgous to that in Proposition 2.37, is left as an exercise for the reader. ●

Example 4.9.
We show that $R_\psi : \mathbb{R}^2 \to \mathbb{R}^2$, rotation about the origin by an angle ψ, is a linear transformation. If we identify $\mathbb{R}^2$ with $\mathbb{C}$, then every point can be written (in polar form) as $(r \cos \theta, r \sin \theta)$, and we have the formula:

$$R_\psi[(r \cos \theta, r \sin \theta)] = (r \cos(\theta + \psi), r \sin(\theta + \psi)).$$

Denote the standard basis of $\mathbb{R}^2$ by e_1, e_2, where

$$e_1 = (1, 0) = (\cos 0, \sin 0)$$

and

$$e_2 = (0, 1) = (\cos \pi/2, \sin \pi/2).$$

Thus,

$$R_\psi(e_1) = R_\psi(\cos 0, \sin 0) = (\cos \psi, \sin \psi),$$

and

$$R_\psi(e_2) = R_\psi(\cos \pi/2, \sin \pi/2) = (\cos(\pi/2 + \psi), \sin(\pi/2 + \psi)).$$

On the other hand, if T is the linear transformation with

$$T(e_1) = (\cos \psi, \sin \psi)$$

and

$$T(e_2) = (-\sin \psi, \cos \psi),$$

then the addition formulas for cosine and sine give

$$
\begin{aligned}
T(r \cos \theta, r \sin \theta) &= T(r \cos \theta e_1) + T(r \sin \theta e_2) \\
&= r \cos \theta (\cos \psi, \sin \psi) + r \sin \theta (-\sin \psi, \cos \psi) \\
&= r[(\cos \theta \cos \psi - \sin \theta \sin \psi, \cos \theta \sin \psi + \sin \theta \cos \psi)] \\
&= r[\cos(\theta + \psi), \sin(\theta + \psi)] \\
&= R_\psi[(r \cos \theta, r \sin \theta)].
\end{aligned}
$$

Therefore, $R_\psi = T$, and so R_ψ is a linear transformation; the matrix of R_ψ is

$$\mu(R_\psi) = \begin{bmatrix} \cos\psi & -\sin\psi \\ \sin\psi & \cos\psi \end{bmatrix}.$$

Notice that $\det(\mu(R_\psi)) = \cos^2\psi + \sin^2\psi = 1 \neq 0$.

It is not difficult to give a geometric proof of this result. The function R_ψ sends any parallelogram having one vertex at the origin into another such parallelogram; that is, $R_\psi(P + Q) = R_\psi(P) + R_\psi(Q)$. That $R_\psi(aP) = aR_\psi(P)$ for all $a \in \mathbb{R}$ essentially follows from R_ψ preserving lengths; we leave the details to the interested reader. ◀

Lemma 4.21. *Let $T : V \to W$ be a linear transformation.*

(i) *If T is nonsingular (i.e., is an isomorphism), then for every basis $X = v_1, v_2, \ldots, v_n$ of V, we have $T(X) = T(v_1), T(v_2), \ldots, T(v_n)$ a basis of W.*

(ii) *Conversely, if there exists some basis $X = v_1, v_2, \ldots, v_n$ of V for which $T(X) = T(v_1), T(v_2), \ldots, T(v_n)$ is a basis of W, then T is nonsingular.*

Proof. (i) Assume that T is nonsingular. To prove that $T(X)$ is linearly independent, suppose that $0 = \sum a_i T(v_i)$. Then $0 = T(\sum a_i v_i)$, and so $\sum a_i v_i \in \ker T = 0$. Therefore, $\sum a_i v_i = 0$; since X is linearly independent, however, we have $a_i = 0$ for all i, as desired. Therefore, $T(X)$ is linearly independent.

To prove that $T(X)$ spans W, let $w \in W$. Since T is surjective, there is $v \in V$ with $w = T(v)$. Since X is a basis of V, we have $v = \sum b_i v_i$; hence, $T(v) = \sum b_i T(v_i)$, as desired.

Therefore, $T(X)$ is a basis of W.

(ii) Conversely, assume that $T(X)$ is a basis of W. To see that T is injective, suppose that $v \in \ker T$, so that $T(v) = 0$. Now $v = \sum b_i v_i$, so that $0 = T(v) = \sum b_i T(v_i)$. As $T(X)$ is linearly independent, we have $b_i = 0$ for all i; therefore, $v = \sum b_i v_i = 0$. hence, Lemma 4.20(ii) shows that T is injective.

To see that T is surjective, take $w \in W$. As $T(X)$ is a basis of W, we have $w = \sum a_i T(v_i) = T(\sum a_i v_i) \in \operatorname{im} T$.

Therefore, T is nonsingular. •

There are other characterizations of nonsingularity. For example, one can define the *determinant* of an $n \times n$ matrix, and it can then be proved that T is nonsingular if and only if $\det(\mu(T)) \neq 0$.

Theorem 4.22.   *If V is an n-dimensional vector space over a field k, then V is isomorphic to k^n.*

Proof.   Choose a basis $v_1, \ldots, v_n$ of V. If $e_1, \ldots, e_n$ is the standard basis of k^n, then Theorem 4.14 says that there is a linear transformation $T : V \rightarrow k^n$ with $T(v_i) = e_i$ for all i; by Lemma 4.21, T is nonsingular.   •

Theorem 4.22 does more than say that every finite-dimensional vector space is essentially the familiar vector space of all n-tuples. It says that a choice of basis in V is tantamount to a choice of coordinates for each vector in V. One wants the freedom to change coordinates because the usual coordinates may not be the most convenient ones for a given problem, as the reader has probably seen (in a calculus course) when rotating axes to simplify the equation of a conic section.

Corollary 4.23.   *Two finite-dimensional vector spaces V and W over a field k are isomorphic if and only if* $\dim(V) = \dim(W)$.

Proof.   Assume that there is a nonsingular $T : V \rightarrow W$. If $X = v_1, \ldots, v_n$ is a basis of V, then Lemma 4.21 says that $T(v_1), \ldots, T(v_n)$ is a basis of W. Therefore, $\dim(W) = |X| = \dim(V)$.

If $n = \dim(V) = \dim(W)$, then there are isomorphisms $T : V \rightarrow k^n$ and $S : W \rightarrow k^n$, by Theorem 4.22. It follows that the composite $S^{-1}T : V \rightarrow W$ is nonsingular.   •

Having classified finite-dimensional vector spaces, the next project in linear algebra is to classify linear transformations. Given a linear transformation $T : V \rightarrow V$, every choice of basis of V gives a matrix $\mu(T)$: that is, $\mu(T)$ depends not only on T but on the basis as well, as we have seen in Example 4.8. One can prove that there is a basis giving the "simplest" matrix $\mu(T)$, called the *rational canonical form* of T, which displays all the important properties of T.

EXERCISES

4.9 Let V and W be vector spaces over a field k, and let $S, T : V \to W$ be linear transformations.

(i) Define functions $S \pm T : V \to W$ by $v \mapsto S(v) \pm T(v)$ for all $v \in V$. Prove that $S + T$ and $S - T$ are linear transformations.

(ii) If $\alpha \in \mathbb{R}$, define $\alpha T : V \to W$ by $v \mapsto \alpha T(v)$ for all $v \in V$. Show that αT is a linear transformation.

(iii) Prove that $\mathcal{L}(V, W)$, the set of all the linear transformations from V to W, is a vector space over k.

(iv) Prove that if V and W are finite-dimensional, then $\dim(\mathcal{L}(V, W)) = \dim(V)\dim(W)$.

(v) The **dual space** V^* of a vector space V over k is defined by

$$V^* = \mathcal{L}(V, k).$$

If $\dim(V) = n$, prove that $\dim(V^*) = n$, and hence that $V^* \cong V$.

4.10 In Chapter 2, we called a 2×2 matrix

$$A = \begin{bmatrix} a & b \\ c & d \end{bmatrix}$$

nonsingular if $\det(A) = ad - bc \neq 0$. If V is a vector space with basis v_1, v_2, define $T : V \to V$ by $T(v_1) = av_1 + bv_2$ and $T(v_2) = cv_1 + dv_2$. Prove that A is a nonsingular matrix if and only if T is a nonsingular linear transformation (i.e., T is an isomorphism).

4.11 (i) If U is a subspace of a vector space V over a field k, define a scalar multiplication on the quotient group V/U by

$$\alpha(v + U) = \alpha v + U,$$

where $\alpha \in k$ and $v \in V$. Prove that this is a single-valued function that makes V/U into a vector space over k (V/U is called a **quotient space**).

(ii) Prove that the **natural map** $\pi : V \to V/U$, given by $v \mapsto v + U$, is a linear transformation with kernel U.

(iii) State and prove the **first isomorphism theorem** for vector spaces.

4.12 If V is a finite-dimensional vector space and U is a subspace, prove that

$$\dim(U) + \dim(V/U) = \dim(V).$$

Applications to Fields

We can now better understand an earlier result.

Example 4.10.
Recall Corollary 3.79: Every finite field K has p^n elements for some prime p. Now K has a prime field, which must be isomorphic to $\mathbb{Z}_p$ for some prime p (because K is finite), and so K is a vector space over $\mathbb{Z}_p$, by Example 4.1(iii). Since K is obviously finite-dimensional, Theorem 4.22 shows that K is isomorphic to $(\mathbb{Z}_p)^n$ for some n, and so $|K| = |(\mathbb{Z}_p)^n| = p^n$. ◄

The main reason we have introduced vector spaces here is that we wish to define the *degree* of a field over a subfield (which is the dimension of the big field viewed as a vector space over the subfield). This notion will be used to resolve the classical Greek problems; in particular, we will see (in the next section) that it is impossible to trisect a 60° angle using only straightedge and compass.

Definition.   If k is a subfield of a field K, then we also say that K is an *extension* of k. We abbreviate this by writing "K/k is an extension."[6]

If K/k is an extension, then K may be regarded as a vector space over k, as in Example 4.1(iii). One says that K is a *finite extension* of k if K is a finite-dimensional vector space over k. The dimension of K, denoted by $[K : k]$, is called the *degree* of K/k.

The next result shows why $[K : k]$ is called the degree. We remind the reader of an earlier notation: If K/k is an extension and $z \in K$, then $k(z)$ is the smallest subfield of K containing k and z. In particular, Theorem 3.75 shows that if $p(x) \in k[x]$ is an irreducible polynomial and $x + (p(x)) \in k[x]/(p(x))$ is denoted by z, then every element in the field $k[x]/(p(x))$ has the form $b_0 + b_1 z + \cdots + b_{n-1} z^{n-1}$, and so $k[x]/(p(x)) = k(z)$.

Proposition 4.24.

(i) *If K/k is a finite extension and $z \in K$, then there is a monic irreducible polynomial $p(x) \in k[x]$ having z as a root.*

(ii) *If k is a field and $p(x) \in k[x]$ is an irreducible polynomial of degree n, then $k(z) = k[x]/(p(x))$ is a field with $[k(z) : k] = n = \deg(p)$.*

---

[6]One pronounces K/k as "K over k"; there should be no confusing this notation with that of a quotient ring, for K is a field and hence it has no proper nonzero ideals.

Proof. (i) If $[K : k] = m$, the list $1, z, z^2, \ldots, z^m$ must be linearly dependent (use Proposition 4.10); that is, there are scalars $a_0, a_1, \ldots, a_m \in k$, not all zero, with $\sum a_i z^i = 0$. We have shown that there is some nonzero polynomial $f(x) = \sum a_i x^i \in k[x]$ having z as a root. By Proposition 3.76, there is an irreducible polynomial $p(x) \in k[x]$ having z as a root. Multiplying by the inverse of the leading coefficient, we may assume that $p(x)$ is monic.

(ii) In Theorem 3.75, we showed that if $z = x + (p(x))$, then every element in $k[x]/(p(x))$ has a unique expression of the form $b_0 + b_1 z + \cdots + b_{n-1} z^{n-1}$, where $b_i \in k$ and $n = \deg(p)$. By Theorem 4.4, $1, z, z^2, \ldots, z^{n-1}$ is a basis of $k[x]/(p(x))$ viewed as a vector space over k, and so $\dim(k[x]/(p(x))) = n$.

Proposition 3.76 gives an isomorphism of fields $\varphi : k[x]/(p(x)) \to k(z)$ with φ preserving k-linear combinations; that is, φ is an isomorphism of vector spaces. Since $[k[x]/(p(x)) : k] = n$, Corollary 4.23 gives $[k(z) : k] = n$. •

For example, $\mathbb{C} = \mathbb{R}(i) \cong \mathbb{R}[x]/(x^2 + 1)$, so that $\mathbb{C}$ is a finite extension of $\mathbb{R}$ of degree 2.

The following formula is quite useful, especially when one is proving a theorem by induction on degrees.

Theorem 4.25. *Let* $k \leq K \leq E$ *be fields, with* K *a finite extension of* k *and* E *a finite extension of* K. *Then* E *is a finite extension of* k, *and*

$$[E : k] = [E : K][K : k]$$

Proof. If $A = a_1, \ldots, a_n$ is a basis of K over k and if $B = b_1, \ldots, b_m$ is a basis of E over K, then it suffices to prove that a list X of all $a_i b_j$ is a basis of E over k.

To see that X spans E, take $e \in E$. Since B is a basis of E over K, there are scalars $\lambda_j \in K$ with $e = \sum_j \lambda_j b_j$. Since A is a basis of K over k, there are scalars $\mu_{ji} \in k$ with $\lambda_j = \sum_i \mu_{ji} a_i$. Therefore, $e = \sum_{ij} \mu_{ji} a_i b_j$, and X spans E over k.

To prove that X is linearly independent over k, assume that there are scalars $\mu_{ji} \in k$ with $\sum_{ij} \mu_{ji} a_i b_j = 0$. If we define $\lambda_j = \sum_i \mu_{ji} a_i$, then $\lambda_j \in K$ and $\sum_j \lambda_j b_j = 0$. Since B is linearly independent over K, it follows that

$$0 = \lambda_j = \sum_i \mu_{ji} a_i$$

for all j. Since A is linearly independent over k, it follows that $\mu_{ji} = 0$ for all j and i, as desired. •

Definition.   Assume that k is a subfield of a field K and that $z \in K$. We call z *algebraic* over k if there is some nonzero polynomial $f(x) \in k[x]$ having z as a root; otherwise, z is called *transcendental* over k.

It is proved in Proposition 4.24 that if K/k is a finite extension, then every $z \in K$ is algebraic over k.

When one says that a real number is transcendental, one usually means that it is transcendental over $\mathbb{Q}$. For example, F. Lindemann (1852–1939) proved, in 1882, that π is transcendental, so that $[\mathbb{Q}(\pi) : \mathbb{Q}]$ is infinite (see A. Baker, *Transcendental Number Theory*, p. 5). Using this fact, we can see that $\mathbb{R}$, viewed as a vector space over $\mathbb{Q}$, is infinite-dimensional. (For a proof of the irrationality of π, a more modest result, we refer the reader to Niven and Zuckerman, *An Introduction to the Theory of Numbers*.)

Proposition 4.26.   *Let K/k be an extension.*

(i) *If $z \in K$, then z is algebraic over k if and only if $k(z)/k$ is finite.*[7]

(ii) *If $z_1, z_2, \dots , z_n \in K$ are algebraic over k, then $k(z_1, z_2, \dots , z_n)/k$ is a finite extension.*

(iii) *If $y, z \in K$ are algebraic over k, then $y + z$ and yz are also algebraic over k.*

(iv) *Define*

$$K_{alg} = \{z \in K : z \text{ is algebraic over } k\}.$$

Then K_{alg} is a subfield of K.

Proof.   (i) If $k(z)/k$ is finite, then Proposition 4.24(i) shows that z is algebraic over k. Conversely, if z is algebraic over k, then Proposition 4.24(ii) shows that $k(z)/k$ is finite.

(ii) We prove this by induction on $n \geq 1$; the base step is part (i). For the inductive step, there is a tower of fields

$$k \leq k(z_1) \leq k(z_1, z_2) \leq \cdots \leq k(z_1, \dots , z_n) \leq k(z_1, \dots , z_{n+1}).$$

By the inductive hypothesis, we have $[k(z_1, \dots , z_n) : k]$ finite; moreover, $[k(z_{n+1}) : k]$ is finite. Indeed, $[k(z_{n+1}) : k] = d$, where d is the degree of

---

[7] Compare this statement, as well as that of part (iii), with Proposition 3.58 concerning algebraic integers.

the monic irreducible polynomial in $k[x]$ having z_{n+1} as a root (by Proposition 4.24). But if z_{n+1} satisfies a polynomial of degree d over k, then it satisfies a polynomial of degree $d' \leq d$ over the larger field $F = k(z_1, \dots, z_n)$. We conclude that

$$[k(z_1, \dots, z_{n+1}) : k(z_1, \dots, z_n)] = [F(z_{n+1}) : F] \leq [k(z_{n+1}) : k].$$

Therefore,

$$[k(z_1, \dots, z_{n+1}) : k] = [F(z_{n+1}) : k] = [F(z_{n+1}) : F][F : k]$$

is finite.

(iii) Now $k(y, z)/k$ is finite, by part (ii). Therefore, $k(y + z) \leq k(y, z)$ and $k(yz) \leq k(y, z)$ are also finite, for any subspace of a finite-dimensional vector space is itself finite-dimensional (Corollary 4.13(ii)). By part (i), both $y + z$ and yz are algebraic over k.

(iv) This follows at once from part (iii).   •

EXERCISES

4.13 In Proposition 4.24, we proved that if K/k is a finite extension and $z \in K$, then there is an irreducible polynomial $p(x) \in k[x]$ having z as a root (of course, we can assume that $p(x)$ is monic). Prove that $p(x)$ is the unique monic irreducible polynomial in $k[x]$ having z as a root.

4.14 Let $k \leq K \leq E$ be fields. Prove that if E is a finite extension of k, then E is a finite extension of K and K is a finite extension of k.

4.15 Let $k \leq F \leq K$ be a tower of fields, and let $z \in K$. Prove that if $k(z)/k$ is finite, then $[F(z)/F] \leq [k(z) : k]$. In particular, $[F(z) : F]$ is finite.

4.2 EUCLIDEAN CONSTRUCTIONS

There are myths in several ancient civilizations in which the gods demand precise solutions of mathematical problems in return for granting relief from catastrophes. We quote from van der Waerden, *Geometry and Algebra in Ancient Civilizations.*

In the dialogue 'Platonikos' of Eratosthenes, a story was told about the problem of doubling the cube. According to this story, as Theon of Smyrna recounts it in his book 'Exposition of mathematical things useful for the reading of Plato', the Delians asked for an oracle in order to be liberated from a plague. The god (Apollo) answered through the oracle that they had to construct an altar twice as large as the existing one without changing its shape. The Delians sent a delegation to Plato, who referred them to the mathematicians Eudoxos and Helikon of Kyzikos.

The altar was cubical in shape, and so the problem involves constructing $\sqrt[3]{2}$. The gods were cruel, for although there is a geometric construction of $\sqrt{2}$ (it is the length of the diagonal of a square with sides of length 1), we are going to prove that it is impossible to construct $\sqrt[3]{2}$ by the methods of euclidean geometry – that is, by using only straightedge and compass. (Actually, the gods were not so cruel, for the Greeks did use other methods. Thus, Menaechmus constructed $\sqrt[3]{2}$ with the intersection of the parabolas $y^2 = 2x$ and $x^2 = y$; this is elementary for us, but it was an ingenious feat when there was no analytic geometry and no algebra. There was also a solution found by Nicomedes.)

There are several other geometric problems handed down from the Greeks. Can one trisect every angle? Can one construct a regular n-gon? Can one "square the circle"; that is, can one construct a square whose area is equal to the area of a given circle?

If we are not careful, then some of these problems appear ridiculously easy. For example, a $60°$ angle can be trisected using a protractor: just find $20°$ and draw the angle. Thus, it is essential to state the problems carefully and to agree on certain ground rules. The Greek problems specify that only two tools are allowed, and each must be used in only one way. Let P and Q be points in the plane; we denote the line segment with endpoints P and Q by PQ, and we denote the length of this segment by $|PQ|$. A *straightedge* is a tool that can draw the line $L(P, Q)$ determined by P and Q; a *compass* is a tool that draws the circle with radius $|PQ|$ and center either P or Q; denote these circles by $C(P; Q)$ or $C(Q; P)$, respectively. Since every construction has only a finite number of steps, we shall be able to define "constructible" points inductively.

What we are calling a *straightedge*, others call a ***ruler***; we use the first term to avoid possible confusion, for one can mark distances on a ruler (as well as draw lines with it). This added function of a ruler turns out to make it a more powerful instrument. For example, Nicomedes solved the Delian problem of doubling the cube using a ruler and compass; both Nicomedes and Archimedes were able to trisect arbitrary angles with these tools (we present Archimedes's proof later in this section). On the other hand, we are going to show that both of these construction are impossible to do using only a straightedge and compass. (Some angles, e.g., 90° and 45°, can be trisected using a straightedge and compass; however, we are saying that there are some angles, e.g., 60°, that can never be so trisected.) When we say *impossible*, we mean what we say; we do not mean that it is merely very difficult. The reader should ponder how one might prove that something is impossible. About 425 B.C., Hippias of Elis was able to square the circle by drawing a certain curve as well as lines and circles. We shall see that this construction is also impossible using only straightedge and compass.

Given the plane, we establish a coordinate system by first choosing two distinct points, A and $\overline{A}$; call the line they determine the *x-axis*. Use a compass to draw the two circles $C(A; \overline{A})$ and $C(\overline{A}; A)$ of radius $|A\overline{A}|$ with centers A and $\overline{A}$, respectively. These two circles intersect in two points; the line they determine is called the *y-axis*; it is the perpendicular bisector of $A\overline{A}$, and it intersects the *x*-axis in a point O, called the *origin*. We define the distance $|OA|$ to be 1. We have introduced coordinates in the plane; in particular, $A = (1, 0)$ and $\overline{A} = (-1, 0)$.

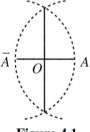

Figure 4.1

Informally, one constructs a new point T from (not necessarily distinct) old points P, Q, R, and S by using the first pair P, Q to draw a line or circle, the second pair R, S to draw a line or circle, and then obtaining T as one of the points of intersection of the two drawn lines, the drawn line and the drawn circle, or the two drawn circles. More generally, a point is called constructible if it is obtained from A and $\overline{A}$ by a finite number of such steps. Given a pair of constructible points, we do *not* assert that every point on the drawn line or the drawn circles they determine is constructible.

Here is the formal discussion. Recall that if P and Q are distinct points in the plane, then $L(P, Q)$ is the line they determine and $C(P; Q)$ is the circle with center P and radius $|PQ|$.

Definition.   Let $E \neq F$ and $G \neq H$ be points in the plane. A point Z is *constructible from* E, F, G, and H if either

(i) $Z \in L(E, F) \cap L(G, H)$, where $L(E, F) \neq L(G, H)$;

(ii) $Z \in L(E, F) \cap C(G; H)$;

(iii) $Z \in C(E; F) \cap C(G; H)$, where $C(E; F) \neq C(G; H)$.

A point Z is **constructible** if $Z = A$ or $Z = \overline{A}$ or if there are points $P_1, \dots, P_n$ with $Z = P_n$ so that, for all $j \geq 1$, the point P_{j+1} is constructible from points in $\{A, \overline{A}, P_1, \dots, P_j\}$.

Example 4.11.
Let us show that $Z = (0, 1)$ is constructible. We have seen, in Figure 4.1, that the points $P_2 = (0, \sqrt{3})$ and $P_3 = (0, -\sqrt{3})$ are constructible, for both lie in $C(A; \overline{A}) \cap C(\overline{A}; A)$, and so the y-axis $L(P_2, P_3)$ can be drawn. Finally,

$$Z = (0, 1) \in L(P_2, P_3) \cap C(O; A). \quad \blacktriangleleft$$

In our discussion, we shall freely use any standard result of euclidean geometry. For example, every angle can be bisected with ruler and compass; i.e., if $(\cos\theta, \sin\theta)$ is constructible, then so is $(\cos\frac{1}{2}\theta, \sin\frac{1}{2}\theta)$.

Definition.   A complex number $z = x + iy$ is **constructible** if the point (x, y) is a constructible point.

Example 4.11 shows that the numbers $1, -1, 0, i\sqrt{3}, -i\sqrt{3}$, and i are constructible numbers.

Lemma 4.27.   *A complex number $z = x + iy$ is constructible if and only if its real part x and its imaginary part y are constructible.*

Proof.   If z is constructible, then a standard euclidean construction draws the vertical line L through (x, y) which is parallel to the y-axis. It follows that x is constructible, for the point $(x, 0)$ is constructible, being the intersection of L and the x-axis. Similarly, the point $(0, y)$ is the intersection of the y-axis and a line through (x, y) which is parallel to the x-axis. It follows that $P = (y, 0)$ is constructible, for it is an intersection point of the x-axis and $C(O; P)$. Hence, y is a constructible number.

Conversely, assume that x and y are constructible numbers; that is, $Q = (x, 0)$ and $P = (y, 0)$ are constructible points. The point $(0, y)$ is constructible, being the intersection of the y-axis and $C(O; P)$. One can draw the vertical line through $(x, 0)$ as well as the horizontal line through $(0, y)$, and (x, y) is the intersection of these lines. Therefore, (x, y) is a constructible point, and so $z = x + iy$ is a constructible number. •

Definition.   We denote by K the subset of $\mathbb{C}$ consisting of all the ***constructible numbers***.

Lemma 4.28.

  (i) *If $K \cap \mathbb{R}$ is a subfield of $\mathbb{R}$, then K is a subfield of $\mathbb{C}$.*

  (ii) *If $K \cap \mathbb{R}$ is a subfield of $\mathbb{R}$ and if $\sqrt{a} \in K$ whenever $a \in K \cap \mathbb{R}$ is positive, then K is closed under square roots.*

Proof.   (i) If $z = a + ib$ and $w = c + id$ are constructible, then $a, b, c, d \in K \cap \mathbb{R}$, by Lemma 4.27. Hence, $a + c, b + d \in K \cap \mathbb{R}$, because $K \cap \mathbb{R}$ is a subfield, and so $(a + c) + i(b + d) \in K$, by Lemma 4.27. Similarly, $zw = (ac - bd) + i(ad + bc) \in K$. If $z \neq 0$, then $z^{-1} = (a/z\overline{z}) - i(b/z\overline{z})$. Now $a, b \in K \cap \mathbb{R}$, by Lemma 4.27, so that $z\overline{z} = a^2 + b^2 \in K \cap \mathbb{R}$, because $K \cap \mathbb{R}$ is a subfield of $\mathbb{C}$. Therefore, $z^{-1} \in K$.

(ii) If $z = a + ib \in K$, then $a, b \in K \cap \mathbb{R}$, by Lemma 4.27, and so $r^2 = a^2 + b^2 \in K \cap \mathbb{R}$, as in part (i). Since r^2 is nonnegative, the hypothesis gives $r \in K \cap \mathbb{R}$ and $\sqrt{r} \in K \cap \mathbb{R}$. Now $z = re^{i\theta}$, so that $e^{i\theta} = r^{-1}z \in K$, because K is a subfield of $\mathbb{C}$. That every angle can be bisected gives $e^{i\theta/2} \in K$, and so $\sqrt{z} = \sqrt{r}e^{i\theta/2} \in K$, as desired.   •

Theorem 4.29.   *The set of all constructible numbers K is a subfield of $\mathbb{C}$ that is closed under square roots and complex conjugation.*

Proof.   For the first two statements, it suffices to prove that the properties of $K \cap \mathbb{R}$ in Lemma 4.28 do hold. Let a and b be constructible reals.

(i) $-a$ is constructible.   If $P = (a, 0)$ is a constructible point, then $(-a, 0)$ is the other intersection of the x-axis and $C(O; P)$.

(ii) $a + b$ is constructible.   Assume that a and b are positive. Let $I = (0, 1)$,

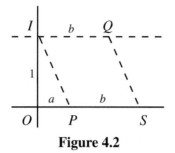

Figure 4.2

$P = (a, 0)$ and $Q = (b, 1)$. Now Q is constructible: it is the intersection of the horizontal line through I and the vertical line through $(b, 0)$ [the latter point is constructible, by hypothesis]. The line through Q parallel to IP intersects the x-axis in $S = (a + b, 0)$, as desired.

To construct $b - a$, let $P = (-a, 0)$ in Figure 4.2. Thus, both $a + b$ and $-a + b$ are constructible; by part (i), both $-a - b$ and $a - b$ are also constructible.

(iii) ab is constructible.   By part (i), we may assume that both a and b are

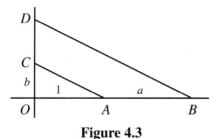

Figure 4.3

positive. In Figure 4.3, $A = (1, 0)$, $B = (1 + a, 0)$, and $C = (0, b)$. Define D to be the intersection of the y-axis and the line through B parallel to AC. Since the triangles $\triangle OAC$ and $\triangle OBD$ are similar,

$$|OB|/|OA| = |OD|/|OC|;$$

hence $(a + 1)/1 = (b + |CD|)/b$, and $|CD| = ab$. Therefore, $b + ab$ is constructible. Since $-b$ is constructible, by part (i), we have $ab = (b+ab)-b$ constructible, by part (ii).

(iv) If $a \neq 0$, then a^{-1} is constructible. Let $A = (1, 0)$, $S = (0, a)$, and

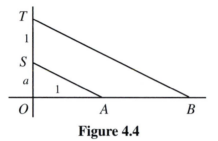

Figure 4.4

$T = (0, 1+a)$. Define B as the intersection of the x-axis and the line through T parallel to AS; thus, $B = (1 + u, 0)$ for some u. Similarity of the triangles $\triangle OSA$ and $\triangle OTB$ gives

$$|OT|/|OS| = |OB|/|OA|.$$

Hence, $(1 + a)/a = (1 + u)/1$, and so $u = a^{-1}$. Therefore, $1 + a^{-1}$ is constructible, and so $(1 + a^{-1}) - 1 = a^{-1}$ is constructible.

(v) If $a \geq 0$, then $\sqrt{a}$ is constructible. Let $A = (1, 0)$ and $P = (1 + a, 0)$;

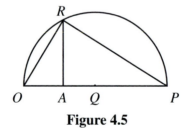

Figure 4.5

construct Q, the midpoint of OP. Define R as the intersection of the circle $C(Q; O)$ with the vertical line through A. The (right) triangles $\triangle AOR$ and $\triangle ARP$ are similar, so that

$$|OA|/|AR| = |AR|/|AP|,$$

and, hence, $|AR| = \sqrt{a}$.

(vi) If $z = a + ib \in K$, then $\bar{z} = a - ib$ is constructible.

By Lemma 4.28, K is a subfield of $\mathbb{C}$. Now $a, b \in K$, by Lemma 4.27, and $i \in K$, by Example 4.11. Therefore, $-bi \in K$, and so $a - ib \in K$.   •

Corollary 4.30.   *If a, b, c are constructible, then the roots of the quadratic $ax^2 + bx + c$ are also constructible.*

Proof.   This follows from the quadratic formula and the theorem.   •

We now consider subfields of $\mathbb{C}$ to enable us to prove an inductive step in the upcoming theorem.

Lemma 4.31.   *Let F be a subfield of $\mathbb{C}$ which contains i and is closed under complex conjugation. Let $z = a + ib$, $w = c + id \in F$, and let $P = (a, b)$ and $Q = (c, d)$.*

(i)   *If $a + ib \in F$, then $a \in F$ and $b \in F$.*

(ii)   *If the equation of $L(P, Q)$ is $y = mx + q$, where $m, q \in \mathbb{R}$, then $m, q \in F$.*

(iii)   *If the equation of $C(P; Q)$ is $(x - a)^2 + (y - b)^2 = r^2$, where $a, b, r \in \mathbb{R}$, then $r^2 \in F$.*

Proof.   (i) If $z = a + ib \in F$, then $a = \frac{1}{2}(z + \bar{z}) \in F$ and $ib = \frac{1}{2}(z - \bar{z}) \in F$; since we are assuming that $i \in F$, we have $b = -i(ib) \in F$.

(ii) If $L(P, Q)$ is not vertical, then its equation is $y - b = m(x - a)$. Now $m = (d - b)/(a - c) \in F$, since $a, b, c, d \in F$, and so $q = -ma + b \in F$.

(iii) The circle $C(P; Q)$ has equation $(x - a)^2 + (y - b)^2 = r^2$, and $r^2 = (c - a)^2 + (d - b)^2 \in F$.   •

Lemma 4.32.   *Let F be a subfield of $\mathbb{C}$ which contains i and is closed under complex conjugation. Let P, Q, R, S be points whose coordinates lie in F, and let $\alpha = u + iv \in \mathbb{C}$. If either*

$$\alpha \in L(P, Q) \cap L(R, S), \quad where \ L(P, Q) \neq L(R, S),$$
$$\alpha \in L(P, Q) \cap C(R; S),$$

or

$$\alpha \in C(P; Q) \cap C(R, S), \quad where \ C(P; Q) \neq C(R; S),$$

then $[F(\alpha) : F] \leq 2$.

Proof. If $L(P, Q)$ is not vertical, then Lemma 4.31(ii) says that $L(P, Q)$ has equation $y = mx + b$, where $m, b \in F$. If $L(P, Q)$ is vertical, then its equation is $x = b$ because $P = (a, b) \in L(P, Q)$, and so $b \in F$, by Lemma 4.31(i). Similarly, $L(R, S)$ has equation $y = nx + c$ or $x = c$, where $m, b, n, c \in F$. Since these lines are not parallel, one can solve the pair of linear equations for (u, v), the coordinates of $\alpha \in L(P, Q) \cap L(R, S)$, and they also lie in F. In this case, therefore, $[F(\alpha) : F] = 1$.

Let $L(P, Q)$ have equation $y = mx + b$ or $x = b$, and let $C(R; S)$ have equation $(x - c)^2 + (y - d)^2 = r^2$; by Lemma 4.31, we have $m, q, r^2 \in F$. Since $\alpha = u + iv \in L(P, Q) \cap C(R; S)$,

$$r^2 = (u - c)^2 + (v - d)^2$$
$$= (u - c)^2 + (mu + q - d)^2,$$

so that u is a root of a quadratic polynomial with coefficients in $F \cap \mathbb{R}$. Hence, $[F(u) : F] \le 2$. Since $v = mu + q$, we have $v \in F(u)$, and, since $i \in F$, we have $\alpha \in F(u)$. Therefore, $\alpha = u + iv \in F(u)$, and so $[F(\alpha) : F] \le 2$.

Let $C(P; Q)$ have equation $(x - a)^2 + (y - b)^2 = r^2$, and let $C(R; S)$ have equation $(x - c)^2 + (y - d)^2 = s^2$. By Lemma 4.31, we have $r^2, s^2 \in F \cap \mathbb{R}$. Since $\alpha \in C(P; Q) \cap C(R; S)$, there are equations

$$(u - a)^2 + (v - b)^2 = r^2 \text{ and } (u - c)^2 + (v - d)^2 = s^2.$$

After expanding, both equations have the form $u^2 + v^2 + \ something = 0$. Setting the *something*'s equal gives an equation of the form $tu + t'v + t'' = 0$, where $t, t', t'' \in F$. Coupling this with the equation of one of the circles returns us to the situation of the second paragraph.   •

Here is the result we have been seeking: an algebraic characterization of a geometric idea.

Theorem 4.33.   *If a complex number z is constructible, then there is a tower of fields*

$$\mathbb{Q} = K_0 \subset K_1 \subset \cdots \subset K_n,$$

where $z \in K_n$ and $[K_{j+1} : K_j] \le 2$ for all j.

Remark.   The converse is also true, so that constructibility of a point $P = (a, b)$ is equivalent to the existence of such a tower for a complex number $z = a + ib$. Therefore, this theorem gives an exact translation from geometry into algebra.   ◄

Proof.   If z is constructible, there is a sequence of points $1, -1, z_1, \ldots, z_n = z$ with each z_j obtainable from $\{1, -1, z_1, \ldots, z_{j-1}\}$; since i is constructible, we may assume that $z_1 = i$. Define

$$K_1 = \mathbb{Q}(z_1) \text{ and } K_{j+1} = K_j(z_{j+1}),$$

where there are points $E, F, G, H \in K_j$ with one of the following:

$$z_{j+1} \in L(E, F) \cap L(G, H);$$
$$z_{j+1} \in L(E, F) \cap C(G; H);$$
$$z_{j+1} \in C(E; F) \cap C(G; H).$$

We may assume, by induction on $j \geq 1$, that K_j is closed under complex conjugation, so that Lemma 4.32 applies to show that $[K_{j+1} : K_j] \leq 2$. Finally, note that K_{j+1} is also closed under complex conjugation, for if z_{j+1} is a root of a quadratic $f(x) \in K_j[x]$, then $\overline{z}_{j+1}$ is the other root of $f(x)$.   •

Corollary 4.34.   *If a complex number z is constructible, then $[\mathbb{Q}(z) : \mathbb{Q}]$ is a power of 2.*

Proof.   This follows from the theorem and Theorem 4.25.   •

Remark.   The converse of this corollary is false; it can be shown that there are nonconstructible numbers z with $[\mathbb{Q}(z) : \mathbb{Q}] = 4$.   ◄

It was proved by G. Mohr in 1672 and, independently, by L. Mascheroni, in 1797, that every geometric construction carried out by straightedge and compass can be done without the straightedge. There is a short proof of this theorem given by M. Hungerbühler in *The American Mathematical Monthly*, 101 (1994), pp. 784–787.

We can now deal with the Greek problems, two of which were solved by P. L. Wantzel (1814–1848) in 1837. The notion of dimension of a vector space was not known at that time; in place of Theorem 4.33, Wantzel proved that if a number is constructible, then it is a root of an irreducible polynomial in $\mathbb{Q}[x]$ of degree 2^n for some n.

Theorem 4.35 (Wantzel).   *It is impossible to duplicate the cube using only straightedge and compass.*

Proof.   The question is whether $z = \sqrt[3]{2}$ is constructible. Since $x^3 - 2$ is irreducible, $[\mathbb{Q}(z) : \mathbb{Q}] = 3$, by Theorem 4.24; but 3 is not a power of 2.   •

Consider how ingenious this proof is. At the beginning of this section, we asked the reader to ponder how one might prove impossibility. The idea here is to translate the problem of constructibility into a statement of arithmetic, and then to show that the existence of a construction produces a contradiction.

A student in one of my classes, imbued with the idea of continual progress through technology, asked me, "Will it ever be possible to duplicate the cube with straightedge and compass?" *Impossible* here is used in its literal sense.

Theorem 4.36 (Wantzel).   *It is impossible to trisect a 60° angle using only straightedge and compass.*

Proof.   We may assume that one side of the angle is on the x-axis, and so the question is whether $z = \cos 20° + i \sin 20°$ is constructible. If z were constructible, then Lemma 4.27 would show that $\cos 20°$ is constructible. Corollary 1.23, the triple angle formula, gives

$$\cos 3\alpha = 4 \cos^3 \alpha - 3 \cos \alpha.$$

Setting $\alpha = 20°$, we have $\cos 3\alpha = \frac{1}{2}$, so that $z = \cos 20°$ is a root of $4x^3 - 3x - \frac{1}{2}$; equivalently, $\cos 20°$ is a root of $f(x) = 8x^3 - 6x - 1 \in \mathbb{Z}[x]$. Now a cubic is irreducible in $\mathbb{Q}[x]$ if and only if it has no rational roots. By Theorem 3.56, the only candidates for rational roots are $\pm 1, \pm\frac{1}{2}, \pm\frac{1}{4}$, and $\pm\frac{1}{8}$; since none of these is a root, as one easily checks, it follows that $f(x)$ is irreducible [alternatively, one can prove irreducibility using Theorem 3.65, for $f(x)$ is irreducible mod 7]. Therefore, $3 = [\mathbb{Q}(z) : \mathbb{Q}]$, by Theorem 4.24(ii), and so $z = \cos 20°$ is not constructible because 3 is not a power of 2.   •

Theorem 4.37 (Archimedes).   *Every angle can be trisected using ruler and compass.*

Remark.   Recall that a *ruler* is a straightedge on which one can mark a distance.   ◄

Proof.   Since it is easy to construct 30°, 60°, and 90°, it suffices to trisect an acute angle α, for if $3\beta = \alpha$, then $3(\beta + 30°) = \alpha + 90°$, $3(\beta + 60°) = \alpha + 180°$, and $3(\beta + 90°) = \alpha + 270°$.

Draw the given angle $\alpha = \angle AOE$, where the origin O is the center of the unit circle. Take a ruler on which the distance 1 has been marked; that is, there are points U and V on the ruler with $|UV| = 1$. There is a chord through A parallel to EF; place the ruler so that this chord is AU. Since α

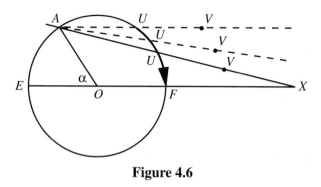

Figure 4.6

is acute, U lies in the first quadrant. Keeping A on the sliding ruler, move the point U down the circle; the ruler intersects the extended diameter EF in some point X with $|UX| > 1$. Continue moving U down the circle, keeping A on the sliding ruler, until the ruler intersects EF in the point V.

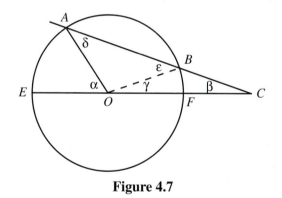

Figure 4.7

Relabel the points as in Figure 4.7, so that $U = B$, $V = C$, and $|BC| = 1$. We claim that $\beta = \angle BCO = \frac{1}{3}\alpha$. Now

$$\alpha = \delta + \beta,$$

because α is an exterior angle of $\triangle AOC$, and hence it is the sum of the two opposite internal angles. Since $\triangle OAB$ is isosceles (OA and OB are radii), $\delta = \varepsilon$, and so

$$\alpha = \varepsilon + \beta.$$

But $\varepsilon = \gamma + \beta = 2\beta$, for it is an exterior angle of the isosceles triangle $\triangle BCO$; therefore,

$$\alpha = 2\beta + \beta = 3\beta. \quad \bullet$$

Theorem 4.38 (*Lindemann*).   *It is impossible to square the circle with straightedge and compass.*

Proof.   The problem is whether one can construct a square whose area is the same as the area of the unit circle. If a side of the square has length z, then one is asking whether $z = \sqrt{\pi}$ is constructible. Now $\mathbb{Q}(\pi)$ is a subspace of $\mathbb{Q}(\sqrt{\pi})$. We have already mentioned that Lindemann proved that π is transcendental (over $\mathbb{Q}$), so that $[\mathbb{Q}(\pi) : \mathbb{Q}]$ is infinite. It follows from Corollary 4.13(ii) that $[\mathbb{Q}(\sqrt{\pi}) : \mathbb{Q}]$ is also infinite. Thus, $[\mathbb{Q}(\sqrt{\pi}) : \mathbb{Q}]$ is surely not a power of 2, and so $\sqrt{\pi}$ is not constructible. $\bullet$

Sufficiency of the following result was discovered, around 1796, by Gauss, when he was still in his teens (he wrote that this result led to his decision to become a mathematician). He claimed necessity as well, but none of his published papers contains a complete proof of it. The first published proof of necessity is due to P. L. Wantzel, in 1837.

Theorem 4.39 (*Gauss-Wantzel*).   *Let p be an odd prime. A regular p-gon is constructible if and only if $p = 2^{2^t} + 1$ for some $t \geq 0$.*

Proof.   We only prove necessity. The problem is whether $z = e^{2\pi i/p}$ is constructible. Now z is a root of the cyclotomic polynomial $\Phi_p(x)$, which is an irreducible polynomial in $\mathbb{Q}[x]$ of degree $p - 1$, by Corollary 3.68.

Since z is constructible, $p - 1 = 2^s$ for some s, so that

$$p = 2^s + 1.$$

We claim that s itself is a power of 2. Otherwise, there is an odd number $k > 1$ with $s = km$. Now k odd implies that -1 is a root of $x^k + 1$; in fact, there is a factorization in $\mathbb{Z}[x]$:

$$x^k + 1 = (x + 1)(x^{k-1} - x^{k-2} + x^{k-3} - \cdots + 1).$$

Thus, setting $x = 2^m$ gives a forbidden factorization of p in $\mathbb{Z}$:

$$\begin{aligned} p = 2^s + 1 &= (2^m)^k + 1 \\ &= [2^m + 1][(2^m)^{k-1} - (2^m)^{k-2} + (2^m)^{k-3} - \cdots + 1]. \quad \bullet \end{aligned}$$

Gauss constructed a regular 17-gon explicitly, a feat the Greeks would have envied. On the other hand, it follows, for example, that it is impossible to construct regular 7-gons, 11-gons, or 13-gons.

Numbers F_t of the form $F_t = 2^{2^t} + 1$ are called **Fermat primes** if they are prime. For $0 \le t \le 4$, one can check that F_t is, indeed, prime; they are

$$3, \ 5, \ 17, \ 257, \ \text{and} \ 65{,}537.$$

It is known that the next few values of t give composite numbers, and it is unknown whether any other Fermat primes exist.

The following result is known.

Theorem.   *A regular n-gon is constructible if and only if n is a product of a power of 2 and distinct Fermat primes.*

Proof.   See Hadlock, *Field Theory and its Classical Problems*, p. 106.   •

4.3 CLASSICAL FORMULAS

Revolutionary events were changing the western world in the early 1500s: the printing press had just been invented; trade with Asia and Africa was flourishing; Columbus had just discovered the New World; and Martin Luther was challenging papal authority. The Reformation and the Renaissance were beginning.

The Italian peninsula was not one country but a collection of city states with many wealthy and cosmopolitan traders. And there were public mathematics contests sponsored by the dukes of the cities. In particular, one of the problems frequently set involved finding roots of a given cubic equation

$$X^3 + bX^2 + cX + d = 0,$$

where a, b, and c were real numbers, usually integers. [Such contests were an old tradition; there are records from 1225 of Leonardo of Pisa (c. 1180– c. 1245), also called Fibonacci, approximating roots of $x^3 + 2x^2 + 10x - 20$ with good accuracy.[8]]

---

[8]Around 1074, Omar Khayyam (1048–1123), a mathematician now more famous for his poetry, used intersections of conic sections to give geometric constructions of roots of cubics.

We must mention that modern notation did not exist at the beginning of the sixteenth century, and so the feat of finding roots involved not only mathematical ingenuity but also an ability to surmount linguistic obstacles. Designating variables by letters was invented in 1591 by F. Viète (1540–1603) who used consonants to denote constants and vowels to denote variables (the modern notation of using letters $a, b, c, \ldots$ at the beginning of the alphabet to denote constants and letters x, y, z at the end of the alphabet to denote variables was introduced by R. Descartes in 1637 in his book *La Géométrie*). The exponential notation $A^2, A^3, A^4, \ldots$ was essentially introduced by J. Hume in 1636 (he used $A^{ii}, A^{iii}, A^{iv}, \ldots$). The symbols $+, -,$ and $\sqrt{\ }$, as well as the symbol $/$ for division, as in a/b, were introduced by J. Widman in 1486. The symbol $\times$ for multiplication was introduced by W. Oughtred in 1631, and the symbol $\div$ for division by J. H. Rahn in 1659. The symbol $=$ was introduced by the Oxford don Robert Recorde in 1557, in his *Whetstone of Wit*:

> And to avoide the tediouse repetition of these woordes: is equal
> to: I will lette as I doe often in woorke use, a paire of paralleles,
> or gemowe lines of one lengthe, thus: $=$, because noe 2 thynges,
> can be moare equalle.

(*Gemowe* is an obsolete word meaning *twin* or, in this case, *parallel*.)

These symbols were not adopted at once, and often there were competing notations. Only in the next century, with the publication of Descartes's *La Géométrie*, did most of this notation become universal in Europe.

Let us return to cubic equations. The lack of a good notation was a great handicap. For example, the cubic equation $X^3 + 2X^2 + 4X - 1 = 0$ would be given, roughly, as follows:

> Take the cube of a thing, add to it twice the square of the thing,
> to this add 4 times the thing, and this must all be set equal to 1.

Complicating matters even more, negative numbers were not accepted; an equation of the form $X^3 - 2X^2 - 4X + 1 = 0$ would only be given in the form $X^3 + 1 = 2X^2 + 4X$. Thus, there were many forms of cubic equations, depending (in our notation) on whether coefficients were positive, negative, or zero.

About 1515, Scipione del Ferro of Bologna discovered a method for finding the roots of several forms of a cubic. Given the competitive context, it was natural for him to keep his method secret. Before his death in 1526, Scipione shared his result with some of his students.

The following history is from the excellent account in J.-P. Tignol, *Galois'*
Theory of Equations.

> In 1535, Niccolò Fontana (c. 1500–1557), nicknamed "Tartaglia"
> ("Stammerer"), from Brescia, who had dealt with some very par-
> ticular cases of cubic equations, was challenged to a problem-
> solving contest by Antonio Maria Fior, a former pupil of Scipi-
> one del Ferro. When he heard that Fior had received the solution
> of cubic equation from his master, Tartaglia threw all his energy
> and skill into the struggle. He succeeded in finding the solution
> just in time to inflict upon Fior a humiliating defeat.
>
> The news that Tartaglia had found the solution of cubic equation
> reached Giralamo Cardano (1501–1576), a very versatile scien-
> tist, who wrote a number of books on a wide variety of subjects,
> including medicine, astrology, astronomy, philosophy, and math-
> ematics. Cardano then asked Tartaglia to give him his solution,
> so that he could include it in a treatise on arithmetic, but Tartaglia
> flatly refused, since he was himself planning to write a book on
> this topic. It turns out that Tartaglia later changed his mind, at
> least partially, since in 1539 he handed on to Cardano the solu-
> tion of $x^3 + qx = r$, $x^3 = qx + r$, and a very brief indication
> of $x^3 + r = qx$ in verses. ... Having received Tartaglia's poem,
> Cardano set to work. Not only did he find justifications for the
> formulas, but he also solved all the other types of cubics. He
> then published his results, giving due credit to Tartaglia and to
> del Ferro, in the epoch-making book *Ars Magna, sive de regulis*
> *algebraicis* (The Great Art, or the Rules of Algebra).

Let us now derive the formulas for the roots of polynomials of low degree.
Consider the quadratic polynomial $f(X) = X^2 + bX + c$. The usual way to
derive the quadratic formula is by "completing the square."

$$X^2 + bX + c = X^2 + bX + \tfrac{1}{4}b^2 + c - \tfrac{1}{4}b^2$$
$$= (X + \tfrac{1}{2}b)^2 + \tfrac{1}{4}(4c - b^2).$$

Therefore, if u is a root of $f(X)$, then

$$u + \tfrac{1}{2}b = \pm\tfrac{1}{2}\sqrt{b^2 - 4c}.$$

We are now going to present a different derivation which generalizes to give the roots of cubic and quartic polynomials.

If U is a root of $f(X) = X^2 + bX + c$, then

$$U^2 + bU + c = 0.$$

If one defines a number u by $U = u - \frac{1}{2}$, then

$$(u - \tfrac{1}{2}b)^2 + b(u - \tfrac{1}{2}b) + c = 0;$$

the linear terms in u cancel, and the equation simplifies to

$$u^2 = \tfrac{1}{4}(b^2 - 4c).$$

It follows that $u = \pm \frac{1}{2}\sqrt{b^2 - 4c}$. One now obtains the familiar quadratic formula:

$$U = u - \tfrac{1}{2}b = \tfrac{1}{2}(-b \pm \sqrt{b^2 - 4c}).$$

The following consequence of the quadratic formula will be used in deriving the cubic formula.

Lemma 4.40.   *Given numbers c and d, there exist numbers α and β with $\alpha + \beta = c$ and $\alpha\beta = d$.*

Proof.   If $d = 0$, choose $\alpha = 0$ and $\beta = c$. If $d \neq 0$, then $\alpha \neq 0$ and we may set $\beta = d/\alpha$. Substituting, $c = \alpha + \beta = \alpha + d/\alpha$, so that

$$\alpha^2 - c\alpha + d = 0.$$

The quadratic formula now shows that such an α exists, as does $\beta = d/\alpha$. (Of course, α and β might be complex numbers.)   •

The proof of the cubic formula will involve replacing a given cubic polynomial by a simpler one.

Definition.   A cubic $f(x) = x^3 + qx + r$ having no x^2 term is called a *reduced cubic*.

Lemma 4.41. *The substitution $X = x - \frac{1}{3}b$ changes*

$$f(X) = X^3 + bX^2 + cX + d$$

into a reduced cubic

$$\tilde{f}(x) = f(x - \tfrac{1}{3}b) = x^3 + qx + r;$$

moreover, if u is a root of $\tilde{f}(x)$, then $u - \frac{1}{3}b$ is a root of $f(X)$.

Proof.   The substitution $X = x - \frac{1}{3}b$ gives

$$
\begin{aligned}
f(X) = f(x - \tfrac{1}{3}b) \\
= (x - \tfrac{1}{3}b)^3 + b(x - \tfrac{1}{3}b)^2 + c(x - \tfrac{1}{3}b) + dx^3 - bx^2 \\
+ (\tfrac{1}{3}b^2)x - \tfrac{1}{27}b^3 + bx^2 - \tfrac{2}{3}b^2x + \tfrac{1}{9}b^3 + cx - \tfrac{1}{3}bc + d.
\end{aligned}
$$

As the x^2 terms cancel, this new polynomial $\tilde{f}(x) = f(x - \frac{1}{3}b)$ is reduced: $\tilde{f}(x) = x^3 + qx + r$, where the coefficients q and r can be found from the last equation by collecting terms.

Finally, if u is a root of $\tilde{f}(x)$, then $0 = \tilde{f}(u) = f(u - \frac{1}{3}b)$; that is, $u - \frac{1}{3}b$ is a root of $f(X)$.   •

Thus, the lemma reduces the problem of finding the roots of a cubic polynomial to the problem of finding the roots of a simpler cubic having no x^2 term.

The "trick" is to write a root u of $x^3 + qx + r$ as

$$u = \alpha + \beta,$$

and then to find α and β. Now

$$
\begin{aligned}
0 &= u^3 + qu + r \\
&= (\alpha + \beta)^3 + q(\alpha + \beta) + r.
\end{aligned}
$$

Note that

$$
\begin{aligned}
(\alpha + \beta)^3 &= \alpha^3 + 3\alpha^2\beta + 3\alpha\beta^2 + \beta^3 \\
&= \alpha^3 + \beta^3 + 3(\alpha^2\beta + \alpha\beta^2) \\
&= \alpha^3 + \beta^3 + 3\alpha\beta(\alpha + \beta) \\
&= \alpha^3 + \beta^3 + 3\alpha\beta u.
\end{aligned}
$$

Therefore, $0 = \alpha^3 + \beta^3 + 3\alpha\beta u + qu + r$, and so

$$0 = \alpha^3 + \beta^3 + u(3\alpha\beta + q) + r. \tag{1}$$

We have already set $\alpha + \beta = u$; by Lemma 4.40, we may impose a second condition

$$\alpha\beta = -q/3 \tag{2}$$

which makes the u term in Eq. (1) go away, leaving

$$\alpha^3 + \beta^3 = -r. \tag{3}$$

Cubing each side of Eq. (2) gives

$$\alpha^3\beta^3 = -q^3/27. \tag{4}$$

Equations (3) and (4) in the two unknowns α^3 and β^3 can be solved, as in Lemma 4.40. Substituting[9] $\beta^3 = -q^3/27\alpha^3$ into Eq. (3) gives

$$\alpha^3 - q^3/27\alpha^3 = -r,$$

which may be rewritten as

$$\alpha^6 + r\alpha^3 - q^3/27 = 0,$$

a quadratic in α^3. The quadratic formula gives

$$\alpha^3 = \tfrac{1}{2}\left(-r + \sqrt{D}\right),$$

where $D = r^2 + 4q^3/27$. Now take a cube root to obtain α.[10] By Eq. (2), $\beta = -q/3v$, and so $u = \alpha + \beta$ has been found.

What are the other two roots? Theorem 3.32 says that if u is a root of a polynomial $f(x)$, then $f(x) = (x - u)g(x)$ for some polynomial $g(x)$. After finding one root $u = \alpha + \beta$, divide $x^3 + qx + r$ by $x - u$, and use the quadratic formula on the quadratic quotient $g(x)$ to find the other two roots. [If γ is a root of $g(x)$, then γ is also a root of $f(x)$, for $g(\gamma) = 0$ and $f(\gamma) = (\gamma - u)g(\gamma) = 0$.]

---

[9]If $\alpha = 0$, then $q = 0$, and the polynomial is $f(x) = x^3 + r = 0$; of course, the roots in this case are the cube roots of $-r$.

[10]The number $z = \tfrac{1}{2}(-r + \sqrt{D})$ might be complex. The easiest way to find a cube root of z is to write it in polar form $se^{i\theta}$, where $s \geq 0$; a cube root is then $\sqrt[3]{s}\, e^{i\theta/3}$.

Here is an explicit formula for the other two roots of $f(x)$ (instead of the method just given for finding them). There are three cube roots of unity, namely, 1, $\omega = -\frac{1}{2} + i\frac{\sqrt{3}}{2}$, and $\omega^2 = -\frac{1}{2} - i\frac{\sqrt{3}}{2}$. It follows that the other cube roots of α^3 are $\omega\alpha$ and $\omega^2\alpha$. If β is the "mate" of α, that is, if $\alpha + \beta$ is a root of $f(x)$, then the mate of $\omega\alpha$ is

$$-q/3\omega\alpha = \beta/\omega = \omega^2\beta,$$

and the mate of $\omega^2\alpha$ is

$$-q/3\omega^2\alpha = \beta/\omega^2 = \omega\beta.$$

Therefore, explicit formulas for the roots of $f(x)$ are $\alpha + \beta$, $\omega\alpha + \omega^2\beta$, and $\omega^2\alpha + \omega\beta$.

We have proved the *cubic formula* (also called **Cardano's formula**).

Theorem 4.42 (Cubic Formula).   *The roots of $x^3 + qx + r$, where $q \neq 0$, are*

$$\alpha + \beta, \qquad \omega\alpha + \omega^2\beta, \qquad and \qquad \omega^2\alpha + \omega\beta,$$

where $\alpha^3 = \frac{1}{2}(-r + \sqrt{D})$, $\beta = -q/3\alpha$, $D = r^2 + \frac{4}{27}q^3$, and $\omega = -\frac{1}{2} + i\frac{\sqrt{3}}{2}$ is a cube root of unity.

Remark.   Since $\alpha\beta = -q/3$, the assumption $q \neq 0$ gives $\alpha \neq 0$, and so $\beta = -q/3\alpha$ makes sense. It is not difficult to show, when $\alpha \neq 0$, that

$$\beta^3 = \frac{1}{2}(-r - \sqrt{D}).$$

If one writes the formula for β in this way, then the cubic formula holds even in the degenerate case $q = 0$; that is, the roots are $\alpha + \beta$, $\omega\alpha + \omega^2\beta$, and $\omega^2\alpha + \omega\beta$.   ◄

Example 4.12.
We find the roots of
$$x^3 - 15x - 126.$$

Since there is no x^2 term, the polynomial is already reduced, and so it is in the form to which the cubic formula applies (were it not reduced, one would first reduce it, as in Lemma 4.41). Here, $q = -15$, $r = -126$, $D = (-126)^2 + 4(-15)^3/27 = 15{,}376$, and $\sqrt{D} = 124$. Hence, $\alpha^3 = \frac{1}{2}[-(-126) + 124] =$

125 and $\alpha = 5$, while $\beta = -q/3\alpha = 15/(3 \cdot 5) = 1$. The roots are thus $\alpha + \beta = 6$, $\omega\alpha + \omega^2\beta = -3 + 2i\sqrt{3}$, and $\omega^2\alpha + \omega\beta = -3 - 2i\sqrt{3}$.

Alternatively, having found $u = 6$ to be a root, the division algorithm gives

$$x^3 - 15x - 126 = (x - 6)(x^2 + 6x + 21),$$

and the quadratic formula gives $-3 \pm 2i\sqrt{3}$ as the roots of the quadratic factor.

The quadratic and cubic formulas are not valid for arbitrary coefficient fields k. For example, if k has characteristic 2 (that is, if the prime field of k is $\mathbb{Z}_2$), then the quadratic formula does not make sense because $\frac{1}{2}$ is not defined (for $2 = 0$). Similarly, the cubic formula (and the quartic formula below) do not apply to polynomials with coefficients in a field of characteristic 2 or a field of characteristic 3 because the formulas involve $\frac{1}{2}$ and $\frac{1}{3}$, neither of which makes sense in these fields. ◀

Here is an application.

Definition. If u, v, and w are the roots of $f(x) = x^3 + qx + r$, let $\Delta = (u - v)(u - w)(v - w)$, and define

$$\Delta^2 = [(u - v)(u - w)(v - w)]^2;$$

the number Δ^2 is called the ***discriminant*** [11] of $f(x)$.

It is natural to consider Δ^2 instead of Δ, for Δ is a number depending not only on the roots but also on the order in which they are listed. Had we listed the roots as u, w, v, for example, then $(u - w)(u - v)(w - v) = -\Delta$, because the factor $w - v = -(v - w)$ has changed sign. Squaring eliminates this difference.

Note that if $\Delta^2 = 0$, then $\Delta = 0$ and the cubic has a repeated root. Can we detect this without first computing the roots? The cubic formula allows us to compute Δ^2 in terms of q and r.

Lemma 4.43.   *The discriminant Δ^2 of $f(x) = x^3 + qx + r$ is*

$$\Delta^2 = -27r^2 - 4q^3 = -27D.$$

---

[11] More generally, if $f(x) = (x - u_1)(x - u_2)\ldots(x - u_n)$ is a polynomial of degree n, then the discriminant of $f(x)$ is defined to be Δ^2, where $\Delta = \prod_{i<j}(u_i - u_j)$ (one takes $i < j$ so that each difference $u_i - u_j$ occurs just once in the product). In particular, the quadratic formula shows that the discriminant of $x^2 + bx + c$ is $b^2 - 4c$.

Proof.   If the roots of $f(x)$ are u, v, and w, then the cubic formula gives

$$u = \alpha + \beta; \qquad v = \omega\alpha + \omega^2\beta; \qquad w = \omega^2\alpha + \omega\beta,$$

where $\omega = -\frac{1}{2} - i\frac{\sqrt{3}}{2}$, $D = r^2 + \frac{4}{27}q^3$, $\alpha = [\frac{1}{2}(-r + \sqrt{D})]^{1/3}$, and $\beta = [\frac{1}{2}(-r - \sqrt{D})]^{1/3}$. One checks easily that:

$$u - v = \alpha + \beta - \omega\alpha - \omega^2\beta = (1 - \omega)(\alpha - \omega^2\beta);$$
$$u - w = \alpha + \beta - \omega^2\alpha - \omega\beta = -\omega^2(1 - \omega)(\alpha - \omega\beta);$$
$$v - w = \omega\alpha + \omega^2\beta - \omega^2\alpha - \omega\beta = \omega(1 - \omega)(\alpha - \beta).$$

Therefore,

$$\Delta = -\omega^3(1 - \omega)^3(\alpha - \beta)(\alpha - \omega\beta)(\alpha - \omega^2\beta).$$

Of course, $-\omega^3 = -1$, while

$$(1 - \omega)^3 = 1 - 3\omega + 3\omega^2 - \omega^3 = -3(\omega - \omega^2).$$

But $\omega = -\frac{1}{2} + i\frac{\sqrt{3}}{2}$ and $\omega^2 = \bar{\omega} = -\frac{1}{2} - i\frac{\sqrt{3}}{2}$, so that $(1 - \omega)^3 = -3(\omega - \omega^2) = -i3\sqrt{3}$, and

$$-\omega^3(1 - \omega)^3 = i3\sqrt{3}.$$

Finally, Exercise 3.74(iii) gives

$$(\alpha - \beta)(\alpha - \omega\beta)(\alpha - \omega^2\beta) = \alpha^3 - \beta^3 = \sqrt{D}.$$

Therefore, $\Delta = i3\sqrt{3}\sqrt{D}$, and

$$\Delta^2 = -27D = -27r^2 - 4q^3. \quad \bullet$$

It follows, for example, that the cubic formula is not needed to see that $x^3 - 3x + 2$ has a repeated root, for $-27r^2 - 4q^3 = 0$.

We are now going to use the discriminant to detect whether the roots of a cubic are all real.

Lemma 4.44. *Every $f(x) \in \mathbb{R}[x]$ of odd degree has a real root.*

Remark. The proof we give assumes that $f(x)$ has a complex root (which follows from the Fundamental Theorem of Algebra). There is a proof using the intermediate value theorem that does not make this assumption [see item iv on page 383]. ◀

Proof. The proof is by induction on $n \geq 0$, where $\deg(f) = 2n + 1$. The base step $n = 0$ is obviously true. Let $n \geq 1$ and let u be a complex root of $f(x)$. If u is real, we are done. Otherwise $u = a + ib$, and Exercise 4.18 shows that the complex conjugate $\bar{u} = a - ib$ is also a root; moreover, $u \neq \bar{u}$ because u is not real. As in the proof of Theorem 3.33, there is a factorization in $\mathbb{C}[x]$:

$$f(x) = (x - u)(x - \bar{u})g(x),$$

and hence there is a factorization in $\mathbb{R}[x]$:

$$\begin{aligned} f(x) &= (x - u)(x - \bar{u})g(x) \\ &= (x^2 - (u + \bar{u})x + u\bar{u})g(x) \\ &= (x^2 - 2ax + a^2 + b^2)g(x). \end{aligned}$$

Now $g(x) = f(x)/(x^2 - 2ax + a^2 + b^2)$ has real coefficients and degree $(2n + 1) - 2 = 2n - 1 = 2(n - 1) + 1$. By induction, $g(x)$, and hence $f(x)$, has a real root. ●

Theorem 4.45. *The roots u, v, w of $x^3 + qx + r$ are real numbers if and only if the discriminant $\Delta^2 \geq 0$; that is, $27r^2 + 4q^3 \leq 0$.*

Proof. If u, v, and w are real numbers, then $\Delta = (u - v)(u - w)(v - w)$ is a real number. Therefore, $-27r^2 - 4q^3 = \Delta^2 \geq 0$, and $27r^2 + rq^3 \leq 0$.

Conversely, assume that $w = s + ti$ is not real (i.e., $t \neq 0$); by Exercise 4.18 below, the complex conjugate of a root is also a root, so we may take $v = s - ti$; by Lemma 4.44, the other root u is real. Now

$$\begin{aligned} \Delta &= (u - s + ti)(u - s - ti)(s - ti - [s + ti]) \\ &= (-2ti)[(u - s)^2 + t^2)]. \end{aligned}$$

Since u, s, and t are real numbers,

$$\begin{aligned} \Delta^2 &= (-2ti)^2[(u - s)^2 + t^2)]^2 \\ &= 4t^2 i^2[(u - s)^2 + t^2)]^2 \\ &= -4t^2[(u - s)^2 + t^2)]^2 \leq 0, \end{aligned}$$

and so $0 \geq \Delta^2 = -27r^2 - 4q^3$. We have shown that if there is a nonreal root, then $27r^2 + 4q^3 \geq 0$; equivalently, if there is no nonreal root (i.e., if all the roots are real), then $27r^2 + 4q^3 \leq 0$.   •

Example 4.13.
Let us try the cubic formula on the polynomial

$$x^3 - 7x + 6 = (x - 1)(x - 2)(x + 3)$$

whose roots are, obviously, 1, 2, and -3. There is no x^2 term, $q = -7, r = 6$, and $D = r^2 + 4q^3/27 = -400/27 < 0$ (notice that $27D = 27r^2 + 4q^3$ is negative, as Theorem 4.45 predicts). The cubic formula gives a messy answer: the roots are

$$\alpha + \beta, \qquad \omega\alpha + \omega^2\beta, \qquad \omega^2\alpha + \omega\beta,$$

where $\alpha^3 = \frac{1}{2}(-6 + \sqrt{-400/27})$ and $\beta^3 = \frac{1}{2}(-6 - \sqrt{-400/27})$. Something strange has happened. There are three curious equations saying that each of 1, 2, and -3 is equal to one of the messy expressions displayed above; thus,

$$\omega \sqrt[3]{\frac{1}{2}\left(-6 + \sqrt{-400/27}\right)} + \omega^2 \sqrt[3]{\frac{1}{2}\left(-6 - \sqrt{-400/27}\right)}$$

is equal to 1, 2, or -3. Aside from the complex cube roots of unity, this expression involves square roots of the negative number $-400/27$.   ◀

Until the Middle Ages, mathematicians had no difficulty in ignoring square roots of negative numbers when dealing with quadratic equations. For example, consider the problem of finding the sides x and y of a rectangle having area A and perimeter p. The equations

$$xy = A \qquad \text{and} \qquad 2x + 2y = p$$

lead to the quadratic equation $2x^2 - px + 2A = 0$, and the quadratic formula gives the roots

$$x = \frac{1}{4}\left(p \pm \sqrt{p^2 - 16A}\right).$$

If $p^2 - 16A \geq 0$, one has found x (and y); if $p^2 - 16A < 0$, one merely says that there is no rectangle whose perimeter and area are in this relation. But the cubic formula does not allow one to discard "imaginary" roots, for we have just seen that an "honest" real and positive root, even a positive integer, can appear in terms of complex numbers.[12] The Pythagoreans in ancient Greece considered *number* to mean positive integer. By the Middle Ages, *number* came to mean positive real number (although there was little understanding of what real numbers are). The importance of the cubic formula in the history of mathematics is that it forced mathematicians to take both complex numbers and negative numbers seriously.

The physicist R. P. Feynman (1918–1988), one of the first winners of the annual Putnam national mathematics competition (and also a Nobel laureate in physics), suggested another possible value of the cubic formula. As mentioned at the beginning of this section, the cubic formula was found in 1515, a time of great change. One of the factors contributing to the Dark Ages was an almost slavish worship of the classical Greek and Roman civilizations. It was believed that that earlier era had been the high point of man's accomplishments; contemporary man was inferior to his forebears (a world view opposite to the modern one of continual progress!). The cubic formula was essentially the first instance of a mathematical formula unknown to the ancients, and so it may well have been a powerful example showing that sixteenth-century man was the equal of his ancestors.

The quartic formula was discovered by Lodovici Ferrari (1522–1565) in the early 1540s; we present the derivation given by Descartes.

Theorem 4.46 (Quartic Formula). *There is a method to compute the four roots of a quartic*

$$X^4 + bX^3 + cX^2 + dX + e.$$

Proof. As with the cubic, the quartic can be simplified, by setting $X = x - \frac{1}{4}b$, to

$$x^4 + qx^2 + rx + s; \tag{5}$$

moreover, if a number u is a root of the second polynomial, then $u = \frac{1}{4}b$ is a root of the first.

Factor the quartic in Eq. (5) into quadratics:

$$x^4 + qx^2 + rs + s = (x^2 + jx + \ell)(x^2 - jx + m) \tag{6}$$

(the coefficient of x in the second factor is $-j$ because the quartic has no x^3 term). If j, ℓ, and m can be found, then the quadratic formula can be used to find the roots of the quartic in Eq. (5).

---

[12]We saw a similar phenomenon in Theorem 1.13: the integer terms of the Fibonacci sequence are given in terms of $\sqrt{5}$.

Expanding the right-hand side of Eq. (6) and equating coefficients of like terms gives the equations

$$\begin{cases} m + \ell - j^2 & = q; \\ j(m - \ell) & = r; \\ \ell m & = s. \end{cases} \tag{7}$$

Adding and subtracting the top two equations in Eqs. (7) yield

$$\begin{cases} 2m & = j^2 + q + r/j; \\ 2\ell & = j^2 + q - r/j. \end{cases} \tag{8}$$

Now substitute these into the bottom equation of Eqs. (7):

$$\begin{aligned} 4s = 4\ell m & = (j^2 + q + r/j)(j^2 + q - r/j) \\ & = (j^2 + q)^2 - r^2/j^2 \\ & = j^4 + 2j^2 q + q^2 - r^2/j^2. \end{aligned}$$

Clearing denominators and transposing gives

$$j^6 + 2qj^4 + (q^2 - 4s)j^2 - r^2 = 0, \tag{9}$$

a cubic equation in j^2. The cubic formula allows one to solve for j^2, and one then finds ℓ and m using Eqs. (8).   •

The quartic formula appeared in Cardano's book, but it was given much less attention there than the cubic formula. The reason is that cubic polynomials have an interpretation as volumes, whereas quartic polynomials have no such obvious justification. We quote Cardano.

> As the first power refers to a line, the square to a surface, and the cube to a solid body, it would be very foolish for us to go beyond this point. Nature does not permit it. Thus, ... , all those matters up to and including the cubic are fully demonstrated, but for the others which we will add, we do not go beyond barely setting out.

Definition. Let E be a field containing a subfield k. An ***automorphism***[13] of E is an isomorphism $\sigma : E \to E$; we say that σ ***fixes*** k if $\sigma(a) = a$ for every $a \in k$.

For example, consider $f(x) = x^2 + 1 \in \mathbb{Q}[x]$; we saw above that a splitting field of $f(x)$ over $\mathbb{Q}$ is $E = \mathbb{Q}(i)$. Complex conjugation $\sigma : a \mapsto \overline{a}$ is an example of an automorphism of E fixing $\mathbb{Q}$.

Proposition 4.48. *Let k be a subfield of a field K, let*

$$f(x) = x^n + a_{n-1}x^{n-1} + \cdots + a_1 x + a_0 \in k[x],$$

and let $E = k(z_1, \ldots, z_n)$ be a splitting field. If $\sigma : E \to E$ is an automorphism fixing k, then σ permutes the roots $z_1, \ldots, z_n$ of $f(x)$ and σ fixes the coefficients of $f(x)$; that is, $\sigma(a_i) = a_i$ for all i.

Proof. If r is a root of $f(x)$, then

$$0 = f(r) = r^n + a_{n-1}r^{n-1} + \cdots + a_1 r + a_0.$$

Applying σ to this equation gives

$$
\begin{aligned}
0 &= \sigma(r)^n + \sigma(a_{n-1})\sigma(r)^{n-1} + \cdots + \sigma(a_1)\sigma(r) + \sigma(a_0) \\
  &= \sigma(r)^n + a_{n-1}\sigma(r)^{n-1} + \cdots + a_1\sigma(r) + a_0,
\end{aligned}
$$

because σ fixes k. Therefore, $\sigma(r)$ is a root of $f(x)$; if Z is the set of all the roots, then $\sigma' : Z \to Z$, where σ' is the restriction $\sigma | Z$. But σ' is injective (because σ is), so that Exercise 2.8 says that σ' is a permutation.

That σ fixes the coefficients of $f(x)$ follows from Eqs. (11), which express each of the coefficients a_i of $f(x)$ in terms of the roots. Each expression, e.g., $a_{n-2} = \sum_{i<j} z_i z_j$, remains unchanged upon permuting the roots.   •

The following technical lemma will be useful.

Lemma 4.49. *Let $E = k(z_1, \ldots, z_n)$. If $\sigma : E \to E$ is an automorphism fixing k and if $\sigma(z_i) = z_i$ for all i, then σ is the identity.*

---

[13]The word *automorphism* is made up of two Greek roots: *auto* meaning "self" and *morph* meaning "shape" or "form." Just as an isomorphism carries one group onto an identical replica, an automorphism carries a group onto itself.

Proof.   We prove the lemma by induction on $n \geq 1$. If $n = 1$, then each $u \in E$ has the form $f(z_1)/g(z_1)$, where $f(x), g(x) \in k[x]$ and $g(z_1) \neq 0$. But σ fixes z_i as well as the coefficients of $f(x)$ and of $g(x)$, so that σ fixes all $u \in E$. For the inductive step, write $K = k(z_1, \ldots, z_{n-1})$, and note that $E = K(z_n)$ [for $K(z_n)$ is the smallest subfield containing k and $z_1, \ldots, z_{n-1}, z_n$]. Having noted this, the inductive step is just a repetition of the base step with k replaced by K.   •

Here is the analog of the symmetry group $\Sigma(\Omega)$ of a polygon Ω.

Definition.   Let k be a subfield of a field E. The ***Galois group*** of E over k, denoted by $\mathrm{Gal}(E/k)$, is the set of all those automorphisms of E that fix k. If $f(x) \in k[x]$, and if $E = k(z_1, \ldots, z_n)$ is a splitting field, then the ***Galois group*** of $f(x)$ over k is defined to be $\mathrm{Gal}(E/k)$.

It is easy to check that $\mathrm{Gal}(E/k)$ is a group with operation composition of functions. This definition is due to E. Artin (1898– 1962), in keeping with his and E. Noether's emphasis on "abstract" algebra. Galois's original version (a group isomorphic to this one) was phrased, not in terms of automorphisms, but in terms of certain permutations of the roots of a polynomial.

Theorem 4.50.   *If $f(x) \in k[x]$ has degree n, then its Galois group $\mathrm{Gal}(E/k)$ is isomorphic to a subgroup of S_n.*

Proof.   Let $X = \{z_1, \ldots, z_n\}$. If $\sigma \in \mathrm{Gal}(E/k)$, then Theorem 4.48 shows that its restriction $\sigma|X$ is a permutation of X; that is, $\sigma|X \in S_X$. Define $\varphi : \mathrm{Gal}(E/k) \to S_X$ by $\varphi : \sigma \mapsto \sigma|X$. To see that φ is a homomorphism, note that both $\varphi(\sigma\tau)$ and $\varphi(\sigma)\varphi(\tau)$ are functions $X \to X$, and hence they are equal if they agree on each $z_i \in X$. But $\varphi(\sigma\tau) : z_i \mapsto (\sigma\tau)(z_i)$, while $\varphi(\sigma)\varphi\tau) : z_i \mapsto \sigma(\tau(z_i))$, and these are the same.

The image of φ is a subgroup of $S_X \cong S_n$. The kernel of φ is the set of all $\sigma \in \mathrm{Gal}(E/k)$ such that σ is the identity permutation on X; that is, σ fixes each of the roots z_i. As σ also fixes k, by definition of the Galois group, Lemma 4.49 gives $\ker \varphi = 1$. Therefore, φ is injective, giving the theorem.   •

If $f(x) = x^2 + 1 \in \mathbb{Q}[x]$, then complex conjugation σ is an automorphism of its splitting field $\mathbb{Q}(i)$, which fixes $\mathbb{Q}$ (it interchanges the roots i and $-i$). Since $\mathrm{Gal}(\mathbb{Q}(i)/\mathbb{Q})$ is a subgroup of the symmetric group S_2, which has order 2, it follows that $\mathrm{Gal}(\mathbb{Q}(i)/\mathbb{Q}) = \langle \sigma \rangle \cong \mathbb{Z}_2$. One should regard the elements of $\mathrm{Gal}(E/k)$ as generalizations of complex conjugation.

That the Galois group of a polynomial does not depend on the choice of splitting field is proven in Theorem 4.52.

The analogy is complete.

Polygon Ω . . . . polynomial $f(x) \in k[x]$
Vertices of Ω . . . . roots of $f(x)$
Plane . . . . splitting field E of $f(x)$
Motion . . . . automorphism fixing k
Symmetry group $\Sigma(\Omega)$ . . . . Galois group $\mathrm{Gal}(E/k)$.

Here is the basic strategy. First, we will translate the classical formulas (giving the roots of $f(x) \in k[x]$) in terms of subfields of a splitting field E over k. Second, this translation into the language of fields will itself be translated into the language of groups: If there is a formula for the roots of $f(x)$, then $\mathrm{Gal}(E/k)$ must be a *solvable* group (which we will soon define). Finally, polynomials of degree at least 5 can have Galois groups that are not solvable. The conclusion is that there are polynomials of degree 5 for which there is no formula, analogous to the classical formulas, giving their roots.

Formulas and Solvability by Radicals

Without further ado, here is the translation of the existence of a formula for the roots of a polynomial in terms of subfields of a splitting field.

Recall that when k is a subfield of a field E, we may also say that E is an *extension* of k, and we may write E/k is an extension.

Definition. A *pure extension* of *type* m is an extension $k(u)/k$, where $u^m \in k$ for some $m \geq 1$. An extension K/k is a *radical extension* if there is a tower of fields

$$k = K_0 \leq K_1 \leq \cdots \leq K_t = K$$

in which each K_{i+1}/K_i is a pure extension.

If $u^m = a \in k$, then $k(u)$ arises from k by adjoining an mth root of a. If $k \leq \mathbb{C}$, there are m different mth roots of a, namely, $u, \omega u, \omega^2 u, \ldots, \omega^{m-1} u$, where $\omega = e^{2\pi i/m}$ is a primitive mth root of unity. Not every subfield k of $\mathbb{C}$ contains all the roots of unity; for example, 1 and -1 are the only roots of unity in $\mathbb{Q}$. Since we seek formulas involving extraction of roots, it will eventually be convenient to assume that k contains appropriate roots of unity.

When we say that there is a *formula* for the roots of a polynomial $f(x)$ analogous to the quadratic, cubic, and quartic formulas, we mean that there is some expression giving the roots of $f(x)$ in terms of the coefficients of $f(x)$. As in the classical formulas, the expression may involve the field operations, constants, and extraction of roots, but it should not involve any other operations involving cosines, definite integrals, or limits, for example. We maintain that a formula as informally described above exists precisely when $f(x)$ is *solvable by radicals*, which we now define.

Definition.   Let $f(x) \in k[x]$ have a splitting field E. We say that $f(x)$ is *solvable by radicals* if there is a radical extension

$$k = K_0 \leq K_1 \leq \cdots \leq K_t$$

with $E \leq K_t$.

Let us illustrate this definition by considering the classical formulas for the polynomials of small degree.

Quadratics

Let $f(x) = x^2 + bx + c$, and let $k = \mathbb{Q}(b, c)$. Define $K_1 = k(u)$, where $u = \sqrt{b^2 - 4c}$. Then K_1 is a radical extension of k, for $u^2 \in k$. Moreover, the quadratic formula implies that K_1 is the splitting field of $f(x)$, and so $f(x)$ is solvable by radicals.

Cubics

Let $f(X) = X^3 + bX^2 + cX + d$, and let $k = \mathbb{Q}(b, c, d)$. The change of variable $X = x - \frac{1}{3}b$ yields a new polynomial $\widetilde{f}(x) = x^3 + qx + r \in k[x]$ having the same splitting field E [for if u is a root of $\widetilde{f}(x)$, then $u - \frac{1}{3}b$ is a root of $f(x)$]. Define $K_1 = k(\sqrt{D})$, where $D = r^2 + 4q^3/27$, and define $K_2 = K_1(\alpha)$, where $\alpha^3 = \frac{1}{2}(-r + \sqrt{D})$. The cubic formula shows that K_2 contains the root $\alpha + \beta$ of $\widetilde{f}(x)$, where $\beta = -q/3\alpha$. Finally, define $K_3 = K_2(\omega)$, where $\omega^3 = 1$. The other roots of $\widetilde{f}(x)$ are $\omega\alpha + \omega^2\beta$ and $\omega^2\alpha + \omega\beta$, both of which lie in K_3, and so $E \leq K_3$.

A splitting field E need not equal K_3, for if all the roots of $f(x)$ are real, then $E \leq \mathbb{R}$, whereas $K_3 \not\leq \mathbb{R}$. An interesting aspect of the cubic formula is the so-called *casus irreducibilis*; the formula for the roots of an irreducible cubic in $\mathbb{Q}[x]$ having all roots real (as in Example 4.13) requires the presence of complex numbers (see Rotman, *Galois Theory*, 2d ed.).

Casus Irreducibilis.   If $f(x) = x^3 + qx + r \in \mathbb{R}[x]$ is an irreducible polynomial having real roots, then any radical extension $K_t/\mathbb{Q}(q, r)$ containing
the splitting field of $f(x)$ is not real; that is, $K_t \not\subset \mathbb{R}$.

Quartics

Let $f(x) = X^4 + bX^3 + cX^2 + dX + e$, and let $k = \mathbb{Q}(b, c, d, e)$. The
change of variable $X = x - \frac{1}{4}b$ yields a new polynomial $\widetilde{f}(x) = x^4 + qx^2 +$
$rx + s \in k[x]$; moreover, the splitting field E of $f(x)$ is equal to the splitting
field of $\widetilde{f}(x)$, for if u is a root of $\widetilde{f}(x)$, then $u - \frac{1}{4}b$ is a root of $f(x)$. Recall
that we factored $\widetilde{f}(x)$:

$$x^4 + qx^2 + rx + s = (x^2 + jx + \ell)(x^2 - jx + m),$$

and that j^2 is a root of a cubic,

$$(j^2)^3 + 2q(j^2)^2 + (q^2 - 4s)j^2 - r^2.$$

Define pure extensions

$$k = K_0 \le K_1 \le K_2 \le K_3,$$

as in the cubic case, so that $j^2 \in K_3$. Define $K_4 = K_3(j)$, and note
that Eqs. (9) give $\ell, m \in K_4$. Finally, define $K_5 = K_4(\sqrt{j^2 - 4\ell})$ and
$K_6 = K_5(\sqrt{j^2 - 4m})$. The quartic formula gives $E \le K_6$ (this tower can
be shortened).

We have seen that quadratics, cubics, and quartics are solvable by radicals.
Conversely, if $f(x)$ is a polynomial that is solvable by radicals, then there is a
formula of the desired kind that expresses its roots in terms of its coefficients.
For suppose that

$$k = K_0 \le K_1 \le \cdots \le K_t$$

is a radical extension with splitting field $E \le K_t$. Let z be a root of $f(x)$.
Now $K_t = K_{t-1}(u)$, where u is an mth root of some element $\alpha \in K_{t-1}$;
hence, z can be expressed in terms of u and K_{t-1}; that is, z can be expressed
in terms of $\sqrt[m]{\alpha}$ and K_{t-1}. But $K_{t-1} = K_{t-2}(v)$, where some power of v lies
in K_{t-2}. Hence, z can be expressed in terms of u, v, and K_{t-2}. Ultimately, z
is expressed by a formula analogous to those of the classical formulas.

Translation into Group Theory

The second stage of the strategy involves investigating the effect of $f(x)$ being solvable by radicals on its Galois group.

Suppose that $k(u)/k$ is a pure extension of type 6; that is, $u^6 \in k$. Now $k(u^3)/k$ is a pure extension of type 2, for $(u^3)^2 = u^6 \in k$, and $k(u)/k(u^3)$ is obviously a pure extension of type 3. Thus, $k(u)/k$ can be replaced by a tower of pure extensions $k \leq k(u^3) \leq k(u)$ of types 2 and 3. More generally, one may assume, given a tower of pure extensions, that each field is of prime type over its predecessor: if $k \leq k(u)$ is of type m, then factor $m = p_t \ldots p_q$, where the p's are (not necessarily distinct) primes, and replace $k \leq k(u)$ by

$$k \leq k(u^{m/p_1}) \leq k(u^{m/p_1 p_2}) \leq \cdots \leq k(u).$$

We are now going to compare different splitting fields of a polynomial over a given field k. Recall Exercise 3.41: If R and S are commutative rings and $\varphi : R \to S$ is a homomorphism, then $\varphi^* : R[x] \to S[x]$, defined by

$$\varphi^* : f(x) = r_0 + r_1 x + r_2 x^2 + \cdots$$
$$\mapsto \varphi(r_0) + \varphi(r_1)x + \varphi(r_2)x^2 + \cdots = f^*(x),$$

is a homomorphism; if φ is an isomorphism, so is φ^*.

The definition of a splitting field E of $f(x) \in k[x]$ was given in terms of some field extension K/k over which $f(x)$ is a product of linear factors. What if K is not given at the outset? For example, suppose that $k = \mathbb{Z}_3$ and $f(x) = x^9 - x \in \mathbb{Z}_3[x]$. Now Kronecker's theorem, Theorem 3.77, does give a field extension $K/\mathbb{Z}_3$ containing all the roots of $f(x)$, but this extension field is not unique, as we saw in Example 3.25. Nevertheless, we are now going to show that, to isomorphism, splitting fields do not depend on the choice of extension field K.

Lemma 4.51. *Let $f(x) \in k[x]$, and let E be a splitting field of $f(x)$ over k. Let $\varphi : k \to k'$ be an isomorphism of fields, let $\varphi^* : k[x] \to k'[x]$ be the isomorphism $g(x) \mapsto g^*(x)$ given by Exercise 3.41, and let E' be a splitting field of $f^*(x)$ over k'. Then there is an isomorphism $\Phi : E \to E'$ extending φ.*

$$
\begin{array}{ccc}
E & \xdashrightarrow{\Phi} & E' \\
| & & | \\
k & \xrightarrow{\varphi} & k'
\end{array}
$$

Proof.   The proof is by induction on $d = [E : k]$. If $d = 1$, then $f(x)$ is a product of linear polynomials in $k[x]$, and it follows easily that $f^*(x)$ is also a product of linear polynomials in $k'[x]$. Thus, we may set $\Phi = \varphi$.

For the inductive step, choose a root z of $f(x)$ in E that is not in k, and let $p(x)$ be the irreducible polynomial in $k[x]$ of which z is a root (Proposition 4.24). Since $z \notin k$, $\deg(p) > 1$; moreover, $[k(z) : k] = \deg(p)$, by Theorem 4.24. Let $p^*(x)$ be the corresponding irreducible polynomial in $k'[x]$, and let z' be a root of $p^*(x)$ in E'.

As in Theorem 3.76, if $I = (p(x))$, there is an isomorphism $\psi : k[x]/I \rightarrow k(z)$ given by $g(x) + I \mapsto g(z)$. Similarly, there is an isomorphism $\theta : k'[x]/I' \rightarrow k'(z')$ given by $g(x) + I' \mapsto g(z')$, where $I' = (p^*(x))$. Consider the composite $\widetilde{\varphi}$

$$k(z) \xrightarrow{\psi^{-1}} k[x]/I \xrightarrow{\varphi'} k'[x]/I' \xrightarrow{\theta} k'(z'),$$

where $\varphi' : g(x) + I \mapsto g^*(x) + I'$. Now $\widetilde{\varphi}$ is an isomorphism (because it is the composite of three isomorphisms); moreover, $\widetilde{\varphi} : z \mapsto z'$ and it extends φ. We may regard $f(x)$ as a polynomial with coefficients in $k(z)$ (for $k \leq k(z)$ implies $k[x] \leq k(z)[x]$). We claim that E is is a splitting field of $f(x)$ over $k(z)$; that is,

$$E = k(z)(z_1, \ldots, z_n),$$

where $z_1, \ldots, z_n$ are the roots of $f(x)$. Clearly,

$$E = k(z_1, \ldots, z_n) \subset k(z)(z_1, \ldots, z_n).$$

For the reverse inclusion, since $z \in E$, we have

$$k(z)(z_1, \ldots, z_n) \subset k(z_1, \ldots, z_n) = E.$$

But $[E : k(z)] < [E : k]$, by Theorem 4.25, so that the inductive hypothesis gives an isomorphism $\Phi : E \rightarrow E'$ that extends $\widetilde{\varphi}$, and hence φ. $\bullet$

Theorem 4.52.   *If k is a field and $f(x) \in k[x]$, then any two splitting fields of $f(x)$ over k are isomorphic.*

Proof.   Let E and E' be splitting fields of $f(x)$ over k. If φ is the identity, then the theorem applies at once. $\bullet$

Notice that this result implies that the Galois group $\mathrm{Gal}(E/k)$ of a polynomial $f(x) \in k[x]$ with splitting field E depends only on $f(x)$ and k, but not upon the choice of E: If $\varphi : E \to E'$ is an isomorphism fixing k, then there is an isomorphism $\mathrm{Gal}(E/k) \to \mathrm{Gal}(E'/k)$ given by $\sigma \mapsto \varphi\sigma\varphi^{-1}$.

It is remarkable that the next theorem was not proved until the 1890s, 60 years after Galois discovered finite fields.

Corollary 4.53 (E. H. Moore).   *Any two finite fields having exactly p^n elements are isomorphic.*

Proof.   If E is a field with $q = p^n$ elements, then Lagrange's theorem applied to the multiplicative group $E^\times$ shows that $a^{q-1} = 1$ for every $a \in E^\times$. It follows that every element of E is a root of $f(x) = x^q - x \in \mathbb{Z}_p[x]$, and so E is a splitting field of $f(x)$ over $\mathbb{Z}_p$. •

E. H. Moore (1862–1932) began his mathematical career as an algebraist, but he did important work in many other parts of mathematics as well; for example, Moore–Smith convergence is named in part after him.

Here is another nice corollary of Lemma 4.51; it says, in our analogy between Galois theory and symmetry of polygons, that irreducible polynomials correspond to regular polygons.

Proposition 4.54.   *Let E/k be a splitting field of some polynomial in $k[x]$, and let $p(x) \in k[x]$ split in E. If $p(x)$ is irreducible in $k[x]$, then $\mathrm{Gal}(E/k)$ acts transitively on the roots of $p(x)$.*

Proof.   Let z and z' be roots of $p(x)$. By Proposition 3.76, there is an isomorphism $\theta : k(z) \to k(z')$ with $\theta(z) = z'$ and which fixes k. Lemma 4.51 shows that θ extends to an automorphism Θ of E which fixes k; that is, $\Theta \in \mathrm{Gal}(E/k)$. •

Here is a key result allowing us to translate solvability by radicals into the language of Galois groups.

Theorem 4.55.   *Let $k \leq K \leq E$ be a tower of fields, let $f(x), g(x) \in k[x]$, let K be a splitting field of $f(x)$, and let E be a splitting field of $g(x)$. Then $\mathrm{Gal}(E/K)$ is a normal subgroup of $\mathrm{Gal}(E/k)$, and*

$$\mathrm{Gal}(E/k)/\mathrm{Gal}(E/K) \cong \mathrm{Gal}(K/k).$$

Proof.   Let $K = k(z_1, \ldots, z_t)$, where $z_1, \ldots, z_t$ are the roots of $f(x)$ in E. If $\sigma \in \mathrm{Gal}(E/k)$, then σ permutes $z_1, \ldots, z_t$, by Theorem 4.48(i) (for σ fixes

k), and so $\sigma(K) \leq K$. Define $\rho : \text{Gal}(E/k) \to \text{Gal}(K/k)$ by $\sigma \mapsto \sigma|K$. It is easy to see, as in the proof of Theorem 4.50, that ρ is a homomorphism and that $\ker \rho = \text{Gal}(E/K)$. It follows that $\text{Gal}(E/K)$ is a normal subgroup of $\text{Gal}(E/k)$. But ρ is surjective: if $\tau \in \text{Gal}(K/k)$, then Lemma 4.51 applies to show that there is $\sigma \in \text{Gal}(E/k)$ extending τ; i.e., $\rho(\sigma) = \sigma|K = \tau$. The first isomorphism theorem completes the proof. •

The next technical result will be needed when we apply Theorem 4.55.

Lemma 4.56.   *Let K be a finite extension of a field k. There is a finite extension E of K that is a splitting field over k of some polynomial in $k[x]$. Moreover, if K is a radical extension of k, then E is also a radical extension of k.*

Proof.   Since E is a finite extension, there are elements $z_1, \ldots, z_m$ with $E = K(z_1, \ldots, z_\ell)$. For each i, Theorem 3.76 gives an irreducible polynomial $p_i(x) \in k[x]$ with $p_i(z_i) = 0$ and a splitting field E of $g(x) = p_1(x) \ldots p_\ell(x)$ containing K.

For each i and each pair of roots z and z' of $p_i(x)$, the proof of the inductive step of Lemma 4.51 provides an isomorphism $\gamma : k(z) \to k(z')$ taking $z \mapsto z'$, for both $k(z)$ and $k(z')$ are isomorphic to $k[x]/(p_i(x))$; by Lemma 4.51, each such γ extends to an isomorphism $\sigma \in \text{Gal}(E/k)$. It follows that E is generated by $\{\sigma(K) : \sigma \in \text{Gal}(E/k)\}$, for every root z_i of $g(x)$ lies in at least one $\sigma(K)$.

If K is a radical extension of k, then $K = k(u_1, \ldots, u_t)$, where each $k(u_1, \ldots, u_{i+1})$ is a pure extension of $k(u_1, \ldots, u_i)$; of course, each $\sigma(K) = k(\sigma(u_1), \ldots, \sigma(u_t))$ is also a radical extension of k. We now show that E is a radical extension of k. Define B_1 to be k with all $\sigma(u_1)$ adjoined, where $\sigma \in \text{Gal}(E/k) = \{1, \sigma, \tau, \ldots\}$; then

$$k \leq k(u_1) \leq k(u_1, \sigma(u_1)) \leq k(u_1, \sigma(u_1), \tau(u_1)) \leq \cdots \leq B_1$$

displays B_1 as a radical extension of k; for example, if u_1^m lies in k, then $\tau(u_1^m) = \tau(u_1)^m$ lies in $\tau(k) = k$, and hence $\tau(u_1)^m$ lies in $k(u_1, \sigma(u_1))$. Define B_2 to be B_1 with all $\sigma(u_2)$ adjoined, where $\sigma \in \text{Gal}(E/k)$:

$$B_1 \leq B_1(u_2) \leq B_1(u_2, \sigma(u_2)) \leq B_1(u_2, \sigma(u_2), \tau(u_2)) \leq \cdots \leq B_2.$$

Now B_2 is a radical extension of B_t; for example, if $u_2' \in k(u_1)$, then $\sigma(u_2') = \sigma(u_2)' \in \sigma(k(u_1)) = k(\sigma(u_1)) \leq B_1(\sigma(u_1))$. Since B_1 is a radical extension of k, it follows that B_2 is a radical extension of k. For each $i \geq 2$, define

B_{i+1} to be B_i with all $\sigma(u_i)$ adjoined. The argument above shows that B_{i+1} is a radical extension of k. Finally, since $E = B_t$, we have shown that E is a radical extension of k. •

We remarked earlier that it would be convenient, at some point, to assume that our fields contain appropriate roots of unity. That time has come.

Lemma 4.57. *Let p be a prime, let k be a field of characteristic 0 which contains the pth roots of unity, and let $k \le k(u)$ be a pure extension of type p. If $u \notin k$, then $\mathrm{Gal}(k(u)/k) \cong \mathbb{Z}_p$.*

Proof. Let $a = u^p \in k$. If $\omega = e^{2\pi i/p}$, then the roots of $f(x) = x^p = a$ are $u, \omega u, \omega^2 u, \ldots, \omega^{p-1} u$; since $\omega \in k$, it follows that $k(u)$ is the splitting field of $f(x)$ over k. If $\sigma \in \mathrm{Gal}(k(u)/k)$, then $\sigma(u) = \omega^i u$ for some i, by Theorem 4.48(i). Define $\varphi : \mathrm{Gal}(k(u)/k) \to \mathbb{Z}_p$ by $\varphi(\sigma) = [i]$. To see that φ is a homomorphism, observe that if $\sigma, \tau \in \mathrm{Gal}(k(u)/k)$, then $\sigma\tau(u) = \sigma(\omega^j u) = \omega^{i+j} u$, so that $\varphi(\sigma\tau) = [i + j] = [i] + [j] = \varphi(\sigma) + \varphi(\tau)$. Now $\ker\varphi = 1$, for if $\varphi(\sigma) = [0]$, then $\sigma(u) = u$; since σ fixes k, by the definition of $\mathrm{Gal}(k(u)/k)$, Lemma 4.49 gives $\sigma = 1$. Finally, we show that φ is a surjection. Since $u \notin k$, the automorphism taking $u \mapsto \omega u$ is not the identity, so that $\mathrm{im}\,\varphi \notin \{0\}$, for it contains [1]. But $\mathbb{Z}_p$, having prime order p, has no subgroups aside from $\{0\}$ and $\mathbb{Z}_p$; as $\mathrm{im}\,\varphi \ne \{0\}$, $\mathrm{im}\,\varphi = \mathbb{Z}_p$. Therefore, φ is an isomorphism. •

Here is the heart of the translation we have been seeking.

Theorem 4.58. *Let*

$$k = K_0 \le K_1 \le K_2 \le \cdots \le K_t$$

be a radical extension of a field k. Assume, for each i, that each K_i is a pure extension of prime type p_i over K_{i-1} and that k contains all the p_ith roots of unity. If K_t is a splitting field over k, then there is a sequence of subgroups

$$\mathrm{Gal}(K_t/k) = G_0 \ge G_1 \ge G_2 \ge \cdots \ge G_t = \{1\},$$

with each G_{i+1} a normal subgroup of G_i and with G_i/G_{i+1} cyclic of prime order.

Proof. Defining $G_i = \mathrm{Gal}(K_t/K_i)$ does give a sequence of subgroups of $\mathrm{Gal}(K_t/k)$. Since $K_1 = k(u)$, where $u^{p_1} \in k$, the assumption that k contains $\omega = e^{2\pi i/p_1}$ shows that K_1 is a splitting field of $x^{p_1} - u^{p_1}$ (for the roots are $u, \omega u, \ldots, \omega^{p_1-1} u$). We may thus apply Theorem 4.55 to see that $G_1 = \mathrm{Gal}(K_t/K_1)$ is a normal subgroup of $G_0 = \mathrm{Gal}(K_t/k)$ and that $G_0/G_1 \cong \mathrm{Gal}(K_1/k) = \mathrm{Gal}(K_1/K_0)$. By Lemma 4.57, $G_0/G_1 \cong \mathbb{Z}_{p_1}$. This argument can be repeated for each i.   •

Definition.   A *normal series* of a group G is a sequence of subgroups

$$G = G_0 \geq G_1 \geq G_2 \geq \cdots \geq G_t = \{1\}$$

with each G_{i+1} a normal subgroup of G_i; the *factor groups* of this series are the quotient groups

$$G_0/G_1, G_1/G_2, \ldots, G_{n-1}/G_n.$$

A group G is called *solvable* if it has a normal series each of whose factor groups has prime order.

In this language, Theorem 4.58 says that $\mathrm{Gal}(K_t/k)$ is a solvable group if k contains appropriate roots of unity and K_t is a radical extension of k.

Example 4.17.

(i) Let us see that S_4 is a solvable group. Consider the chain of subgroups

$$S_4 \geq A_4 \geq \mathbf{V} \geq W \geq \{1\},$$

where $\mathbf{V}$ is the four-group and W is any subgroup of $\mathbf{V}$ of order 2. Note, since $\mathbf{V}$ is abelian, that W is a normal subgroup of $\mathbf{V}$. Now $|S_4/A_4| = |S_4|/|A_4| = 24/12 = 2$, $|A_4/\mathbf{V}| = |A_4|/|\mathbf{V}| = 12/4 = 3$, $|\mathbf{V}/W| = |\mathbf{V}|/|W| = 4/2 = 2$, and $|W/\{1\}| = |W| = 2$. Since each factor group has prime order, S_4 is solvable.

(ii) We now show that S_n is not a solvable group for all $n \geq 5$.

In Exercise 2.106, we saw, for all $n \geq 5$, that A_n is the only proper non-trivial normal subgroup of S_n (the key fact in the proof is that A_n is a simple group). It follows that S_n has only one normal series, namely,

$$S_n > A_n > \{1\}$$

(this is not quite true; another normal series is $S_n > A_n \geq A_n > \{1\}$, which repeats a term; of course, this repetition only contributes the new factor group $A_n/A_n = \{1\}$). But the factor groups of this normal series are $S_n/A_n \cong \mathbb{Z}_2$ and $A_n/\{1\} \cong A_n$, and the latter group is not of prime order. Therefore, S_n is not a solvable group for $n \geq 5$.   ◄

Lemma 4.59.   *Let k be a field of characteristic 0 which contains any desired roots of unity, and assume that $f(x) \in k[x]$ has splitting field E over k. If $f(x)$ is solvable by radicals, then its Galois group $\mathrm{Gal}(E/k)$ is a quotient of a solvable group.*

Proof.   There is a tower of pure extensions of prime type

$$k = K_0 \leq K_1 \leq K_2 \leq \cdots \leq K_t$$

with $E \leq K_t$; moreover, Lemma 4.56 allows us to assume that K_t is also a splitting field of some polynomial in $k[x]$. Since E is a splitting field, if $\sigma \in \mathrm{Gal}(K_t/k)$, then $\sigma|E \in \mathrm{Gal}(E/k)$, and so $\rho : \sigma \mapsto \sigma|E$ is a homomorphism $\mathrm{Gal}(K_t/k) \to \mathrm{Gal}(E/k)$. Finally, Lemma 4.51 shows that ρ is surjective, because K_t is a splitting field over k.   •

Theorem 4.60.   *Every quotient G/N of a solvable group G is itself a solvable group.*

Remark.   One can also prove that every subgroup of a solvable group is itself a solvable group.   ◄

Proof.   By the first isomorphism theorem for groups, which says that quotient groups are isomorphic to homomorphic images, it suffices to prove that if $f : G \to H$ is a surjection (for some group H), then H is a solvable group.

Let $G = G_0 \geq G_1 \geq G_2 \geq \cdots \geq G_t = \{1\}$ be a sequence of subgroups as in the definition of solvable group. Then

$$H = f(G_0) \geq f(G_1) \geq f(G_2) \geq \cdots \geq f(G_t) = \{1\}$$

is a sequence of subgroups of H. If $f(x_{i+1}) \in f(G_{i+1})$ and $u_i \in f(G_i)$, then $u_i = f(x_i)$ and $u_i f(x_{i+1}) u_i^{-1} = f(x_i) f(x_{i+1}) f(x_i)^{-1} = f(x_i x_{i+1} x_i^{-1}) \in f(G_i)$, because $G_{i+1} \triangleleft G_i$; that is, $f(G_{i+1})$ is a normal subgroup of $f(G_i)$. The map $\varphi : G_i \to f(G_i)/f(G_{i+1})$, defined by $x_i \mapsto f(x_i)f(G_{i+1})$, is a surjection, for it is the composite of the surjections $G_i \to f(G_i)$ and the natural map $f(G_i) \to f(G_i)/f(G_{i+1})$. Since $G_{i+1} \leq \ker\varphi$, this map induces a surjection $G_i/G_{i+1} \to f(G_i)/f(G_{i+1})$, namely, $x_i G_{i+1} \mapsto f(x_i)f(G_{i+1})$. Now G_i/G_{i+1} is cyclic of prime order, so that its quotient $f(G_i)/f(G_{i+1})$ is a cyclic group of order 1 or order a prime. Thus, deleting any repetitions if necessary, $H = f(G)$ has a series in which all the quotient groups are cyclic of prime order; therefore, H is a solvable group.   •

Here is the main criterion.

Theorem 4.61 (*Galois*).   *Let k be a field of characteristic 0 which contains any desired roots of unity (in particular, if k is an extension of $\mathbb{C}$), and assume that $f(x) \in k[x]$ has splitting field E over k. If $f(x)$ is solvable by radicals, then its Galois group $\mathrm{Gal}(E/K)$ is a solvable group.*

Proof.   By Lemma 4.59, $\mathrm{Gal}(E/k)$ is a quotient of a solvable group and, by Theorem 4.60, any quotient of a solvable group is itself solvable.   •

Further analysis allows one to eliminate the hypothesis that k contains desired roots of unity; Theorem 4.61 holds for any extension k of $\mathbb{Q}$, including $\mathbb{Q}$ itself. Moreover, the converse (also proved by Galois) is true: If the Galois group of a polynomial $f(x) \in k[x]$ is a solvable group, where k is an extension field of $\mathbb{Q}$, then $f(x)$ is solvable by radicals.

In 1827, Abel proved a theorem that says, in group-theoretic language not known to him, that if the Galois group of a polynomial $f(x)$ is commutative, then $f(x)$ is solvable by radicals. This is why abelian groups are so called. It can be shown that every finite abelian group is solvable, and so Abel's theorem is a special case of Galois's.

As we remarked earlier, every subgroup of a solvable group is itself solvable. It now follows from Example 4.17(i) that every subgroup of S_4 is a solvable group. Since S_2 and S_3 are subgroups of S_4, Theorem 4.50 shows that the Galois group of every quadratic, cubic, and quartic polynomial is a solvable group. Thus, the converse of Galois's theorem shows that each such polynomial is solvable by radicals (of course, we already know this because we have proved the classical formulas).

We now complete the discussion by showing that the general quintic polynomial has a Galois group that is not solvable.

Theorem 4.62 (*Abel-Ruffini*).   *The general quintic*

$$f(x) = (x - y_1)(x - y_2)(x - y_3)(x - y_4)(x - y_5)$$

is not solvable by radicals.

Proof.   In Example 4.16 on page 365, we saw that if $E = \mathbb{C}(y_1, \ldots, y_5)$ is the field of all rational functions in 5 variables $y_1, \ldots, y_5$ with coefficients in $\mathbb{C}$, and if $k = \mathbb{C}(a_0, \ldots, a_5)$, where the a_i are the coefficients of $f(x)$, then E is the splitting field of $f(x)$ over k. As k is an extension field of $\mathbb{C}$, it contains all the roots of unity.

We claim that S_5 is (isomorphic to) a subgroup of $\mathrm{Gal}(E/k)$. Recall Exercise 3.46(ii)(i): If A and R are domains and $\varphi : A \to R$ is an isomorphism, then $[a, b] \mapsto [\varphi(a), \varphi(b)]$ is an isomorphism $\mathrm{Frac}(A) \to \mathrm{Frac}(R)$.

If $\sigma \in S_n$, then there is an isomorphism $\tilde{\sigma}$ of $\mathbb{C}[y_1, \ldots, y_5]$ with itself defined by $f(y_1, \ldots, y_5) \mapsto f(y_{\sigma 1}, \ldots, y_{\sigma 5})$; that is, $\tilde{\sigma}$ just permutes the variables of a polynomial in several variables. By Exercise 3.46(ii)(i), $\tilde{\sigma}$ extends to an automorphism σ^* of E, for $E = \text{Frac}(\mathbb{C}[y_1, \ldots, y_n])$. Equations (11) show that σ^* fixes k, and so $\sigma^* \in \text{Gal}(E/k)$. Using Lemma 4.49, it is easy to see that $\sigma \mapsto \sigma^*$ is an isomorphism of S_5 with a subgroup of $\text{Gal}(E/k)$; in fact, Theorem 4.50 shows that $\text{Gal}(E/k) \cong S_5$. Therefore, $\text{Gal}(E/k)$ is not a solvable group, and Theorem 4.61 shows that $f(x)$ is not solvable by radicals.   •

We have proved that there is no generalization of the classical formulas to polynomials of degree 5. As we remarked earlier, it can be shown that Theorem 4.61 and its converse can be generalized as follows: If $f(x) \in \mathbb{Q}[x]$ has splitting field E, then $f(x)$ is solvable by radicals if and only if $\text{Gal}(E/\mathbb{Q})$ is a solvable group. One can then show that the specific polynomial $f(x) = x^5 - 4x + 2$ has Galois group isomorphic to S_5, and so $f(x)$ is not solvable by radicals.

Let us record two facts that can be used in the Exercises below.

Fact I.   If $f(x) \in \mathbb{Q}[x]$ has degree n, then Theorem 4.50 says there is an injection $\varphi : G \to S_n$. If $f(x)$ has discriminant D, then $\text{im } \varphi \le A_n$ if and only if $\sqrt{D}$ is rational.

Fact II.   If $f(x) \in \mathbb{Q}[x]$ has degree n and Galois group $G = \text{Gal}(E/\mathbb{Q})$, then $|G| = [E : \mathbb{Q}]$. Consequently, if $f(x)$ is irreducible, then $n \mid |G|$.

EXERCISES

4.25 Prove that $\mathbb{Z}_3[x]/(x^3 - x^2 - 1) \cong \mathbb{Z}_3[x]/(x^3 - x^2 + x + 1)$.

4.26 Is GF(4) a subfield of GF(8)?

4.27 Let k be a field of characteristic p.

    (i) Prove that the function $F : k \to k$, defined by $F : a \mapsto a^p$, is an automorphism fixing the prime field $\mathbb{Z}_p$. Conclude that $F \in \text{Gal}(k/\mathbb{Z}_p)$. (The automorphism F is called the **Frobenius automorphism**.)

    (ii) Prove that $F : k \to k$ is an injection.

    (iii) Prove that if k is finite, then every $a \in k$ has a pth root; that is, there is $b \in k$ with $b^p = a$.

4.28   (i) If α is a generator of $\text{GF}(p^n)^\times$, prove that $\text{GF}(p^n) = \mathbb{Z}_p(\alpha)$.

(ii) Prove that the irreducible polynomial $p(x) \in \mathbb{Z}_p[x]$ of α has degree n.

(iii) Prove that if $G = \text{Gal}(\text{GF}(p^n)/\mathbb{Z}_p)$, then $|G| \leq n$.

(iv) Prove that $\text{Gal}(\text{GF}(p^n)/\mathbb{Z}_p)$ is cyclic of order n with generator the Frobenius F.

4.29 Prove that the following statements are equivalent for a quadratic $f(x) = ax^2 + bx + c \in \mathbb{Q}[x]$.

(i) $f(x)$ is irreducible.

(ii) $\sqrt{b^2 - 4ac}$ is irrational.

(iii) $\text{Gal}(\mathbb{Q}(\sqrt{b^2 - 4ac}), \mathbb{Q})$ has order 2.

4.30 Prove that if $f(x) \in \mathbb{Q}[x]$ has a rational root a, then its Galois group is the same as the Galois group of $f(x)/(x - a)$.

4.31 Let $f(x) \in \mathbb{Q}[x]$ be an irreducible cubic with Galois group G.

(i) Prove that if $f(x)$ has exactly one real root, then $G \cong S_3$.

(ii) Assume that $f(x)$ has three real roots. Prove that if $\sqrt{D}$ is rational, where D is the discriminant of $f(x)$, then $G \cong A_3 \cong \mathbb{Z}_3$, and if $\sqrt{D}$ is irrational, then $G \cong S_3$.

4.32   (i) Find the Galois group of $f(x) = x^3 - 2 \in \mathbb{Q}[x]$.

(ii) Find the Galois group of $f(x) = x^3 - 4x + 2 \in \mathbb{Q}[x]$.

(iii) Find the Galois group of $f(x) = x^3 - x + \frac{1}{3} \in \mathbb{Q}[x]$.

4.33   (i) If k is a field and $f(x) \in k[x]$ has derivative $f'(x)$, prove that either $f'(x) = 0$ or $\deg(f') < \deg(f)$.

(ii) If k is a field of characteristic 0, prove that an irreducible polynomial $p(x) \in k[x]$ has no repeated roots; that is, if E is the splitting field of $p(x)$, then there is no $a \in E$ with $(x - a)^2 \mid p(x)$ in $E[x]$.

4.34 Let k be a field of characteristic p.

(i) Prove that if $f(x) = \sum_i a_i x^i \in k[x]$, then $f'(x) = 0$ if and only if the only nonzero coefficients are those a_i with $p \mid i$.

(ii) If k is finite and $f(x) = \sum_i a_i x^i \in k[x]$, prove that $f'(x) = 0$ if and only if there is $g(x) \in k[x]$ with $f(x) = g(x)^p$.

(iii) Prove that if k is a finite field, then every irreducible polynomial $p(x) \in k[x]$ has no repeated roots.

4.35 If $k = \mathbb{Z}_p(x)$, the field of rational functions over $\mathbb{Z}_p$, prove that there is an irreducible $p(t) \in k[t]$ having repeated roots.

4.5 EPILOG

Further investigation of these ideas is the subject of Galois theory, which studies the relation between extension fields and their Galois groups. Aside from its intrinsic interest, Galois theory is used extensively in algebraic number theory.

We recall some notation. If E is a field with subfields L and M, then $L \vee M$ denotes the smallest subfield containing L and M – that is, the intersection of all the subfields of E containing $L \cup M$. Similarly, if G is a group with subgroups H and K, then $H \vee K$ denotes the smallest subgroup of G containing H and K.

Definition.   A polynomial $f(x) \in k[x]$ is *separable* if its irreducible factors have no repeated roots.

We have seen that k is separable if it has characteristic 0 [Exercise 4.33(ii)] or if it is finite [Exercise 4.34(iii)]. On the other hand, there are inseparable polynomials, as we have seen in Exercise 4.35.

Definition.   Let E/F be a field extension with Galois group $G = \text{Gal}(E/F)$. If $H \leq G$, then the *fixed field* E^H is defined by

$$E^H = \{u \in E : \sigma(u) = u \text{ for all } \sigma \in H\}.$$

The following theorems are proved. First, one has a characterization of extension fields that are splitting fields.

Theorem.   *Let E/F be a field extension with Galois group $G = \text{Gal}(E/F)$. Then the following statements are equivalent.*

(i) *E is a splitting field of some separable polynomial $f(x) \in F[x]$.*

(ii) *Every irreducible $p(x) \in F[x]$ having one root in E is separable and all its roots are in E; that is, $p(x)$ splits in $E[x]$.*

(iii) *$F = E^G$.*

Definition.   A field extension E/F is a *Galois extension* if it satisfies any of the equivalent conditions in this theorem.

The following theorem shows that there is an intimate connection between the intermediate fields B in a Galois extension E/F (that is, subfields with $F \leq B \leq E$) and the subgroups of the Galois group.

Theorem (Fundamental Theorem of Galois Theory). *Let E/F be a finite Galois extension with Galois group $G = \mathrm{Gal}(E/F)$.*

(i) *The function $H \mapsto E^H$ is a bijection from the family of all intermediate fields to the subgroups of the $\mathrm{Gal}(E/F)$ which reverses inclusions:*

$$H \leq K \text{ if and only if } E^K \leq E^H.$$

(ii) *For every intermediate field B and every $H \leq G$,*

$$E^{\mathrm{Gal}(E/B)} = B \text{ and } \mathrm{Gal}(E/E^H) = H.$$

(iii)

$$E^{H \vee K} = E^H \cap E^K;$$
$$E^{H \cap K} = E^H \vee E^K;$$
$$\mathrm{Gal}(E/(B \vee C) = \mathrm{Gal}(E/B) \cap \mathrm{Gal}(E/C);$$
$$\mathrm{Gal}(E/(B \cap C) = \mathrm{Gal}(E/B) \vee \mathrm{Gal}(E/C).$$

(iv)

$$[B : F] = [G : \mathrm{Gal}(E/B)] \text{ and } [G : H] = [E^H : F].$$

(v) *B/F is a Galois extension if and only if $\mathrm{Gal}(E/B)$ is a normal subgroup of G.*

Here are some corollaries.

Theorem (Theorem of the Primitive Element). *If E/F is a Galois extension, then there is $\alpha \in E$ with $E = F(\alpha)$.*

Theorem. *The Galois field $GF(p^n)$ has exactly one subfield of order p^d for every divisor d of n, and no others.*

Theorem. *If E/F is a Galois extension whose Galois group is abelian, then every intermediate field is a Galois extension.*

We are now going to prove the Fundamental Theorem of Algebra, first proved by Gauss (1799) (at one point, we will invoke the Sylow theorem that we will prove in Chapter 5). Assume that $\mathbb{R}$ satisfies a weak form of the intermediate value theorem: if $f(x) \in \mathbb{R}[x]$ and there exist $a, b \in \mathbb{R}$ such that $f(a) > 0$ and $f(b) < 0$, then $f(x)$ has a real root. Here are some preliminary consequences.

(i) *Every positive real r has a real square root.*

If $f(x) = x^2 - r$, then $f(1 + r) > 0$ and $f(0) < 0$.

(ii) *Every quadratic $g(x) \in \mathbb{C}[x]$ has a complex root.*

First, every complex number z has a complex square root. Write z in polar form: $z = re^{i\theta}$, where $r \geq 0$, and $\sqrt{z} = \sqrt{r}e^{i\theta/2}$. It follows that the quadratic formula can give the (complex) roots of $g(x)$.

(iii) *The field $\mathbb{C}$ has no extensions of degree 2.*

Such an extension would contain an element whose irreducible polynomial is a quadratic in $\mathbb{C}[x]$, and item (ii) shows that no such polynomial exists.

(iv) *Every $f(x) \in \mathbb{R}[x]$ having odd degree has a real root.*

Let $f(x) = a_0 + a_1 x + \ldots + a_{n-1}x^{n-1} + x^n \in \mathbb{R}[x]$. Define $t = 1 + \sum |a_i|$. Now $|a_i| \leq t - 1$ for all i, and

$$
\begin{aligned}
|a_0 + a_1 t + \ldots + a_{n-1}t^{n-1}| &\leq (t-1)[1 + t + \ldots + t^{n-1}] \\
&= t^n - 1 \\
&< t^n.
\end{aligned}
$$

It follows that $f(t) > 0$ (for any not necessarily odd n), because the sum of the early terms is dominated by t^n. When n is odd, $f(-t) < 0$, for

$$(-t)^n = (-1)^n t^n < 0,$$

and so the same estimate as above now shows that $f(-t) < 0$.

(v) *There is no field extension $E/\mathbb{R}$ of odd degree > 1.*

If $\alpha \in E$, then its irreducible polynomial must have even degree, by item (iv), so that $[\mathbb{R}(\alpha) : \mathbb{R}]$ is even. Hence $[E : \mathbb{R}] = [E : \mathbb{R}(\alpha)][\mathbb{R}(\alpha) : \mathbb{R}]$ is even.

Theorem 4.63 (Fundamental Theorem of Algebra).   *If $f(x) \in \mathbb{C}[x]$ has degree $n \geq 1$, then $f(x)$ has a factorization*

$$f(x) = c(x - \alpha_1) \cdots (x - \alpha_n),$$

where $c, \alpha_1, \ldots, \alpha_n \in \mathbb{C}$.

Proof.   We show that $f(x)$ has a complex root. If $f(x) = \sum a_i x^i \in \mathbb{C}[x]$, define $\overline{f(x)} = \sum \overline{a_i} x^i$, where $\overline{a_i}$ is the complex conjugate of a_i. If $f(x)\overline{f(x)} = \sum c_k x^k$, then $c_k = \sum_{i+j=k} a_i \overline{a_j}$; hence, $\overline{c_k} = c_k$, and so $f(x)\overline{f(x)} \in \mathbb{R}[x]$. Since $f(x)$ has a complex root if and only if $f(x)\overline{f(x)}$ has a complex root, it suffices to prove that every real polynomial has a complex root.

Let $p(x)$ be an irreducible polynomial in $\mathbb{R}[x]$, and let $E/\mathbb{R}$ be a splitting field of $(x^2 + 1)p(x)$ which contains $\mathbb{C}$. Since $\mathbb{R}$ has characteristic 0, $E/\mathbb{R}$ is a Galois extension; let G be its Galois group. If $|G| = 2^m k$, where k is odd, then G has a subgroup H of order 2^m, by the Sylow theorem (see Chapter 5); let $B = E^H$ be the corresponding intermediate field. Now the degree $[B : \mathbb{R}]$ is equal to the index $[G : H] = k$. But we have seen above that $\mathbb{R}$ has no extension of odd degree > 1; hence $k = 1$ and G is a 2-group. By Proposition 2.77, the subgroup $\mathrm{Gal}(E/\mathbb{C})$ of G (corresponding to $\mathbb{C}$) has a subgroup of index 2 provided $|\mathrm{Gal}(E/\mathbb{C})| > 1$; its corresponding intermediate field is an extension of $\mathbb{C}$ of degree 2, and this contradicts item (iii) above. We conclude that $\mathrm{Gal}(E/\mathbb{C}) = \{1\}$ and $E = \mathbb{C}$.

The result now follows by induction on $n \geq 1$.   •

5

Groups II

5.1 FINITE ABELIAN GROUPS

We continue our study of groups by considering finite abelian groups; as is customary, these groups are written additively. We are going to prove that every finite abelian group is a direct sum of (finitely many) cyclic groups, and so we begin by considering direct sums.

Definition. The **external direct sum** of two abelian groups S and T is the abelian group $S \times T$ whose underlying set is the cartesian product of S and T and whose operation is given by $(s, t) + (s', t') = (s + s', t + t')$.

If S and T are subgroups of an abelian group G, then G is the **internal direct sum**, denoted by $G = S \oplus T$, if $S + T = G$ (that is, each $a \in G$ can be written $a = s + t$, where $s \in S$ and $t \in T$), and $S \cap T = \{0\}$.

It is routine to check that both the internal and external direct sums are (abelian) groups.

The following is the additive version of Proposition 2.61.

Proposition 5.1. *If S and T are subgroups of an abelian group G, then $G = S \oplus T$ if and only if every $a \in G$ has a unique expression of the form $a = s + t$, where $s \in S$ and $t \in T$.*

Proof. By hypothesis, $G = S + T$, so that each $a \in G$ has an expression of the form $a = s + t$ with $s \in S$ and $t \in T$. To see that this expression is unique, suppose also that $a = s' + t'$, where $s' \in S$ and $t' \in T$. Then

$s + t = s' + t'$ gives $s - s' = t' - t \in S \cap T = \{0\}$. Therefore, $s = s'$ and $t = t'$, as desired.

Conversely, if $x \in S \cap T$ and $x \neq 0$, then x has two expressions of the form $s + t$, namely, $x = x + 0$ and $x = 0 + x$. Since expressions are unique, we must have $x = 0$. •

The next result shows that there is no essential difference between internal and external direct sums.

Corollary 5.2. *If S and T are subgroups of an abelian group G which generate their internal direct sum $S \oplus T$, then $S \oplus T \cong S \times T$.*

Conversely, given abelian groups S and T, define subgroups S' and T' of the external direct sum $S \times T$ by

$$S' = \{(s, 0) : s \in S\} \quad and \quad T' = \{(0, t) : t \in T\};$$

then $S \times T \cong S' \oplus T'$.

Proof. Define $f : S \oplus T \to S \times T$ as follows. If $a \in S \oplus T$, then the proposition says that there is a unique expression of the form $a = s + t$, and so $f : a \mapsto (s, t)$ is a well-defined function. It is routine to check that f is an isomorphism.

Conversely, if we define $g : S \times T \to S' \oplus T'$ by $g : (s, t) \mapsto (s, 0) + (0, t)$, then it is also routine to check that g is an isomorphism. •

From now on, we shall use the notation $S \oplus T$ to denote either version of the direct sum, because our point of view is almost always internal.

Definition. If $S_1, S_2, \ldots, S_n, \ldots$ are abelian groups, define the *finite direct sum* $S_1 \oplus S_2 \oplus \cdots \oplus S_n$ by induction on $n \geq 2$:

$$S_1 \oplus S_2 \oplus \cdots \oplus S_{n+1} = [S_1 \oplus S_2 \oplus \cdots \oplus S_n] \oplus S_{n+1}.$$

If $S_1, S_2, \ldots, S_n$ are subgroups of an abelian group G, when is the subgroup they generate, $\langle S_1, S_2, \ldots, S_n \rangle$, equal to their direct sum? A common mistake is to say that it is enough to assume that $S_i \cap S_j = \{0\}$ for all $i \neq j$, but the following example shows that this is not enough.

Example 5.1.
Let V be a 2-dimensional vector space over a field k, and let x, y be a basis. It is easy to check that the intersection of any two of the subspaces $\langle x \rangle$, $\langle y \rangle$, and $\langle x + y \rangle$ is $\{0\}$. On the other hand, we do not have $V = \langle x \rangle \oplus \langle y \rangle \oplus \langle x + y \rangle$ because $[\langle x \rangle \oplus \langle y \rangle] \cap \langle x + y \rangle \neq \{0\}$. ◀

Proposition 5.3.   *Let $G = S_1 + S_2 + \cdots + S_n$, where the S_i are subgroups; that is, each $a \in G$ has an expression of the form*

$$a = s_1 + s_2 + \cdots + s_n,$$

where $s_i \in S_i$ for all i. Then the following conditions are equivalent.

(i) $G = S_1 \oplus S_2 \oplus \cdots \oplus S_n$.

(ii) *Every $a \in G$ has a unique expression of the form $a = s_1 + s_2 + \cdots + s_n$, where $s_i \in S_i$ for all i.*

(iii) *For each i,*

$$S_i \cap (S_1 + S_2 + \cdots + \widehat{S_i} + \cdots + S_n) = \{0\},$$

where $\widehat{S_i}$ means that the term S_i is omitted from the sum.

Proof.   (i) $\Rightarrow$ (ii) The proof is by induction on $n \geq 2$. The base step is Proposition 5.1. For the inductive step, define $T = S_1 + S_2 + \cdots + S_n$, so that $G = T \oplus S_{n+1}$. If $a \in G$, then a has a unique expression of the form $a = t + s_{n+1}$, where $t \in T$ and $s_{n+1} \in S_{n+1}$ (by the proposition). But the inductive hypothesis says that t has a unique expression of the form $t = s_1 + \cdots + s_n$, where $s_i \in S_i$ for all i, as desired.

(ii) $\Rightarrow$ (iii) Suppose that

$$x \in S_i \cap (S_1 + S_2 + \cdots + \widehat{S_i} + \cdots + S_n).$$

Then $x = s_i \in S_i$ and $s_i = \sum_{j \neq i} s_j$, where $s_j \in S_j$. Unless all the $s_j = 0$, the element 0 has two distinct expressions: $0 = -s_j + \sum_{j \neq i} s_j$ and $0 = 0 + 0 + \cdots + 0$. Therefore, all $s_j = 0$ and $x = s_i = 0$.

(iii) $\Rightarrow$ (i) The proof is by induction. Since $S_{n+1} \cap (S_1 + S_2 + \cdots + S_n) = \{0\}$, we have $G = S_{n+1} \oplus (S_1 + S_2 + \cdots + S_n)$. By the inductive hypothesis,

$$S_1 + S_2 + \cdots + S_n = S_1 \oplus S_2 \oplus \cdots \oplus S_n. \quad \bullet$$

Example 5.2.
If V is an n-dimensional vector space over a field k, and if $v_1, \ldots, v_n$ is a basis, then

$$V = \langle v_1 \rangle \oplus \langle v_2 \rangle \oplus \cdots \oplus \langle v_n \rangle.$$

Since $v_1, \ldots, v_n$ is a basis, each $v \in V$ has a unique expression of the form $v = s_1 + \cdots + s_n$, where $s_i = r_i v_i \in \langle v_i \rangle$.   ◄

It will be convenient to analyze groups "one prime at a time."

Definition.   If p is a prime, then an abelian group G is *p-primary* if, for each $a \in G$, there is $n \geq 1$ with $p^n a = 0$.

If G is any abelian group, then its *p-primary component* is

$$G_p = \{a \in G : p^n a = 0 \text{ for some } n \geq 1\}.$$

If we do not want to specify the prime p, we may write that an abelian group is *primary* (instead of p-primary). It is clear that primary components are subgroups.

Theorem 5.4 (Primary Decomposition).

(i) *Every finite abelian group G is a direct sum of its p-primary components:*

$$G = \sum_p G_p.$$

(ii) *Two finite abelian groups G and G' are isomorphic if and only if $G_p \cong G'_p$ for every prime p.*

Proof.   (i) Let $x \in G$ be nonzero, and let its order be d. By the Fundamental Theorem of Arithmetic, there are distinct primes $p_1, \ldots, p_n$ and positive exponents $e_1, \ldots, e_n$ with

$$d = p_1^{e_1} \cdots p_n^{e_n}.$$

Define $r_i = d/p_i^{e_i}$, so that $p_i^{e_i} r_i = d$. It follows that $r_i x \in G_{p_i}$ for each i. But the gcd d of $r_1, \ldots, r_n$ is 1 (the only possible prime divisors of d are $p_1, \ldots, p_n$, but no p_i is a common divisor because $p_i \nmid r_i$); hence, there are integers $s_1, \ldots, s_n$ with $1 = \sum_i s_i r_i$. Therefore,

$$x = \sum_i s_i r_i x \in G_{p_1} + \cdots + G_{p_n}.$$

Write $H_i = G_{p_1} + G_{p_2} + \cdots + \widehat{G_{p_i}} + \cdots + G_{p_n}$. By Proposition 5.3, it suffices to prove that if

$$x \in G_{p_i} \cap H_i,$$

then $x = 0$. Since $x \in G_{p_i}$, we have $p_i^\ell x = 0$ for some $\ell \geq 0$; since $x \in H_i$, we have $ux = 0$, where $u = \prod_{j \neq i} p_j^{g_j}$. But p_i^ℓ and u are relatively prime, so there exist integers s and t with $1 = sp_i^\ell + tu$. Therefore,

$$x = (sp_i^\ell + tu)x = sp_i^\ell x + tux = 0.$$

(ii) If $f : G \to G'$ is a homomorphism, then $f(G_p) \le G'_p$ for every prime p, for if $p^\ell a = 0$, then $0 = f(p^\ell a) = p^\ell f(a)$. If f is an isomorphism, then $f^{-1} : G' \to G$ is also an isomorphism (so that $f^{-1}(G'_p) \le G_p$ for all p). It follows that each restriction $f|G_p : G_p \to G'_p$ is an isomorphism, with inverse $f^{-1}|G'_p$.

Conversely, if there are isomorphisms $f_p : G_p \to G'_p$ for all p, then there is an isomorphism $\varphi : \sum_p G_p \to \sum_p G'_p$ given by $\sum_p a_p \mapsto \sum_p f_p(a_p)$.   •

Recall the notation: if G is an abelian group and m is an integer, then

$$mG = \{ma : a \in G\}.$$

The next type of subgroup will play an important role.

Definition.   Let p be a prime and let G be a p-primary abelian group.[1] A subgroup $S \le G$ is a **pure**[2] **subgroup** if, for all $n \ge 0$,

$$S \cap p^n G = p^n S.$$

The inclusion $S \cap p^n G \ge p^n S$ is true for every subgroup $S \le G$, and so it is only the reverse inclusion $S \cap p^n G \le p^n S$ that is significant. It says that if $s \in S$ satisfies an equation $s = p^n a$ for some $a \in G$, then there exists $s' \in S$ with $s = p^n s'$.

Example 5.3.
(i) Every direct summand S of G is a pure subgroup. If $G = S \oplus T$ and $(s, 0) = p^n(u, v)$, where $u \in S$ and $t \in T$, then it is clear that $(s, 0) = p^n(u, 0)$. (The converse, Every pure subgroup S is a direct summand, is true when S is finite, but it may be false when S is infinite.)

(ii) If $G = \langle a \rangle$ is a cyclic group of order p^2, where p is a prime, then $S = \langle pa \rangle$ is not a pure subgroup of G, for if $s = pa \in S$, then there is no element $s' \in S$ with $s = pa = ps'$.   ◄

---

[1] If G is not a primary group, then a pure subgroup $S \le G$ is defined to be a subgroup which satisfies $S \cap mG = mS$ for all $m \in \mathbb{Z}$ (see Exercises 5.1 and 5.2).

[2] A polynomial equation is called **pure** if it has the form $x^n = a$; pure subgroups are defined in terms of such equations, and they are probably so called because of this.

Lemma 5.5. *If p is a prime and G is a finite p-primary abelian group, then G has a nonzero pure cyclic subgroup.*

Proof. Let $G = \langle x_1, \ldots, x_q \rangle$. The order of x_i is p^{n_i} for all i, because G is p-primary. If $x \in G$, then $x = \sum_i a_i x_i$, where $a_i \in \mathbb{Z}$, so that if ℓ is the largest of the n_i, then $p^\ell x = 0$. Now choose any $y \in G$ of largest order p^ℓ (for example, y could be one of the y_i). We claim that $S = \langle y \rangle$ is a pure subgroup of G.

Suppose that $s \in S$, so that $s = mp^t y$, where $t \geq 0$ and $p \nmid m$, and let

$$s = p^n a$$

for some $a \in G$. If $t \geq n$, define $s' = mp^{t-n} y \in S$, and note that

$$p^n s' = p^n mp^{t-n} y = mp^t y = s.$$

If $t < n$, then

$$p^\ell a = p^{\ell-n} p^n a = p^{\ell-n} s = p^{\ell-n} mp^t y = mp^{\ell-n+t} y.$$

But $p \nmid m$ and $\ell - n + t < \ell$, because $-n + t < 0$, and so $p^\ell a \neq 0$. This contradicts y having largest order, and so this case cannot occur. •

Proposition 5.6. *If G is an abelian group and p is a prime, then G/pG is a vector space over $\mathbb{Z}_p$ which is finite-dimensional when G is finite.*

Proof. If $[r] \in \mathbb{Z}_p$ and $a \in G$, define scalar multiplication

$$[r](a + pG) = ra + pG.$$

This formula is well-defined, for if $k \equiv r \bmod p$, then $k = r + pm$ for some integer m, and so

$$ka + pG = ra + pma + pG = ra + pG,$$

because $pma \in pG$. It is now routine to check that the axioms for a vector space do hold.

If G is finite, then so is G/pG, and it is clear that G/pG has a finite basis. •

Definition.   If p is a prime and G is a finite p-primary abelian group, then

$$d(G) = \dim(G/pG).$$

Observe that d is additive over direct sums,

$$d(G \oplus H) = d(G) + d(H),$$

for Proposition 2.60 gives

$$(G \oplus H)/p(G \oplus H) = (G \oplus H)/(pG \oplus pH)$$
$$\cong (G/pG) \oplus (H/pH).$$

The dimension of the left side is $d(G \oplus H)$ and the dimension of the right side is $d(G) + d(H)$, for the union of a basis of G/pG and a basis of H/pH is a basis of $(G/pG) \oplus (H/pH)$.

The abelian groups G with $d(G) = 1$ are easily characterized.

Lemma 5.7.   *If G is a p-primary abelian group, then $d(G) = 1$ if and only if G is cyclic.*

Proof.   If G is cyclic, then so is any quotient of G; in particular, G/pG is cyclic, and so $\dim(G/pG) = 1$.

Conversely, if $G/pG = \langle z + pG \rangle$, then $G/pG \cong \mathbb{Z}_p$. Since $\mathbb{Z}_p$ is a simple group, the correspondence theorem says that pG is a maximal subgroup of G; we claim that pG is the only maximal subgroup of G. If $L \le G$ is any maximal subgroup, then $G/L \cong \mathbb{Z}_p$, for G/L is a simple abelian group of order a power of p, hence has order p (by Proposition 2.78, the abelian simple groups are precisely the cyclic groups of prime order). Thus, if $a \in G$, then $p(a + L) = 0$ in G/L, so that $pa \in L$; hence $pG \le L$. But pG is maximal, and so $pG = L$. It follows that every proper subgroup of G is contained in pG (for every proper subgroup is contained in some maximal subgroup). In particular, if $\langle z \rangle$ is a proper subgroup of G, then $\langle z \rangle \le pG$, contradicting $z + pG$ being a generator of G/pG. Therefore, $G = \langle z \rangle$, and so G is cyclic.   •

Lemma 5.8.   *Let G be a finite p-primary abelian group.*

(i) *If $S \le G$, then $d(G/S) \le d(G)$.*

(ii) *If S is a pure subgroup of G, then*

$$d(G) = d(S) + d(G/S).$$

Proof.   (i) By the correspondence theorem, $p(G/S) = (pG + S)/S$, so that

$$(G/S)/p(G/S) = (G/S)/[(pG + S)/S] \cong G/(pG + S),$$

by the third isomorphism theorem. Since $pG \leq pG + S$, there is a surjective homomorphism (of vector spaces over $\mathbb{Z}_p$),

$$G/pG \to G/(pG + S),$$

namely, $g + pG \mapsto g + (pG + S)$. Hence, $\dim(G/pG) \geq \dim(G/(pG+S))$; that is, $d(G) \geq d(G/S)$.

(ii) We now analyze $(pG + S)/pG$, the kernel of $G/pG \to G/(pG + S)$. By the second isomorphism theorem,

$$(pG + S)/pG \cong S/(S \cap pG).$$

Since S is a pure subgroup, $S \cap pG = pS$. Therefore,

$$(pG + S)/pG \cong S/pS,$$

and so $\dim[(pG + S)/pG] = d(S)$. But if W is a subspace of a finite-dimensional vector space V, then $\dim(V) = \dim(W) + \dim(V/W)$, by Exercise 4.12. Hence, if $V = G/pG$ and $S = (pG + S)/pG$, we have

$$d(G) = d(S) + d(G/S). \quad \bullet$$

Theorem 5.9 (Basis Theorem).   *Every finite abelian group G is a direct sum of primary cyclic groups.*

Proof.   By the primary decomposition, Theorem 5.4, we may assume that G is p-primary for some prime p. We prove that G is a direct sum of cyclic groups by induction on $d(G) \geq 1$. The base step is easy, for Lemma 5.7 shows that G must be cyclic in this case.

   To prove the inductive step, we begin by using Lemma 5.5 to find a nonzero pure cyclic subgroup $S \leq G$. By Lemma 5.8, we have

$$d(G/S) = d(G) - d(S) = d(G) - 1 < d(G).$$

By induction, G/S is a direct sum of cyclic groups, say,

$$G/S = \sum_{i=1}^{q} \langle \bar{x}_i \rangle,$$

where $\bar{x}_i = x_i + S$.

Let $x \in G$ and let $\bar{x}$ have order p^ℓ, where $\bar{x} = x + S$. We claim that there is $z \in G$ with $z + S = \bar{x} = x + S$ such that

$$\text{order } z = \text{ order } (\bar{x}).$$

Now x has order p^n, where $n \geq \ell$. But $p^\ell(x+S) = p^\ell \bar{x} = 0$ in G/S, so there is some $s \in S$ with $p^\ell x = s$. By purity, there is $s' \in S$ with $p^\ell x_i = p^\ell s'$. If we define $z = x - s'$, then $z + S = x + S$ and $p^\ell z = 0$. Hence, if $m\bar{x} = 0$ in G/S, then $p^\ell \mid m$, and so $mz = 0$ in G.

For each i, choose $z_i \in G$ with $z_i + S = \bar{x}_i = x_i + S$ and with order $z_i = \text{order } \bar{x}_i$; let $T = \langle z_1, \dots, z_q \rangle$. Now $S+T = G$, because G is generated by S and the z_i's. To see that $G = S \oplus T$, it now suffices to prove that $S \cap T = \{0\}$. If $y \in S \cap T$, then $y = \sum_i m_i z_i$, where $m_i \in \mathbb{Z}$. Now $y \in S$, and so $\sum_i m_i \bar{x}_i = 0$ in G/S. Since this is a direct sum, each $m_i \bar{x}_i = 0$; after all, for each i,

$$-m_i \bar{x}_i = \sum_{j \neq i} m_j \bar{x}_j \in \langle \bar{x}_i \rangle \cap \left(\langle \bar{x}_1 \rangle \right) + \cdots + \widehat{\langle \bar{x}_i \rangle} + \cdots + \langle \bar{x}_q \rangle \right) = \{0\}.$$

Therefore, $m_i z_i = 0$ for all i, and hence $y = 0$.   •

When are two finite abelian groups G and G' isomorphic? By the basis theorem, such groups are direct sums of cyclic groups, and so one's first guess is that $G \cong G'$ if they have the same number of cyclic summands of each type. But this hope is dashed by Theorem 2.62, which says that if m and n are relatively prime, then $\mathbb{Z}_{mn} \cong \mathbb{Z}_m \times \mathbb{Z}_n$; for example, $\mathbb{Z}_6 \cong \mathbb{Z}_2 \times \mathbb{Z}_3$. Thus, we retreat and try to count *primary* cyclic summands. But how can we do this? As in the Fundamental Theorem of Arithmetic, we must ask whether there is some kind of unique factorization theorem here.

Before stating the next lemma, recall that we have defined

$$d(G) = \dim(G/pG).$$

In particular, $d(pG) = \dim(pG/p^2G)$ and, more generally,

$$d(p^nG) = \dim(p^nG/p^{n+1}G).$$

Lemma 5.10. *Let G be a finite p-primary abelian group, where p is a prime, and let $G = \sum_j C_j$, where each C_j is cyclic. If b_n is the number of summands C_j having order p^n, then there is some $t \geq 1$ with*

$$d(p^n G) = b_{n+1} + b_{n+2} + \cdots + b_t.$$

Proof. Let B_n be the direct sum of all C_j, if any, with order p^n. Thus,

$$G = B_1 \oplus B_2 \oplus \cdots \oplus B_t$$

for some t. Now

$$p^n G = p^n B_{n+1} \oplus \cdots \oplus p^n B_t,$$

because $p^n B_j = \{0\}$ for all $j \leq n$. Similarly,

$$p^{n+1} G = p^{n+1} B_{n+2} \oplus \cdots \oplus p^{n+1} B_t.$$

Now Proposition 2.60 shows that $p^n G / p^{n+1} G$ is isomorphic to

$$\left[p^n B_{n+1} / p^{n+1} B_{n+1} \right] \oplus \left[p^n B_{n+2} / p^{n+1} B_{n+2} \right] \oplus \cdots \oplus \left[p^n B_t / p^{n+1} B_t \right].$$

Since d is additive over direct sums, we have

$$d(p^n G) = b_{n+1} + b_{n+2} + \cdots + b_t. \quad \bullet$$

The numbers b_n can now be described in terms of G.

Definition. If G is a finite p-primary abelian group, where p is a prime, then

$$U_p(n, G) = d(p^n G) - d(p^{n+1} G).$$

Lemma 5.10 shows that

$$d(p^n G) = b_{n+1} + \cdots + b_t$$

and

$$d(p^{n+1} G) = b_{n+2} + \cdots + b_t,$$

so that $U_p(n, G) = b_{n+1}$.

Theorem 5.11.   *If p is a prime, then any two decompositions of a finite p-primary abelian group G into direct sums of cyclic groups have the same number of cyclic summands of each type. More precisely, for each $n \geq 0$, the number of cyclic summands having order p^{n+1} is $U_p(n, G)$.*

Proof.   By the basis theorem, there exist cyclic subgroups C_i with $G = \sum_i C_i$. The lemma shows, for each $n \geq 0$, that the number of C_i having order p^{n+1} is $U_p(n, G)$, a number that is defined without any mention of the given decomposition of G into a direct sum of cyclics. Thus, if $G = \sum_j D_j$ is another decomposition of G, where each D_j is cyclic, then the number of D_j having order p^{n+1} is also $U_p(n, G)$, as desired.   •

Corollary 5.12.   *If G and G' are finite p-primary abelian groups, then $G \cong G'$ if and only if $U_p(n, G) = U_p(n, G')$ for all $n \geq 0$.*

Proof.   If $\varphi : G \to G'$ is an isomorphism, then $\varphi(p^n G) = p^n G'$ for all $n \geq 0$, and hence it induces isomorphisms of the $\mathbb{Z}_p$-vector spaces $p^n G / p^{n+1} G \cong p^n G' / p^{n+1} G'$ for all $n \geq 0$. Hence, their dimensions are the same; that is, $U_p(n, G) = U_p(n, G')$.

Conversely, assume that $U_p(n, G) = U_p(n, G')$ for all $n \geq 0$. If $G = \sum_i C_i$ and $G' = \sum_j C'_j$, where the C_i and C'_j are cyclic, then Lemma 5.10 shows that there are the same number of summands of each type, and so it is a simple matter to construct an isomorphism $G \to G'$.   •

Definition.   If G is a p-primary abelian group, then the ***elementary divisors*** of G are the numbers p^{n+1}, each repeated with multiplicity $U_p(n, G)$.

If G is a finite abelian group, then its ***elementary divisors*** are the elementary divisors of all its primary components.

Theorem 5.13 (Fundamental Theorem of Finite Abelian Groups).   *Two finite abelian groups G and G' are isomorphic if and only if they have the same elementary divisors; that is, any two decompositions of G and G' as direct sums of primary cyclic groups have the same number of summands of each order.*

Proof.   By the primary decomposition, Theorem 5.4(ii), $G \cong G'$ if and only if, for each prime p, their primary components are isomorphic: $G_p \cong G'_p$. The result now follows from Theorem 5.11.   •

396 Groups II Ch. 5

For example, the elementary divisors of an elementary abelian group of order 8 are $(2, 2, 2)$, and the elementary divisors of $\mathbb{Z}_6$ are $(2, 3)$. The elementary divisors of $\mathbb{Z}_2 \oplus \mathbb{Z}_2 \oplus \mathbb{Z}_4 \oplus \mathbb{Z}_8$ are $(2, 2, 4, 8)$.

The results of this section can be generalized from finite abelian groups to finitely generated abelian groups, where an abelian group G is *finitely generated* if there are finitely many elements $a_1, \dots, a_n \in G$ so that every $x \in G$ is a linear combination of them: $x = \sum_i m_i a_i$, where $m_i \in \mathbb{Z}$ for all i. The basis theorem generalizes: Every finitely generated abelian group G is a direct sum of cyclic groups, each of which is a finite primary group or an infinite cyclic group; the fundamental theorem also generalizes: Given two decompositions of G into a direct sum of cyclic groups (as in the basis theorem), the number of cyclic summands of each type is the same in both decompositions. The basis theorem is no longer true for abelian groups that are not finitely generated; for example, the additive group $\mathbb{Q}$ of rational numbers is not a direct sum of cyclic groups.

EXERCISES

5.1 Let G be an abelian group, not necessarily primary. Define a subgroup $S \le G$ to be a *pure subgroup* if, for all $m \in \mathbb{Z}$,

$$S \cap mG = mS.$$

Prove that if G is a p-primary abelian group, where p is a prime ideal, then a subgroup $S \le G$ is pure as just defined if and only if $S \cap p^n G = p^n S$ for all $n \ge 0$ (the definition in the text).

5.2 Let G be a possibly infinite abelian group.

(i) Prove that every direct summand S of G is a pure subgroup.

Define the *torsion*[3] *subgroup* tG of G as

$$tG = \{a \in G : a \text{ has finite order}\}.$$

(ii) Prove that tG is a pure subgroup of G. (There exist abelian groups G whose torsion subgroup tG is not a direct summand; hence, a pure subgroup need not be a direct summand.)

---

[3]This terminology comes from algebraic topology. To each space X, one assigns a sequence of abelian groups, called *homology groups*, and if X is "twisted," then there are elements of finite order in some of these groups.

(iii) Prove that G/tG is an abelian group in which every nonzero element has infinite order.

5.3   (i) If G and H are finite abelian groups, prove, for all primes p and all $n \geq 0$, that

$$U_p(n, G \oplus H) = U_p(n, G) + U_p(n, H),$$

(ii) If A, B, and C are finite abelian groups, prove that $A \oplus B \cong A \oplus C$ implies $B \cong C$.

(iii) If A and B are finite abelian groups, prove that $A \oplus A \cong B \oplus B$ implies $A \cong B$.

5.4 If n is a positive integer, then a ***partition of*** n is a sequence of positive integers $i_1 \leq i_2 \leq \cdots \leq i_r$ with $i_1 + i_2 + \cdots + i_r = n$. If p is a prime, prove that the number of abelian groups of order p^n, to isomorphism, is equal to the number of partitions of n.

5.5 To isomorphism, how many abelian groups are there of order 288?

5.6 Prove the Fundamental Theorem of Arithmetic by applying the Fundamental Theorem of Finite Abelian Groups to $G = \mathbb{Z}_n$.

5.7   (i) If G is a finite abelian group, define

$$\nu_k(G) = \text{the number of elements in } G \text{ of order } k.$$

Prove that two finite abelian groups G and G' are isomorphic if and only if $\nu_k(G) = \nu_k(G')$ for all integers k.

(ii) Give an example of two nonisomorphic finite nonabelian groups G and G' for which $\nu_k(G) = \nu_k(G')$ for all integers k.

5.8 Prove that the additive group $\mathbb{Q}$ is not a direct sum: $\mathbb{Q} \ncong A \oplus B$, where A and B are nonzero subgroups.

5.2 THE SYLOW THEOREMS

We return to nonabelian groups, and so we revert to the multiplicative notation. The Sylow theorems give an analog for finite nonabelian groups of the primary decomposition for finite abelian groups.

Recall that a group $G \neq \{1\}$ is called *simple* if it has no normal subgroups other than $\{1\}$ and G itself. We saw, in Proposition 2.78, that the abelian simple groups are precisely the cyclic groups $\mathbb{Z}_p$ of prime order p, and we saw, in Theorem 2.83, that A_n is a nonabelian simple group for all $n \geq 5$. In fact, A_5 is the nonabelian simple group of smallest order. How can one prove that a nonabelian group G of order less than $60 = |A_5|$ is not simple? Let us solve Exercise 2.105: If G is a group of order $|G| = mp$, where p is prime and $1 < m < p$, then G is not simple. By Cauchy's theorem, G contains an element x of order p, and so G has a (cyclic) subgroup $H = \langle x \rangle$ of order p, hence of index m. By Theorem 2.67, there is a homomorphism $\varphi : G \to S_m$ with $\ker \varphi \leq H$. Since G is simple, however, it has no proper normal subgroups; hence $\ker \varphi = \{1\}$ and φ is an injection; that is, $G \cong \varphi(G) \leq S_m$, and so $|G|$ divides $m!$. But $p \mid |G|$, by Lagrange's theorem, and so $p \mid m!$. This contradicts Euclid's lemma, for $m! = \prod_{i=1}^{m} i$ and p divides none of the factors.

This exercise shows that many of the numbers less than 60 are not orders of simple groups. On the other hand, it does not eliminate 36, for example. If we could find a larger subgroup H of G such that $|H|$ is a multiple of a prime p and $p \nmid m = [G : H]$, then the exercise could be generalized to groups G of order $|G| = m|H|$. What proper subgroups H of G do we know to play the role of $H = \langle x \rangle$? The center $Z(G)$ of a group G is a possible candidate, but this subgroup might not be proper or it might be trivial: if G is abelian, then $Z(G) = G$; if $G = S_3$, then $Z(G) = \{1\}$. Hence, $Z(G)$ cannot be used to generalize the exercise.

The first book on Group Theory, *Traités des Substitutions et des Équations Algébriques*, by C. Jordan, was published in 1870 (more than half of it is devoted to Galois Theory, then called the Theory of Equations). At about the same time, but too late for publication in Jordan's book, three fundamental theorems were discovered. In 1868, E. Schering proved the Basis Theorem: Every finite abelian group is a direct product of cyclic groups, each of prime power order; in 1870, L. Kronecker, unaware of Schering's proof, also proved this result. In 1878, G. Frobenius and L. Stickelberger proved the Fundamental Theorem of Finite Abelian Groups. In 1872, L. Sylow showed, for every finite group G and every prime p, that if p^k is the largest power of p dividing $|G|$, then G has a subgroup of order p^k. A Sylow subgroup can be used to generalize Exercise 2.105.

Recall that a *p-group* is a finite group G of order p^k for some $k \geq 0$. (When working wholly in the context of abelian groups, as we were in the last section, one calls G a *p*-primary group.)

Definition.   Let p be a prime. A ***Sylow p-subgroup*** of a finite group G is a maximal p-subgroup P.

Maximality means that if Q is a p-subgroup of G and $P \leq Q$, then $P =$

Q. Sylow p-subgroups always exist: indeeed, we now show that if S is any p-subgroup of G (perhaps $S = \{1\}$), then there exists a Sylow p-subgroup P containing S. If there is no p-subgroup strictly containing S, then S itself is a Sylow p-subgroup. Otherwise, there is a p-subgroup P_1 with $S < P_1$. If P_1 is maximal, it is Sylow, and we are done. Otherwise, there is some p-subgroup P_2 with $P_1 < P_2$. This procedure of producing larger and larger p-subgroups P_i must end after a finite number of steps, for $|P_i| \leq |G|$ for all i; the largest P_i must, therefore, be a Sylow p-subgroup.

Definition. If H is a subgroup of a group G, then a ***conjugate*** of H is a subgroup of G of the form

$$aHa^{-1} = \{aha^{-1} : h \in H\},$$

where $a \in G$.

Since conjugation $h \mapsto aha^{-1}$ is an injection $H \to G$ with image aHa^{-1}, it follows that conjugate subgroups of G are isomorphic. For example, in S_3, all cyclic subgroups of order 2 are conjugate (for their generators are conjugate).

The ideas of group actions are going to be used, and so it is a good idea to review the notions of *orbit* and *stabilizer* that we discussed in Chapter 2.

Definition. If X is a set and G is a group, then G ***acts*** on X if, for each $g \in G$, there is a function $\alpha_g : X \to X$, such that

(i) for $g, h \in G$, $\alpha_g \circ \alpha_h = \alpha_{gh}$;

(ii) $\alpha_1 = 1_X$, the identity function.

If G acts on X, we usually write gx instead of $\alpha_g(x)$.

Definition. If G acts on X and $x \in X$, then the ***orbit*** of x, denoted by $\mathcal{O}(x)$, is the subset of X

$$\mathcal{O}(x) = \{gx : g \in G\} \subset X;$$

the ***stabilizer*** of x, denoted by G_x, is

$$G_x = \{g \in G : gx = x\} \leq G.$$

Recall Theorem 2.71: If G acts on a set X and $x \in X$, then

$$|\mathcal{O}(x)| = [G : G_x] :$$

the size of the orbit $\mathcal{O}(x)$ is the index of the stabilizer G_x in G.

In particular, a group G acts on $X = \mathbf{Sub}(G)$, the set of all its subgroups, by conjugation: if $a \in G$, then a acts by $H \mapsto aHa^{-1}$, where $H \leq G$. The orbit of a subgroup H consists of all its conjugates; the stabilizer of H consists of $\{a \in G : aHa^{-1} = H\}$. This last subgroup has a name.

Definition.   If H is a subgroup of a group G, then the ***normalizer*** of H in G is the subgroup

$$N_G(H) = \{a \in G : aHa^{-1} = H\}.$$

Of course, $H \lhd N_G(H)$, and so the quotient group $N_G(H)/H$ is defined.

Proposition 5.14.   *If H is a subgroup of a finite group G, then the number of conjugates of H in G is $[G : N_G(H)]$.*

Proof.   This is a special case of Theorem 2.71: the orbit size is the index of the stabilizer.   •

Lemma 5.15.   *Let P be a Sylow p-subgroup of a finite group G.*

(i) *Every conjugate of P is also a Sylow p-subgroup of G.*

(ii) *$|N_G(P)/P|$ is prime to p.*

(iii) *If $a \in G$ has order some power of p and if $aPa^{-1} = P$, then $a \in P$.*

Proof.   (i) If $a \in G$, then aPa^{-1} is a p-subgroup of G; if it is not a maximal such, then there is a p-subgroup Q with $aPa^{-1} < Q$. Hence, $P < a^{-1}Qa$, contradicting the maximality of P.

(ii) If p divides $|N_G(P)/P|$, then Cauchy's theorem shows that $N_G(P)/P$ contains an element aP of order p, and hence $N_G(P)/P$ contains a subgroup $S^* = \langle aP \rangle$ of order p. By the correspondence theorem (Theorem 2.58), there is a subgroup S with $P \leq S \leq N_G(P)$ such that $S/P \cong S^*$. But S is a p-subgroup of $N_G(P) \leq G$ (by Exercise 2.89) strictly larger than P, and this contradicts the maximality of P. We conclude that p does not divide $|N_G(P)/P|$.

(iii) By the definition of normalizer, the element a lies in $N_G(P)$. If $a \notin P$, then the coset aP is a nontrivial element of $N_G(P)/P$ having order some power of p; in light of part (ii), this contradicts Lagrange's theorem.   •

Since every conjugate of a Sylow p-subgroup is a Sylow p-subgroup, it is reasonable to let G act by conjugation on the Sylow p-subgroups.

Theorem 5.16 (*Sylow*).   *Let G be a finite group, and let P be a Sylow p-subgroup of G for some prime p.*

  (i) *Every Sylow p-subgroup is conjugate to P.*

  (ii) *If there are r Sylow p-subgroups, then r is a divisor of G and*

$$r \equiv 1 \bmod p.$$

Proof.   Let $X = \{P_1, \ldots, P_r\}$ be the set of all the conjugates of P, where we have denoted P by P_1. If Q is any Sylow p-subgroup of G, then Q acts on X by conjugation: if $a \in Q$, then it sends

$$P_i = g_i P g_i^{-1} \mapsto a \left(g_i P g_i^{-1} \right) a^{-1} = (a g_i) P (a g_i)^{-1} \in X.$$

By Corollary 2.72, the number of elements in any orbit is a divisor of $|Q|$; that is, every orbit has size some power of p (because Q is a p-group). If there is an orbit of size 1, then there is some P_i with $a P_i a^{-1} = P_i$ for all $a \in Q$. By Lemma 5.15, we have $a \in P_i$ for all $a \in Q$; that is, $Q \le P_i$. But Q, being a Sylow p-subgroup, is a maximal p-subgroup of G, and so $Q = P_i$. In particular, if $Q = P_1$, then every orbit has size an honest power of p except one, the orbit consisting of P_1 alone. We conclude that $|X| = r \equiv 1 \bmod p$.

Suppose now that there is some Sylow p-subgroup Q that is not a conjugate of P; thus, $Q \ne P_i$ for any i. Again, we let Q act on X, and again, we ask if there is an orbit of size 1, say, $\{P_j\}$. As in the previous paragraph, this implies $Q = P_j$, contrary to our present assumption that $Q \notin X$. Hence, there are no orbits of size 1, which says that each orbit has size an honest power of p. It follows that $|X| = r$ is a multiple of p; that is, $r \equiv 0 \bmod p$, which contradicts the congruence $r \equiv 1 \bmod p$. Therefore, no such Q can exist, and so all Sylow p-subgroups are conjugate to P. Finally, since all Sylow p-subgroups are conjugate, we have $r = [G : N_G(P)]$, and so r is a divisor of $|G|$.   •

Corollary 5.17.   *A finite group G has a unique Sylow p-subgroup P, for some prime p, if and only if $P \lhd G$.*

Proof.   Assume that P, a Sylow p-subgroup of G, is unique. For each $a \in G$, the conjugate $a P a^{-1}$ is also a Sylow p-subgroup; by uniqueness, $a P a^{-1} = P$ for all $a \in G$, and so $P \lhd G$.

Conversely, assume that $P \lhd G$. If Q is any Sylow p-subgroup, then $Q = a P a^{-1}$ for some $a \in G$; but $a P a^{-1} = P$, by normality, and so $Q = P$.   •

The following result gives the order of a Sylow subgroup.

Theorem 5.18 (Sylow).   *If G is a finite group of order $p^e m$, where p is a prime and $p \nmid m$, then every Sylow p-subgroup P of G has order p^e.*

Proof.   We first show that $p \nmid [G : P]$. Now

$$[G : P] = [G : N_G(P)][N_G(P) : P].$$

The first factor, $[G : N_G(P)] = r$, is the number of conjugates of P in G, and we know that $r \equiv 1 \bmod p$; hence, p does not divide $[G : N_G(P)]$. The second factor is $[N_G(P) : P] = |N_G(P)/P|$; this, too, is not divisible by p, by Lemma 5.15. Therefore, p does not divide $[G : P]$, by Euclid's lemma.

Now $|P| = p^k$ for some $k \le e$, and so

$$[G : P] = |G|/|P| = p^e m / p^k = p^{e-k} m.$$

Since p does not divide $[G : P]$, we must have $k = e$; that is, $|P| = p^e$.   •

This last result was used, in Chapter 4, to prove the Fundamental Theorem of Algebra.

Example 5.4.
(i) Let $G = S_4$. Now $|S_4| = 24 = 2^3 3$. Thus, a Sylow 2-subgroup of S_4 has order 8. We have seen, in Exercise 2.96, that S_4 contains a copy of the dihedral group D_8 consisting of the symmetries of a square. The Sylow theorem says that all subgroups of order 8 are conjugate, hence isomorphic, to D_8. Moreover, the number r of Sylow 2-subgroups is a divisor of 24 congruent to 1 mod 2; that is, r is an odd divisor of 24. Since $r \ne 1$ (see Exercise 5.9), there are exactly 3 Sylow 2-subgroups.

(ii) If G is a finite abelian group, then a Sylow p-subgroup is just its p-primary component (since G is abelian, every subgroup is normal, and so there is a unique Sylow p-subgroup for every prime p).   ◄

Here is a second proof of the last Sylow theorem, due to Wielandt.

Theorem 5.19.   *If G is a finite group of order $p^e m$, where p is a prime and $p \nmid m$, then G has a subgroup of order p^e.*

Proof.   If X is the family of all those subsets of G having exactly p^e elements, then $|X| = \binom{n}{p^e}$; by Exercise 1.63, $p \nmid |X|$. Now G acts on X: define gB, for $g \in G$ and $B \in X$, by

$$gB = \{gb : b \in B\}.$$

If p divides $|\mathcal{O}(B)|$ for every $B \in X$, then p is a divisor of $|X|$, for X is the disjoint union of orbits, by Proposition 2.70. As $p \nmid |X|$, there exists a subset B with $|B| = p^e$ and with $|\mathcal{O}(B)|$ not divisible by p. If G_B is the stabilizer of this subset B, then Theorem 2.71 gives $[G : G_B] = |\mathcal{O}(B)|$, and so $|G| = |G_B| \cdot |\mathcal{O}(B)|$. Since $p^e \mid |G|$ and $p \nmid \mathcal{O}(B)|$, repeated application of Euclid's lemma gives $p^e \mid |G_B|$. Therefore, $p^e \leq |G_B|$.

To prove the reverse inequality, choose an element $b \in B$ and define a function $\tau : G_B \to B$ by $g \mapsto gb$. Note that $\tau(g) = gb \in gB = B$, for $g \in G_B$, the stabilizer of B. If $g, h \in G_B$ and $h \neq g$, then $\tau(h) = hb \neq gb = \tau(g)$; that is, τ is an injection. We conclude that $|G_B| \leq |B| = p^e$, and so G_B is a subgroup of G of order p^e.   •

We can now generalize Exercise 2.105 and its solution.

Lemma 5.20.   *There is no nonabelian simple group G of order $|G| = p^e m$, where p is prime and $1 \leq m < p^e$.*

Proof.   If $m = 1$, then $|G| = p^e$ for some prime p, and G is a p-group. Theorem 2.75 says that G has a nontrivial center, $Z(G)$. Now $Z(G)$ is always normal, so that G is not simple if $Z(G)$ is a proper subgroup. If $Z(G) = G$, then G is abelian, and Proposition 2.78 shows that G is not simple unless $|G| = p$.

Suppose now that $m > 1$ and that G is simple. By Sylow's theorem, G contains a subgroup P of order p^e, hence of index m. By Theorem 2.67, there exists a homomorphism $\varphi : G \to S_m$ with $\ker \varphi \leq H$. Since G is simple, however, it has no proper normal subgroups; hence $\ker \varphi = \{1\}$ and φ is an injection; that is, $G \cong \varphi(G) \leq S_m$, and so $|G| \mid m!$. But $p \mid |G|$, by Lagrange's theorem, and so $p \mid m!$. This contradicts Euclid's lemma, for $m! = \prod_{i=1}^{m} i$ and p divides none of the factors.   •

Proposition 5.21.   *There are no nonabelian simple groups of order less than 60.*

Proof.   The reader may now check that each integer n between 2 and 59, with the exception of $n = 30$, has a factorization of the form $n = p^e m$, where p is prime and $1 \leq m < p^e$. By the lemma, there is no nonabelian simple group of any such order.

Only the case $n = 30$ remains; assume there is a simple group G of this order. Let P be a Sylow 5-subgroup of G, so that $|P| = 5$. The number r_5 of conjugates of P is a divisor of 30 and $r_5 \equiv 1 \bmod 5$. Now $r_5 \neq 1$

lest $P \lhd G$, so that $r_5 = 6$. By Lagrange's theorem, the intersection of any two of these is trivial (intersections of Sylow subgroups can be more complicated; see Exercise 5.10). There are 4 nonidentity elements in each of these subgroups, and so there are $6 \times 4 = 24$ nonidentity elements in their union. Similarly, the number r_3 of Sylow 3-subgroups of G is 10 (for $r_3 \neq 1$, r_3 is a divisor of 30, and $r_3 \equiv 1 \bmod 3$). There are 2 nonidentity elements in each of these subgroups, and so the union of these subgroups has 20 nonidentity elements. We have exceeded the number of elements in G, and so G cannot be simple. •

The "converse" of Lagrange's theorem is false: If G is a finite group of order n, and if $d \mid n$, then G may not have a subgroup of order d. For example, we proved, in Proposition 2.40, that the alternating group A_4 is a group of order 12 having no subgroup of order 6.

Proposition 5.22. *Let G be a finite group. If p is a prime and if p^k divides $|G|$, then G has a subgroup of order p^k.*

Proof. If $|G| = p^e m$, where $p \nmid m$, then a Sylow p-subgroup P of G has order p^e. Hence, if p^k divides $|G|$, then p^k divides $|P|$. By Proposition 2.77, P has a subgroup of order p^k; *a fortiori*, G has a subgroup of order p^k. •

What examples of p-groups have we seen? Of course, cyclic groups of order p^n are p-groups, as is any direct product of copies of these. By the Fundamental Theorem, this describes all (finite) abelian p-groups. The only nonabelian examples we have seen so far are the dihedral groups D_{2n} (which are 2-groups when n is a power of 2) and the quaternions $\mathbf{Q}$ of order 8 (of course, for every 2-group A, the direct products $D_8 \times A$ and $\mathbf{Q} \times A$ are also nonabelian 2-groups). Here are some new examples.

Definition. If k is a field, then an $n \times n$ **unitriangular** matrix over k is an upper triangular matrix each of whose diagonal terms is 1. Define $\mathrm{UT}(n, k)$ to be the set of all $n \times n$ unitriangular matrices over k.

Proposition 5.23. $\mathrm{UT}(n, k)$ *is a subgroup of* $\mathrm{GL}(n, k)$.

Proof. If $A \in \mathrm{UT}(n, k)$, then $A = E + N$, where N is *strictly* upper triangular; that is, N is an upper triangular matrix having only 0's on its diagonal. Note that the sum and product of strictly upper triangular matrices is again strictly upper triangular.

Let $e_1, \ldots, e_n$ be the standard basis of k^n. If N is strictly upper triangular, define $T: k^n \rightarrow k^n$ by $T(e_i) = Ne_i$, where e_i is regarded as a column matrix. Now T satisfies the equations, for all i,

$$T(e_1) = 0 \quad \text{and} \quad T(e_{i+1}) \in \langle e_1, \ldots, e_i \rangle.$$

It is easy to see, by induction on i, that

$$T^i(e_j) = 0 \text{ for all } j \leq i.$$

It follows that $T^n = 0$ and, hence, that $N^n = 0$. Thus, if $A \in \text{UT}(n, k)$, then $A = E + N$, where $N^n = 0$.

We can now show that $\text{UT}(n, k)$ is a subgroup of $\text{GL}(n, k)$. First of all, if A is unitriangular, then it is nonsingular. In analogy to the power series expansion $1/(1+x) = 1-x+x^2-x^3+\cdots$, we try $B = E-N+N^2-N^3+\cdots$ as the inverse of $A = E + N$ (we note that the matrix power series stops after $n - 1$ terms because $N^n = 0$), The reader may now check that this works: $BA = E = AB$; therefore, A is nonsingular. Moreover, since N is strictly upper triangular, so is $-N + N^2 - N^3 + \cdots$, so that A^{-1} is also unitriangular. Finally, $(E + N)(E + M) = E + (N + M + NM)$ is unitriangular, and so $\text{UT}(n, k)$ is a subgroup of $\text{GL}(n, k)$.   •

Notation.   We are going to denote the finite field with q elements by $\mathbb{F}_q$ [instead of by $\text{GF}(q)$].

Proposition 5.24.   *Let* $q = p^e$, *where p is a prime. For each $n \geq 2$,* $\text{UT}(n, \mathbb{F}_q)$ *is a p-group of order* $q^{\binom{n}{2}} = q^{n(n-1)/2}$.

Proof.   By Exercise 1.9, the number of entries in an $n \times n$ unitriangular matrix lying strictly above the diagonal is $\binom{n}{2}$. Since each of these entries can be any element of $\mathbb{F}_q$, there are exactly $q^{\binom{n}{2}}$ $n \times n$ unitriangular matrices over $\mathbb{F}_q$, and so this is the order of $\text{UT}(n, \mathbb{F}_q)$.   •

Recall Exercise 2.41: If G is a group and $x^2 = 1$ for all $x \in G$, then G is abelian. We now ask whether a group G satisfying $x^p = 1$ for all $x \in G$, where p is an odd prime, must also be abelian.

Proposition 5.25.   *If p is an odd prime, then there exists a nonabelian group G of order p^3 with $x^p = 1$ for all $x \in G$.*

Proof.   If $G = \mathrm{UT}(3, \mathbb{F}_p)$, then $|G| = p^3$. If $A \in G$, then $A = E + N$, where $N^p = 0$ and $EN = N = NE$. Hence,

$$A^p = (E + N)^p = E^p + N^p = E. \quad \bullet$$

Theorem 5.26.   *Let $\mathbb{F}_q$ denote the finite field with q elements. Then*

$$|\mathrm{GL}(n, \mathbb{F}_q)| = (q^n - 1)(q^n - q)(q^n - q^2) \cdots (q^n - q^{n-1}).$$

Proof.   Let V be an n-dimensional vector space over $\mathbb{F}_q$. We show first that there is a bijection $\Phi : \mathrm{GL}(n, \mathbb{F}_q) \to \mathcal{B}$, where $\mathcal{B}$ is the set of all bases of V. Choose, once for all, a basis $e_1, \ldots, e_n$ of V. If $T \in \mathrm{GL}(n, \mathbb{F}_q)$, define

$$\Phi(T) = Te_1, \ldots, Te_n.$$

By Lemma 4.21, $\Phi(T) \in \mathcal{B}$ because T, being nonsingular, carries a basis into a basis. But Φ is a bijection, for given a basis $v_1, \ldots, v_n$, there is a unique linear transformation S, necessarily nonsingular (by Lemma 4.21), with $Se_i = v_i$ for all i (by Theorem 4.14).

Our problem now is to count the number of bases $v_1, \ldots, v_n$ of V. There are q^n vectors in V, and so there are $q^n - 1$ candidates for v_1 (the zero vector is not a candidate). Having chosen v_1, we see that the candidates for v_2 are those vectors not in $\langle v_1 \rangle$, the subspace spanned by v_1; there are thus $q^n - q$ candidates for v_2. More generally, having chosen a linearly independent list $v_1, \ldots, v_i$, then v_{i+1} can be any vector not in $\langle v_1, \ldots, v_i \rangle$. Thus, there are $q^n - q^i$ candidates for v_{i+1}. The result follows by induction on n. $\bullet$

Theorem 5.27.   *If p is a prime and $q = p^m$, then the unitriangular group $\mathrm{UT}(n, \mathbb{F}_q)$ is a Sylow p-subgroup of $\mathrm{GL}(n, \mathbb{F}_q)$.*

Proof.   Since $q^n - q^i = q^i(q^{n-i} - 1)$, the highest power of p dividing $|\mathrm{GL}(n, \mathbb{F}_q)|$ is

$$qq^2q^3 \cdots q^{n-1} = q^{\binom{n}{2}}.$$

But $|\mathrm{UT}(n, \mathbb{F}_q)| = q^{\binom{n}{2}}$, and so it must be a Sylow p-subgroup. $\bullet$

Corollary 5.28.   *If p is a prime, then every finite p-group G is isomorphic to a subgroup of the unitriangular group $\mathrm{UT}(m, \mathbb{F}_p)$ for some m.*

Proof.   We show first, for every $m \geq 1$, that the symmetric group S_m can be imbedded in $\mathrm{GL}(m, k)$, where k is a field. Let V be an m-dimensional vector space over k, and let $v_1, \ldots, v_m$ be a basis of V. Define a function $\varphi : S_m \to \mathrm{GL}(V)$ by $\sigma \mapsto T_\sigma$, where $T_\sigma : v_i \mapsto v_{\sigma(i)}$ for all i. It is easy to see that φ is an injective homomorphism.

By Cayley's theorem, G can be imbedded in S_G; hence, G can be imbedded in $\mathrm{GL}(m, \mathbb{F}_p)$, where $m = |G|$. Now G is contained in some Sylow p-subgroup P of $\mathrm{GL}(m, \mathbb{F}_p)$, for every p-subgroup lies in some Sylow p-subgroup. Since all Sylow p-subgroups are conjugate, there is $a \in \mathrm{GL}(m, \mathbb{F}_p)$ with $P = a\left(\mathrm{UT}(m, \mathbb{F}_p)\right) a^{-1}$. Therefore,

$$G \cong aGa^{-1} \leq a^{-1}Pa \leq \mathrm{UT}(m, \mathbb{F}_p). \quad \bullet$$

A natural question is to find the Sylow subgroups of symmetric groups. This can be done, and the answer is in terms of a construction called *wreath product*.

EXERCISES

5.9 Prove that S_4 has more than one Sylow 2-subgroup.

5.10 Give an example of a finite group G having 3 Sylow p-subgroups (for some prime p) P, Q and R such that $P \cap Q = \{1\}$ and $P \cap R \neq \{1\}$.

5.11 **(Frattini Argument).** Let K be a normal subgroup of a finite group G. If P is a Sylow p-subgroup of K for some prime p, prove that

$$G = KN_G(P),$$

where $KN_G(P) = \{ab : a \in K \text{ and } b \in N_G(P)\}$.

5.12 Prove that $\mathrm{UT}(3, 2) \cong \mathbf{Q}$.

5.13 Show that a Sylow 2-subgroup of S_6 is isomorphic to $D_8 \times \mathbb{Z}_2$.

5.14 Let Q be a normal p-subgroup of a finite group G. Prove that $Q \leq P$ for every Sylow p-subgroup P of G.

5.15 For each prime divisor p of the order of a finite group G, choose a Sylow p-subgroup Q_p. Prove that $G = \left\langle \bigcup_p Q_p \right\rangle$.

5.16   (i) Let G be a finite group and let P be a Sylow p-subgroup of G. If $H \triangleleft G$, prove that HP/H is a Sylow p-subgroup of G/H and $H \cap P$ is a Sylow p-subgroup of H.

(ii) Let P be a Sylow p-subgroup of a finite group G. Give an example of a subgroup H of G with $H \cap P$ not a Sylow p-subgroup of H.

5.17 Prove that a Sylow 2-subgroup of A_5 has exactly 5 conjugates.

5.18 Prove that there are no simple groups of order 300, 312, 616, or 1000.

5.19 Prove that if every Sylow subgroup of a finite group G is normal, then G is the direct product of its Sylow subgroups.

5.20 For any group G, prove that if $H \triangleleft G$, then $Z(H) \triangleleft G$.

5.3 THE JORDAN-HÖLDER THEOREM

Galois introduced groups to investigate polynomials in $k[x]$, where k is a field of characteristic 0, and he saw that such a polynomial is solvable by radicals if and only if its Galois group is a solvable group. Solvable groups are an interesting family of groups in their own right, and we now examine them a bit more.

Definition.   A *normal series* of a group G is a finite sequence of subgroups $G = G_0, G_1, G_2, \ldots, G_n = \{1\}$ with

$$G = G_0 \geq G_1 \geq G_2 \geq \cdots \geq G_n = \{1\}$$

and $G_{i+1} \triangleleft G_i$ for all i. The *factor groups* of the series are the groups G_0/G_1, $G_1/G_2, \ldots, G_{n-1}/G_n$, and the *length* of the series is the number of strict inclusions (equivalently, the length is the number of nontrivial factor groups).

A group G is *solvable* if it has a normal series whose factor groups are cyclic of prime order.

We begin with a technical result that generalizes the second isomorphism theorem, for we will want to compare different normal series of a group.

Lemma 5.29 (Zassenhaus Lemma).   *Given four subgroups $A \triangleleft A^*$ and $B \triangleleft B^*$ of a group G, then $A(A^* \cap B) \triangleleft A(A^* \cap B^*)$, $B(B^* \cap A) \triangleleft B(B^* \cap A^*)$, and there is an isomorphism*

$$\frac{A(A^* \cap B^*)}{A(A^* \cap B)} \cong \frac{B(B^* \cap A^*)}{B(B^* \cap A)}.$$

Remark.   The isomorphism is symmetric in the sense that the right side is obtained from the left by interchanging the symbols A and B.   ◄

Proof.   Since $A \lhd A^*$, we have $A \lhd A^* \cap B^*$, and so

$$A \cap B^* = A \cap (A^* \cap B^*) \lhd A^* \cap B^*,$$

by the second isomorphism theorem; similarly,

$$A^* \cap B \lhd A^* \cap B^*.$$

Therefore, the subset D, defined by

$$D = (A \cap B^*)(A^* \cap B),$$

is a normal subgroup of $A^* \cap B^*$, because it is generated by two normal subgroups.

Using the symmetry in the remark, it suffices to show that there is an isomorphism

$$\frac{A(A^* \cap B^*)}{A(A^* \cap B)} \to \frac{A^* \cap B^*}{D}.$$

Define $\varphi : A(A^* \cap B^*) \to (A^* \cap B^*)/D$ by

$$\varphi : ax \mapsto xD,$$

where $a \in A$ and $x \in A^* \cap B^*$. Note that φ is single-valued, for if $ax = a'x'$, where $a' \in A$ and $x' \in A^* \cap B^*$, then

$$(a')^{-1}a = x'x^{-1} \in A \cap (A^* \cap B^*) = A \cap B^* \leq D.$$

It is routine to check that φ is surjective and that $\ker \varphi = A(A^* \cap B)$, and so the first isomorphism theorem completes the proof.   •

The reader should check that the Zassenhaus lemma implies the second isomorphism theorem: If S and T are subgroups of a group G with $T \lhd G$, then $TS/T \cong S/(S \cap T)$; set $A^* = G$, $A = T$, $B^* = S$, and $B = S \cap T$.

Definition.   A *composition series* is a series all of whose nontrivial factor groups are simple. The factor groups of a composition series are called *composition factors* of G.

A group need not have a composition series; for example, the abelian group $\mathbb{Z}$ has no composition series. However, every finite group does have a composition series.

Proposition 5.30.   *Every finite group G has a composition series.*

Proof.   If the proposition is false, let G be a least criminal; that is, G is a finite group of smallest order which does not have a composition series. Now G is not simple, otherwise $G > \{1\}$ is a composition series. Hence, G has a proper normal subgroup H; we may assume that H is a maximal normal subgroup, so that G/H is a simple group. But $|H| < |G|$, so that H does have a composition series: say, $H = H_0 > H_1 > \cdots > \{1\}$, and $G > H_0 > H_1 > \cdots > \{1\}$ is a composition series for G, a contradiction.   •

A group G is solvable if it has a normal series with factor groups cyclic of prime order. As cyclic groups of prime order are simple groups, a normal series as in the definition of solvable group is a composition series, and so composition factors of G are cyclic groups of prime order.

Here are two composition series of $G = \langle a \rangle$, a cyclic group of order 30 (note that normality of subgroups is automatic because G is abelian). The first is

$$G = \langle a \rangle \geq \langle a^2 \rangle \geq \langle a^6 \rangle \geq \{1\};$$

the factor groups of this series are $\langle a \rangle / \langle a^2 \rangle \cong \mathbb{Z}_2$, $\langle a^2 \rangle / \langle a^6 \rangle \cong \mathbb{Z}_3$, and $\langle a^6 \rangle / \{1\} \cong \langle a^6 \rangle \cong \mathbb{Z}_5$. Another normal series is

$$G = \langle a \rangle \geq \langle a^5 \rangle \geq \langle a^{10} \rangle \geq \{1\};$$

the factor groups of this series are $\langle a \rangle / \langle a^5 \rangle \cong \mathbb{Z}_5$, $\langle a^5 \rangle / \langle a^{10} \rangle \cong \mathbb{Z}_2$, and $\langle a^{10} \rangle / \{1\} \cong \langle a^{10} \rangle \cong \mathbb{Z}_3$. Notice that the same factor groups arise, although the order in which they arise is different. We will see that this phenomenon always occurs: different composition series of the same group have the same factor groups. This is the *Jordan-Hölder theorem*, and the next definition makes its statement more precise.

Definition.   Two normal series of a group G are *equivalent* if there is a bijection between the sets of nontrivial factor groups of each so that corresponding factor groups are isomorphic.

The Jordan-Hölder theorem says that any two composition series of a group are equivalent. It will be more efficient to prove a more general theorem, due to Schreier.

Definition.   A *refinement* of a normal series is a normal series $G = N_0, N_1, \ldots, N_k = \{1\}$ having the original series as a subsequence.

In other words, a refinement of a normal series is a new normal series obtained from the original by inserting more subgroups.

Notice that a composition series admits only insignificant refinements; one can merely repeat terms (if G_i/G_{i+1} is simple, then it has no proper nontrivial normal subgroups and, hence, there is no intermediate subgroup L with $G_i > L > G_{i+1}$ and $L \lhd G_i$). More precisely, any refinement of a composition series is equivalent to the original composition series.

Theorem 5.31 (Schreier Refinement Theorem).   *Any two series $G = G_0$, $G_1, \ldots, G_n = \{1\}$ and $G = N_0, N_1, \ldots, N_k = \{1\}$ of a group G have equivalent refinements.*

Proof.   We insert a copy of the second series between each pair of adjacent terms in the first series. In more detail, for each $i \geq 0$, define

$$G_{ij} = G_{i+1}(G_i \cap N_j)$$

(this is a subgroup because $G_{i+1} \lhd G_i$). Note that

$$G_{i0} = G_{i+1}(G_i \cap N_0) = G_{i+1}G_i = G_i,$$

because $N_0 = G$, and that

$$G_{ik} = G_{i+1}(G_i \cap N_k) = G_{i+1},$$

because $N_k = \{1\}$. Therefore, the series of G_{ij} is a refinement of the series of G_i:

$$\cdots \geq G_i = G_{i0} \geq G_{i1} \geq G_{i2} \geq \cdots \geq G_{ik} = G_{i+1} \geq \cdots .$$

Similarly, there is a refinement of the second series arising from subgroups N_{pq}, where

$$N_{pq} = N_{p+1}(N_p \cap G_q).$$

Both refinements have nk terms. For each i, j, the Zassenhaus lemma applied to the four subgroups $G_{i+1} \lhd G_i$ and $N_{j+1} \lhd N_j$ gives an isomorphism

$$\frac{G_{i+1}(G_i \cap N_j)}{G_{i+1}(G_i \cap N_{j+1})} \cong \frac{N_{j+1}(N_j \cap G_i)}{N_{j+1}(N_j \cap G_{i+1})};$$

that is,

$$G_{ij}/G_{ij+1} \cong N_{ji}/N_{ji+1}.$$

The association $G_{ij}/G_{ij+1} \mapsto N_{ji}/N_{ji+1}$ is a bijection showing that the two refinements are equivalent.   •

Theorem 5.32 (Jordan-Hölder[4] Theorem).   *Any two composition series of a group G are equivalent. In particular, the length of a composition series, if one exists, is an invariant of G.*

Proof.   As we remarked above, any refinement of a composition series is equivalent to the original composition series. It now follows from Schreier's theorem that any two composition series are equivalent.   •

Here is a new proof of the Fundamental Theorem of Arithmetic.

Corollary 5.33.   *Every integer $n \geq 2$ has a factorization into primes, and the prime factors are uniquely determined by n.*

Proof.   Since the abelian group $\mathbb{Z}_n$ is finite, it has a composition series; let $S_1, \dots, S_t$ be the factor groups. Now an abelian group is simple if and only if it is of prime order, by Proposition 2.78; since $n = |\mathbb{Z}_n|$ is the product of the orders of the factor groups (see Exercise 5.24), we have proved that n is a product of primes. Moreover, the Jordan-Hölder theorem gives the uniqueness of the (prime) orders of the factor groups.   •

Example 5.5.
Let $G = \mathrm{GL}(2, \mathbb{F}_4)$ be the general linear group of all 2×2 nonsingular matrices with entries in the field $\mathbb{F}_4$ with 4 elements. Now $\det \colon G \to (\mathbb{F}_4)^\times$, where $(\mathbb{F}_4)^\times \cong \mathbb{Z}_3$ is the multiplicative group of nonzero elements of $\mathbb{F}_4$. Since $\ker \det = \mathrm{SL}(2, \mathbb{F}_4)$, the special linear group consisting of those matrices of determinant 1, there is a normal series

$$G = \mathrm{GL}(2, \mathbb{F}_4) \geq \mathrm{SL}(2, \mathbb{F}_4) \geq \{1\}.$$

The factor groups of this series are $\mathbb{Z}_3$ and $\mathrm{SL}(2, \mathbb{F}_4)$. It is true that $\mathrm{SL}(2, \mathbb{F}_4)$ is simple, and so this series is a composition series. We cannot yet conclude that G is not solvable, for the definition of solvability requires that there be some composition series, not necessarily this one, having factor groups of prime order. However, the Jordan-Hölder theorem says that if one composition series of G has all its factor groups of prime order, then so does every other composition series. We may now conclude that $\mathrm{GL}(2, \mathbb{F}_4)$ is not a solvable group.   ◄

We now discuss the import of the Jordan-Hölder theorem in group theory.

---

[4]In 1868, C. Jordan proved that the orders of the factor groups of a composition series depend only on G and not upon the composition series; in 1889, O. Hölder proved that the factor groups themselves, to isomorphism, do not depend upon the composition series.

Definition.   If G is a group and $K \lhd G$, then one calls G an *extension* of K by G/K.

With this terminology, Exercise 2.89 says that an extension of one p-group by another p-group is itself a p-group, while Proposition 5.35 (proved below) says that an extension of one solvable group by another solvable group is itself solvable.

The study of extensions involves the inverse question: How much of G can be recovered from a normal subgroup K and the quotient $Q = G/K$? For example, we do know that if K and Q are finite, then $|G| = |K||Q|$.

Example 5.6.
(i) The direct product $K \times Q$ is an extension of K by Q (and $K \times Q$ is an extension of Q by K).

(ii) Both S_3 and $\mathbb{Z}_6$ are extensions of $\mathbb{Z}_3$ by $\mathbb{Z}_2$. On the other hand, $\mathbb{Z}_6$ is an extension of $\mathbb{Z}_2$ by $\mathbb{Z}_3$, but S_3 is not, for S_3 contains no normal subgroup of order 2.   ◀

We have just seen, for any given pair of groups, that an extension of one by the other always exists (the direct product), but it may not be unique to isomorphism. Hence, if we view an extension of K by Q as a "product" of K and Q, then this product is not single-valued. The *extension problem* is to classify all possible extensions of a given pair of groups K and Q.

Suppose that a group G has a normal series

$$G = K_0 \geq K_1 \geq K_2 \geq \cdots \geq K_{n-1} \geq K_n = \{1\}$$

with factor groups $Q_1, \ldots, Q_n$, where

$$Q_i = K_{i-1}/K_i$$

for all $i \geq 1$. Now $K_n = \{1\}$, so that $K_{n-1} = Q_n$, but something more interesting occurs next: $K_{n-2}/K_{n-1} = Q_{n-1}$, so that K_{n-2} is an extension of K_{n-1} by Q_{n-1}. If we could solve the extension problem, then we could recapture K_{n-2} from K_{n-1} and Q_{n-1} – that is, from Q_n and Q_{n-1}. Next, observe that $K_{n-3}/K_{n-2} = Q_{n-2}$, so that K_{n-3} is an extension of K_{n-2} by Q_{n-2}. If we could solve the extension problem, then we could recapture K_{n-3} from K_{n-2} and Q_{n-2}; that is, we could recapture K_{n-3} from Q_n, Q_{n-1}, and Q_{n-2}. Climbing up the composition series in this way, we end with $G = K_0$ being recaptured from $Q_n, Q_{n-1}, \ldots, Q_1$. Thus, G is a "product" of the factor groups. If the normal series is a composition series, then the Jordan-Hölder theorem says that the factors in this product (that is, the composition factors of G) are uniquely determined by G. Therefore, we could survey all finite groups if we knew the finite simple groups and if we could solve the extension problem. Now all the finite simple groups were classified in the 1980s; this theorem is one of the deepest theorems in mathematics; it does no less than give a complete list of every finite simple group (along with interesting properties of them). In a sense, the extension problem has also been solved. There is a solution to the extension problem, due to Schreier, which gives all possible multiplication tables for extensions; this study leads to **cohomology of groups** and, ultimately, to an upper bound for the number of possible Gs. On the other hand, no one knows a way, given K and Q, to compute the number of nonisomorphic extensions G of K by Q.

We now pass from general groups whose composition factors are arbitrary simple groups to solvable groups whose composition factors are cyclic groups of prime order (they are simple in every sense of the word). Even though solvable groups arose in determining those polynomials which are solvable by radicals, there are purely group-theoretic theorems about solvable groups making no direct reference to Galois theory and polynomials. For example, a theorem of P. Hall generalizes the Sylow theorems as follows: If G is a solvable group of order ab, where a and b are relatively prime, then G contains a subgroup of order a; moreover, any two such subgroups are conjugate. A theorem of Burnside says that if $|G| = p^m q^n$, where p and q are prime, then G is solvable. The remarkable **Feit-Thompson theorem** states that every group of odd order must be solvable.

In Chapter 4, we proved that every quotient of a solvable group is solvable (Theorem 4.60); we now prove that solvability is inherited by subgroups.

Lemma 5.34. *Every subgroup H of a solvable group G is itself a solvable group.*

Proof. Since G is solvable, there is a sequence of subgroups

$$G = G_0 \geq G_1 \geq G_2 \geq \cdots \geq G_t = \{1\}$$

with G_i normal in G_{i-1} and G_{i-1}/G_i cyclic, for all i. Consider the sequence of subgroups

$$H = H \cap G_0 \geq H \cap G_1 \geq H \cap G_2 \geq \cdots \geq H \cap G_t = \{1\}.$$

If $h \in H \cap G_i$ and $g \in H \cap G_{i-1}$, then $ghg^{-1} \in H \cap G_i$ because G_i is normal in G_{i-1}. Finally, the second isomorphism theorem, Theorem 2.55, gives

$$(H \cap G_{i-1})/(H \cap G_i) = (H \cap G_{i-1})/[(H \cap G_{i-1}) \cap G_i]$$
$$\cong G_i(H \cap G_{i-1})/G_i.$$

But the last (quotient) group is a subgroup of G_{i-1}/G_i. Since every sub-group of a cyclic group is itself cyclic, by Exercise 2.50, it follows that $(H \cap G_{i-1})/(H \cap G_i)$ is cyclic for every i. Therefore, H is a solvable group.   •

Here is another way to manufacture solvable groups.

Proposition 5.35.   *Every extension of one solvable group by another solv-able group is itself solvable: if $H \lhd G$ and if both H and G/H are solvable groups, then G is solvable.*

Proof.   Since G/H is solvable, there is a normal series

$$G/H \geq K_1^* \geq K_2^* \geq \cdots K_m^* = \{1\}$$

having factor groups of prime order. By the correspondence theorem for groups, there are subgroups K_i of G,

$$G \geq K_1 \geq K_2 \geq \cdots \geq K_m = H,$$

with $K_i/H = K_i^*$ and $K_{i+1} \lhd K_i$ for all i. By the third isomorphism theorem,

$$K_i^*/K_{i+1}^* \cong K_i/K_{i+1}$$

for all i, and so K_i/K_{i+1} is cyclic of prime order for all i.
   Since H is solvable, there is a normal series

$$H \geq H_1 \geq H_2 \geq \cdots H_q = \{1\}$$

having factor groups of prime order. Splice these two series together,

$$G \geq K_1 \geq K_2 \geq \cdots \geq K_m \geq H_1 \geq H_2 \geq \cdots H_q = \{1\},$$

to obtain a normal series of G having factor groups of prime order.   •

Corollary 5.36.   *If H and K are solvable groups, then $H \times K$ is solvable.*

Proof.   If $G = H \times K$, then $H \lhd G$ and $G/H \cong K$.   •

Corollary 5.37.   *Every finite p-group G is solvable.*

Proof.   If G is abelian, then G is solvable. Otherwise, its center, $Z(G)$, is a proper nontrivial normal abelian subgroup, by Theorem 2.75. Now $Z(G)$ is solvable, because it is abelian, and $G/Z(G)$ is solvable, by induction on $|G|$, and so G is solvable, by the proposition.   •

It follows, of course, that a direct product of finite p-groups is solvable.

Definition.   A *commutator* in a group G is an element of the form

$$[x, y] = xyx^{-1}y^{-1}.$$

If X and Y are subsets of a group G, then $[X, Y]$ is defined by

$$[X, Y] = \langle [x, y] : x \in X \text{ and } y \in Y \rangle.$$

In particular, the *commutator subgroup* G' of a group G is

$$G' = [G, G],$$

the subgroup generated by all the commutators[5].

It is clear that two elements x and y in a group G commute if and only if their commutator $[x, y]$ is 1. The next proposition generalizes this observation.

Proposition 5.38.   *Let G be a group.*

(i) *The commutator subgroup G' is a normal subgroup of G, and G/G' is abelian.*

(ii) *If $H \lhd G$ and G/H is abelian, then $G' \leq H$.*

Proof.   (i) The inverse of a commutator $xyx^{-1}y^{-1}$ is itself a commutator, namely, $yxy^{-1}x^{-1}$; that is, $[x, y]^{-1} = [y, x]$. Therefore, each element of G' is a product of commutators. But any conjugate of a commutator (and hence, a product of commutators) is another such:

$$a[x, y]a^{-1} = a(xyx^{-1}y^{-1})a^{-1}$$
$$= axa^{-1}aya^{-1}ax^{-1}a^{-1}ay^{-1}a^{-1}$$
$$= [axa^{-1}, aya^{-1}].$$

---

[5]The subset consisting of all the commutators need not be closed under products, and so the *set* of all commutators may not be a subgroup. The smallest group in which a product of two commutators is not a commutator has order 96. Also, see Carmichael's exercise on page 418.

Therefore, $G' \lhd G$.

If $aG', bG' \in G/G'$, then

$$aG'bG'(aG')^{-1}(bG')^{-1} = aba^{-1}b^{-1}G' = [a, b]G' = G',$$

and so G/G' is abelian.

(ii) Suppose that $H \lhd G$ and G/H is abelian. If $a, b \in G$, then $aHbH = bHaH$; that is, $abH = baH$, and so $b^{-1}a^{-1}ba \in H$. As every commutator has the form $b^{-1}a^{-1}ba$, we have $G' \leq H$.   •

Example 5.7.

(i) A group G is abelian if and only if $G' = \{1\}$.

(ii) If G is a simple group, then $G' = \{1\}$ or $G' = G$, for G' is a normal subgroup. The first case occurs when G has prime order; the second case occurs otherwise. In particular, $(A_n)' = A_n$ for all $n \geq 5$.

(iii) We show that $(S_n)' = A_n$ for all $n \geq 5$. Let $\alpha, \beta \in S_n$, so that α is a product of, say, ℓ transpositions and β is a product of m transpositions. It follows that the commutator $[\alpha, \beta] = \alpha\beta\alpha^{-1}\beta^{-1}$ is a product of $2(\ell + m)$ transpositions; that is, $[\alpha, \beta]$ is even, and so $[\alpha, \beta] \in A_n$. Hence, $(S_n)' \leq A_n$. For the reverse inclusion, note that $(S_n)' \cap A_n \lhd A_n$, so that the simplicity of A_n gives this intersection trivial or A_n. Clearly, $(S_n)' \cap A_n \neq \{(1)\}$, and so $A_n \leq (S_n)'$.   ◄

Let us iterate the formation of the commutator subgroup.

Definition.   The *derived series* of G is

$$G = G^{(0)} \geq G^{(1)} \geq G^{(2)} \geq \cdots \geq G^{(i)} \geq G^{(i+1)} \geq \cdots,$$

where $G^{(0)} = G$, $G^{(1)} = G'$, and, more generally, $G^{(i+1)} = (G^{(i)})' = [G^{(i)}, G^{(i)}]$ for all $i \geq 0$.

It is easy to prove, by induction, that $G^{(i)} \lhd G$ for every i; it follows that $G^{(i+1)} \lhd G^{(i)}$, and so the derived series is a normal series. The derived series can be used to give a characterization of solvability: G is solvable if and only if the derived series reaches $\{1\}$.

Proposition 5.39.

(i) *A finite group G is solvable if and only if it has a normal series with abelian factor groups.*

(ii) *A finite group G is solvable if and only if there is some n with*

$$G^{(n)} = \{1\}.$$

Proof.   (i) If G is solvable, then it has a normal series whose factor groups G_i/G_{i+1} are all cyclic of prime order, hence are abelian.

Conversely, if G has a normal series with abelian factor groups, then the factor groups of any refinement are also abelian. In particular, the factor groups of a composition series of G, which exists because G is finite, are abelian simple groups; hence, they are cyclic of prime order, and so G is solvable.

(ii) Assume that G is solvable, so there is a normal series

$$G \geq G_1 \geq G_2 \geq \cdots \geq G_n = \{1\}$$

whose factor groups G_i/G_{i+1} are abelian. We show, by induction on $i \geq 0$, that $G^{(i)} \leq G_i$. Since $G^{(0)} = G = G_0$, the base step is obviously true. For the inductive step, since G_i/G_{i+1} is abelian, Proposition 5.38 gives $(G_i)' \leq G_{i+1}$. On the other hand, the inductive hypothesis gives $G^{(i)} \leq G_i$, which implies that

$$G^{(i+1)} = (G^{(i)})' \leq (G_i)' \leq G_{i+1}.$$

In particular, $G^{(n)} \leq G_n = \{1\}$, which is what we wished to show.

Conversely, if $G^{(n)} = \{1\}$, then the derived series is a normal series (a normal series must end with $\{1\}$) with abelian factor groups, and so part (i) gives G solvable.   •

For example, the derived series of $G = S_4$ is easily seen to be

$$S_4 > A_4 > \mathbf{V} > \{(1)\}.$$

Our earlier definition of solvability applies only to finite groups, whereas the characterization in the proposition makes sense for all groups, possibly infinite. Nowadays, most authors define a group to be solvable if its derived series reaches $\{1\}$ after a finite number of steps; with this new definition, every abelian group is solvable, whereas it is easy to see that abelian groups are solvable (in the original definition of the term) if and only if they are finite. In the exercises, the reader will be asked to prove, using the criterion in Proposition 5.39, that subgroups and quotient groups of solvable groups are also solvable (in the new, generalized, sense).

There are other interesting classes of groups defined in terms of normal series. One of the most interesting such consists of the *nilpotent* groups, The finite nilpotent groups are characterized as those groups which are direct products of their Sylow subgroups.

EXERCISES

5.21 Let p be an odd prime and let G be a nonabelian group of order p^3 and with $x^p = 1$ for all $x \in G$. Prove that $Z(G) = G'$.

5.22 Find the commutator subgroup $(S_n)'$ for $n \leq 4$.

5.23 Prove that if H is a subgroup of a group G and $G' \leq H$, then $H \lhd G$.

5.24 If G is a finite group and

$$G = G_0 \geq G_1 \geq \cdots \geq G_n = \{1\}$$

is a normal series, prove that the order of G is the product of the orders of the factor groups:

$$|G| = \prod_{i=0}^{n-1} |G_i/G_{i+1}|.$$

5.25 This exercise asks for new proof of theorems already proved in the text.

   (i) Prove that if $H \leq G$, then $H^{(i)} \leq G^{(i)}$ for all i. Conclude, using Proposition 5.39, that every subgroup of a solvable group is solvable.

   (ii) Prove that if $f : G \to K$ is a surjective homomorphism, then

$$f(G^{(i)}) \leq K^{(i)}$$

   for all i. Conclude, using Proposition 5.39, that every quotient of a solvable group is also solvable.

   (iii) For every group G, prove, by double induction, that

$$G^{(m+n)} = (G^{(m)})^{(n)}.$$

   (iv) Prove, using Proposition 5.39, that if $H \lhd G$ and both H and G/H are solvable, then G is solvable.

5.26 Let p and q be primes.

   (i) Prove that every group of order pq is solvable.

   (ii) Prove that every group G of order p^2q is solvable.

5.27 Show that the Feit-Thompson theorem: Every finite group of odd order is solvable, is equivalent to: Every nonabelian finite simple group has even order.

5.28   (i) Prove that the infinite cyclic group $\mathbb{Z}$ does not have a composition series.

(ii) Prove that an abelian group G has a composition series if and only if G is finite.

5.29 Prove that if G is a finite group and $H \lhd G$, then there is a composition series of G one of whose terms is H.

5.30   (i) Prove that if S and T are solvable subgroups of a group G and $S \lhd G$, then ST is a solvable subgroup of G.

(ii) If G is a finite group, define $\mathcal{S}(G)$ to be the subgroup of G generated by all normal solvable subgroups of G. Prove that $\mathcal{S}(G)$ is the unique maximal normal solvable subgroup of G and that $G/\mathcal{S}(G)$ has no nontrivial normal solvable subgroups.

5.31   (i) Prove that the dihedral groups D_{2n} are solvable.

(ii) Give a composition series for D_{2n}.

5.4 PRESENTATIONS

How can one describe a group? By Cayley's theorem, a finite group G is (isomorphic to) a subgroup of the symmetric group S_n, where $n = |G|$, and so G can be defined as the subgroup of S_n generated by certain permutations. An example of this kind of construction occurs in the following exercise from Carmichael's group theory book:

Let G be the subgroup of S_{16} generated by the following permutations:

$$(a\ c)(b\ d); \quad (e\ g)(f\ h);$$
$$(i\ k)(j\ \ell); \quad (m\ o)(n\ p)$$
$$(a\ c)(e\ g)(i\ k); \quad (a\ b)(c\ d)(m\ o);$$
$$(e\ f)(g\ h)(m\ n)(o\ p); \quad (i\ j)(k\ \ell).$$

Prove that $|G| = 256$, $|G'| = 16$,

$$\alpha = (i\ k)(j\ \ell)(m\ o)(n\ p) \in G',$$

but α is not a commutator.

A second way of describing a group is by replacing S_n by $GL(n, k)$ for some $n \geq 2$ and some field k [remember that all $n \times n$ permutation matrices form a subgroup of $GL(n, k)$ isomorphic to S_n, and so every group of order n can be imbedded in $GL(n, k)$]. We have already described some groups in terms of matrices; for example, we defined the quaternion group $\mathbf{Q}$ in this way. For relatively small groups, descriptions in terms of permutations or matrices are useful, but when n is large, such descriptions are cumbersome.

One can also describe groups as being generated by elements subject to certain relations. For example, the dihedral group D_{2n} could be described as a group of order $2n$ that can be generated by two elements a and b, such that $a^n = 1 = b^2$ and $bab = a^{-1}$. Consider the following definition.

Definition.   The group of *generalized quaternions* $\mathbf{Q}_n$, where $n \geq 3$, is a group of order 2^n that is generated by two elements a and b such that

$$a^{2^{n-1}} = 1, \quad bab^{-1} = a^{-1}, \text{ and } b^2 = a^{2^{n-2}}.$$

When $n = 3$, this is the group $\mathbf{Q}$ of order 8. An obvious defect in this definition is that the existence of such a group is left in doubt; for example, is there such a group of order 16? Notice that it is not enough to find a group $G = \langle a, b \rangle$ in which $a^8 = 1$, $bab^{-1} = a^{-1}$, and $b^2 = a^4$. For example, the group $G = \langle a, b \rangle$ in which $a^2 = 1$ and $b = 1$ (which is, of course, cyclic of order 2) satisfies all of the equations.

It was W. von Dyck, in the 1880s, who invented *free groups* in order to make such descriptions rigorous.

Here is a modern definition of a free group.

Definition.   If X is a subset of a group F, then F is a *free group* with *basis* X if, for every group G and every function $f : X \to G$, there exists a unique homomorphism $\varphi : F \to G$ with $\varphi(x) = f(x)$ for all $x \in X$.

This definition is modeled on a fundamental result in linear algebra, Theorem 4.14, which is the reason why matrices describe linear transformations.

Theorem.   *Let $X = v_1, \ldots, v_n$ be a basis of a vector space V. If W is a vector space and $u_1, \ldots, u_n$ is a list in W, then there exists a unique linear transformation $T : V \to W$ with $T(v_i) = u_i$ for all i.*

We may draw a diagram of this theorem after we note that giving a list $u_1, \ldots, u_n$ of vectors in W is the same thing as giving a function $f : X \to W$, where $f(v_i) = u_i$; after all, a function $f : X \to W$ is determined by its values on $v_i \in X$.

If we knew that free groups exist, then we could define $\mathbf{Q}_n$ as follows. Let F be the free group with basis $X = \{x, y\}$, let R be the normal subgroup of F generated by $\{x^{2^{n-1}}, yxy^{-1}x, y^{-2}x^{2^{n-2}}\}$, and define $\mathbf{Q}_n = F/R$. It is clear that F/R is a group generated by two elements $a = xR$ and $b = yR$ which satisfy the relations in the definition; what is not clear is that F/R has order 2^n, and this needs proof (see Proposition 5.46).

The first question, then, is whether free groups exist. The idea of the construction is simple and natural, but checking the details is a bit fussy. We begin by describing the ingredients of a free group.

Let X be a nonempty set, and let X^{-1} be a disjoint replica of X; that is, X and X^{-1} are disjoint and there is a bijection $X \to X^{-1}$, which we denote by $x \mapsto x^{-1}$. Define the **alphabet** on X to be

$$X \cup X^{-1}.$$

If n is a positive integer, we define a **word** on X of **length** $n \geq 1$ to be a function $w : \{1, 2, \ldots, n\} \to X \cup X^{-1}$. In practice, we shall write a word w of length n as follows:

$$w = x_1^{e_1} \cdots x_n^{e_n},$$

where $x_i \in X$, $e_i = \pm 1$, and $w(i) = x_i^{e_i}$. The length n of a word w will be denoted by $|w|$. The **empty word**, denoted by 1, is a new symbol; the length of the empty word is defined to be 0.

The definition of equality of functions reads here as follows. If $u = x_1^{e_1} \cdots x_n^{e_n}$ and $v = y_1^{d_1} \cdots y_m^{d_m}$ are words, where $x_i, y_j \in X$ for all i, j, then $u = v$ if and only if $m = n$, $x_i = y_i$, and $e_i = d_i$ for all i; thus, every word has a unique spelling.

Definition. A *subword* of a word $w = x_1^{e_1} \cdots x_n^{e_n}$ is either the empty word or a word of the form $u = x_r^{e_r} \cdots x_s^{e_s}$, where $1 \leq r \leq s \leq n$.

A word w on X is *reduced* if $w = 1$ or if w has no subwords of the form xx^{-1} or $x^{-1}x$, where $x \in X$.

The *inverse* of a word $w = x_1^{e_1} \cdots x_n^{e_n}$ is $w^{-1} = x_n^{-e_n} \cdots x_1^{-e_1}$.

Any two words on X can be multiplied.

Definition. If $u = x_1^{e_1} x_2^{e_2} \cdots x_n^{e_n}$ and $v = y_1^{d_1} \cdots y_m^{d_m}$ are words on X, then their *juxtaposition* is the word

$$uv = x_1^{e_1} \cdots x_n^{e_n} y_1^{d_1} \cdots y_m^{d_m}.$$

If 1 is the empty word, then $1v = v$ and $u1 = u$.

The basic idea underlying the construction of a free group F with basis X is quite simple. The elements of F are the words on X, the operation is juxtaposition, the identity is the empty word 1, and the inverse of a word in the group is its inverse as defined above. But there is a problem: we want $x^{-1}x = 1$, but this is not so; $x^{-1}x$ has length 2, not length 0. We can try to remedy this by restricting the elements of F to be *reduced* words on X; but, even if u and v are reduced, their juxtaposition uv may not be reduced. Of course, one can reduce uv, that is, do all the cancellation to convert uv into a reduced word, but now it is tricky to prove associativity. We solve this problem as follows. Since words such as $zx^{-1}xyzx^{-1}$ and $zyzx^{-1}$, for example, must be identified, it is reasonable to impose an equivalence relation on the set of all the words on X. If we define the elements of F to be the equivalence classes, then associativity can be proved without much difficulty, and it turns out that there is a unique reduced word in each equivalence class. Therefore, one can regard the elements of F as reduced words and the product of two elements as their juxtaposition followed by reduction.

The casual reader may accept the existence of free groups as described above and proceed to Proposition 5.43 on page 428; here are the details for everyone else.

Definition. Let A and B be words on X, possibly empty, and let $w = AB$. An *elementary operation* is either an *insertion*, which changes $w = AB$ to $Aaa^{-1}B$ for some $a \in X \cup X^{-1}$, or a *deletion* of a subword of w of the form aa^{-1}. We write

$$w \to w'$$

to denote w' arising from w by an elementary operation. Two words u and v on X are **equivalent**, denoted by $u \sim v$, if there are words $u = w_1, w_2, \dots, w_n = v$ and elementary operations

$$u = w_1 \to w_2 \to \cdots \to w_n = v.$$

Denote the equivalence class of a word w by $[w]$.

Note that $xx^{-1} \sim 1$ and $x^{-1}x \sim 1$.

We construct free groups in two stages.

Definition. A **semigroup** is a set having an associative operation; a **monoid** is a semigroup S having an identity element 1; that is, $1s = s = s1$ for all $s \in S$. If S and S' are semigroups, then a **homomorphism** is a function $f : S \to S'$ such that $f(xy) = f(x)f(y)$; if S and S' are monoids, then a **homomorphism** $f : S \to S'$ must also satisfy $f(1) = 1$.

Of course, every group is a monoid, and a homomorphism between groups is a homomorphism of them *qua* monoids.

Example 5.8.

(i) The natural numbers $\mathbb{N}$ form a commutative monoid under addition.

(ii) A direct product of monoids is again a monoid (with cooordinatewise operation). In particular, the set $\mathbb{N}^n$ of all n-tuples of natural numbers is a commutative additive monoid. ◄

Here is an example of a noncommutative monoid.

Lemma 5.40. *Let X be a set, and let $\mathcal{W}(X)$ be the set of all words on X (if $X = \varnothing$, then $\mathcal{W}(X)$ consists of only the empty word).*

(i) *$\mathcal{W}(X)$ is a monoid under juxtaposition.*

(ii) *If $u \sim v$ and $u' \sim v'$, then $uv \sim u'v'$.*

(iii) *If G is a group and $f : X \to G$ is a function, then there is a homomorphism $\widetilde{f} : \mathcal{W}(X) \to G$ extending f such that $w \sim w'$ implies $\widetilde{f}(w) = \widetilde{f}(w')$ in G.*

Proof. (i) Associativity of juxtaposition is obvious once we note that there is no cancellation in $\mathcal{W}(X)$.

(ii) The elementary operations that take u to u', when applied to the word uv, give a chain taking uv to $u'v$; the elementary operations that take v to

v', when applied to the word $u'v$, give a chain taking $u'v$ to $u'v'$. Hence, $uv \sim u'v'$.

(iii) If $w = x_1^{e_1} \cdots x_n^{e_n}$, then define

$$\widetilde{f}(w) = f(x_1)^{e_1} f(x_2)^{e_2} \cdots f(x_n)^{e_n}.$$

That w has a unique spelling shows that $\widetilde{f}$ is a well-defined function, and it is obvious that $\widetilde{f} : \mathcal{W}(X) \to G$ is a homomorphism.

Let $w \sim w'$. We prove, by induction on the number of elementary operations in a chain from w to w', that $\widetilde{f}(w) = \widetilde{f}(w')$ in G. Consider the deletion $w = Aaa^{-1}B \to AB$, where A and B are subwords of w. That $\widetilde{f}$ is a homomorphism gives

$$\widetilde{f}(Aaa^{-1}B) = \widetilde{f}(A)\widetilde{f}(a)\widetilde{f}(a)^{-1}\widetilde{f}(B).$$

But

$$\widetilde{f}(A)\widetilde{f}(a)\widetilde{f}(a)^{-1}\widetilde{f}(B) = \widetilde{f}(A)\widetilde{f}(B) \text{ in } G,$$

because there is cancellation in the group G, so that $\widetilde{f}(Aaa^{-1}B) = \widetilde{f}(AB)$. A similar argument holds for insertions.   •

The next proposition will be used to prove that each element in a free group has a normal form.

Proposition 5.41.   *Every word w on a set X is equivalent to a unique reduced word.*

Proof.   If $X = \varnothing$, then there is only one word on X, the empty word 1, and 1 is reduced.

If $X \neq \varnothing$, we show first that there exists a reduced word equivalent to w. If w has no subword of the form aa^{-1}, where $a \in X \cup X^{-1}$, then w is reduced. Otherwise, delete the first such pair, producing a new word w_1, which may be empty, with $|w_1| < |w|$. Now repeat: if w_1 is reduced, stop; if there is a subword of w_1 of the form aa^{-1}, then delete it, producing a shorter word w_2. Since the lengths are strictly decreasing, this process ends with a reduced word which is equivalent to w.

To prove uniqueness, suppose, on the contrary, that u and v are distinct reduced words and there is a chain of elementary operations

$$u = w_1 \to w_2 \to \cdots \to w_n = v;$$

we may assume that n is minimal. Since u and v are both reduced, the first elementary operation is an insertion, while the last elementary operation is a deletion, and so there must be a first deletion, say, $w_i \to w_{i+1}$. Thus, the elementary operation $w_{i-1} \to w_i$ inserts aa^{-1} while the elementary operation $w_i \to w_{i+1}$ deletes bb^{-1}, where $a, b \in X \cup X^{-1}$.

There are three cases. If the subwords aa^{-1} and bb^{-1} of w_i coincide, then $w_{i-1} = w_{i+1}$, for w_{i+1} is obtained from w_{i-1} by first inserting aa^{-1} and then deleting it; hence, the chain

$$u = w_1 \to w_2 \to \cdots \to w_{i-1} = w_{i+1} \to \cdots \to w_n = v$$

is shorter than the original shortest chain. The second case has aa^{-1} and bb^{-1} overlapping subwords of w_i; this can happen in two ways. One way is

$$w_i = Aaa^{-1}b^{-1}C,$$

where A, C are subwords of w_i and $a^{-1} = b$; hence, $a = b^{-1}$ and

$$w_i = Aaa^{-1}aC.$$

Therefore, $w_{i-1} = AaC$, because we are inserting aa^{-1}, and $w_{i+1} = AaC$, because we are deleting $bb^{-1} = a^{-1}a$. Thus, $w_{i-1} = w_{i+1}$, and removing w_i gives a shorter chain. The second way an overlap can happen is $w_i = Aa^{-1}aa^{-1}C$, where $b^{-1} = a$. As in the first way, this leads to $w_{i-1} = w_{i+1}$. Finally, suppose that the subwords aa^{-1} and bb^{-1} do not overlap: $w_i = Aaa^{-1}Bbb^{-1}C$ and $w_{i+1} = Aaa^{-1}BC$. Now bb^{-1} became a subword of w_i by an earlier insertion: there is some $j < i$ with $w_{j-1} = XY$ and $w_j = Xbb^{-1}Y$ (since only insertions occur before the ith elementary operation, this subword could only have been changed to $bcc^{-1}b^{-1}$). Schematically, the subchain $w_{j-1} \to \cdots \to w_{i+1}$ looks like

$$XY \to Xbb^{-1}Y \to \cdots \to ABbb^{-1}C \to Aaa^{-1}Bbb^{-1}C \to Aaa^{-1}BC.$$

But we can shorten this chain by not inserting bb^{-1}:

$$XY \to \cdots \to ABC \to Aaa^{-1}BC.$$

In all cases, we are able to shorten the shortest chain, and so no such chain can exist. •

Theorem 5.42.   *If X is a set, then the set F of all equivalence classes of words on X with operation $[u][v] = [uv]$ is a free group with basis X.*

Moreover, every element in F has a normal form: for each $[u] \in F$, there is a unique reduced word w with $[u] = [w]$.

Proof. If $X = \varnothing$, then $\mathcal{W}(\varnothing)$ consists only of the empty word 1, and so $F = \{1\}$. The reader may show that this is, indeed, a free group on $\varnothing$.

Assume now that $X \neq \varnothing$. We have already seen, in Lemma 5.40(ii), that juxtaposition is compatible with the equivalence relation, and so the operation on F is well defined. The operation is associative, because of associativity in $\mathcal{W}(X)$:

$$
\begin{aligned}
[u]([v][w]) &= [u][vw] \\
&= [u(vw)] \\
&= [(uv)w] \\
&= [uv][w] \\
&= ([u][v])[w].
\end{aligned}
$$

The identity is the class $[xx^{-1}] = [1]$, the inverse of $[w]$ is $[w^{-1}]$, and so F is a group.

If $[w] \in F$, then

$$
[w] = [x_1^{e_1} \cdots x_n^{e_n}] = [x_1^{e_1}][x_2^{e_2}] \cdots [x_n^{e_n}],
$$

so that F is generated by X (if we identify each $x \in X$ with $[x]$). It follows from Proposition 5.41 that for every $[w]$, there is a unique reduced word u with $[w] = [u]$.

To prove that F is free with basis X, suppose that $f : X \to G$ is a function, where G is a group. Define $\varphi : F \to G$ by

$$
\varphi : [x_1^{e_1}][x_2^{e_2}] \cdots [x_n^{e_n}] \mapsto f(x_1)^{e_1} f(x_2)^{e_2} \cdots f(x_n)^{e_n},
$$

where $x_1^{e_1} \cdots x_n^{e_n}$ is reduced. Uniqueness of the reduced expression of a word shows that φ is a well-defined function (which obviously extends f). Notice the relation of φ to the homomorphism $\tilde{f} : \mathcal{W}(X) \to G$ in Lemma 5.40: when w is reduced,

$$
\varphi([w]) = \tilde{f}(w)
$$

It remains to prove that φ is a homomorphism (if so, it is the unique homomorphism extending f, because the subset X generates F). Let $[u], [v] \in F$, where u and v are reduced words, and let $uv \sim w$, where w is reduced. Now

$$\varphi([u][v]) = \varphi([w]) = \tilde{f}(w),$$

because w is reduced, and

$$\varphi([u])\varphi([v]) = \tilde{f}(u)\tilde{f}(v),$$

because u and v are reduced. Finally, $\tilde{f}(u)\tilde{f}(v) = \tilde{f}(w)$, by Lemma 5.40(iii), and so $\varphi([u][v]) = \varphi([u])\varphi([v])$.   •

We have proved, for every set X, that there exists a free group that is free with basis X. Moreover, the elements of a free group F on X may be regarded as reduced words and the operation may be regarded as juxtaposition followed by reduction; brackets are no longer used, and the elements $[w]$ of F are written as w.

The free group F with basis X that we have just constructed is generated by X. Are any two free groups with basis X isomorphic?

Proposition 5.43.

   (i) *Let X_1 be a basis of a free group F_1 and let X_2 be a basis of a free group F_2. If there is a bijection $f : X_1 \to X_2$, then there is an isomorphism $\varphi : F_1 \to F_2$ extending f.*

   (ii) *If F is a free group with basis X, then F is generated by X.*

Proof.   (i) The diagram below, in which the vertical arrows are inclusions, will help the reader to follow the proof.

$$
\begin{array}{ccc}
F_1 & \xrightarrow{\ \varphi\ } & F_2 \\
\uparrow & & \uparrow \\
X_1 & \underset{f^{-1}}{\overset{f}{\rightleftarrows}} & X_2.
\end{array}
$$

We may regard f as having target F_2, because $X_2 \subset F_2$; since F_1 is a free group with basis X_1, there is a homomorphism $\varphi_1 : F_1 \to F_2$ extending f. Similarly, there exists a homomorphism $\varphi_2 : F_2 \to F_1$ extending f^{-1}. The composite $\varphi_2\varphi_1 : F_1 \to F_1$ is thus a homomorphism extending 1_X. But the identity 1_{F_1} also extends 1_X, so that uniqueness of the extension gives $\varphi_2\varphi_1 = 1_{F_1}$. In the same way, we see that the other composite $\varphi_1\varphi_2 = 1_{F_2}$, and so φ_1 is an isomorphism.

(ii) Let there be a bijection $f: X_1 \to X$ for some set X_1. If F_1 is the free group with basis X_1 constructed in Theorem 5.42, then X_1 generates F_1. By part (i), there is an isomorphism $\varphi: F_1 \to F$ with $\varphi(X_1) = X$. But if X generates F, then $\varphi(X_1)$ generates $\operatorname{im} \varphi$; that is, X generates F.   •

Remark.   There is an analog for free groups of the dimension of a vector space: any two bases of a free group F have the same number of elements. This number is called the ***rank*** of F, and one can prove that two free groups are isomorphic if and only if they have the same rank.   ◄

Corollary 5.44.   *Every group G is a quotient of a free group.*

Proof.   Let X be a set for which there exists a bijection $f: X \to G$ (for example, one could take X to be the underlying set of G and $f = 1_G$), and let F be the free group with basis X. There exists a homomorphism $\varphi: F \to G$ extending f, and φ is surjective because f is. Therefore, $G \cong F/\ker \varphi$.   •

Let us return to describing groups.

Definition.   A *presentation* of a group G is an ordered pair

$$G = (X \mid R),$$

where F is a free group with basis X, R is a set of words on X, and $F/N \cong G$, where N is the normal subgroup of F generated by R.

One calls the set X **generators**[6] and the set R **relations**.

Corollary 5.44 says that every group has a presentation.

Definition.   A group G is *finitely generated* if it has a presentation $(X|R)$ with X finite. A group G is called *finitely presented* if it has a presentation $(X|R)$ in which both X and R are finite.

It is easy to see that a group G is finitely generated if and only if there exists a finite subset $A \subset G$ with $G = \langle A \rangle$. There do exist finitely generated groups that are not finitely presented.

---

[6]The term *generators* is now being used in a generalized sense, for X is not a subset of G. The subset $\{xN : x \in X\}$ of $G = F/N$ does generate G in the usual sense.

Remark.   One of the most fundamental results about free groups is the Nielsen-Schreier theorem: Every subgroup S of a free group F is free. One consequence of the proof of this theorem is that a subgroup of a finitely generated group need not be finitely generated. In fact, if F is a free group of rank $m \geq 2$, then its commutator subgroup F' is a free group which is not finitely generated. It follows, in particular, that F contains a free subgroup of rank $n > m$. Thus, there exist nonisomorphic finitely generated groups, namely, free groups of ranks m and n, respectively, each of which is isomorphic to a subgroup of the other.   ◄

Example 5.9.

(i) A group has many presentations. For example, $G = \mathbb{Z}_6$ has presentations

$$(x \mid x^6)$$

as well as

$$(a, b \mid a^3, b^2, aba^{-1}b^{-1}).$$

A fundamental problem is how to determine whether two presentations give isomorphic groups. It can be proved that no algorithm can exist that solves this problem.

(ii) The free group with basis X has a presentation

$$(X \mid \varnothing).$$

Indeed, a free group is so called precisely because it has a presentation with no relations.   ◄

A word on notation.   Often, one writes the relations in a presentation as equations. Thus, the relations

$$a^3, \quad b^2, \quad aba^{-1}b^{-1}$$

in the second presentation of $\mathbb{Z}_6$ may also be written

$$a^3 = 1, \quad b^2 = 1, \quad ab = ba.$$

If r is a word on $x_1, \dots, x_n$, we may write $r = r(x_1, \dots, x_n)$. If H is a group and $h_1, \dots, h_n \in H$, then $r(h_1, \dots, h_n)$ denotes the element in H obtained from r by replacing each x_i by h_i.

The next, elementary, result is quite useful.

Theorem 5.45 (von Dyck's Theorem).   *Let a group G have a presentation*

$$G = (x_1, \ldots, x_n \mid r_j, \ j \in J),$$

where $r_j = r_j(x_1, \ldots, x_n)$. If H is a group with $H = \langle h_1, \ldots, h_n \rangle$ and if $r_j(h_1, \ldots, h_n) = 1$ in H for all $j \in J$, then there is a surjective homomorphism $G \to H$ with $x_i R \mapsto h_i$ for all i.

Proof.   If F is the free group with basis $\{x_1, \ldots, x_n\}$, then there is a homomorphism $\varphi : F \to H$ with $\varphi(x_i) = h_i$ for all i. Since $r_j(h_1, \ldots, h_n) = 1$ in H for all $j \in J$, we have $r_j \in \ker \varphi$ for all $j \in J$. Therefore, φ induces a homomorphism $G = F/\ker \varphi \to H$ with $x_i R \mapsto h_i$ for all i.   •

The next proposition will show how von Dyck's theorem enters into the analysis of presentations, but we begin with the construction of a concrete group of matrices.

Example 5.10.
We are going to construct a group H_n that is a good candidate to be the generalized quaternion group $\mathbf{Q}_n$ for $n \geq 3$ on page 421. Consider the complex matrices

$$A = \begin{bmatrix} 0 & \omega \\ \omega & 0 \end{bmatrix} \text{ and } B = \begin{bmatrix} 0 & 1 \\ -1 & 0 \end{bmatrix},$$

where ω is a primitive 2^{n-1}th root of unity, and let $H_n = \langle A, B \rangle \leq \mathrm{GL}(2, \mathbb{C})$. We claim that A and B satisfy the relations in the definition of the generalized quaternion group. For all $i \geq 1$,

$$A^{2^i} = \begin{bmatrix} \omega^{2^i} & 0 \\ 0 & \omega^{2^i} \end{bmatrix},$$

so that $A^{2^{n-1}} = E$; indeed, A has order 2^{n-1}. Moreover,

$$B^2 = \begin{bmatrix} -1 & 0 \\ 0 & -1 \end{bmatrix} = A^{2^{n-2}} \text{ and } BAB^{-1} = \begin{bmatrix} 0 & -\omega \\ -\omega & 0 \end{bmatrix} = A^{-1}.$$

Notice that A and B do not commute; hence, $B \notin \langle A \rangle$, and so the cosets $\langle A \rangle$ and $B \langle A \rangle$ are distinct. Since A has order 2^{n-1}, it follows that

$$|H_n| \geq |\langle A \rangle \cup B \langle A \rangle| = 2^{n-1} + 2^{n-1} = 2^n.$$

The next theorem will show that $|H_n| = 2^n$.   ◄

Proposition 5.46. *For all $n \geq 3$, the generalized quaternion groups $\mathbf{Q}_n$ exist.*

Proof. Let G_n be the group defined by the presentation

$$G_n = \left(a, b \mid a^{2^{n-1}} = 1, bab^{-1} = a^{-1}, b^2 = a^{2^{n-2}}\right).$$

The group G_n satisfies all the requirements in the definition of the generalized quaternions with one possible exception: we do not yet know that its order is 2^n. By von Dyck's theorem, there is a surjective homomorphism $G_n \to H_n$, where H_n is the group just constructed in Example 5.10. Hence, $|G_n| \geq 2^n$.

On the other hand, the cyclic subgroup $\langle a \rangle$ in G_n has order at most 2^{n-1}, because $a^{2^{n-1}} = 1$. The relation $bab^{-1} = a^{-1}$ implies that $\langle a \rangle \lhd G_n = \langle a, b \rangle$, so that $G_n / \langle a \rangle$ is generated by the image of b. Finally, the relation $b^2 = a^{2^{n-2}}$ shows that $|G_n / \langle a \rangle| \leq 2$. Hence,

$$|G_n| \leq |\langle a \rangle||G_n / \langle a \rangle| \leq 2^{n-1} \cdot 2 = 2^n.$$

Therefore, $|G_n| = 2^n$, and so $G_n \cong \mathbf{Q}_n$. •

It now follows that the group H_n in Example 5.10 is isomorphic to $\mathbf{Q}_n$.

A concrete construction of the dihedral group D_{2n} was given in Exercise 2.73, and we can use that group, as in the proof just given, to give a presentation.

Proposition 5.47. *The dihedral group D_{2n} has a presentation*

$$D_{2n} = (a, b \mid a^n, b^2, bab = a^{-1}).$$

Proof. Let D_{2n} denote the group defined by the presentation, and let C_n be the group of order $2n$ constructed in Exercise 2.73. By von Dyck's theorem, there is a surjective homomorphism $f \colon D_{2n} \to C_n$, and so $|D_{2n}| \geq 2n$. To see that f is an isomorphism, we prove the reverse inequality. The cyclic sugroup $\langle a \rangle$ in D_{2n} has order at most n, because $a^n = 1$. The relation $bab^{-1} = a^{-1}$ implies that $\langle a \rangle \lhd D_{2n} = \langle a, b \rangle$, so that $D_{2n} / \langle a \rangle$ is generated by the image of b. Finally, the relation $b^2 = 1$ shows that $|D_{2n} / \langle a \rangle| \leq 2$. Hence,

$$|D_{2n}| \leq |\langle a \rangle||D_{2n} / \langle a \rangle| \leq 2n.$$

Therefore, $|D_{2n}| = 2n$, and so $D_{2n} \cong C_n$. •

In Chapter 2, we classified the groups of order 7 or less. Since groups of prime order are cyclic, it was only a question of classifying the groups of orders 4 and 6. The proof we gave, in Proposition 2.69, that every nonabelian group of order 6 is isomorphic to S_3 was rather complicated, analyzing the representation of a group on the cosets of a cyclic subgroup. Here is a proof in the present spirit.

Proposition 5.48. *If G is a nonabelian group of order 6, then $G \cong S_3$.*

Proof. As in the proof of Proposition 2.69, G must contain elements a and b of orders 3 and 2, respectively. Now $\langle a \rangle \lhd G$, because it has index 2, and so either $bab^{-1} = a$ or $bab^{-1} = a^{-1}$. The first possibility cannot occur, because G is not abelian. Therefore, G satisfies the conditions in the presentation of $D_6 \cong S_3$, and so von Dyck's theorem gives a surjective homomorphism $D_6 \to G$. Since both groups have the same order, this map must be an isomorphism. •

We can now classify the groups of order 8.

Theorem 5.49. *Every group G of order 8 is isomorphic to*

$$D_8, \quad \mathbf{Q}, \quad \mathbb{Z}_8, \quad \mathbb{Z}_4 \oplus \mathbb{Z}_2, \quad or \quad \mathbb{Z}_2 \oplus \mathbb{Z}_2 \oplus \mathbb{Z}_2.$$

Moreover, no two of the displayed groups are isomorphic.

Proof. If G is abelian, then the basis theorem shows that G is a direct sum of cyclic groups, and the fundamental theorem shows that the only such groups are those listed. Therefore, we may assume that G is not abelian.

Now G cannot have an element of order 8, lest it be cyclic, hence abelian; moreover, not every nonidentity element can have order 2, lest G be abelian, by Exercise 2.41. We conclude that G must have an element a of order 4; hence, $\langle a \rangle$ has index 2, and so $\langle a \rangle \lhd G$. Choose $b \in G$ with $b \notin \langle a \rangle$; note that $G = \langle a, b \rangle$ because $\langle a \rangle$ has index 2, hence is a maximal subgroup. Now $b^2 \in \langle a \rangle$, because $G/\langle a \rangle$ is a group of order 2, and so $b^2 = a^i$, where $0 \le i \le 3$. We cannot have $b^2 = a$ or $b^2 = a^3 = a^{-1}$ lest b have order 8. Therefore, either

$$b^2 = a^2 \quad or \quad b^2 = 1.$$

Furthermore, $bab^{-1} \in \langle a \rangle$, by normality, and so $bab^{-1} = a$ or $bab^{-1} = a^{-1}$ (for bab^{-1} has the same order as a). Now $bab^{-1} = a$ says that a and b

commute, which implies that G is abelian. We conclude that $bab^{-1} = a^{-1}$. Therefore, there are only two possibilities:

$$a^4 = 1, \quad b^2 = a^2, \quad \text{and} \quad bab^{-1} = a^{-1},$$

or

$$a^4 = 1, \quad b^2 = 1, \quad \text{and} \quad bab^{-1} = a^{-1}.$$

By the lemma, the first equations give relations of a presentation for $\mathbf{Q}$, while Proposition 5.47 shows that the second equations give relations of a presentation of D_8. By von Dyck's theorem, there is a surjective homomorphism $\mathbf{Q} \to G$ or $D_8 \to G$; as $|G| = 8$, however, this homomorphism must be an isomorphism.

Finally, Exercise 2.76 shows that $\mathbf{Q}$ and D_8 are not isomorphic (for example, $\mathbf{Q}$ has a unique element of order 2 while D_8 has several such elements). •

The reader may continue this classification of the groups of small order. Here are the results. By Corollary 2.76, every group of order p^2, where p is a prime, is abelian, and so every group of order 9 is abelian; by the fundamental theorem of finite abelian groups, there are only two such groups: $\mathbb{Z}_9$ and $\mathbb{Z}_3 \times \mathbb{Z}_3$. If p is a prime, then every group of order $2p$ is either cyclic or dihedral (see Exercise 5.39). Thus, there are only two groups of order 10 and only two groups of order 14. There are 5 groups of order 12, two of which are abelian. The nonabelian groups of order 12 are $D_{12} \cong S_3 \times \mathbb{Z}_2$, A_4, and a group having the presentation

$$\left(a, b \mid a^6, b^2 = a^3 = (ab)^2 \right).$$

A group of order pq, where $p < q$ are primes and $p \nmid q$, must be cyclic, and so there is only one group of order 15. There are 15 nonisomorphic groups of order 16, and so this is a good place to stop.

We are at the very beginning of a rich subject called ***combinatorial group theory***, which investigates how much one can say about a group, given a presentation of it. A finitely presented group G has a ***solvable word problem*** if it has a presentation $G = (X \mid R)$ for which there exists an algorithm to determine whether an arbitrary word w on X is equal to the identity element in G. One of the most remarkable results was proved in the late 1950s by P. S. Novikov and W. W. Boone, independently: there exists a finitely presented group G that does not have a solvable word problem.

Finding presentations of known groups, as we have just done for $\mathbf{Q}_n$ and D_{2n}, is an interesting problem; an excellent reference for such questions is the book by Coxeter and Moser.

Another problem is whether a group defined by a presentation is finite or infinite. For example, ***Burnside's problem*** asks whether a finitely generated group G of finite exponent n, that is, $x^n = 1$ for all $x \in G$, must be finite [he had proved that every such group G which happens to be a subgroup of $GL(n, \mathbb{C})$ is finite]. The answer in general, however, is negative; such a group can be infinite. This was first proved, for n odd and large, by P. S. Novikov and S. I. Adyan in a long and complicated paper; using geometric methods (called *van Kampen diagrams*), A. Yu. Olshanskii gave a much shorter and simpler proof; finally, S. V. Ivanov was able to complete the solution of the problem by showing that the presented group can be infinite when n is even and large; thus, if n is any sufficiently large integer, there are infinite finitely generated groups G with $x^n = 1$ for all $x \in G$.

Another geometric technique involves the ***Cayley graph***, which is a graph depending on a given presentation; one can prove that a group is free if and only if it has a Cayley graph that is a tree.

Finally, the interaction between presentations and algorithms is both theoretical and practical. On the practical side, nowadays, many efficient algorithms have been implemented. The first was ***coset enumeration***, which computes the order of a group G, defined by a presentation, provided that $|G|$ is finite (unfortunately, there can be no algorithm to determine, in advance, whether G is finite). A theorem of G. Higman states that a finitely generated group G can be imbedded as a subgroup of a finitely presented group H if and only if G is *recursively presented*; that is, there is a presentation of G whose relations can be given by an algorithm.

There are other directions in group theory. In the 1980s, in an amazing collective effort, all finite simple groups were classified. There is now a list of every finite simple group, and many important properties of each of them is known. Many questions about arbitrary finite groups can be reduced to problems about simple groups. Thus, the classification theorem can be used by checking, one by one, whether each simple group on the list satisfies the desired result.

Another important direction is representation theory–the systematic study of homomorphisms of a group into groups of nonsingular matrices. One of the first applications of this theory is a theorem of Burnside: Every group of order $p^m q^n$, where p and q are primes, must be solvable.

EXERCISES

5.32 Let F be a free group with basis X and let $A \subset X$. Prove that if N is the normal subgroup of F generated by A, then F/N is a free group.

5.33 Let F be a free group.
   (i) Prove that F has no elements (other than 1) of finite order.
   (ii) Prove that $Z(F) = \{1\}$, where $Z(F)$ is the center of F,

5.34 (i) Prove that $\mathbb{Z}$ is a free group with one generator.
   (ii) Prove that a free group is solvable if and only if it is infinite cyclic.

5.35 If G is a finitely generated group and n is a positive integer, prove that G has only finitely many subgroups of index n.

5.36 Prove that if G is a finite group generated by two elements a, b having order 2, then $G \cong D_{2n}$ for some $n \geq 2$.

5.37 Prove that every finite group has a presentation which has a finite number of generators and a finite number of relations.

5.38 (i) Prove that the generalized quaternion group $\mathbf{Q}_n$ has a unique element of order 2, namely, $\langle b^2 \rangle$, and this subgroup is the center $Z(\mathbf{Q}_n)$.
   (ii) Prove that $\mathbf{Q}_n/Z(\mathbf{Q}_n) \cong D_{2 \cdot 2^{n-2}}$.

5.39 If p is a prime, prove that every group G of order $2p$ is either cyclic or isomorphic to D_{2p}.

6

Commutative Rings II

6.1 PRIME IDEALS AND MAXIMAL IDEALS

Our main interest in this chapter is the study of polynomials in several variables. One sees in analytic geometry that polynomials correspond to geometric figures; for example, $f(x, y) = x^2/a^2 + y^2/b^2 - 1$ is intimately related to an ellipse in the plane $\mathbb{R}^2$. But there is a very strong connection between the rings $k[x_1, \ldots, x_n]$, where k is a field, and the geometry of subsets of k^n going far beyond this. Given a set of polynomials $f_1, \ldots, f_t$ of n variables, call the subset $V \subset k^n$ consisting of their common zeros a *variety*. Of course, one can study varieties because solutions of systems of polynomial equations (an obvious generalization of systems of linear equations) are intrinsically interesting. On the other hand, some systems are more interesting than others. Investigating a problem often leads to a parametrization of its solutions by a variety, and so understanding the variety and its properties, e.g., irreducibility, dimension, genus, singularities, and so forth, leads to an understanding of the original problem. For example, once calculus became established, Leibniz raised the question of determining those functions that could be integrated explicitly in terms of "elementary functions:" algebraic combinations of polynomials, trigonometric and inverse trigonometric functions, exponentials, and logarithms. In 1694, John Bernoulli conjectured that the integrand arising from the arclength of an ellipse could not be so integrated. Such integrands arise, not only in arclength problems, but also in finding periods of pendulums and problems in mechanics, for example. Some of the best mathematicians – Euler, Legendre, Abel, Jacobi, Gauss, Weierstrass, Riemann, Poincaré –

investigated such *elliptic integrals* and their associated *elliptic functions*. A key observation is that elliptic functions are *doubly periodic*, and this leads to a subgroup Λ of the additive group $\mathbb{C}$ whose (two) generators are these periods. The quotient group $\mathbb{C}/\Lambda$ is a torus (i.e., a doughnut), which is a surface. Using complex coordinates instead of real ones, however, we see that it is a (complex) *elliptic curve*. Geometric properties of this curve are crucial to the understanding of elliptic functions. The interplay between $k[x_1, \ldots, x_n]$ and varieties has evolved into what is nowadays called *Algebraic Geometry*, and this chapter may be regarded as an introduction to this subject.

As usual, it is simpler to begin by looking at the more general setting – in this case, commutative rings – before getting involved with polynomial rings. A great deal of the number theory we have presented involves divisibility: given two integers a and b, when does $a \mid b$; that is, when is b a multiple of a? This question translates into a question about principal ideals, for $a \mid b$ if and only if $(a) \subset (b)$. We now introduce two especially interesting types of ideal: *prime ideals*, which are related to Euclid's lemma, and *maximal ideals*.

Let us begin with the analog of Theorem 2.58, the correspondence theorem for groups.

Proposition 6.1 (Correspondence Theorem for Rings). *If I is a proper ideal in a commutative ring R, then there is an inclusion-preserving bijection φ from the set of all intermediate ideals J containing I, that is, $I \subset J \subset R$, to the set of all the ideals in R/I, given by*

$$\varphi: J \mapsto \pi(J) = J/I = \{a + I : a \in J\},$$

where $\pi: R \to R/I$ is the natural map.

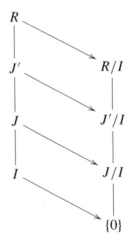

Proof.   If one forgets its multiplication, the commutative ring R is only an additive abelian group and its ideal I is only a (normal) subgroup. The correspondence theorem for groups, Theorem 2.58, now applies, and it gives an inclusion-preserving bijection

$$\Phi : \{\text{all subgroups of } R \text{ containing } I\} \to \{\text{all subgroups of } R/I\},$$

where $\Phi(J) = \pi(J) = J/I$.

If J is an ideal, then $\Phi(J)$ is also an ideal, for if $r \in R$ and $a \in J$, then $ra \in J$, and so

$$(r + I)(a + I) = ra + I \in J/I.$$

Let φ be the restriction of Φ to the set of intermediate ideals; φ is an injection because Φ is a bijection. To see that φ is surjective, let J^* be an ideal in R/I. Now $\pi^{-1}(J^*)$ is an intermediate ideal in R [it contains $I = \pi^{-1}(\{0\})$], and $\varphi(\pi^{-1}(J^*)) = \pi(\pi^{-1}(J^*)) = J^*$, by Lemma 2.57.   •

In practice, the correspondence theorem is invoked, tacitly, by saying that every ideal in the quotient ring R/I has the form J/I for some unique ideal J with $I \leq J \leq R$.

Example 6.1.
Let $I = (m)$ be a nonzero ideal in $\mathbb{Z}$. Every ideal J in $\mathbb{Z}$ is principal, say $J = (a)$, and it is easy to see that $(m) \leq (a)$ if and only if $a \mid m$. By the correspondence theorem, every ideal in $\mathbb{Z}_m$ has the form $([a])$ for some divisor a of m.   ◄

Definition.   An ideal I in a commutative ring R is called a ***prime ideal*** if it is a proper ideal, that is, $I \neq R$, and $ab \in I$ implies $a \in I$ or $b \in I$.

Example 6.2.
Recall that a commutative ring R is a domain if and only if $ab = 0$ in R implies $a = 0$ or $b = 0$. Thus, the ideal $\{0\}$ in R is a prime ideal if and only if R is a domain.   ◄

Example 6.3.

We claim that the prime ideals in $\mathbb{Z}$ are precisely the ideals (p), where either $p = 0$ or p is a prime. Since m and $-m$ generate the same principal ideal, we may restrict our attention to nonnegative generators. If $p = 0$, then the result follows from our example above, for $\mathbb{Z}$ is a domain. If $p > 0$, we show first that (p) is a proper ideal; otherwise, $1 \in (p)$, and there would be an integer a with $ap = 1$, a contradiction. Next, if $ab \in (p)$, then $p \mid ab$. By Euclid's lemma, either $p \mid a$ or $p \mid b$; that is, either $a \in (p)$ or $b \in (p)$. Therefore, (p) is a prime ideal.

Conversely, if $m > 1$ is not a prime, then it has a factorization $m = ab$ with $0 < a < m$ and $0 < b < m$. Thus, neither a nor b is a multiple of m, hence neither lies in (m), and so (m) is not a prime ideal. ◄

The proof in the example works in more generality.

Proposition 6.2.   *If k is a field, then a nonzero polynomial $p(x) \in k[x]$ is irreducible if and only if $(p(x))$ is a prime ideal.*

Proof.   Suppose that $p(x)$ is irreducible. First, (p) is a proper ideal; otherwise, $R = (p)$ and hence $1 \in (p)$, so there is a polynomial $f(x)$ with $1 = p(x)f(x)$. But $p(x)$ has degree at least 1, whereas

$$0 = \deg(1) = \deg(pf) = \deg(p) + \deg(f) \geq \deg(p) \geq 1.$$

This contradiction shows that (p) is a proper ideal.

Second, if $ab \in (p)$, then $p \mid ab$, and so Euclid's lemma in $k[x]$ gives $p \mid a$ or $p \mid b$. Thus, $a \in (p)$ or $b \in (p)$. It follows that (p) is a prime ideal.

Conversely, suppose that $p(x)$ is not irreducible; there is thus a factorization

$$p(x) = a(x)b(x)$$

with $\deg(a) < \deg(p)$ and $\deg(b) < \deg(p)$. As every nonzero polynomial $g(x) \in (p)$ has the form $g(x) = d(x)p(x)$ for some $d(x) \in k[x]$, we have $\deg(g) \geq \deg(p)$; it follows that neither $a(x)$ nor $b(x)$ lies in (p), and so (p) is not a prime ideal.   •

Proposition 6.3.   *A proper ideal I in R is a prime ideal if and only if R/I is a domain.*

Proof.   Let I be a prime ideal. Since I is a proper ideal, we have $1 \notin I$ and so $1 + I \neq 0 + I$ in R/I. If $0 = (a + I)(b + I) = ab + I$, then $ab \in I$. Since I is a prime ideal, either $a \in I$ or $b \in I$; that is, either $a + I = 0$ or $b + I = 0$. Hence, R/I is a domain. The converse is just as easy.   •

We are going to change notation. If I is an ideal in a commutative ring R, we will write $I \leq R$, instead of $I \subset R$, and we may write $I < R$ if I is a proper ideal. More generally, if I and J are ideals, we will write $I \leq J$, instead of $I \subset J$, and we may write $I < J$ if $I \leq J$ and $I \neq J$.

Here is a second interesting type of ideal.

Definition.   A proper ideal I in a commutative ring R is a ***maximal ideal*** if there is no ideal J with $I < J < R$.

Thus, if I is a maximal ideal in a commutative ring R and if J is a proper ideal with $I \leq J$, then $I = J$.

The prime ideals in the polynomial ring $k[x_1, \dots, x_n]$ can be quite complicated, but when k is algebraically closed, we shall see that every maximal ideal has the form $(x_1 - a_1, \dots, x_n - a_n)$ for some point $(a_1, \dots, a_n) \in k^n$.

We may restate Example 3.10(ii) in the present language.

Lemma 6.4.   *The ideal $\{0\}$ is a maximal ideal in a commutative ring R if and only if R is a field.*

Proof.   It is shown in Example 3.10(ii) that every nonzero ideal I in R is equal to R itself if and only if every nonzero element in R is a unit. That is, $\{0\}$ is a maximal ideal if and only if R is a field.   •

Proposition 6.5.   *A proper ideal I in a commutative ring R is a maximal ideal if and only if R/I is a field.*

Proof.   The correspondence theorem for rings shows that I is a maximal ideal if and only if R/I has no ideals other than $\{0\}$ and R/I itself; Lemma 6.4 shows that this property holds if and only if R/I is a field.   •

Corollary 6.6.   *Every maximal ideal I in a commutative ring R is a prime ideal.*

Proof.   If I is a maximal ideal, then R/I is a field. Since every field is a domain, R/I is a domain, and so I is a prime ideal.   •

Example 6.4.
The converse of the last corollary is false. For example, consider the principal ideal (x) in $\mathbb{Z}[x]$. By Exercise 3.80, we have

$$\mathbb{Z}[x]/(x) \cong \mathbb{Z};$$

since $\mathbb{Z}$ is a domain, (x) is a prime ideal; since $\mathbb{Z}$ is not a field, (x) is not a maximal ideal.

It is not difficult to exhibit a proper ideal J strictly containing (x); let

$$J = \{f(x) \in \mathbb{Z}[x]:\ f(x) \text{ has even constant term}\}.$$

Since $\mathbb{Z}[x]/J \cong \mathbb{Z}_2$, which is a field, it follows that J is a maximal ideal containing (x). ◀

Example 6.5.
Let k be a field, and let $a = (a_1, \ldots, a_n) \in k^n$. Define the *evaluation map* $e_a: k[x_1, \ldots, x_n] \to k$ by

$$e_a:\ f(x_1, \ldots, x_n) \mapsto f(a) = f(a_1, \ldots, a_n).$$

It is easy to see that e_a is a surjective ring homomorphism, and so $\ker e_a$ is a maximal ideal. Now $(x_1 - a_n, \ldots, x_n - a_n) \leq \ker e_a$. In Exercise 6.5(i), however, we shall see that $(x_1 - a_n, \ldots, x_n - a_n)$ is a maximal ideal, and so it must be equal to $\ker e_a$. ◀

The converse of Corollary 6.6 is true when R is a PID.

Theorem 6.7.   *If R is a principal ideal domain, then every nonzero prime ideal I is a maximal ideal.*

Proof.   Assume there is a proper ideal J with $I \leq J$. Since R is a PID, $I = (a)$ and $J = (b)$ for some $a, b \in R$. Now $a \in J$ implies that $a = rb$ for some $r \in R$, and so $rb \in I$; but I is a prime ideal, so that $r \in I$ or $b \in I$. If $r \in I$, then $r = sa$ for some $s \in R$, and so $a = rb = sab$. Since R is a domain, $1 = sb$, and Exercise 3.27 gives $J = (b) = R$, contradicting J being a proper ideal. If $b \in I$, then $J \leq I$, and so $J = I$. Therefore, I is a maximal ideal. •

We can now give a second proof of Proposition 3.73.

Corollary 6.8.   *If k is a field and $p(x) \in k[x]$ is irreducible, then the quotient ring $k[x]/(p(x))$ is a field.*

Proof.   Since $p(x)$ is irreducible, the principal ideal $I = (p(x))$ is a nonzero prime ideal; since $k[x]$ is a PID, I is a maximal ideal, and so $k[x]/I$ is a field. •

Does every commutative ring R contain a maximal ideal? The (positive) answer to this question involves *Zorn's lemma*, a theorem equivalent to the Axiom of Choice, which is usually discussed in a sequel course (but see Corollary 6.21).

EXERCISES

6.1   (i) Find all the maximal ideals in $\mathbb{Z}$.

   (ii) Find all the maximal ideals in $k[x]$, where k is a field.

6.2   (i) Give an example of a commutative ring containing two prime ideals P and Q for which $P \cap Q$ is not a prime ideal.

   (ii) If $P_1 \supseteq P_2 \supseteq \cdots P_n \supseteq P_{n+1} \supseteq \cdots$ is a decreasing sequence of prime ideals in a commutative ring R, prove that $\bigcap_{n \geq 1} P_n$ is a prime ideal.

6.3 Let $f : R \to S$ be a ring homomorphism.

   (i) If Q is a prime ideal in S, prove that $f^{-1}(Q)$ is a prime ideal in R. Conclude, in the correspondence theorem, that if J/I is a prime ideal in R/I, where $I \leq J \leq R$, then J is a prime ideal in R.

   (ii) Give an example to show that if P is a prime ideal in R, then $f(P)$ need not be a prime ideal in S.

6.4   (i) If k is a field and $a \in k$, recall that the *evaluation map* $e_a : k[x] \to k$ is defined by $e_a : f(x) \mapsto f(a)$; we saw, in Example 3.8(iv), that e_a is a ring homomorphism. Prove that the kernel of any evaluation map is a maximal ideal in $k[x]$.

   (ii) If k is an algebraically closed field, prove that the function

$$k \to \{\text{maximal ideals in } k[x]\},$$

given by $a \mapsto (x - a)$, the principal ideal in $k[x]$ generated by $x - a$, is a bijection.

6.5   (i) Let k be a field, and let $a_1, \ldots, a_n \in k$. Prove that $(x_1 - a_1, \ldots, x_n - a_n)$ is a maximal ideal in $k[x_1, \ldots, x_n]$.

   (ii) Prove that if $x_i - b \in (x_1 - a_1, \ldots, x_n - a_n)$ for some i, where $b \in k$, then $b = a_i$.

   (iii) Prove that $\mu : k^n \to \{\text{maximal ideals in } k[x_1, \ldots, x_n]\}$, given by

$$\mu : (a_1, \ldots, a_n) \mapsto (x_1 - a_1, \ldots, x_n - a_n),$$

is an injection, and give an example of a field k for which μ is not a surjection.

6.6 Prove that if P is a prime ideal in a commutative ring R and if $r^n \in P$ for some $r \in R$ and $n \geq 1$, then $r \in P$.

6.7 Prove that the ideal $(x^2 - 2, y^2 + 1, z)$ in $\mathbb{Q}[x, y, z]$ is a proper ideal.

6.8 Call a nonempty subset S of a commutative ring R **multiplicatively closed** if $0 \notin S$ and, if $s, s' \in S$, then $ss' \in S$. Prove that an ideal I which is maximal with the property that $I \cap S = \emptyset$ is a prime ideal. (The existence of such an ideal I can be proved using Zorn's lemma.)

6.9   (i) If I and J are ideals in a commutative ring R, define

$$IJ = \left\{ \sum_\ell a_\ell b_\ell : a_\ell \in I \text{ and } b_\ell \in J \right\}.$$

Prove that IJ is an ideal in R and that $IJ \leq I \cap J$.

(ii) Let $R = k[x, y]$, where k is a field and let $I = (x, y) = J$. Prove that $I^2 = IJ < I \cap J = I$.

6.10 Let P be a prime ideal in a commutative ring R. If there are ideals I and J in R with $IJ \leq P$, prove that $I \leq P$ or $J \leq P$.

6.11 If I and J are ideals in a commutative ring R, define

$$(I : J) = \{r \in R : rJ \subset I\}.$$

(i) Prove that $(I : J)$ is an ideal.

(ii) Let R be a domain and let $a, b \in R$, where $b \neq 0$. If $I = (ab)$ and $J = (b)$, prove that $(I : J) = (a)$.

6.12 Let I and J be ideals in a commutative ring R.

(i) Prove that there is an injection $R/(I \cap J) \to R/I \times R/J$ given by $\varphi : r \mapsto (r + I, r + J)$

(ii) Call I and J **coprime** if $I + J = R$. Prove that the ring homomorphism $\varphi : R/(I \cap J) \to R/I \times R/J$ is a surjection if I and J are coprime.

(iii) Generalize the Chinese Remainder Theorem as follows. Let R be a commutative ring and let $I_1, \dots, I_n$ be pairwise coprime ideals; that is, I_i and I_j are coprime for all $i \neq j$. Prove that if $a_1, \dots, a_n \in R$, then there exists $r \in R$ with $r + I_i = a_i + I_i$ for all i.

6.13 Recall that a Boolean ring is a commutative ring R for which $a^2 = a$ for all $a \in R$. Prove that every prime ideal in a Boolean ring is a maximal ideal.

6.14 A commutative ring R is called a **local ring** if it has a unique maximal ideal.

(i) If p is a prime, prove that

$$\{a/b \in \mathbb{Q} : p \nmid b\}$$

is a local ring.

(ii) If R is a local ring with unique maximal ideal M, prove that $a \in R$ is a unit if and only if $a \notin M$.

6.2 UNIQUE FACTORIZATION

We have proved unique factorization theorems in $\mathbb{Z}$ and in $k[x]$, where k is a field. In fact, we have proved a common generalization of these two results: every euclidean ring has unique factorization. Our aim now is to generalize this result, first to general PID's, and then to $R[x]$, where R is a ring having unique factorization. It will then follow that there is unique factorization in the ring $k[x_1, \ldots , x_n]$ of all polynomials in several variables over a field k. One immediate consequence is that any two polynomials in several variables have a gcd.

We begin by generalizing some earlier definitions.

Definition.   Elements a and b in a commutative ring R are **associates** if there exists a unit $u \in R$ with $b = ua$.

For example, in $\mathbb{Z}$, the units are ± 1, and so the only associates of an integer m are $\pm m$; in $k[x]$, where k is a field, the units are the nonzero constants, and so the only associates of a polynomial $f(x) \in k[x]$ are the polynomials $uf(x)$, where $u \in k$ and $u \neq 0$. The only units in $\mathbb{Z}[x]$ are ± 1, and so the only associates of a polynomial $f(x) \in \mathbb{Z}[x]$ are $\pm f(x)$.

Consider two principal ideals (a) and (b) in a commutative ring R. It is easy to see that the following are equivalent: $a \in (b)$; $a = rb$ for some $r \in R$; $(a) \leq (b)$. We can say more when R is a domain.

Proposition 6.9.   *Let R be a domain and let $a, b \in R$.*

(i) *$a \mid b$ and $b \mid a$ if and only if a and b are associates.*

(ii) *The principal ideals (a) and (b) are equal if and only if a and b are associates.*

Proof.   (i) This is Proposition 3.9.

(ii) If $(a) = (b)$, then $(a) \leq (b)$ and $(b) \leq (a)$; hence, $a \in (b)$ and $b \in (a)$. Thus, $a \mid b$ and $b \mid a$; by part (i), a and b are associates. The converse is easy, and one does not need to assume that R is a domain to prove it. $\bullet$

The notions of prime number in $\mathbb{Z}$ or irreducible polynomial in $k[x]$, where k is a field, have a common generalization.

Definition.   A element p in a commutative ring R is **irreducible** if it is neither 0 nor a unit and if its only factors are units or associates of p.

For example, the irreducibles in $\mathbb{Z}$ are the numbers $\pm p$, where p is a prime, and the irreducibles in $k[x]$, where k is a field, are the irreducible polynomials $p(x)$; that is, $\deg(p) \geq 1$ and $p(x)$ has no factorization $p(x) = f(x)g(x)$ where $\deg(f) < \deg(p)$ and $\deg(g) < \deg(g)$. This characterization of irreducible polynomial does not persist in rings $R[x]$ when R is not a field. For example, in $\mathbb{Z}[x]$, the polynomial $f(x) = 2x + 2$ cannot be factored into two polynomials, each having degree smaller than $\deg(f) = 1$, yet $f(x)$ is not irreducible, for in the factorization $2x + 2 = 2(x + 1)$, neither 2 nor $x + 1$ is a unit.

Here is the definition we have been seeking.

Definition.   A domain R is a **unique factorization domain (UFD)** if

(i) every $r \in R$, neither 0 nor a unit, is irreducible or is a product of irreducibles;

(ii) if $p_1 \cdots p_m = q_1 \cdots q_n$, where p_i and q_j are irreducible, then $m = n$ and there is a permutation $\sigma \in S_n$ with p_i and $q_{\sigma(i)}$ associates for all i.

When we proved that $\mathbb{Z}$ and $k[x]$, for k a field, have unique factorization into irreducibles, we did not mention associates because, in each case, irreducible elements were always replaced by favorite choices of associates: in $\mathbb{Z}$, *positive* irreducibles, i.e., primes, are chosen; in $k[x]$, *monic* irreducible polynomials are chosen. The reader should see, for example, that the statement: "$\mathbb{Z}$ is a UFD" is just a restatement of the Fundamental Theorem of Arithmetic.

The proof that every PID is a UFD uses a new idea: chains of ideals.

Lemma 6.10.   *Let R be a PID.*

(i) *There is no infinite strictly ascending chain of ideals*

$$I_1 < I_2 < \cdots < I_n < I_{n+1} < \cdots .$$

(ii) *If $r \in R$ is neither 0 nor a unit, then r is irreducible or a product of irreducibles.*

Proof.   (i) If, on the contrary, an infinite strictly ascending chain exists, then define $J = \bigcup_{n=1}^{\infty} I_n$. We claim that J is an ideal. If $a \in J$, then $a \in I_n$ for some n; if $r \in R$, then $ra \in I_n$, because I_n is an ideal; hence, $ra \in J$. If $a, b \in J$, then there are ideals I_n and I_m with $a \in I_n$ and $b \in I_m$; since the chain is ascending, we may assume that $I_n \subset I_m$, and so $a, b \in I_m$. As I_m is an ideal, $a - b \in I_m$ and, hence, $a - b \in J$. Therefore, J is an ideal.

Since R is a PID, we have $J = (d)$ for some $d \in J$. Now d got into J by being in I_n for some n. Hence

$$J = (d) \le I_n < I_{n+1} \le J,$$

and this is a contradiction.

(ii) If r is a divisor of an element $a \in R$, then $a = rs$; r is called a **proper divisor** of a if neither r nor s is a unit. We first show that if r is a proper divisor of a, then $(a) < (r)$. By Proposition 6.9, $(a) \le (r)$ and, if the inequality is not strict, then a and r are associates. In the latter case, there is a unit $u \in R$ with $a = ur$, and this contradicts r being a proper divisor of a.

Call a nonzero nonunit $a \in R$ *good* if it is irreducible or a product of irreducibles; otherwise, call *a* bad. We must show that there are no bad elements. If a is bad, it is not irreducible, and so $a = rs$, where both r and s are proper divisors. But the product of good elements is good, and so at least one of the factors, say r, is bad. The first paragraph shows that $(a) < (r)$. It follows, by induction, that there exists a sequence $a = a_1, r = a_2, \dots, a_n, \dots$ of bad elements with each a_{n+1} a proper divisor of a_n, and this sequence yields a strictly ascending chain

$$(a_1) < (a_2) < \cdots < (a_n) < (a_{n+1}) < \cdots,$$

contradicting part (i) of this lemma.   •

Proposition 6.11.   *Let R be a domain in which every $r \in R$, neither 0 nor a unit, is irreducible or a product of irreducibles. Then R is a UFD if and only if (p) is a prime ideal in R for every irrreducible element $p \in R$.*

Proof.   Assume that R is a UFD. If $a, b \in R$ and $ab \in (p)$, then there is $r \in R$ with

$$ab = rp.$$

Factor each of a, b, and r into irreducibles; by unique factorization, the left side of the equation must involve an associate of p. This associate arose as a factor of a or b, and hence $a \in (p)$ or $b \in (p)$.

The proof of the converse is merely an adaptation of the proof of the Fundamental Theorem of Arithmetic. Assume that

$$p_1 \cdots p_m = q_1 \cdots q_n,$$

where the p_i's and the q_j's are irreducible elements. We prove, by induction on $\max\{m, n\} \geq 1$, that $n = m$ and the q's can be reindexed so that q_i and p_i are associates for all i. The base step $\max\{m, n\} = 1$ has $p_1 = q_1$, and the result is obviously true. For the inductive step, the given equation shows that $p_1 \mid q_1 \cdots q_n$. By hypothesis, (p_1) is a prime ideal (which is the analog of Euclid's lemma), and so there is some q_j with $p_1 \mid q_j$. But q_j, being irreducible, has no divisors other than units and associates, so that q_j and p_1 are associates: $q_j = up_1$ for some unit u. Canceling p_1 from both sides, we have $p_2 \cdots p_m = uq_1 \cdots \widehat{q_j} \cdots q_n$. By the inductive hypothesis, $m-1 = n-1$ (so that $m = n$), and, after possible reindexing, q_i and p_i are associates for all i. •

Example 6.6.
We claim that $f(x, y) = x^2 + y^2 - 1 \in k[x, y]$ is irreducible, where k is a field. Write $Q = k(y) = \mathrm{Frac}(k[y])$, and view $f(x, y) \in Q[x]$. Now the quadratic $g(x) = x^2 + (y^2 - 1)$ is irreducible in $Q[x]$ if and only if it has no roots in $Q = k(y)$, and this is so, by Exercise 3.58.

It follows from Proposition 6.11 that $(x^2 + y^2 - 1)$ is a prime ideal because it is generated by an irreducible polynomial. ◀

Theorem 6.12.   *If R is a PID, then R is a UFD. In particular, every euclidean ring is a UFD.*

Proof.   In view of the last two results, it suffices to prove that (p) is a prime ideal whenever p is irreducible. Suppose that $p \mid ab$; we must show that $p \mid b$ or $p \mid a$. The subset

$$I = \{sb + tp : s, t \in R\}$$

is an ideal in R and, hence, $I = (d)$ because R is a PID. Now $b, p \in I$, so that $d \mid p$ and $d \mid b$. Since p is irreducible, either d is an associate of p or d is a unit. In the first case, $d = up$ for some unit u, and so $d \mid b$ implies $p \mid b$. In the second case, d is a unit. Now $d = sb + tp$, and so $da = sab + tap$. Since $p \mid ab$, we have $p \mid da$. But da is an associate of a, and so $p \mid a$. It follows that (p) is a prime ideal. •

Recall that the notion of gcd can be defined in any commutative ring.

Definition.   Let R be a commutative ring and let $a_1, \ldots, a_n \in R$. A *common divisor* of $a_1, \ldots, a_n$ is an element $c \in R$ with $c \mid a_i$ for all i. A *greatest* *common divisor* or *gcd* of $a_1, \ldots, a_n$ is a common divisor d with $c \mid d$ for every common divisor c.

Even in the familiar examples of $\mathbb{Z}$ and $k[x]$, gcd's are not unique unless an extra condition is imposed. For example, if d is a gcd of a pair of integers in $\mathbb{Z}$, as defined above, then $-d$ is also a gcd. To force gcd's to be unique, one defines nonzero gcd's in $\mathbb{Z}$ to be positive; similarly, in $k[x]$, where k is a field, one imposes the condition that nonzero gcd's are monic polynomials. In a general PID, however, elements may not have favorite associates.

If R is a domain, then it is easy to see that if d and d' are gcd's of elements $a_1, \ldots, a_n$, then $d \mid d'$ and $d' \mid d$. It follows from Proposition 6.9 that d and d' are associates and, hence, that $(d) = (d')$. Thus, gcd's are not unique, but they all generate the same principal ideal.

In Exercise 3.71, we saw that there exist domains R containing a pair of elements having no gcd. However, the idea in Proposition 1.44 carries over to show that gcd's do exist in UFD's.

Proposition 6.13. *If R is a UFD, then the gcd of any set of elements $a_1, \ldots, a_n$ exists.*

Proof. It suffices to prove that the gcd of two elements a and b exists, for we leave as an exercise that if a gcd of two elements always exists, then a gcd of any finite number of elements also exists.

There are units u and v and distinct irreducibles $p_1, \ldots, p_t$ with

$$a = u p_1^{e_1} p_2^{e_2} \cdots p_t^{e_t}$$

and

$$b = v p_1^{f_1} p_2^{f_2} \cdots p_t^{f_t},$$

where $e_i \geq 0$ and $f_i \geq 0$ for all i. It is easy to see that if $c \mid a$, then the factorization of c into irreducibles is $c = w p_1^{g_1} p_2^{g_2} \cdots p_t^{g_t}$, where w is a unit and $g_i \leq e_i$ for all i. Thus, c is a common divisor of a and b if and only if $g_i \leq m_i$ for all i, where

$$m_i = \min\{e_i, f_i\}.$$

It is now clear that $p_1^{m_1} p_2^{m_2} \cdots p_t^{m_t}$ is a gcd of a and b.   •

We caution the reader that we have not proved that a gcd of elements $a_1, \ldots, a_n$ is a linear combination of them; indeed, this may not be true (see Exercise 6.20).

Definition.   Elements $a_1, \dots, a_n$ in a UFD R are called *relatively prime* if all their gcd's are units – that is, if every common divisor of $a_1, \dots, a_n$ is a unit.

We are now going to prove that if R is a UFD, then so is $R[x]$. This theorem was essentially found by Gauss, and the proof uses ideas in the proof of Gauss's theorem, Theorem 3.63. It will follow that $k[x_1, \dots, x_n]$ is a UFD whenever k is a field.

Definition.   A polynomial $f(x) = a_n x^n + \dots + a_1 x + a_0 \in R[x]$, where R is a UFD, is called *primitive* if its coefficients are relatively prime; that is, the only common divisors of $a_n, \dots, a_1, a_0$ are units.

Observe that if $f(x)$ is not primitive, then there exists an irreducible $q \in R$ that divides each of its coefficients: if the gcd is a nonunit d, then take for q any irreducible factor of d.

Example 6.7.

We now show, in a UFD R, that every irreducible $p(x) \in R[x]$ of positive degree is primitive. If not, then there is an irreducible $q \in R$ with $p(x) = qg(x)$; note that $\deg(q) = 0$ because $q \in R$. Since $p(x)$ is irreducible, its only factors are units and associates, and so q must be an associate of $p(x)$. But every unit in $R[x]$ has degree 0, i.e., is a constant (for $uv = 1$ implies $\deg(u) + \deg(v) = \deg(1) = 0$); hence, associates in $R[x]$ have the same degree. Therefore, q is not an associate of $p(x)$, because the latter has positive degree.   ◄

We begin with a generalization of Gauss's lemma.

Lemma 6.14.   *If R is a UFD and $f(x), g(x) \in R[x]$ are both primitive, then their product $f(x)g(x)$ is also primitive.*

Proof.   Let $f(x) = \sum a_i x^i$, $g(x) = \sum b_j x^j$, and $f(x)g(x) = \sum c_k x^k$. If $f(x)g(x)$ is not primitive, then there is an irreducible p that divides every c_k. Since $f(x)$ is primitive, at least one of its coefficients is not divisible by p; let a_i be the first such; similarly, let b_j be the first coefficient of $g(x)$ that is not divisible by p. The definition of multiplication of polynomials gives

$$a_i b_j = c_{i+j} - (a_0 b_{i+j} + \dots + a_{i-1} b_{j+1} + a_{i+1} b_{j-1} + \dots + a_{i+j} b_0).$$

Each term on the right side is divisible by p, and so p divides $a_i b_j$. As p divides neither a_i nor b_j, however, this contradicts Proposition 6.11, which says that (p) is a prime ideal.   •

Remark.   Here is a less computational proof of this lemma. If $\pi : R \rightarrow R/(p)$ is the natural map $\pi : a \mapsto a + (p)$, then Exercise 3.41 shows that the function $\tilde{\pi} : R[x] \rightarrow (R/(p))[x]$, which replaces each coefficient c of a polynomial by $\pi(c)$, is a ring homomorphism. Now the hypothesis that a polynomial $h(x) \in R[x]$ is not primitive says there is some irreducible p such that all the coefficients of $\tilde{\pi}(h)$ are 0 in $R/(p)$; that is, $\tilde{\pi}(h) = 0$ in $(R/(p))[x]$. Thus, if the product $f(x)g(x)$ is not primitive, there is some irreducible p with $0 = \tilde{\pi}(fg) = \tilde{\pi}(f)\tilde{\pi}(g)$ in $(R/(p))[x]$. Since (p) is a prime ideal, $R/(p)$ is a domain, and hence $(R/(p))[x]$ is also a domain. But, neither $\tilde{\pi}(f)$ nor $\tilde{\pi}(g)$ is 0 in $(R/(p))[x]$, because f and g are primitive, and this contradicts $(R/(p))[x]$ being a domain. ◄

Definition.   If R is a UFD and $f(x) = a_n x^n + \cdots + a_1 x + a_0 \in R[x]$, define $c(f) \in R$ to be a gcd of $a_n, \dots, a_1, a_0$; one calls $c(f)$ the ***content*** of $f(x)$.

Note that the content of a polynomial $f(x)$ is not unique, but that any two contents of $f(x)$ are associates.

It is obvious that if $b \in R$ and $b \mid f(x) \in R[x]$, where R is a UFD, then b is a common divisor of the coefficients of $f(x)$, and so $b \mid c(f)$.

Lemma 6.15.   *Let R be a UFD.*

(i) *Every nonzero $f(x) \in R[x]$ has a factorization*

$$f(x) = c(f)f^*(x),$$

where $c(f) \in R$ and $f^*(x) \in R[x]$ is primitive.

(ii) *This factorization is unique in the sense that if $f(x) = dg^*(x)$, where $d \in R$ and $g^*(x) \in R[x]$ is primitive, then d and $c(f)$ are associates and $f^*(x)$ and $g^*(x)$ are associates.*

(iii) *Let $g^*(x), f(x) \in R[x]$. If $g^*(x)$ is primitive and $g^*(x) \mid bf(x)$, where $b \in R$, then $g^*(x) \mid f(x)$.*

Proof.   (i) If $f(x) = a_n x^n + \cdots + a_1 x + a_0$ and $c(f)$ is the content of f, then there are factorizations $a_i = c(f)b_i$ in R for $i = 0, 1, \dots, n$; if we define $f^*(x) = b_n x^n + \cdots + b_1 x + b_0$, then it is easy to see that $f^*(x)$ is primitive and $f(x) = c(f)f^*(x)$.

(ii) To prove uniqueness, suppose that $f(x) = dg^*(x)$ is a second such factorization, as in the statement. Now $c(f)f^*(x) = f(x) = dg^*(x)$, so that, in

$Q[x]$, where $Q = \text{Frac}(R)$, we have $f^*(x) = [d/c(f)]g^*(x)$. Exercise 6.16 allows us to write $d/c(f)$ in lowest terms: $d/c(f) = u/v$, where u and v are relatively prime elements of R. The equation $vf^*(x) = ug^*(x)$ holds in $R[x]$; equating like coefficients, v is a common divisor of each coefficient of $ug^*(x)$. Since u and v are relatively prime, Exercise 6.17 gives v a common divisor of the coefficients of $g^*(x)$. But $g^*(x)$ is primitive, and so v is a unit. A similar argument shows that u is a unit. Therefore, $d/c(f) = u/v$ is a unit in R, call it w, and $d = wc(f)$; that is, d and $c(f)$ are asssociates and, hence, $g^*(x) = f^*(x)$ are associates.

(iii) Since $g^*(x) \mid bf(x)$, there is $h(x) \in R[x]$ with $bf(x) = g^*(x)h(x)$. By part (i), we have

$$h(x) = c(h)h^*(x) \text{ and } f(x) = c(f)f^*(x),$$

where h^* and f^* are primitive. Therefore,

$$bc(f)f^*(x) = c(h)g^*(x)h^*(x).$$

Now $g^*(x)h^*(x)$ is primitive, by Lemma 6.14, and so the uniqueness in part (ii) gives a unit $u \in R$ with $c(h) = ubc(f)$. Therefore,

$$bf(x) = g^*(x)c(h)h^*(x) = g^*(x)[ubc(f)h^*(x)].$$

Canceling b gives $f(x) = g^*(x)h'(x)$, where $h'(x) = uc(f)h^*(x) \in R[x]$; that is, $g^*(x) \mid f(x)$.   •

Theorem 6.16 (*Gauss*).   *If R is a UFD, then $R[x]$ is also a UFD.*

Proof.   In this proof, the phrase "product of irreducibles" means "irreducible or a product of irreducibles."

We show first, by induction on $\deg(f)$, that every $f(x) \in R[x]$, neither zero nor a unit, is a product of irreducibles. If $\deg(f) = 0$, then $f(x)$ is a constant, hence lies in R. Since R is a UFD, f is a product of irreducibles. If $\deg(f) > 0$, then $f(x) = c(f)f^*(x)$, where $c(f) \in R$ and $f^*(x)$ is primitive. Now $c(f)$ is either a unit or a product of irreducibles, by the base step. If $f^*(x)$ is irreducible, we are done. Otherwise, $f^*(x) = g(x)h(x)$, where neither g nor h is a unit. Since $f^*(x)$ is primitive, however, neither g nor h is a constant; therefore, each of these has degree less than $\deg(f^*) = \deg(f)$, and so each is a product of irreducibles, by the inductive hypothesis.

Proposition 6.11 now applies: $R[x]$ is a UFD if $(p(x))$ is a prime ideal for every irreducible $p(x) \in R[x]$; that is, if $p \mid fg$, then $p \mid f$ or $p \mid g$. Let us assume that $p(x) \nmid f(x)$.

Case (i). Suppose that $\deg(p) = 0$. Write

$$f(x) = c(f)f^*(x) \text{ and } g(x) = c(g)g^*(x),$$

where $c(f), c(g) \in R$, and $f^*(x), g^*(x)$ are primitive. Now $p \mid fg$, so that $p \mid c(f)c(g)f^*(x)g^*(x)$. Since $f^*(x)g^*(x)$ is primitive, we must have $c(f)c(g)$ an associate of $c(fg)$, by Lemma 6.15(ii). However, if $p \mid f(x)g(x)$, then p divides each coefficient of fg; that is, p is a common divisor of all the coefficients of fg, and hence $p \mid c(fg) = c(f)c(g)$ in R, which is a UFD. But Proposition 6.11 says that (p) is a prime ideal in R, and so $p \mid c(f)$ or $p \mid c(g)$. If $p \mid c(f)$, then $p \mid c(f)f^*(x) = f(x)$, a contradiction. Therefore, $p \mid c(g)$ and, hence, $p \mid g(x)$, as desired.

Case (ii). Suppose that $\deg(p) > 0$. Let

$$(p, f) = \{s(x)p(x) + t(x)f(x) : s(x), t(x) \in R[x]\};$$

of course, (p, f) is an ideal containing $p(x)$ and $f(x)$. Choose $m(x) \in (p, f)$ of minimal degree. If $Q = \text{Frac}(R)$ is the fraction field of R, then the division algorithm in $Q[x]$ gives polynomials $q'(x), r'(x) \in Q[x]$ with $f(x) = m(x)q'(x) + r'(x)$, where either $r'(x) = 0$ or $\deg(r') < \deg(m)$. Clearing denominators, there are polynomials $q(x), r(x) \in R[x]$ and a constant $b \in R$ with

$$bf(x) = q(x)m(x) + r(x),$$

where $r(x) = 0$ or $\deg(r) < \deg(m)$. Since $m \in (p, f)$, there are polynomials $s(x), t(x) \in R[x]$ with $m(x) = s(x)p(x) + t(x)f(x)$; hence

$$r = bf - qm = bf - q(sp + tf)$$
$$= (b - tq)f - spq \in (p, f).$$

Since m has minimal degree in (p, f), we must have $r = 0$; that is, $bf(x) = m(x)q(x)$, and so $bf(x) = c(m)m^*(x)q(x)$. But $m^*(x)$ is primitive, and $m^*(x) \mid bf(x)$, so that $m^*(x) \mid f(x)$, by Lemma 6.15(iii). A similar argument, replacing $f(x)$ by $p(x)$, gives $m^*(x) \mid p(x)$. Since $p(x)$ is irreducible, its only factors are units and associates. If $m^*(x)$ were an associate of $p(x)$, then the equation

$$m(x) = c(m)m^*(x) = s(x)p(x) + t(x)f(x)$$

would give $p(x) \mid f(x)$, contrary to the hypothesis. Hence, $m^*(x)$ must be a unit; that is, $m(x) = c(m) \in R$, and so (p, f) contains the nonzero constant $c(m)$. Now $c(m) = sp + tf$, and so

$$c(m)g(x) = s(x)p(x)g(x) + t(x)f(x)g(x).$$

Since $p(x) \mid f(x)g(x)$, we have $p(x) \mid c(m)g(x)$. But $p(x)$ is primitive, because it is irreducible, and so $p(x) \mid g(x)$, by Lemma 6.15(iii). This completes the proof.   •

It follows from Proposition 6.13 that if R is a UFD, then gcd's exist in $R[x]$.

Corollary 6.17.   *If k is a field, then $k[x_1, \dots , x_n]$ is a UFD.*

Proof.   The proof is by induction on $n \geq 1$. We proved, in Chapter 3, that the polynomial ring $k[x_1]$ in one variable is a UFD. For the inductive step, recall that $k[x_1, \dots , x_n, x_{n+1}] = R[x_{n+1}]$, where $R = k[x_1, \dots , x_n]$. By induction, R is a UFD, and by Theorem 6.16, so is $R[x_{n+1}]$.   •

The theorem of Gauss, Theorem 3.63, can be generalized.

Corollary 6.18.   *Let R be a UFD, let $Q = \text{Frac}(R)$, and let $f(x) \in R[x]$. If*

$$f(x) = G(x)H(x) \text{ in } Q[x],$$

then there is a factorization

$$f(x) = g(x)h(x) \text{ in } R[x],$$

where $\deg(g) = \deg(G)$ and $\deg(h) = \deg(H)$. Therefore, if $f(x)$ does not factor into polynomials of smaller degree in $R[x]$, then $f(x)$ is irreducible in $Q[x]$.

Proof.   By Lemma 6.15, there is a factorization

$$f(x) = c(G)c(H)G^*(x)H^*(x) \text{ in } Q[x],$$

where $G^*(x)$, $H^*(x) \in R[x]$ are primitive polynomials. But $c(G)c(H) = c(f)$, by Lemma 6.15(ii). Since $c(f) \in R$, there is a factorization $f(x) = g(x)h(x)$ in $R[x]$, where $g(x) = c(f)G^*(x)$ and $h(x) = H^*(x)$.   •

Irreducibility of a polynomial in several variables is more difficult to determine than irreducibility of a polynomial of one variable, but here is one criterion.

Corollary 6.19. *Let k be a field and let $f(x_1, \ldots, x_n)$ be a primitive polynomial in $R[x_n]$, where $R = k[x_1, \ldots, x_{n-1}]$. If f cannot be factored into two polynomials of lower degree in $R[x_n]$, then f is irreducible in $k[x_1, \ldots, x_n]$.*

Proof. Let us write $f(x_1, \ldots, x_n) = F(x_n)$ if we wish to view f as a polynomial in $R[x_n]$ (of course, the coefficients of F are polynomials in $k[x_1, \ldots, x_{n-1}]$). Suppose that $F(x_n) = G(x_n)H(x_n)$; by hypothesis, the degrees of G and H (in x_n) cannot both be less than $\deg(F)$, and so one of them, say, G, has degree 0. It follows, because F is primitive, that G is a unit in $k[x_1, \ldots, x_{n-1}]$. Therefore, $f(x_1, \ldots, x_n)$ is irreducible in $R[x_n] = k[x_1, \ldots, x_n]$.   •

Of course, the corollary applies to any variable x_i, not just to x_n.

Example 6.8.
If

$$f(x_1, \ldots, x_n) = x_n g(x_1, \ldots, x_{n-1}) + h(x_1, \ldots, x_{n-1}),$$

where $(g, h) = 1$, then f is primitive; since f is linear in x_n, it does not factor over $k(x_1, \ldots, x_{n-1})$, and so it is irreducible in $k[x_1, \ldots, x_n]$. For example, $xy^2 + z$ is an irreducible polynomial in $k[x, y, z]$ because it is a primitive polynomial that is linear in x.   ◄

EXERCISES

6.15 In any commutative ring R, prove that if a gcd of any two elements always exists, then a gcd of any finite number of elements also exists.

6.16 Let R be a UFD and let $Q = \operatorname{Frac}(R)$ be its fraction field. Prove that each nonzero $a/b \in Q$ has an expression in lowest terms; that is, a and b are relatively prime.

6.17 Let R be a UFD, and let $a, b, c \in R$. If a and b are relatively prime, and if $a \mid bc$, prove that $a \mid c$.

6.18 If R is a domain, prove that the only units in $R[x_1, \ldots, x_n]$ are units in R.

6.19 If R is a UFD and $f(x), g(x) \in R[x]$, prove that $c(fg)$ and $c(f)c(g)$ are associates.

6.20   (i) Prove that x and y are relatively prime in $k[x, y]$, where k is a field.

   (ii) Prove that 1 is not a linear combination of x and y in $k[x, y]$.

6.21 Prove that $\mathbb{Z}[x_1, \ldots, x_n]$ is a UFD for all $n \geq 1$.

6.22 Let k be a field and let $f(x_1, \ldots, x_n) \in k[x_1, \ldots, x_n]$ be a primitive polynomial in $R[x_n]$, where $R = k[x_1, \ldots, x_{n-1}]$. If f is either quadratic or cubic in x_n, prove that f is irreducible in $k[x_1, \ldots, x_n]$ if and only if f has no roots in $k(x_1, \ldots, x_{n-1})$.

6.23 (***Eisenstein's criterion***) Let R be a UFD with $Q = \mathrm{Frac}(R)$, and let $f(x) = a_0 + a_1 x + \cdots + a_n x^n \in R[x]$. Prove that if there is an irreducible element $p \in R$ with $p \mid a_i$ for all $i < n$ but with $p \nmid a_n$ and $p^2 \nmid a_0$, then $f(x)$ is irreducible in $Q[x]$.

6.24 Prove that

$$f(x, y) = xy^3 + x^2 y^2 - x^5 y + x^2 + 1$$

is an irreducible polynomial in $\mathbb{R}[x, y]$.

6.3 NOETHERIAN RINGS

One of the most important properties of $k[x_1, \ldots, x_n]$, when k is a field, is that every ideal in it can be generated by a finite number of elements. This property is intimately related to chains of ideals, which we have already seen in the course of proving that PID's are UFD's (I apologize for so many acronyms).

Definition. A commutative ring R satisfies the *ACC*, the *ascending chain condition*, if every ascending chain of ideals

$$I_1 \leq I_2 \leq \cdots \leq I_n \leq \cdots$$

stops; that is, the sequence is constant from some point on: there is an integer N with $I_N = I_{N+1} = I_{N+2} = \cdots$.

The proof of Lemma 6.10 shows that every PID satisfies the ACC.
Here is an important type of ideal.

Definition. An ideal I in a commutative ring R is called *finitely generated* if there are finitely many elements $a_1, \ldots, a_n \in I$ with

$$I = \left\{ \sum_i r_i a_i : r_i \in R \text{ for all } i \right\};$$

that is, every element in I is a linear combination of the a_i's. One writes

$$I = (a_1, \dots, a_n)$$

and calls I the ***ideal generated by*** $a_1, \dots, a_n$. A set of generators $a_1, \dots, a_n$ of an ideal I is sometimes called a ***basis*** of I (although this is a weaker notion than that of a basis of a vector space).

Every ideal I in a PID can be generated by one element, and so I is finitely generated.

Proposition 6.20.   *The following conditions are equivalent for a commutative ring R.*

(i) *R has the ACC.*

(ii) *R satisfies the **maximum condition**: Every nonempty family $\mathcal{F}$ of ideals in R has a maximal element; that is, there is some $I_0 \in \mathcal{F}$ with $I \leq I_0$ for all $I \in \mathcal{F}$.*

(iii) *Every ideal in R is finitely generated.*

Proof.   (i) $\Rightarrow$ (ii): Let $\mathcal{F}$ be a family of ideals in R, and assume that $\mathcal{F}$ has no maximal element. Choose $I_1 \in \mathcal{F}$. Since I_1 is not a maximal element, there is $I_2 \in \mathcal{F}$ with $I_1 < I_2$. Now I_2 is not a maximal element in $\mathcal{F}$, and so there is $I_3 \in \mathcal{F}$ with $I_2 < I_3$. Continuing in this way, we can construct an ascending chain of ideals in R that does not stop, contradicting the ACC.

(ii) $\Rightarrow$ (iii): Let I be an ideal in R, and define $\mathcal{F}$ to be the family of all the finitely generated ideals contained in I; of course, $\mathcal{F} \neq \varnothing$. By hypothesis, there exists a maximal element $M \in \mathcal{F}$. Now $M \leq I$ because $M \in \mathcal{F}$. If $M < I$, then there is $a \in I$ with $a \notin M$. The ideal

$$J = \{m + ra : m \in M \text{ and } r \in R\} \leq I$$

is finitely generated, and so $J \in \mathcal{F}$; but $M < J$, and this contradicts the maximality of M. Therefore, $M = I$, and so I is finitely generated.

(iii) $\Rightarrow$ (i): Assume that every ideal in R is finitely generated, and let

$$I_1 \leq I_2 \leq \cdots \leq I_n \leq \cdots$$

be an ascending chain of ideals in R. As in the proof of Lemma 6.10, we show that $J = \bigcup_n I_n$ is an ideal. If $a \in J$, then $a \in I_n$ for some n; if $r \in R$, then $ra \in I_n$, because I_n is an ideal; hence, $ra \in J$. If $a, b \in J$, then there are ideals I_n and I_m with $a \in I_n$ and $b \in I_m$; since the chain is ascending, we may assume that $I_n \subset I_m$, and so $a, b \in I_m$. As I_m is an ideal, $a - b \in I_m$ and, hence, $a - b \in J$. Therefore, J is an ideal.

By hypothesis, there are elements $a_i \in J$ with $J = (a_1, \ldots, a_q)$. Now a_i got into J by being in I_{n_i} for some n_i. If N is the largest n_i, then $I_{n_i} \leq I_N$ for all i; hence, $a_i \in I_N$ for all i, and so

$$J = (a_1, \ldots, a_q) \leq I_N \leq J.$$

It follows that if $n \geq N$, then $J = I_N \leq I_n \leq J$, so that $I_n = J$; therefore, the chain stops, and R has the ACC.   •

We now give a name to a commutative ring which satisfies any of the three equivalent conditions in the proposition.

Definition.   A commutative ring R is called ***noetherian***[1] if every ideal in R is finitely generated.

The maximum condition is related to Zorn's lemma.

Definition.   A ***partially ordered set*** is a nonempty set X equipped with a relation $x \preceq y$ such that, for all $x, y, z \in X$, we have

(i) ***reflexivity***: $x \preceq x$;

(ii) ***antisymmetry***: if $x \preceq y$ and $y \preceq x$, then $x = y$;

(iii) ***transitivity***: if $x \preceq y$ and $y \preceq z$, then $x \preceq z$.

An element u in a partially ordered set X is called a ***maximal element*** if there is no $x \in X$ with $u \preceq x$ and $u \neq x$.

If A is a set, then the family $P(A)$ of all the subsets of A is a partially ordered set if one defines $U \preceq V$ to mean $U \subset V$, where U and V are subsets of A; the family $P(A)^*$, consisting of all the proper subsets of A, is also a partially ordered set (more generally, every nonempty subset of a partially ordered set is itself a partially ordered set). Another example is the real numbers $\mathbb{R}$, with $x \preceq y$ meaning $x \leq y$. There are some partially ordered sets, e.g., $P(A)^*$, having many maximal elements, and there are some partially ordered sets, e.g., $\mathbb{R}$, having no maximal elements. Zorn's lemma is a condition on a partially ordered set which guarantees that it has at least one maximal element.

---

[1]This name honors Emmy Noether (1882–1935), who introduced chain conditions in 1921.

There is usually no need for Zorn's lemma when dealing with noetherian rings, for the maximum condition guarantees the existence of a maximal element in any nonempty family $\mathcal{F}$ of ideals.

Corollary 6.21. *If I is an ideal in a noetherian ring R, then there exists a maximal ideal M in R containing I. In particular, every noetherian ring has maximal ideals.*[2]

Proof. Let $\mathcal{F}$ be the family of all those proper ideals in R which contain I; note that $\mathcal{F} \neq \varnothing$ because $I \in \mathcal{F}$. Since R is noetherian, the maximum condition gives a maximal element M in $\mathcal{F}$. We must still show that M is a maximal ideal in R (that is, that M is actually a maximal element in the larger family $\mathcal{F}'$ consisting of all the proper ideals in R). Suppose there is a proper ideal J with $M \leq J$. Then $I \leq J$, and so $J \in \mathcal{F}$; therefore, maximality of M gives $M = J$, and so M is a maximal ideal in R.   •

Here is one way to construct a new noetherian ring from an old one.

Corollary 6.22. *If R is a noetherian ring and J is an ideal in R, then R/J is also noetherian.*

Proof. If A is an ideal in R/I, then the correspondence theorem provides an ideal J in R with $J/I = A$. Since R is noetherian, the ideal J is finitely generated, say, $J = (b_1, \ldots , b_n)$, and so $A = J/I$ is also finitely generated (by the cosets $b_1 + I, \ldots , b_n + I$). Therefore, R/I is noetherian.   •

The following anecdote is well known. Around 1890, Hilbert proved the famous Hilbert basis theorem, showing that every ideal in $\mathbb{C}[x_1, \ldots , x_n]$ is finitely generated. As we will see, the proof is nonconstructive in the sense that it does not give an explicit set of generators of an ideal. It is reported that when P. Gordan, one of the leading algebraists of the time, first saw Hilbert's proof, he said, "This is not mathematics, but theology!" On the other hand, Gordan said, in 1899 when he published a simplified proof of Hilbert's theorem, "I have convinced myself that theology also has its advantages."

---

[2]This corollary is true without assuming R is noetherian, but the proof of the general result needs Zorn's lemma.

Theorem 6.23 (Hilbert Basis Theorem).   *If R is a commutative noetherian ring, then $R[x]$ is also noetherian.*

Proof.   Let I be an ideal in $R[x]$. We are going to use the leading coefficients of polynomials in I to produce a finite, though extravagantly large, generating set of I.

We begin by setting up notation. If $h(x) \in R[x]$ is not the zero polynomial, then

$$h(x) = a_d x^d + \text{ lower terms,}$$

where its leading coefficient is $a_d \neq 0$; let us write $\mathrm{LC}(h) = a_d$. Define

$$L = \{0\} \cup \{\mathrm{LC}(h) \colon h(x) \in I\},$$

and, for $j \geq 0$, define

$$L_j = \{0\} \cup \{\mathrm{LC}(h) \colon h(x) \in I \text{ and } \deg(h) \leq j\}.$$

It is easy to see that each L_j is an ideal in R. Since R is noetherian, these ideals are finitely generated:

$$L_j = (r_{j1}, r_{j2}, \dots, r_{jt(j)}) \quad \text{and} \quad L = (c_1, c_2, \dots, c_k).$$

The ideals L_j and L are comprised of leading coefficients; for each c_i, choose $f_i(x) \in I$ with $\mathrm{LC}(f_i) = c_i$, and let

$$N = \max\{\deg(f_1), \deg(f_2), \dots, \deg(f_k)\}.$$

For each $j < N$, choose polynomials $g_{j1}(x), g_{j2}(x), \dots, g_{jt(j)}(x)$ in I with degrees $\leq j$ and leading coefficients $r_{j1}, r_{j2}, \dots$, respectively.

The proof begins now. Define J to be the ideal in $R[x]$ generated by all the f's and g_{ji}'s, where $j < N$. It is clear that J is a finitely generated ideal and that $J \leq I$. We claim that the reverse inclusion holds as well, so that $J = I$ and, hence, I is finitely generated.

We are going to prove that if $h(x) \in I$, then there exists $u(x) \in J$, so that either $h(x) - u(x) = 0$ or $\deg(h - u) < \deg(h)$. Assume that

$$h(x) = a_d x^d + \text{ lower terms.}$$

Step (i): $d < N$.

Now $a_d \in L_d = (r_{d1}, r_{d2}, \dots)$, so that $a_d = \sum_i s_i r_{di}$, where $s_i \in R$ for all i. Define $u(x) = \sum s_i x^{d - \deg(g_{di})} g_{di}(x)$; the exponent $d - \deg(g_{di}) \geq 0$ because $\deg(g_{di}) \leq d$. Of course, $u(x) \in J$, and either $h(x) - u(x) = 0$ or $\deg(h - u) < \deg(h) = d$.

Step (ii): $d \geq N$.

Now $a_d \in L = (c_1, c_2, \dots, c_k)$, so that $a_d = \sum_\ell t_\ell c_\ell$, where $t_\ell \in R$. Define $u(x) = \sum_\ell t_\ell x^{d - \deg(f_\ell)} f_\ell(x)$; the exponent $d - \deg(f_\ell) \geq 0$ because $d \geq N \geq \deg(f_\ell)$. Of course, $u(x)$ lies in J and either $h(x) - u(x) = 0$ or $\deg(h - u) < \deg(h)$.

To complete the proof, it suffices to show that $I \leq J$; that is, if $h(x) \in I$, then $h(x) \in J$. Suppose first that $\deg(h) < N$. There exists $u_1(x) \in J$ with $h(x) - u_1(x) = 0$ or $\deg(h - u_1) < \deg(h)$. Since $J \leq I$, we have $h(x) - u_1(x) \in I$, and so this procedure can be repeated: there is $u_2(x) \in J$ with $h(x) - u_1(x) - u_2(x) = 0$ or $\deg(h - u_1 - u_2) < \deg(h - u_1)$. Since degrees decrease strictly, there is some n with $h - u_1 - \cdots - u_n = 0$, and so $h(x) \in J$, as desired.   •

Let us review the proof of the Hilbert basis theorem in the special case when R is a field. In this case, the only ideals are $\{0\}$ and $(1) = R$. Thus, if I is a nonzero ideal in $R[x]$, then, using the notation of the proof of Theorem 6.23, the ideal $L = (1)$, and so only one polynomial $f_1(x) \in I$ is chosen, and it is monic. The ideals L_j, where $j \leq N = \deg(f_1)$, are $\{0\}$ if there are no polynomials in I of degree $\leq j$, and $L_j = (1)$ otherwise. If L_m is the first nonzero ideal, then $r_{m1} = 1$ and $g_{m1}(x)$ is the monic polynomial in I of least degree. The proof says that $I = (g_{m1}(x))$, a fact we have known since Chapter 1.

Corollary 6.24.

(i) *If k is a field, then $k[x_1, \dots, x_n]$ is noetherian.*

(ii) *The ring $\mathbb{Z}[x_1, \dots, x_n]$ is noetherian.*

(iii) *For any ideal I in $k[x_1, \dots, x_n]$, where $k = \mathbb{Z}$ or k is a field, the quotient ring $k[x_1, \dots, x_n]/I$ is noetherian.*

Proof.   The proofs of the first two items are by induction on $n \geq 1$, using the theorem, while the proof of item (iii) follows from Corollary 6.22.   •

EXERCISES

6.25 Let m be a positive integer, and let X be the set of all its (positive) divisors. Prove that X is a partially ordered set if one defines $a \preceq b$ to mean $a \mid b$.

6.26 Prove that the ring $\mathcal{F}(\mathbb{R})$ of Example 3.3(i) on page 210 is not a noetherian ring.

6.27 Prove that if R is a noetherian ring, then the ring of formal power series $R[[x]]$ is also a noetherian ring.

6.28 Let

$$S^2 = \{(a, b, c) \in \mathbb{R}^3 : a^2 + b^2 + c^2 = 1\}$$

be the unit sphere in $\mathbb{R}^3$, and let

$$I = \{f(x, y, z) \in \mathbb{R}[x, y, z] : f(a, b, c) = 0 \text{ for all } (a, b, c) \in S^2\}.$$

Prove that I is a finitely generated ideal in $\mathbb{R}[x, y, z]$.

6.29 If R and S are noetherian rings, prove that their direct product $R \times S$ is also a noetherian ring.

6.30 If R is a ring that is also a vector space over a field k, then R is called a *k-algebra* if

$$(\alpha u)v = \alpha(uv) = u(\alpha v)$$

for all $\alpha \in k$ and $u, v \in R$. Prove that every finite-dimensional k-algebra is a noetherian ring.

6.4 VARIETIES

Analytic geometry gives pictures of equations. For example, we picture a function $f : \mathbb{R} \to \mathbb{R}$ as its graph, which consists of all the ordered pairs $(a, f(a))$ in the plane; that is, f is the set of all the solutions $(a, b) \in \mathbb{R}^2$ of

$$g(x, y) = y - f(x) = 0.$$

We can also picture equations that are not graphs of functions. For example, the set of all the zeros of the polynomial

$$h(x, y) = x^2 + y^2 - 1$$

is the unit circle. One can also picture simultaneous solutions in $\mathbb{R}^2$ of several polynomials of two variables, and, indeed, one can picture simultaneous solutions in $\mathbb{R}^n$ of several polynomials of n variables.

Notation.   Let k be a field and let k^n denote the set of all n-tuples

$$k^n = \{a = (a_1, \ldots, a_n): a_i \in k \text{ for all } i\}.$$

The polynomial ring $k[x_1, \ldots, x_n]$ in several variables may be denoted by $k[X]$, where X is the abbreviation:

$$X = (x_1, \ldots, x_n).$$

In particular, $f(X) \in k[X]$ may abbreviate $f(x_1, \ldots, x_n) \in k[x_1, \ldots, x_n]$.

In what follows, we regard polynomials $f(x_1, \ldots, x_n) \in k[x_1, \ldots, x_n]$ as functions of n variables $k^n \to k$. Here is the precise definition.

Definition.   A polynomial $f(X) \in k[X]$ determines a ***polynomial function*** $f^\flat: k^n \to k$ in the obvious way: if $(a_1, \ldots, a_n) \in k^n$, then

$$f^\flat: (a_1, \ldots, a_n) \mapsto f(a_1, \ldots, a_n).$$

The next proposition generalizes Corollary 3.35 from one variable to several variables.

Proposition 6.25.   *Let k be an infinite field and let $k[X] = k[x_1, \ldots, x_n]$. If $f(X), g(X) \in k[X]$ satisfy $f^\flat = g^\flat$, then $f(x_1, \ldots, x_n) = g(x_1, \ldots, x_n)$.*

Proof.   The proof is by induction on $n \geq 1$; the base step is Corollary 3.35. For the inductive step, write

$$f(X, y) = \sum_i p_i(X)y^i \text{ and } g(X, y) = \sum_i q_i(X)y^i,$$

where X denotes $(x_1, \ldots, x_n)$. If $f^\flat = g^\flat$, then we have $f(a, \alpha) = g(a, \alpha)$ for every $a \in k^n$ and every $\alpha \in k$. For fixed $a \in k^n$, define $F_a(y) = \sum_i p_i(a)y^i$ and $G_a(y) = \sum_i q_i(a)y^i$. Since both $F_a(y)$ and $G_a(y)$ are in $k[y]$, the base step gives $p_i(a) = q_i(a)$ for all $a \in k^n$. By the inductive hypothesis, $p_i(X) = q_i(X)$ for all i, and hence

$$f(X, y) = \sum_i p_i(X)y^i = \sum_i q_i(X)y^i = g(X, t),$$

as desired.   ●

As a consequence of this last proposition, we drop the $f^\flat$ notation and identify polynomials with their polynomial functions when k is infinite.

Definition.   If $f(X) \in k[X] = k[x_1, \ldots, x_n]$ and $f(a) = 0$, where $a \in k^n$, then a is called a *zero* of $f(X)$. [If $f(x)$ is a polynomial in one variable, then a zero of $f(x)$ is also called a root of $f(x)$.]

Proposition 6.26.   *If k is an algebraically closed field and $f(X) \in k[X]$ is not a constant, then $f(X)$ has a zero.*

Proof.   We prove the result by induction on $n \geq 1$, where $X = (x_1, \ldots, x_n)$. The base step follows at once from our assuming that $k^1 = k$ is algebraically closed. As in the previous proof, write

$$f(X, y) = \sum_i g_i(X) y^i.$$

For each $a \in k^n$, define $f_a(y) = \sum_i g_i(a) y^i$. If $f(X, y)$ has no zeros, then each $f_a(y) \in k[y]$ has no zeros, and the base step says that $f_a(y)$ is a nonzero constant for all $a \in k^n$. Thus, $g_i(a) = 0$ for all $i > 0$ and all $a \in k^n$. By Proposition 6.25, which applies because algebraically closed fields are infinite, $g_i(X) = 0$ for all $i > 0$, and so $f(X, y) = g_0(X) y^0 = g_0(X)$. By the inductive hypothesis, $g_0(X)$ is a nonzero constant, and the proof is complete.   •

We now give some general definitions describing solution sets of polynomials.

Definition.   If $F \subset k[X] = k[x_1, \ldots, x_n]$, then the *variety* [3,4] defined by F

---

[3] There is some disagreement about the usage of this term. Some call this an *affine variety*, in contrast to the analogous *projective variety*. Some insist that varieties should be *irreducible*, which we will define later in this section.

[4] The term *variety* arose as a translation by E. Beltami (inspired by Gauss) of the German term *Mannigfaltigkeit* used by Riemann; nowadays, this term is usually translated as *manifold*. The following correspondence, from Aldo Brigaglia to Steven Kleiman, contains more details.

"I believe the usage of the word *varietà* by Italian geometers arose from the (unpublished) Italian translation of Riemann's Habilitationsvortrag, which was later translated into French by J. Hoüel and published in the Italian journal *Annali*. Indeed, Beltrami wrote to Hoüel on 8 January, 1869:

> J'ai traduit *Mannigfaltigkeit* par *varietà*, dans le sens de *multitudo variarum rerum*...

And later, on 14 February, 1869, he wrote

> Je croirais toujours convenable de traduire *Mannigfaltigkeit* par *variété*: j'ai remarqué que Gauss, dans ses Mémoires sur les résidus biquadratiques appelle en latin *varietas* la même chose qui, dans les comptes-rendus rédigés par

is

$$\text{Var}(F) = \{a \in k^n : f(a) = 0 \text{ for every } f(X) \in F\};$$

thus, $\text{Var}(F)$ consists of all those $a \in k^n$ which are zeros of every $f(X) \in F$.

Example 6.9.
(i) Here is a variety defined by two equations.

$$\text{Var}(x, y) = \{(a, b) \in k^2 : x = 0 \text{ and } y = 0\}.$$

Thus,

$$\text{Var}(x, y) = x\text{-axis} \cup y\text{-axis}.$$

(ii) Here is an example in higher-dimensional space. Let A be an $m \times n$ matrix with entries in k. A system of m equations in n unknowns,

$$AX = B,$$

where B is an $n \times 1$ column matrix, defines a variety, $\text{Var}(AX = B)$, which is a subset of k^n. Of course, $AX = B$ is really shorthand for a set of m linear equations in n variables, and $\text{Var}(AX = B)$ is usually called the **solution set** of the system $AX = B$; when this system is homogeneous, that is, when $B = 0$, then $\text{Var}(AX = 0)$ is a subspace of k^n, called the **solution space** of the system.   ◄

The next result shows that, as far as varieties are concerned, one may just as well assume the subsets F of $k[X]$ are ideals of $k[X]$.

Proposition 6.27.

(i) *If $F \subset G \subset k[X]$, then $\text{Var}(G) \subset \text{Var}(F)$.*

(ii) *If $F \subset k[X]$ and $I = (F)$ is the ideal generated by F, then*

$$\text{Var}(F) = \text{Var}(I).$$

---

lui même en allemand dans les *Gelehrte Anzeige*, est désignée par *Mannig-faltigkeit*.

The correspondence of Beltrami and Hoüel can be found in the beautiful book *La découverte de la géométrie non euclidienne sur la pseudosphère: les lettres d'Eugenio Beltrami à Jules Hoüel (1868–1881)*, edited by L. Boi, L. Giacardi, and R. Tazzioli, and published by Blanchard, Paris, 1998."

Proof.   (i) If $a \in \text{Var}(G)$, then $g(a) = 0$ for all $g(X) \in G$; since $F \subset G$, it follows, in particular, that $f(a) = 0$ for all $f(X) \in F$.

(ii) Since $F \subset (F) = I$, we have $\text{Var}(I) \subset \text{Var}(F)$, by part (i). For the reverse inclusion, let $a \in \text{Var}(F)$, so that $f(a) = 0$ for every $f(X) \in F$. If $g(X) \in I$, then $g(X) = \sum_i r_i f_i(X)$, where $r_i \in k$ and $f_i(X) \in F$; hence, $g(a) = \sum_i r_i f_i(a) = 0$ and $a \in \text{Var}(I)$.   •

It follows that not every subset of k^n is a variety. For example, if $n = 1$, then $k[x]$ is a PID. Hence, if F is a subset of $k[x]$, then $(F) = (g(x))$ for some $g(x) \in k[x]$, and so

$$\text{Var}(F) = \text{Var}((F)) = \text{Var}((g(x))) = \text{Var}(g(x)).$$

But $g(x)$ has only a finite number of roots, and so $\text{Var}(F)$ is finite. If k is algebraically closed, then it is an infinite field, and so most subsets of $k^1 = k$ are not varieties.

In spite of our wanting to draw pictures in the plane, there is a major defect with $k = \mathbb{R}$: some polynomials have no zeros. For example, $f(x) = x^2 + 1$ has no real roots, and so $\text{Var}(x^2 + 1) = \varnothing$. More generally, $g(x_1, \dots, x_n) = x_1^2 + \cdots + x_n^2 + 1$ has no zeros in $\mathbb{R}^n$, and so $\text{Var}(g(X)) = \varnothing$. Since we are dealing with (not necessarily linear) polynomials, it is a natural assumption to want all their zeros available. For polynomials in one variable, this amounts to saying that k is algebraically closed and, in light of Proposition 6.26, we know that $\text{Var}(f(X)) \neq \varnothing$ for every nonconstant $f(X) \in k[X]$ if k is algebraically closed. Of course, varieties are of interest for all fields k, but it makes more sense to consider the simplest case before trying to understand more complicated problems. On the other hand, many of the first results below are valid for any field k. Thus, we will state the hypothesis needed for each proposition, but the reader should realize that the most important case is when k is algebraically closed.

Here are some elementary properties of Var.

Proposition 6.28.   *Let k be a field.*

(i) $\text{Var}(x_1, x_1 - 1) = \varnothing$ *and* $\text{Var}(0) = k^n$, *where 0 is the zero polynomial.*

(ii) *If I and J are ideals in $k[X]$, then*

$$\text{Var}(IJ) = \text{Var}(I \cap J) = \text{Var}(I) \cup \text{Var}(J),$$

where $IJ = \left\{\sum_i f_i(X)g_i(X): f_i(X) \in I \text{ and } g_i(X) \in J\right\}$.

(iii) *If $\{I_\ell : \ell \in L\}$ is a family of ideals in $k[X]$, then*

$$\mathrm{Var}\left(\sum_\ell I_\ell\right) = \bigcap_\ell \mathrm{Var}(I_\ell),$$

where $\sum_\ell I_\ell$ is the set of all finite sums of the form $r_{\ell_1} + \cdots + r_{\ell_q}$ with $r_{\ell_i} \in I_{\ell_i}$.

Proof.   (i) If $a = (a_1, \dots, a_n) \in \mathrm{Var}(x_1, x_1 - 1)$, then $a_1 = 0$ and $a_1 = 1$; plainly, there are no such points a, and so $\mathrm{Var}(x_1, x_1 - 1) = \varnothing$. That $\mathrm{Var}(0) = k^n$ is clear, for every point a is a zero of the zero polynomial.

(ii) Since $IJ \leq I \cap J$, it follows that $\mathrm{Var}(IJ) \supset \mathrm{Var}(I \cap J)$; since $IJ \subset I$, it follows that $\mathrm{Var}(IJ) \supset \mathrm{Var}(I)$. Hence,

$$\mathrm{Var}(IJ) \supset \mathrm{Var}(I \cap J) \supset \mathrm{Var}(I) \cup \mathrm{Var}(J).$$

To complete the proof, it suffices to show that $\mathrm{Var}(IJ) \subset \mathrm{Var}(I) \cup \mathrm{Var}(J)$. If $a \notin \mathrm{Var}(I) \cup \mathrm{Var}(J)$, then there exist $f(X) \in I$ and $g(X) \in J$ with $f(a) \neq 0$ and $g(a) \neq 0$. But $f(X)g(X) \in IJ$ and $(fg)(a) = f(a)g(a) \neq 0$, because $k[X]$ is a domain. Therefore, $a \notin \mathrm{Var}(IJ)$, as desired.

(iii) For each ℓ, the inclusion $I_\ell \leq \sum_\ell I_\ell$ gives $\mathrm{Var}\left(\sum_\ell I_\ell\right) \subset \mathrm{Var}(I_\ell)$, and so

$$\mathrm{Var}\left(\sum_\ell I_\ell\right) \subset \bigcap_\ell \mathrm{Var}(I_\ell).$$

For the reverse inclusion, if $g(X) \in \sum_\ell I_\ell$, then there are finitely many ℓ with $g(X) = \sum_\ell h_\ell f_\ell$, where $h_\ell \in k[X]$ and $f_\ell(X) \in I_\ell$. Therefore, if $a \in \bigcap_\ell \mathrm{Var}(I_\ell)$, then $f_\ell(a) = 0$ for all ℓ, and so $g(a) = 0$; that is, $a \in \mathrm{Var}\left(\sum_\ell I_\ell\right)$.   •

Definition.   A *topology* on a set X is a family $\mathcal{F}$ of subsets of X, called *closed sets*[5], which satisfy the following axioms:

(i) $\varnothing \in \mathcal{F}$ and $X \in \mathcal{F}$;

(ii) if $F_1, F_2 \in \mathcal{F}$, then $F_1 \cup F_2 \in \mathcal{F}$; that is, the union of two closed sets is closed;

(iii) if $\{F_\ell : \ell \in L\} \subset \mathcal{F}$, then $\bigcap_\ell F_\ell \in \mathcal{F}$; that is, any intersection of closed sets is also closed.

---

[5]One can also define a topology by specifying its *open subsets*, which are the complements of closed sets.

Proposition 6.28 shows that the family of all varieties is a topology on k^n; it is called the **Zariski topology**, and it is very useful in the deeper study of $k[X]$. The usual topology on $\mathbb{R}$ has many closed sets; for example, every closed interval is a closed set. In contrast, in the Zariski topology on $\mathbb{R}$, every closed set (aside from $\mathbb{R}$) is finite.

Given an ideal I in $k[X]$, we have just defined its variety $\mathrm{Var}(I) \subset k^n$. We now reverse direction: given a subset $A \subset k^n$, we assign an ideal in $k[X]$ to it; in particular, we assign an ideal to every variety.

Definition. If $A \subset k^n$, define its **coordinate ring** $k[A]$ to be the commutative ring of all polynomial functions $f : A \to k$ under pointwise operations.

The polynomial $f(x_1, \dots, x_n) = x_i \in k[X]$, when regarded as a polynomial function, is defined by

$$x_i : (a_1, \dots, a_n) \mapsto a_i;$$

that is, x_i picks out the ith coordinate of a point in k^n. The reason for the name coordinate ring is that if $a \in V$, then $(x_1(a), \dots, x_n(a))$ describes a.

There is an obvious ring homomorphism res: $k[X] \to k[A]$, given by $f(X) \mapsto f|A$, and the kernel of this restriction map is an ideal in $k[X]$.

Definition. If $A \subset k^n$, define

$$\mathrm{Id}(A) = \{f(X) \in k[X] = k[x_1, \dots, x_n] : f(a) = 0 \text{ for every } a \in A\}.$$

The Hilbert basis theorem tells us that $\mathrm{Id}(A)$ is always a finitely generated ideal.

Proposition 6.29. *If $A \subset k^n$, then there is an isomorphism*

$$k[X]/\mathrm{Id}(A) \cong k[A].$$

Proof.   The restriction map res: $k[X] \to k[A]$ is a surjection with kernel $\mathrm{Id}(A)$, and so the result follows from the first isomorphism theorem. Note that two polynomials agreeing on A lie in the same coset of $\mathrm{Id}(A)$.   •

Although the definition of $\mathrm{Var}(F)$ makes sense for any subset F of $k[X]$, it is most interesting when F is an ideal. Similarly, although the definition of $\mathrm{Id}(A)$ makes sense for any subset A of k^n, it is most interesting when A is a variety. After all, varieties are comprised of solutions of (polynomial) equations, which is what we care about.

Proposition 6.30.   *Let k be a field.*

(i) $\mathrm{Id}(\varnothing) = k[X]$ *and, if k is algebraically closed,* $\mathrm{Id}(k^n) = \{0\}$.

(ii) *If* $A \subset B$ *are subsets of* k^n, *then* $\mathrm{Id}(B) \leq \mathrm{Id}(A)$.

(iii) *If* $\{A_\ell : \ell \in L\}$ *is a family of subsets of* k^n, *then*

$$\mathrm{Id}\left(\bigcup_\ell A_\ell\right) = \bigcap_\ell \mathrm{Id}(A_\ell).$$

Proof.   (i) If $f(X) \in \mathrm{Id}(A)$ for some subset $A \subset k^n$, then $f(a) = 0$ for all $a \in A$; hence, if $f(X) \notin \mathrm{Id}(A)$, then there exists $a \in A$ with $f(a) \neq 0$. In particular, if $A = \varnothing$, every $f(X) \in k[X]$ must lie in $\mathrm{Id}(\varnothing)$, for there are no elements $a \in \varnothing$. Therefore, $\mathrm{Id}(\varnothing) = k[X]$.

We prove that $\mathrm{Id}(k^n) = \{0\}$ by induction on $n \geq 1$. The base step $n = 1$ is true, for if $f(x) \in k[x]$ lies in $\mathrm{Id}(k)$, then every element in k is a root of $f(x)$. But algebraically closed fields are infinite, and a nonzero polynomial has only a finite number of roots. If $f(x_1, \ldots, x_{n+1}) \in k[x_1, \ldots, x_{n+1}]$, then collecting terms involving the powers of x_{n+1} gives $g_i(X) \in k[X] = k[x_1, \ldots, x_n]$ with

$$f(X, x_{n+1}) = \sum_{i \geq 0} g_i(X) x_{n+1}^i.$$

If $f \in \mathrm{Id}(k^{n+1})$, then

$$f(\alpha, a) = \sum_{i \geq 0} g_i(\alpha) a^i = 0$$

for all $\alpha \in k^n$ and $a \in k$. For each fixed α, define

$$h_\alpha(x_{n+1}) = \sum_{i \geq 0} g_i(\alpha) x_{n+1}^i \in k[x_{n+1}].$$

Since $h_\alpha(a) = 0$ for all $a \in k$, the base step gives $h_\alpha(x_{n+1}) = 0$; that is, $g_i(\alpha) = 0$ for all i and all α. As $g_i(X) \in k[X] = k[x_1, \ldots, x_n]$, the inductive hypothesis gives $g_i(X) = 0$ for all i, and so $f(X, x_{n+1}) = 0$, as desired.

(ii) If $f(X) \in \mathrm{Id}(B)$, then $f(b) = 0$ for all $b \in B$; in particular, $f(a) = 0$ for all $a \in A$, because $A \subset B$, and so $f(X) \in \mathrm{Id}(A)$.

(iii) Since $A_\ell \subset \bigcup_\ell A_\ell$, we have $\mathrm{Id}(A_\ell) \supset \mathrm{Id}(\bigcup_\ell A_\ell)$ for all ℓ; hence, $\bigcap_\ell \mathrm{Id}(A_\ell) \supset \mathrm{Id}(\bigcup_\ell A_\ell)$. For the reverse inclusion, suppose that $f(X) \in \bigcap_\ell \mathrm{Id}(A_\ell)$; that is, $f(a_\ell) = 0$ for all ℓ and all $a_\ell \in A_\ell$. If $b \in \bigcup_\ell A_\ell$, then $b \in A_\ell$ for some ℓ, and hence $f(b) = 0$; therefore, $f(X) \in \mathrm{Id}(\bigcup_\ell A_\ell)$.   •

One would like to have a formula for $\mathrm{Id}(A \cap B)$. Certainly, $\mathrm{Id}(A \cap B) = \mathrm{Id}(A) \cup \mathrm{Id}(B)$ is not correct, for the union of two ideals is almost never an ideal.

The next idea arises in characterizing those ideals of the form $\mathrm{Id}(V))$ when V is a variety.

Definition.   If I is an ideal in a commutative ring R, then its **radical**, denoted by $\sqrt{I}$, is

$$\sqrt{I} = \{r \in R : r^m \in I \text{ for some integer } m \geq 1\}.$$

An ideal I is called a **radical ideal** [6] if

$$\sqrt{I} = I.$$

Exercise 6.32 asks you to prove that $\sqrt{I}$ is an ideal. It is easy to see that $I \leq \sqrt{I}$, and so an ideal I is a radical ideal if and only if $\sqrt{I} \leq I$. For example, every prime ideal P is a radical ideal, for if $f^n \in P$, then $f \in P$. Here is an example of an ideal that is not radical. Let $b \in k$ and let $I = ((x-b)^2)$. Now I is not a radical ideal, for $(x-b)^2 \in I$ while $x - b \notin I$.

Definition.   An element a in a commutative ring R is called **nilpotent** if there is some $n \geq 1$ with $a^n = 0$.

Note that I is a radical ideal in a commutative ring R if and only if R/I has no nilpotent elements (of course, we mean that R/I has no *nonzero* nilpotent elements).

Proposition 6.31.   *If an ideal* $I = \mathrm{Id}(A)$ *for some* $A \subset k^n$, *then it is a radical ideal. Hence, the coordinate ring* $k[A]$ *has no nilpotent elements.*

Proof.   Since $I \leq \sqrt{I}$ is always true, it suffices to check the reverse inclusion. By hypothesis, $I = \mathrm{Id}(A)$ for some $A \subset k^n$; hence, if $f \in \sqrt{I}$, then $f^m \in \mathrm{Id}(A)$; that is, $f(a)^m = 0$ for all $a \in A$. But the values of $f(a)^m$ lie in the field k, and so $f(a)^m = 0$ implies $f(a) = 0$; that is, $f \in \mathrm{Id}(A) = I$.   •

Proposition 6.32.

   (i) *If* I *and* J *are ideals, then* $\sqrt{I \cap J} = \sqrt{I} \cap \sqrt{J}$.

   (ii) *If* I *and* J *are radical ideals, then* $I \cap J$ *is a radical ideal.*

---

[6]This term is appropriate, for if $r^m \in I$, then its mth root r also lies in I.

Proof. (i) If $f \in \sqrt{I \cap J}$, then $f^m \in I \cap J$ for some $m \geq 1$. Hence, $f^m \in I$ and $f^m \in J$, and so $f \in \sqrt{I}$ and $f \in \sqrt{J}$; that is, $f \in \sqrt{I} \cap \sqrt{J}$.

For the reverse inclusion, assume that $f \in \sqrt{I} \cap \sqrt{J}$, so that $f^m \in I$ and $f^q \in J$. We may assume that $m \geq q$, and so $f^m \in I \cap J$; that is, $f \in \sqrt{I \cap J}$.

(ii) If I and J are radical ideals, then $I = \sqrt{I}$ and $J = \sqrt{J}$ and

$$I \cap J \leq \sqrt{I \cap J} = \sqrt{I} \cap \sqrt{J} = I \cap J. \quad \bullet$$

We are now going to prove Hilbert's *Nullstellensatz* for $\mathbb{C}[X]$. The reader will see that the proof we will give generalizes to any uncountable algebraically closed field. The theorem is actually true for all algebraically closed fields, and so the proof here does not, alas, cover the algebraic closures of the rational function fields $\mathbb{Q}(x)$ or $\mathbb{Z}_p(x)$, for example, which are countable.

Lemma 6.33. *Let k be a field and let $\varphi \colon k[X] \to k$ be a surjective ring homomorphism which fixes k pointwise. If $J = \ker \varphi$, then* $\mathrm{Var}(J) \neq \varnothing$.

Proof. Let $\varphi(x_i) = a_i \in k$ and let $a = (a_1, \ldots, a_n) \in k^n$. If $f(X) = \sum_{\alpha_1, \ldots, \alpha_n} c_{\alpha_1, \ldots, \alpha_n} x_1^{\alpha_1} \cdots x_n^{\alpha_n} \in k[X]$, then

$$\varphi(f(X)) = \sum_{\alpha_1, \ldots, \alpha_n} c_{\alpha_1, \ldots, \alpha_n} \varphi(x_1)^{\alpha_1} \cdots \varphi(x_n)^{\alpha_n}$$

$$= \sum_{\alpha_1, \ldots, \alpha_n} c_{\alpha_1, \ldots, \alpha_n} a_1^{\alpha_1} \cdots a_n^{\alpha_n}$$

$$= f(a_1, \ldots, a_n).$$

Hence, $\varphi(f(X)) = f(a) = \varphi(f(a))$, because $f(a) \in k$. It follows that $f(X) - f(a) \in J$ for every $f(X)$. Now if $f(X) \in J$, then $f(a) \in J$. But $f(a) \in k$, and, since J is a proper ideal, it contains no nonzero constants. Therefore, $f(a) = 0$ and $a \in \mathrm{Var}(J)$. $\quad \bullet$

The next proof will use a bit of cardinality.

Theorem 6.34 (Weak Nullstellensatz[7] over $\mathbb{C}$). *If $f_1(X), \ldots, f_t(X) \in \mathbb{C}[X]$, then the ideal $I = (f_1, \ldots, f_t)$ is a proper ideal in $\mathbb{C}[X]$ if and only if* $\mathrm{Var}(f_1, \ldots, f_t) \neq \varnothing$.

---

[7]A translation from German is "Locus-of-zeros theorem."

Proof.   One direction is clear: if $\text{Var}(I) \neq \varnothing$, then I is a proper ideal, because $\text{Var}(\mathbb{C}[X]) = \varnothing$.

For the converse, suppose that I is a proper ideal. By Corollary 6.21, there is a maximal ideal M containing I, and so $K = \mathbb{C}[X]/M$ is a field. It is plain that the natural map $\mathbb{C}[X] \to \mathbb{C}[X]/M = K$ carries $\mathbb{C}$ to itself, so that $K/\mathbb{C}$ is an extension field; hence, K is a vector space over $\mathbb{C}$. Now $\mathbb{C}[X]$ has countable dimension, for a basis consists of all the monic monomials; it follows that $\dim_{\mathbb{C}}(K)$ is countable (possibly finite).

Suppose that K is a proper extension of $\mathbb{C}$; that is, there is some $t \in K$ with $t \notin \mathbb{C}$. Since $\mathbb{C}$ is algebraically closed, t cannot be algebraic over $\mathbb{C}$, and so it is transcendental. Consider the subset B of K,

$$B = \{1/(t - c) : c \in \mathbb{C}\}$$

(note that $t - c \neq 0$ because $t \notin \mathbb{C}$). The set B is uncountable, for it is indexed by the uncountable set $\mathbb{C}$. We claim that B is linearly independent over $\mathbb{C}$; if so, then we will have contradicted the fact that $\dim_{\mathbb{C}}(K)$ is countable. If B is linearly dependent, there are nonzero $a_1, \ldots, a_r \in \mathbb{C}$ with $\sum_{i=1}^{r} a_i/(t - c_i) = 0$. Clearing denominators, we have a polynomial $h(t) \in \mathbb{C}[t]$:

$$h(t) = \sum_i a_i(t - c_1) \cdots \widehat{(t - c_i)} \cdots (t - c_r) = 0;$$

Now $h(c_1) = a_2(c_1 - c_2) \cdots (c_1 - c_r) \neq 0$, so that $h(t)$ is not the zero polynomial. But this contradicts t being transcendental, and we conclude that $K = \mathbb{C}$. Lemma 6.33 now applies to show that $\text{Var}(M) \neq \varnothing$. But $\text{Var}(M) \subset \text{Var}(I)$, and this completes the proof.   •

Consider the special case of this theorem for $I = (f(x)) \leq \mathbb{C}[x]$, where $f(x)$ is not a constant. To say that $\text{Var}(f) \subset \mathbb{C}$ is nonempty is to say that $f(x)$ has a complex root. Thus, the weak Nullstellensatz is a generalization to several variables of the Fundamental Theorem of Algebra.

Theorem 6.35.   *If k is an (uncountable) algebraically closed field, then every maximal ideal M in $k[x_1, \ldots, x_n]$ has the form*

$$M = (x_1 - a_1, \ldots, x_n - a_n),$$

where $a = (a_1, \ldots, a_n) \in k^n$, and so there is a bijection between k^n and the maximal ideals in $k[x_1, \ldots, x_n]$.

Proof.   Since $k[X]/M \cong k$, Lemma 6.33 gives $\text{Var}(M) \neq \varnothing$. As in the proof of that lemma, there are constants $a_i \in k$ with $x_i + M = a_i + M$ for all i, and so $x_i - a_i \in M$. Therefore, there is an inclusion of ideals

$$(x_1 - a_1, \ldots , x_n - a_n) \le M.$$

But $(x_1 - a_1, \ldots , x_n - a_n)$ is a maximal ideal, by Exercise 6.5(i), and so $M = (x_1 - a_1, \ldots , x_n - a_n)$.   •

The following proof of Hilbert's Nullstellensatz uses the "Rabinowitch trick" of imbedding a polynomial ring in n variables into a polynomial ring in $n + 1$ variables. Again, uncountability is not needed, and we assume it only because our proof of the weak Nullstellensatz uses this hypothesis.

Theorem 6.36 (Nullstellensatz).   *Let k be an (uncountable) algebraically closed field. If I is an ideal in $k[X]$, then $\text{Id}(\text{Var}(I)) = \sqrt{I}$. Thus, f vanishes on $\text{Var}(I)$ if and only if $f^m \in I$ for some $m \ge 1$.*

Proof.   The inclusion $\text{Id}(\text{Var}(I)) \ge \sqrt{I}$ is obviously true, for if $f^m(a) = 0$ for some $m \ge 1$ and all $a \in \text{Var}(I)$, then $f(a) = 0$ for all a, because $f(a) \in k$.

For the converse, assume that $h \in \text{Id}(\text{Var}(I))$, where $I = (f_1, \ldots , f_t)$; that is, if $f_i(a) = 0$ for all i, where $a \in k^n$, then $h(a) = 0$. We must show that some power of h lies in I. Of course, we may assume that h is not the zero polynomial. Let us regard

$$k[x_1, \ldots , x_n] \le k[x_1, \ldots , x_n, y];$$

thus, every $f_i(x_1, \ldots , x_n)$ is regarded as a polynomial in $n + 1$ variables that does not depend on the last variable y. We claim that the polynomials

$$f_1, \ldots , f_t, 1 - yh$$

in $k[x_1, \ldots , x_n, y]$ have no common zeros. If $(a_1, \ldots , a_n, b) \in k^{n+1}$ is a common zero, then $a = (a_1, \ldots , a_n) \in k^n$ is a common zero of $f_1, \ldots , f_t$, and so $h(a) = 0$. But now $1 - bh(a) = 1 \neq 0$. The weak Nullstellensatz now applies to show that the ideal $(f_1, \ldots , f_t, 1 - yh)$ in $k[x_1, \ldots , x_n, y]$ is not a proper ideal. Therefore, there are $g_1, \ldots , g_{t+1} \in k[x_1, \ldots , x_n, y]$ with

$$1 = f_1 g_1 + \cdots + f_t g_t + (1 - yh)g_{t+1}.$$

Now make the substitution $y = 1/h$, so that the last term involving g_{t+1} vanishes. Writing the polynomials $g_i(X, y)$ more explicitly: $g_i(X, y) = \sum_{j=0}^{d_i} u_j(X) y^j$, so that $g_i(X, h^{-1}) = \sum_{j=0}^{d_i} u_j(X) h^{-j}$, we see that

$$h^{d_i} g_i(X, h^{-1}) \in k[X].$$

Therefore, if $m = \max\{d_1, \ldots, d_t\}$, then

$$h^m = (h^m g_1) f_1 + \cdots + (h^m g_t) f_t \in I. \quad \bullet$$

We continue the study of the operators Var and Id.

Proposition 6.37. *Let k be a field.*

(i) *For every subset $A \subset k^n$,*

$$\mathrm{Var}(\mathrm{Id}(A)) \supset A.$$

(ii) *For every ideal $I \leq k[X]$.*

$$\mathrm{Id}(\mathrm{Var}(I)) \supset I.$$

(iii) *If V is a variety of k^n, then $\mathrm{Var}(\mathrm{Id}(V)) = V$.*

Proof. (i) This result is almost a tautology. If $a \in A$, then $f(a) = 0$ for all $f(X) \in \mathrm{Id}(A)$. But every $f(X) \in \mathrm{Id}(A)$ annihilates A, by definition of $\mathrm{Id}(A)$, and so $a \in \mathrm{Var}(\mathrm{Id}(A))$. Therefore, $\mathrm{Var}(\mathrm{Id}(A)) \supset A$.

(ii) Again, one merely looks at the definitions. If $f(X) \in I$, then $f(a) = 0$ for all $a \in \mathrm{Var}(I)$; hence, $f(X)$ is surely one of the polynomials annihilating $\mathrm{Var}(I)$.

(iii) If V is a variety, then $V = \mathrm{Var}(J)$ for some ideal J in $k[X]$. Now

$$\mathrm{Var}(\mathrm{Id}(\mathrm{Var}(J))) \supset \mathrm{Var}(J),$$

by part (i). Also, part (ii) gives $\mathrm{Id}(\mathrm{Var}(J)) \supset J$, and applying Proposition 6.27(i) gives the reverse inclusion

$$\mathrm{Var}(\mathrm{Id}(\mathrm{Var}(J))) \subset \mathrm{Var}(J).$$

Therefore, $\mathrm{Var}(\mathrm{Id}(\mathrm{Var}(J))) = \mathrm{Var}(J)$; that is, $\mathrm{Var}(\mathrm{Id}(V)) = V$. $\quad \bullet$

Corollary 6.38.

(i) *If V_1 and V_2 are varieties and $\mathrm{Id}(V_1) = \mathrm{Id}(V_2)$, then $V_1 = V_2$.*

(ii) *If I_1 and I_2 are radical ideals and $\mathrm{Var}(I_1) = \mathrm{Var}(I_2)$, then $I_1 = I_2$.*

Proof.   (i) If $\mathrm{Id}(V_1) = \mathrm{Id}(V_2)$, then $\mathrm{Var}(\mathrm{Id}(V_1)) = \mathrm{Var}(\mathrm{Id}(V_2))$; by Proposition 6.37(iii), we have $V_1 = V_2$.

(ii) If $\mathrm{Var}(I_1) = \mathrm{Var}(I_2)$, then $\mathrm{Id}(\mathrm{Var}(I_1)) = \mathrm{Id}(\mathrm{Var}(I_2))$. By the Nullstellensatz, $\sqrt{I_1} = \sqrt{I_2}$. Since I_1 and I_2 are radical ideals, by hypothesis, we have $I_1 = I_2$.   •

Can a variety be decomposed into simpler subvarieties?

Definition.   A variety V is ***irreducible*** if it is not a union of two proper subvarieties; that is, $V \neq W' \cup W''$, where both W' and W'' are varieties that are proper subsets of V.

Proposition 6.39.   *Every variety V in k^n is a union of finitely many irreducible subvarieties:*

$$V = V_1 \cup V_2 \cup \cdots \cup V_m.$$

Proof.   Call a variety $W \in k^n$ *good* if it is irreducible or a union of finitely many irreducible subvarieties; otherwise, call W *bad*. We must show that there are no bad varieties. If W is bad, it is not irreducible, and so $W = W' \cup W''$, where both W' and W'' are proper subvarieties. But a union of good varieties is good, and so at least one of W' and W'' is bad; say, W' is bad, and rename it $W' = W_1$. Repeat this construction for W_1 to get a bad subvariety W_2. It follows by induction that there exists a strictly descending sequence

$$W > W_1 > \cdots > W_n > \cdots$$

of bad subvarieties. Since the operator Id reverses inclusions, there is a strictly increasing chain of ideals

$$\mathrm{Id}(W) < \mathrm{Id}(W_1) < \cdots < \mathrm{Id}(W_n) < \cdots$$

[the inclusions are strict because of Corollary 6.38(i)], and this contradicts the Hilbert basis theorem. We conclude that every variety is good.   •

Irreducible varieties have a nice characterization.

Proposition 6.40.   *A variety V in k^n is irreducible if and only if $\mathrm{Id}(V)$ is a prime ideal in $k[X]$. Hence, the coordinate ring $k[V]$ of an irreducible variety V is a domain.*

Proof.   Assume that V is an irreducible variety. It suffices to show that if $f_1(X), f_2(X) \notin \mathrm{Id}(V)$, then $f_1(X)f_2(X) \notin \mathrm{Id}(V)$. Define, for $i = 1, 2$,

$$W_i = V \cap \mathrm{Var}(f_i(X)).$$

Note that each W_i is a subvariety of V, for it is the intersection of two varieties; moreover, since $f_i(X) \notin \mathrm{Id}(V)$, there is some $a_i \in V$ with $f_i(a_i) \neq 0$, and so W_i is a proper subvariety of V. Since V is irreducible, we cannot have $V = W_1 \cup W_2$. Thus, there is some $b \in V$ which is not in $W_1 \cup W_2$; that is, $f_1(b) \neq 0 \neq f_2(b)$. Therefore, $f_1(b)f_2(b) \neq 0$, hence $f_1(X)f_2(X) \notin \mathrm{Id}(V)$, and so $\mathrm{Id}(V)$ is a prime ideal.

Conversely, assume that $\mathrm{Id}(V)$ is a prime ideal. Suppose that $V = V_1 \cup V_2$, where V_1 and V_2 are subvarieties. If $V_2 < V$, then we must show that $V = V_1$. Now

$$\mathrm{Id}(V) = \mathrm{Id}(V_1) \cap \mathrm{Id}(V_2) \geq \mathrm{Id}(V_1)\,\mathrm{Id}(V_2);$$

the equality is given by Proposition 6.30, and the inequality is given by Exercise 6.9(i). Since $\mathrm{Id}(V)$ is a prime ideal, Exercise 6.9(ii) says that $\mathrm{Id}(V_1) \leq \mathrm{Id}(V)$ or $\mathrm{Id}(V_2) \leq \mathrm{Id}(V)$. But $V_2 < V$ implies $\mathrm{Id}(V_2) > \mathrm{Id}(V)$, and we conclude that $\mathrm{Id}(V_1) \leq \mathrm{Id}(V)$. But the reverse inequality $\mathrm{Id}(V_1) \geq \mathrm{Id}(V)$ holds as well, because $V_1 \subset V$, and so $\mathrm{Id}(V_1) = \mathrm{Id}(V)$. Therefore, $V_1 = V$, by Corollary 6.38, and so V is irreducible. $\bullet$

We now consider whether the irreducible subvarieties in the decomposition of a variety into a union of irreducible varieties are uniquely determined. There is one obvious way to arrange nonuniqueness. Suppose that there are two prime ideals $P < Q$ in $k[X]$ (for example, $(x) < (x, y)$ are such prime ideals in $k[x, y]$). Now $\mathrm{Var}(Q) < \mathrm{Var}(P)$, so that if $\mathrm{Var}(P)$ is a subvariety of a variety V, say, $V = \mathrm{Var}(P) \cup V_2 \cup \cdots \cup V_m$, then $\mathrm{Var}(Q)$ can be one of the V_i or it can be left out.

Definition.   A decomposition $V = V_1 \cup \cdots \cup V_m$ is an ***irredundant union*** if no V_i can be omitted; that is, for all i,

$$V \neq V_1 \cup \cdots \cup \widehat{V_i} \cup \cdots \cup V_m.$$

Proposition 6.41.   *Every variety V is an irredundant union of irreducible subvarieties*

$$V = V_1 \cup \cdots \cup V_m;$$

moreover, the irreducible subvarieties V_i are uniquely determined by V.

Proof.   By Proposition 6.39, V is a union of finitely many irreducible subvarieties; say, $V = V_1 \cup \cdots \cup V_m$. If m is chosen minimal, then this union must be irredundant.

We now prove uniqueness. Suppose that $V = W_1 \cup \cdots \cup W_s$ is an irredundant union of irreducible subvarieties. Let $X = \{V_1, \ldots, V_m\}$ and let $Y = \{W_1, \ldots, W_s\}$; we shall show that $X = Y$. If $V_i \in X$, we have

$$V_i = V_i \cap V = \bigcup_j (V_i \cap W_j).$$

Now $V_i = V_i \cap W_j \neq \varnothing$ for some j; since V_i is irreducible, there is only one such W_j. Therefore, $V_i = V_i \cap W_j$, and so $V_i \subset W_j$. The same argument applied to W_j shows that there is exactly one V_ℓ with $W_j \subset V_\ell$. Hence,

$$V_i \subset W_j \subset V_\ell.$$

Since the union $V_1 \cup \cdots \cup V_m$ is irredundant, we must have $V_i = V_\ell$, and so $V_i = W_j = V_\ell$; that is, $V_i \in Y$ and $X \subset Y$. The reverse inclusion is proved in the same way.   •

Definition.   An intersection $I = J_1 \cap \cdots \cap I_m$ is an **irredundant intersection** if no J_i can be omitted; that is, for all i,

$$I \neq J_1 \cap \cdots \cap \widehat{J_i} \cap \cdots \cap J_m.$$

Corollary 6.42.   *Every radical ideal J in $k[X]$ is an irredundant intersection of prime ideals,*

$$J = P_1 \cap \cdots \cap P_m;$$

moreover, the prime ideals P_i are uniquely determined by J.

Proof.   Since J is a radical ideal, there is a variety V with $J = \mathrm{Id}(V)$. Now V is an irredundant union of irreducible subvarieties,

$$V = V_1 \cup \cdots \cup V_m,$$

so that

$$J = \mathrm{Id}(V) = \mathrm{Id}(V_1) \cap \cdots \cap \mathrm{Id}(V_m).$$

By Proposition 6.40, V_i irreducible implies $\mathrm{Id}(V_i)$ is prime, and so J is an intersection of prime ideals. This is an irredundant intersection, for if there is ℓ with $J = \mathrm{Id}(V) = \bigcap_{j \neq \ell} \mathrm{Id}(V_j)$, then

$$V = \mathrm{Var}(\mathrm{Id}(V)) = \bigcup_{j \neq \ell} \mathrm{Var}(\mathrm{Id}(V_j)) = \bigcup_{j \neq \ell} V_j,$$

contradicting the given irredundancy of the union.

Uniqueness is proved similarly. If $J = \mathrm{Id}(W_1) \cap \cdots \cap \mathrm{Id}(W_s)$, where each $\mathrm{Id}(W_i)$ is a prime ideal (hence is a radical ideal), then each W_i is an irreducible variety. Applying Var expresses $V = \mathrm{Var}(\mathrm{Id}(V)) = \mathrm{Var}(J)$ as an irredundant union of irreducible subvarieties, and the uniqueness of this decomposition gives the uniquess of the prime ideals in the intersection.   •

Here are some natural problems arising as one investigates these ideas further. First, what is the dimension of a variety? There are several candidates, and it turns out that prime ideals are the key. If V is a variety, then its dimension is the length of a longest chain of prime ideals in its coordinate ring $k[V]$ (which, by the correspondence theorem, is the length of a longest chain of prime ideals above $\mathrm{Id}(V)$ in $k[X]$). Another problem involves intersections. If $\mathrm{Var}(f)$ is a curve arising from a polynomial of degree d, how many points lie in the intersection of V with a straight line? *Bézout's theorem* says there should be d points, but one must be careful. First, one must demand that the coefficient field be algebraically closed, lest $\mathrm{Var}(f) = \varnothing$ cause a problem. But there may also be multiple roots, and so some intersections may have to be counted with a certain *multiplicity* in order to have Bézout's theorem hold. Defining multiplicities for intersections of higher dimensional varieties is very subtle.

It turns out to be more convenient to work in a larger **projective space** arising from k^n by adjoining a "hyperplane at infinity." For example, in our discussion of projective planes in Chapter 3, we adjoined a line at infinity. To distinguish it from projective space, one calls k^n **affine space**, for it consists of the "finite points" – that is, not the points at infinity. If one studies varieties in projective space, now defined as zeros of a set of *homogeneous* polynomials, then it is often the case that many separate affine cases become part of one simpler projective formula. Indeed, Bézout's theorem is an example of this phenomenon.

Finally, there is a deep analogy between manifolds in differential geometry and varieties. A *manifold* is a subspace of $\mathbb{R}^n$ which is a union of open replicas of euclidean space. For example, a torus T (i.e., a doughnut) is a subspace of $\mathbb{R}^3$, and each point of T has a neighborhood looking like an open disk (which is homeomorphic to the plane). One says that T is "locally euclidean;" it is obtained by gluing copies of $\mathbb{R}^2$ together in a coherent way. A variety V can be viewed as its coordinate ring $k[V]$, and neighborhoods of its points can be described "locally," using local rings!, and these local rings form what is called a *sheaf*. If one glues sheaves together, one obtains a *scheme*, and schemes seem to be the best way to study varieties. Two of the most prominent mathematicians involved in this circle of ideas are A. Grothendieck and J.-P. Serre.

EXERCISES

6.31 Prove that an element a in a commutative ring R is nilpotent if and only $1 + a$ is a unit.

6.32 If I is an ideal in a commutative ring R, prove that its radical, $\sqrt{I}$, is an ideal.

6.33 If R is a commutative ring, then its **nilradical** nil(R) is defined to be the intersection of all the prime ideals in R. Prove that nil(R) is the set of all the nilpotent elements in R:

$$\mathrm{nil}(R) = \{r \in R : r^m = 0 \text{ for some } m \geq 1\}.$$

6.34   (i) Show that $x^2 + y^2$ is irreducible in $\mathbb{R}[x, y]$, and conclude that $(x^2 + y^2)$ is a prime, hence radical, ideal in $\mathbb{R}[x, y]$.

    (ii) Prove that $\mathrm{Var}(x^2 + y^2) = \{(0, 0)\}$.

    (iii) Prove that $\mathrm{Id}(\mathrm{Var}(x^2 + y^2)) > (x^2 + y^2)$, and conclude that the radical ideal $(x^2 + y^2)$ in $\mathbb{R}[x, y]$ is not of the form $\mathrm{Id}(V)$ for some variety V. Conclude that the Nullstellensatz may fail in $k[X]$ if k is not algebraically closed.

    (iv) Prove that $(x^2 + y^2) = (x + iy) \cap (x - iy)$ in $\mathbb{C}[x, y]$.

    (v) Prove that $\mathrm{Id}(\mathrm{Var}(x^2 + y^2)) = (x^2 + y^2)$ in $\mathbb{C}[x, y]$.

6.35 Prove that if k is an (uncountable) algebraically closed field, then every radical ideal in $k[X]$ is an irredundant intersection of prime ideals.

6.36 Prove that if k is an (uncountable) algebraically closed field and $f_1, \ldots, f_t \in k[X]$, then $\mathrm{Var}(f_1, \ldots, f_t) = \varnothing$ if and only if there are $h_1, \ldots, h_t \in k[X]$

such that

$$1 = \sum_{i=1}^{t} h_i(X) f_i(X).$$

6.37 Let R be a commutative ring, and let $\mathrm{Spec}(R)$ denote the set of all the prime ideals in R. If I is an ideal in R, define

$$\mathrm{cl}(I) = \{\text{all the prime ideals in } R \text{ containing } I\}.$$

Prove the following.

   (i) $\mathrm{cl}(\{0\}) = \mathrm{Spec}(R)$.
   (ii) $\mathrm{cl}(R) = \varnothing$.
   (iii) $\mathrm{cl}\left(\bigcup_\ell I_\ell\right) = \bigcap_\ell \mathrm{cl}(I_\ell)$.
   (iv) $\mathrm{cl}(I \cap J) = \mathrm{cl}(IJ) = \mathrm{cl}(I) \cup \mathrm{cl}(J)$.

Conclude that the set of all $\mathrm{cl}(I)$, where I varies over the ideals in R, are the closed sets of a topology on $\mathrm{Spec}(R)$ (it is also called the *Zariski topology*).

6.38 Prove that an ideal P in $\mathrm{Spec}(R)$ is closed (that is, the one-point set $\{P\}$ is a closed set in the Zariski topology) if and only if P is a maximal ideal.

6.39 Let $f : R \to S$ be a ring homomorphism, and define $f^* : \mathrm{Spec}(S) \to \mathrm{Spec}(R)$ by $f^*(Q) = f^{-1}(Q)$, where Q is any prime ideal in S. Prove that if $G = \mathrm{cl}(I)$ is a closed subset of $\mathrm{Spec}(R)$, then $(f^*)^{-1}(G)$ is a closed subset of $\mathrm{Spec}(S)$. (In the terminology of point-set topology, this says that f^* is a continuous function.)

6.5 GRÖBNER BASES

Given two polynomials $f(x), g(x) \in k[x]$ with $g(x) \neq 0$, where k is a field, when is $g(x)$ a divisor of $f(x)$? The division algorithm gives unique polynomials $q(x), r(x) \in k[x]$ with

$$f(x) = q(x)g(x) + r(x),$$

where $r = 0$ or $\deg(r) < \deg(g)$, and $g \mid f$ if and only if the remainder $r = 0$. Let us look at this formula from a different point of view. To say that $g \mid f$ is to say that $f \in (g)$, the principal ideal generated by $g(x)$. Thus, the remainder r is the obstruction to f lying in this ideal; that is, $f \in (g)$ if and only if $r = 0$.

Consider a more general problem. Given polynomials

$$f(x), g_1(x), \ldots, g_m(x) \in k[x],$$

where k is a field, when is $d(x) = \gcd\{g_1(x), \ldots, g_m(x)\}$ a divisor of f? The euclidean algorithm finds d, and the division algorithm determines whether $d \mid f$. From another viewpoint, the two classical algorithms combine to give an algorithm determining whether $f \in (g_1, \ldots, g_m) = (d)$.

We now ask whether there is an algorithm in $k[x_1, \ldots, x_n] = k[X]$ to determine, given $f(X), g_1(X), \ldots, g_m(X) \in k[X]$, whether $f \in (g_1, \ldots, g_m)$. A division algorithm in $k[X]$ should be an algorithm yielding

$$r(X), a_1(X), \ldots, a_m(X) \in k[X],$$

with $r(X)$ unique, such that

$$f = a_1 g_1 + \cdots + a_m g_m + r.$$

Since $(g_1, \ldots, g_m)$ consists of all the linear combinations of the g's, such a generalized division algorithm would say again that the remainder r is the obstruction: $f \in (g_1, \ldots, g_m)$ if and only if $r = 0$.

We are going to show that both the division algorithm and the euclidean algorithm can be extended to polynomials in several variables. Even though these results are elementary, they were discovered only recently, in 1965, by B. Buchberger. Algebra has always dealt with algorithms, but the power and beauty of the axiomatic method has dominated the subject ever since Cayley and Dedekind in the second half of the nineteenth century. After the invention of the transistor in 1948, high-speed calculation became a reality, and old complicated algorithms, as well as new ones, could be implemented; a higher order of computing had entered algebra. Most likely, the development of computer science is a major reason why generalizations of the classical algorithms, from polynomials in one variable to polynomials in several variables, are only now being discovered. This is a dramatic illustration of the impact of external ideas on mathematics.

Generalized Division Algorithm

The most important feature of the division algorithm in $k[x]$ is that the remainder $r(x)$ has small degree. Without the inequality $\deg(r) < \deg(g)$, the result would be virtually useless; after all, given any $Q(x) \in k[x]$, there is an equation

$$f(x) = Q(x)g(x) + [f(x) - Q(x)g(x)].$$

Now polynomials in several variables are sums of monomials $cx_1^{\alpha_1} \cdots x_n^{\alpha_n}$, where $c \in k$ and $\alpha_i \geq 0$ for all i. Here are two degrees that one can assign to a monomial.

Definition. The *multidegree* of a monomial $cx_1^{\alpha_1} \cdots x_n^{\alpha_n} \in k[x_1, \dots, x_n]$, where $c \in k$ is nonzero and $\alpha_i \geq 0$ for all i, is the n-tuple $\alpha = (\alpha_1, \dots, \alpha_n)$; its *total degree* is the sum $|\alpha| = \alpha_1 + \cdots + \alpha_n$.

When dividing $f(x)$ by $g(x)$ in $k[x]$, one usually arranges the monomials in $f(x)$ in descending order, according to degree:

$$f(x) = c_n x^n + c_{n-1} x^{n-1} + \cdots + c_2 x^2 + c_1 x + c_0.$$

Consider a polynomial in several variables:

$$f(X) = f(x_1, \dots, x_n) = \sum c_{(\alpha_1, \dots, \alpha_n)} x_1^{\alpha_1} \cdots x_n^{\alpha_n}.$$

We will abbreviate $(\alpha_1, \dots, \alpha_n)$ to α and $x_1^{\alpha_1} \cdots x_n^{\alpha_n}$ to X^α, so that $f(X)$ can be written more compactly as

$$f(X) = \sum_\alpha c_\alpha X^\alpha.$$

Our aim is to arrange the monomials involved in $f(X)$ in a reasonable way, and we do this by ordering their multidegrees.

In Example 5.8(ii), we saw that the set $\mathbb{N}^n$, consisting of all the n-tuples $\alpha = (\alpha_1, \dots, \alpha_n)$ of natural numbers, is a monoid under addition:

$$(\alpha_1, \dots, \alpha_n) + (\beta_1, \dots, \beta_n) = (\alpha_1 + \beta_1, \dots, \alpha_n + \beta_n).$$

This monoid operation is related to the multiplication of monomials:

$$X^\alpha X^\beta = X^{\alpha+\beta}.$$

Recall that a *partially ordered set* is a nonempty set X equipped with a relation $\preceq$ which is reflexive, antisymmetric, and transitive. Of course, we may write $x \prec y$ if $x \prec y$ and $x \neq y$, and we may write $y \succeq x$ (or $y \succ x$) instead of $x \preceq y$ (or $x \prec y$).

Definition.   A partially ordered set X is **well ordered** if every nonempty subset $S \subset X$ contains a smallest element; that is, there exists $s_0 \in S$ with $s_0 \preceq s$ for all $s \in S$.

For example, the Least Integer Axiom says that the natural numbers $\mathbb{N}$ with the usual inequality $\leq$ is well ordered.

Proposition 6.43.   *Let X be a well ordered set.*

  (i) *If $x, y \in X$, then either $x \preceq y$ or $y \preceq x$.*

  (ii) *Every strictly decreasing sequence is finite.*

Proof.   (i) The subset $S = \{x, y\}$ has a smallest element, which must be either x or y. In the first case, $x \preceq y$; in the second case, $y \preceq x$.

(ii) Assume that there is an infinite strictly decreasing sequence, say,

$$x_1 \succ x_2 \succ x_3 \succ \cdots .$$

Since X is well ordered, the subset S consisting of all the x_i has a smallest element, say, x_n. But $x_{n+1} \prec x_n$, a contradiction.   •

The second property of well ordered sets will be used in showing that an algorithm eventually stops. In the proof of the division algorithm for polynomials in one variable, for example, we associated a natural number to each step: the degree of the remainder. Moreover, if the algorithm does not stop at a given step, then the natural number associated to the next step – the degree of its remainder – is strictly smaller. Since the natural numbers are well ordered by the usual inequality $\leq$, this strictly decreasing sequence of natural numbers must be finite; that is, the algorithm must stop after a finite number of steps.

We are interested in orderings of multidegrees that are compatible with multiplication of monomials – that is, with addition in the monoid $\mathbb{N}^n$.

Definition.   A *monomial order* is a well order of $\mathbb{N}^n$ such that

$$\alpha \preceq \beta \quad \text{implies} \quad \alpha + \gamma \preceq \beta + \gamma$$

for all $\alpha, \beta, \gamma \in \mathbb{N}^n$.

A monomial order will be used as follows. If $X = (x_1, \ldots, x_n)$, then we define $X^\alpha \preceq X^\beta$ in case $\alpha \preceq \beta$; that is, monomials are ordered according to their multidegrees.

Definition.   If $\mathbb{N}^n$ is equipped with a monomial order, then every $f(X) \in k[X] = k[x_1, \dots, x_n]$ can be written with its largest term first, followed by its other, smaller, terms in descending order:

$$f(X) = c_\alpha X^\alpha + \text{ lower terms.}$$

Define its **leading term** to be $\mathrm{LT}(f) = c_\alpha X^\alpha$ and its **Degree** to be $\mathrm{Deg}(f) = \alpha$. Call $f(X)$ **monic** if $\mathrm{LT}(f) = X^\alpha$; that is, if $c_\alpha = 1$.

There are many examples of monomial orders, but we shall give only the two most popular ones.

Definition.   The **lexicographic order** on $\mathbb{N}^n$ is defined by $\alpha \preceq_{\text{lex}} \beta$ in case $\alpha = \beta$ or the first nonzero coordinate in $\beta - \alpha$ is positive.[8]

The term *lexicographic* refers to the standard ordering in a dictionary. If $\alpha \prec_{\text{lex}} \beta$, then they agree for the first $i - 1$ coordinates (where $i \geq 1$), that is, $\alpha_1 = \beta_1, \dots, \alpha_{i-1} = \beta_{i-1}$, and there is strict inequality: $\alpha_i < \beta_i$. For example, the following German words are increasing in lexicographic order (the letters are ordered $a < b < c < \cdots < z$):

ausgehen

ausladen

auslagen

auslegen

bedeuten

Proposition 6.44.   *The lexicographic order $\preceq_{\text{lex}}$ on $\mathbb{N}^n$ is a monomial order.*

Proof.   First, we show that the lexicographic order is a partial order. The relation $\preceq_{\text{lex}}$ is reflexive, for its definition shows that $\alpha \preceq_{\text{lex}} \alpha$. To prove antisymmetry, assume that $\alpha \preceq_{\text{lex}} \beta$ and $\beta \preceq_{\text{lex}} \alpha$. If $\alpha \neq \beta$, there is a first coordinate, say the ith, where they disagree. For notation, we may assume that $\alpha_i < \beta_i$. But this contradicts $\beta \preceq_{\text{lex}} \alpha$. To prove transitivity, suppose that $\alpha \prec_{\text{lex}} \beta$ and $\beta \prec_{\text{lex}} \gamma$ (it suffices to consider strict inequality). Now $\alpha_1 = \beta_1, \dots, \alpha_{i-1} = \beta_{i-1}$ and $\alpha_i < \beta_i$. Let γ_p be the first coordinate with $\beta_p < \gamma_p$. If $p < i$, then

$$\gamma_1 = \beta_1 = \alpha_1, \dots, \gamma_{p-1} = \beta_{p-1} = \alpha_{p-1}, \alpha_p = \beta_p < \gamma_p;$$

---

[8]The difference $\beta - \alpha$ may not lie in $\mathbb{N}^n$, but it does lie in $\mathbb{Z}^n$.

if $p \geq i$, then

$$\gamma_1 = \beta_1 = \alpha_1, \ldots, \gamma_{i-1} = \beta_{i-1} = \alpha_{i-1}, \ \alpha_i < \beta_i = \gamma_i.$$

In either case, the first nonzero coordinate of $\gamma - \alpha$ is positive; that is, $\alpha \prec_{\text{lex}} \gamma$.

Next, we show that the lexicographic order is a well order. If S is a nonempty subset of $\mathbb{N}^n$, define

$$C_1 = \{\text{all first coordinates of } n\text{-tuples in } S\},$$

and define δ_1 to be the smallest number in C_1 (note that C_1 is a nonempty subset of the well ordered set $\mathbb{N}$). Define

$$C_2 = \{\text{all second coordinates of } n\text{-tuples } (\delta_1, \alpha_2, \ldots, \alpha_n) \in S\}.$$

Since $C_2 \neq \varnothing$, it contains a smallest number, δ_2. Similarly, for all $i < n$, define C_{i+1} as all the $(i+1)$th coordinates of those n-tuples in S whose first i coordinates are $(\delta_1, \delta_2, \ldots, \delta_i)$, and define δ_{i+1} to be the smallest number in C_{i+1}. By construction, the n-tuple $\delta = (\delta_1, \delta_2, \ldots, \delta_n)$ lies in S; moreover, if $\alpha = (\alpha_1, \alpha_2, \ldots, \alpha_n) \in S$, then

$$\alpha - \delta = (\alpha_1 - \delta_1, \alpha_2 - \delta_2, \ldots, \alpha_n - \delta_n)$$

has all its coordinates nonnegative. Hence, if $\alpha \neq \delta$, then its first nonzero coordinate is positive, and so $\delta \prec_{\text{lex}} \alpha$. Therefore, the lexicographic order is a well order.

Assume that $\alpha \preceq_{\text{lex}} \beta$; we claim that

$$\alpha + \gamma \preceq_{\text{lex}} \beta + \gamma$$

for all $\gamma \in \mathbb{N}$. If $\alpha = \beta$, then $\alpha + \gamma = \beta + \gamma$. If $\alpha \prec_{\text{lex}} \beta$, then the first nonzero coordinate of $\beta - \alpha$ is positive. But

$$(\beta + \gamma) - (\alpha + \gamma) = \beta - \alpha,$$

and so $\alpha + \gamma \prec_{\text{lex}} \beta + \gamma$. Therefore, $\preceq_{\text{lex}}$ is a monomial order. ●

In the lexicographic order, $x_1 \prec x_2 \prec x_3 \prec \cdots$, for

$$(1, 0, \ldots, 0) \prec (0, 1, 0, \ldots, 0) \prec (0, 0, 1, 0, \ldots, 0) \prec \cdots.$$

Any permutation of the variables $x_{\sigma(1)}, \ldots, x_{\sigma(n)}$ yields a different lexicographic order on $\mathbb{N}^n$.

Remark.   If X is any well ordered set with order $\preceq$, then the lexicographic order on X^n can be defined by $a = (a_1, \ldots, a_n) \preceq_{\text{lex}} b = (b_1, \ldots, b_n)$ in case $a = b$ or if they first disagree in the ith coordinate and $a_i \prec b_i$. It is a simple matter to generalize Proposition 6.44 by replacing $\mathbb{N}$ with X.   ◄

In Lemma 5.40 we constructed, for any set X, a monoid $\mathcal{W}(X)$: its elements are the empty word together with all the words $x_1^{e_1} \cdots x_p^{e_p}$ on a set X, where $p \geq 1$ and $e_i = \pm 1$ for all i; its operation is juxtaposition. In contrast to $\mathbb{N}^n$, in which all words have length n, the monoid $\mathcal{W}(X)$ has words of different lengths. Of more interest here is the submonoid $\mathcal{W}^+(X)$ consisting of all the "positive" words on X:

$$\mathcal{W}^+(X) = \{x_1 \cdots x_p \in \mathcal{W}(X) \colon x_i \in X\}.$$

Corollary 6.45.   *If X is a well ordered set, then $\mathcal{W}^+(X)$ is well ordered in the lexicographic order (which we also denote by $\preceq_{\text{lex}}$).*

Proof.   We will only give a careful definition of the lexicographic order here; the proof that it is a well order is left to the reader. First, define $1 \preceq_{\text{lex}} w$ for all $w \in \mathcal{W}^+(X)$. Next, given words $u = x_1 \cdots x_p$ and $v = y_1 \cdots y_q$ in $\mathcal{W}^+(X)$, make them the same length by adjoining 1's at the end of the shorter word, and rename them u' and v' in $\mathcal{W}^+(X)$. If $m \geq \max\{p, q\}$, we may regard $u', v', \in X^m$, and we define $u \preceq_{\text{lex}} v$ if $u' \preceq_{\text{lex}} v'$ in X^m. (This is the word order commonly used in dictionaries, where a blank precedes any letter: for example, *muse* precedes *museum*).   ●

Example 6.10.

Given a monomial order on $\mathbb{N}^n$, each polynomial $f(X) = \sum_\alpha c_\alpha X^\alpha \in k[X]$ $= k[x_1, \ldots, x_n]$ can be written with the multidegrees of its terms in descending order: $\alpha_1 \succ \alpha_2 \succ \cdots \succ \alpha_p$. Write

$$\text{multiword}(f) = \alpha_1 \cdots \alpha_p \in \mathcal{W}^+(\mathbb{N}^n).$$

Let $c_\beta X^\beta$ be a nonzero term in $f(X)$, let $g(X) \in k[X]$ have $\text{Deg}(g) \prec \beta$, and write

$$f(X) = h(X) + c_\beta X^\beta + \ell(X),$$

where $h(X)$ is the sum of all terms in $f(X)$ of multidegree $\succ \beta$ and $\ell(X)$ is the sum of all terms in $f(X)$ of multidegree $\prec \beta$. We claim that

$$\text{multiword}(f(X) - c_\beta X^\beta + g(X)) \prec_{\text{lex}} \text{multiword}(f) \text{ in } \mathcal{W}^+(X).$$

The sum of the terms in $f(X) - c_\beta X^\beta + g(X)$ with multidegree $\succ \beta$ is $h(X)$, while the sum of the lower terms is $\ell(X) + g(X)$. But $\mathrm{Deg}(\ell + g) \prec \beta$, by Exercise 6.43. Therefore, the initial terms of $f(X)$ and $f(X) - c_\beta X^\beta + g(X)$ agree, while the next term of $f(X) - c_\beta X^\beta + g(X)$ has multidegree $\prec \beta$, and this proves the claim.

Since $\mathcal{W}^+(\mathbb{N}^n)$ is well ordered, it follows that any sequence of steps of the form $f(X) \to f(X) - c_\beta X^\beta + g(X)$, where $c_\beta X^\beta$ is a nonzero term of $f(X)$ and $\mathrm{Deg}(g) \prec \beta$, must be finite. ◄

Here is the second popular monomial order.

Definition.   The *degree-lexicographic order* on $\mathbb{N}^n$ is defined by $\alpha \preceq_{\mathrm{dlex}} \beta$ in case $\alpha = \beta$ or

$$|\alpha| = \sum_{i=1}^n \alpha_i < \sum_{i=1}^n \beta_i = |\beta|,$$

or, if $|\alpha| = |\beta|$, then the first nonzero coordinate in $\beta - \alpha$ is positive.

In other words, given $\alpha = (\alpha_1, \dots, \alpha_n)$ and $\beta = (\beta_1, \dots, \beta_n)$, first check total degrees: if $|\alpha| < |\beta|$, then $\alpha \preceq_{\mathrm{dlex}} \beta$; if there is a tie, that is, if α and β have the same total degree, then order them lexicographically. For example, $(1, 2, 3, 0) \prec_{\mathrm{dlex}} (0, 2, 5, 0)$ and $(1, 2, 3, 4) \prec_{\mathrm{dlex}} (1, 2, 5, 2)$.

Proposition 6.46.   *The degree-lexicographic order $\preceq_{\mathrm{dlex}}$ is a monomial order on $\mathbb{N}^n$.*

Proof.   It is routine to show that $\preceq_{\mathrm{dlex}}$ is a partial order on $\mathbb{N}^n$. To see that it is a well order, let S be a nonempty subset of $\mathbb{N}^n$. The total degrees of elements in S form a nonempty subset of $\mathbb{N}$, and so there is a smallest such, say, t. The nonempty subset of all $\alpha \in S$ having total degree t has a smallest element, because the degree-lexicographic order $\preceq_{\mathrm{dlex}}$ coincides with the lexicographic order $\preceq_{\mathrm{lex}}$ on this subset. Therefore, there is a smallest element in S in the degree-lexicographic order.

Assume that $\alpha \preceq_{\mathrm{dlex}} \beta$ and $\gamma \in \mathbb{N}^n$. Now $|\alpha + \gamma| = |\alpha| + |\gamma|$, so that $|\alpha| = |\beta|$ implies $|\alpha + \gamma| = |\beta + \gamma|$ and $|\alpha| < |\beta|$ implies $|\alpha + \gamma| < |\beta + \gamma|$; in the latter case, Proposition 6.44 shows that $\alpha + \gamma \preceq_{\mathrm{dlex}} \beta + \gamma$. •

The next proposition shows, with respect to a monomial order, that polynomials in several variables behave like polynomials in a single variable.

Proposition 6.47. *Let $\preceq$ be a monomial order on $\mathbb{N}^n$, and let $f(X), g(X),$ $h(X) \in k[X] = k[x_1, \ldots, x_n]$.*

(i) *If $\mathrm{Deg}(f) = \mathrm{Deg}(g)$, then $\mathrm{LT}(g) \mid \mathrm{LT}(f)$.*

(ii) $\mathrm{LT}(hg) = \mathrm{LT}(h)\mathrm{LT}(g)$.

(iii) *If $\mathrm{Deg}(f) = \mathrm{Deg}(hg)$, then $\mathrm{LT}(g) \mid \mathrm{LT}(f)$.*

Proof. (i) If $\mathrm{Deg}(f) = \alpha = \mathrm{Deg}(g)$, then $\mathrm{LT}(f) = cX^\alpha$ and $\mathrm{LT}(g) = dX^\alpha$. Hence, $\mathrm{LT}(g) \mid \mathrm{LT}(f)$ [and also $\mathrm{LT}(f) \mid \mathrm{LT}(g)$].

(ii) Let $h(X) = bX^\gamma +$ lower terms and let $h(X) = cX^\beta +$ lower terms, so that $\mathrm{LT}(h) = cX^\gamma$ and $\mathrm{LT}(g) = bX^\beta$. Clearly, $cbX^{\gamma+\beta}$ is a nonzero term of $h(X)g(X)$. To see that it is the leading term, let $c_\mu X^\mu$ be a term of $h(X)$ with $\mu \prec \gamma$, and let $b_\nu X^\nu$ be a term of $g(X)$ with $\nu \prec \beta$. Now $\mathrm{Deg}(c_\mu X^\mu b_\nu X^\nu) = \mu + \nu$; since $\preceq$ is a monomial order, we have $\mu + \nu \prec \gamma + \nu \prec \gamma + \beta$. Thus, $cbX^{\gamma+\beta}$ is the term in $h(X)g(X)$ with largest multidegree.

(iii) Since $\mathrm{Deg}(f) = \mathrm{Deg}(hg)$, part (i) gives $\mathrm{LT}(hg) \mid \mathrm{LT}(f)$ and part (ii) gives $\mathrm{LT}(h)\mathrm{LT}(g) = \mathrm{LT}(hg)$; hence, $\mathrm{LT}(g) \mid \mathrm{LT}(f)$.   •

Definition.   Let $\preceq$ be a monomial order on $\mathbb{N}^n$ and let $f(X), g(X) \in k[X] = k[x_1, \ldots, x_n]$. If there is a nonzero term $c_\beta X^\beta$ in $f(X)$ with $\mathrm{LT}(g) \mid c_\beta X^\beta$ and

$$h(X) = f(X) - \frac{c_\beta X^\beta}{\mathrm{LT}(g)} g(X),$$

then the ***reduction*** $f \xrightarrow{g} h$ is the replacement of f by h.

Reduction is precisely the usual step involved in long division of polynomials in one variable. Of course, a special case of reduction is when $c_\beta X^\beta = \mathrm{LT}(f)$.

Proposition 6.48. *Let $\preceq$ be a monomial order on $\mathbb{N}^n$, let $f(X), g(X) \in k[X] = k[x_1, \ldots, x_n]$, and assume that $f \xrightarrow{g} h$; that is, there is a nonzero term $c_\beta X^\beta$ in $f(X)$ with $\mathrm{LT}(g) \mid c_\beta X^\beta$ and $h(X) = f(X) - \frac{c_\beta X^\beta}{\mathrm{LT}(g)} g(X)$.*
If $\beta = \mathrm{Deg}(f)$, i.e., if $c_\beta X^\beta = \mathrm{LT}(f)$, then either

$$h(X) = 0 \quad or \quad \mathrm{Deg}(h) \prec \mathrm{Deg}(f);$$

if $\beta \prec \text{Deg}(f)$, then $\text{Deg}(h) = \text{Deg}(f)$. In either case,

$$\text{Deg}\left(\frac{c_\beta X^\beta}{\text{LT}(g)}g(X)\right) \preceq \text{Deg}(f).$$

Proof.   Let us write

$$f(X) = \text{LT}(f) + c_\kappa X^\kappa + \text{lower terms};$$

since $c_\beta X^\beta$ is a term of $f(X)$, we have $\beta \preceq \text{Deg}(f)$. If $\text{LT}(g) = a_\gamma X^\gamma$, so that $\text{Deg}(g) = \gamma$, let us write

$$g(X) = a_\gamma X^\gamma + a_\lambda X^\lambda + \text{lower terms}.$$

Hence,

$$
\begin{aligned}
h(X) &= f(X) - \frac{c_\beta X^\beta}{\text{LT}(g)}g(X) \\
&= f(X) - \frac{c_\beta X^\beta}{\text{LT}(g)}\Big[\text{LT}(g) + a_\lambda X^\lambda + \cdots\Big] \\
&= \Big[f(X) - c_\beta X^\beta\Big] - \frac{c_\beta X^\beta}{\text{LT}(g)}\Big[a_\lambda X^\lambda + \cdots\Big].
\end{aligned}
$$

Now $\text{LT}(g) \mid c_\beta X^\beta$ says that $\beta - \gamma \in \mathbb{N}^n$. We claim that

$$\text{Deg}\left(-\frac{c_\beta X^\beta}{\text{LT}(g)}\Big[a_\lambda X^\lambda + \cdots\Big]\right) = \lambda + \beta - \gamma \prec \beta.$$

The inequality holds, for $\lambda \prec \gamma$ implies $\lambda + (\beta - \gamma) \prec \gamma + (\beta - \gamma) = \beta$. To see that $\lambda + \beta - \gamma$ is the Degree, it suffices to show that $\lambda + \beta - \gamma = \text{Deg}\left(-\frac{c_\beta X^\beta}{\text{LT}(g)}a_\lambda X^\lambda\right)$ is the largest multidegree occurring in $-\frac{c_\beta X^\beta}{\text{LT}(g)}\Big[a_\lambda X^\lambda + \cdots\Big]$. But if $a_\eta X^\eta$ is a lower term in $g(X)$, i.e., $\eta \prec \lambda$, then $\preceq$ being a monomial order gives $\eta + (\beta - \gamma) \prec \lambda + (\beta - \gamma)$, as desired.

If $h(X) \neq 0$, then Exercise 6.43 below gives

$$\text{Deg}(h) \preceq \max\left\{\text{Deg}\left(f(X) - c_\beta X^\beta\right), \text{Deg}\left(-\frac{c_\beta X^\beta}{\text{LT}(g)}\Big[a_\lambda X^\lambda + \cdots\Big]\right)\right\}.$$

Now if $\beta = \text{Deg}(f)$, then $c_\beta X^\beta = \text{LT}(f)$,

$$f(X) - c_\beta X^\beta = f(X) - \text{LT}(f) = c_\kappa X^\kappa + \text{lower terms},$$

and, hence, $\mathrm{Deg}(f(X) - c_\beta X^\beta) = \kappa \prec \mathrm{Deg}(f)$. Therefore, $\mathrm{Deg}(h) \prec \mathrm{Deg}(f)$ in this case. If $\beta \prec \mathrm{Deg}(f)$, then $\mathrm{Deg}(f(X) - c_\beta X^\beta) = \mathrm{Deg}(f)$, while $\mathrm{Deg}\left(-\frac{c_\beta X^\beta}{\mathrm{LT}(g)}\left[a_\lambda X^\lambda + \cdots\right]\right) \prec \beta \prec \mathrm{Deg}(f)$, and so $\mathrm{Deg}(h) = \mathrm{Deg}(f)$ in this case.

The last inequality is clear, for

$$\frac{c_\beta X^\beta}{\mathrm{LT}(g)} g(X) = c_\beta X^\beta + \frac{c_\beta X^\beta}{\mathrm{LT}(g)}\left[a_\lambda X^\lambda + \cdots\right].$$

Since $\mathrm{Deg}\left(-\frac{c_\beta X^\beta}{\mathrm{LT}(g)}\left[a_\lambda X^\lambda + \cdots\right]\right) \prec \beta$, we see that

$$\mathrm{Deg}\left(\frac{c_\beta X^\beta}{\mathrm{LT}(g)} g(X)\right) = \beta \preceq \mathrm{Deg}(f). \quad \bullet$$

Definition.   Let $\{g_1, \ldots, g_m\}$, where $g_i = g_i(X) \in k[X]$. A polynomial $r(X)$ is **reduced mod** $\{g_1, \ldots, g_m\}$ if either $r(X) = 0$ or no $\mathrm{LT}(g_i)$ divides any nonzero term of $r(X)$.

Here is the division algorithm for polynomials in several variables. Because the algorithm requires the "divisor polynomials" $\{g_1, \ldots, g_m\}$ to be used in a specific order (after all, an algorithm must give explicit directions), we will be using an m-tuple of polynomials instead of a subset of polynomials. We use the notation $[g_1, \ldots, g_m]$ for the m-tuple whose ith entry is g_i, because the usual notation $(g_1, \ldots, g_m)$ would be confused with the notation for the ideal $(g_1, \ldots, g_m)$ generated by the g_i.

Theorem 6.49 (Division Algorithm in $k[X]$).   *Let $\preceq$ be a monomial order on $\mathbb{N}^n$, and let $k[X] = k[x_1, \ldots, x_n]$. If $f(X) \in k[X]$ and $G = [g_1(X), \ldots, g_m(X)]$ is an m-tuple of polynomials in $k[X]$, then there is an algorithm giving polynomials $r(X), a_1(X), \ldots, a_m(X) \in k[X]$ with*

$$f = a_1 g_1 + \cdots + a_m g_m + r,$$

where r is reduced mod $\{g_1, \ldots, g_m\}$, and

$$\mathrm{Deg}(a_i g_i) \preceq \mathrm{Deg}(f) \text{ for all } i.$$

Proof.   Once a monomial order is chosen, so that leading terms are defined, the algorithm is a straightforward generalization of the division algorithm in one variable. First, reduce mod g_1 as many times as possible, then reduce mod g_2, then reduce again mod g_1, etc. Here is a pseudocode describing the algorithm more precisely.

$$\text{Input: } f(X) = \sum_\beta c_\beta X^\beta, \quad [g_1, \ldots, g_m]$$
$$\text{Output: } r, a_1, \ldots, a_m$$
$$r := f; \quad a_i := 0$$
$$\text{WHILE } f \text{ is not reduced mod } \{g_1, \ldots, g_m\} \text{ DO}$$
$$\text{select smallest } i \text{ with } \text{LT}(g_i) \mid c_\beta X^\beta \text{ for some } \beta$$
$$f - [c_\beta X^\beta / \text{LT}(g_i)]g_i := f$$
$$a_i + [c_\beta X^\beta / \text{LT}(g_i)] := a_i$$
$$\text{END WHILE}$$

At each step $h_j \xrightarrow{g_i} h_{j+1}$ of the algorithm, we have multiword(h_j) $\succ_{\text{lex}}$ multiword(h_{j+1}) in $\mathcal{W}^+(\mathbb{N}^n)$, by Example 6.10, and so the algorithm does stop, because $\prec_{\text{lex}}$ is a well order on $\mathcal{W}^+(\mathbb{N}^n)$. Obviously, the output $r(X)$ is reduced mod $\{g_1, \ldots, g_m\}$, for if it has a term divisible by some $\text{LT}(g_i)$, then one further reduction is possible.

Finally, each term of $a_i(X)$ has the form $c_\beta X^\beta / \text{LT}(g_i)$ for some intermediate output $h(X)$ (as one sees in the pseudocode). It now follows from Proposition 6.48 that $\text{Deg}(a_i g_i) \preceq \text{Deg}(f)$.   •

Definition.   Given a monomial order on $\mathbb{N}^n$, a polynomial $f(X) \in k[X]$, and an m-tuple $G = [g_1, \ldots, g_m]$, we call the output $r(X)$ of the division algorithm the ***remainder of f mod G***.

Note that the remainder r of f mod G is reduced mod $\{g_1, \ldots, g_m\}$ and $f - r \in I = (g_1, \ldots, g_m)$. The algorithm requires that G be an m-tuple, because of the command

$$\text{select smallest } i \text{ with } \text{LT}(g_i) \mid c_\beta X^\beta \text{ for some } \beta$$

specifying the order of reductions. The next example shows that the remainder may depend not only on the set of polynomials $\{g_1, \ldots, g_m\}$ but also on the ordering of the coordinates in the m-tuple $G = [g_1, \ldots, g_m]$. That is, if $\sigma \in S_m$ is a permutation and $G_\sigma = [g_{\sigma(1)}, \ldots, g_{\sigma(m)}]$, then the remainder r_σ of f mod G_σ may not be the same as the remainder r of f mod G. Even worse, it is possible that $r \neq 0$ and $r_\sigma = 0$, so that the remainder mod G is not the obstruction to f being in the ideal $(g_1, \ldots, g_m)$.

Example 6.11.

Let $f(x, y, z) = x^2 y^2 + xy$, and let $G = [g_1, g_2, g_3]$, where

$$g_1 = y^2 + z^2$$
$$g_2 = x^2 y + yz$$
$$g_3 = z^3 + xy.$$

We use the degree-lexicographic order on $\mathbb{N}^3$. Now $y^2 = \mathrm{LT}(g_1) \mid \mathrm{LT}(f) = x^2 y^2$, and so $f \xrightarrow{g_1} h$, where

$$h = f - \frac{x^2 y^2}{y^2}(y^2 + z^2) = -x^2 z^2 + xy.$$

The polynomial $-x^2 z^2 + xy$ is reduced mod G, because neither $-x^2 z^2$ nor xy is divisible by any of the leading terms $\mathrm{LT}(g_1) = y^2$, $\mathrm{LT}(g_2) = x^2 y$, or $\mathrm{LT}(g_3) = z^3$.

On the other hand, let us apply the division algorithm using the 3-tuple $G' = [g_2, g_1, g_3]$. The first reduction gives $f \xrightarrow{g_2} h'$, where

$$h' = f - \frac{x^2 y^2}{x^2 y}(x^2 y + yz) = -y^2 z + xy.$$

Now h' is not reduced, and reducing mod g_1 gives

$$h' - \frac{-y^2 z}{y^2}(y^2 + z^2) = z^3 + xy.$$

But $z^3 + xy = g_3$, and so $z^3 + xy \xrightarrow{g_3} 0$.

Thus, the remainder depends on the ordering of the divisor polynomials g_i in the m-tuple.

For a simpler example of different remainders (but with neither remainder being 0), see Exercise 6.42. ◀

The dependence of the remainder on the order of the g_i in the m-tuple $G = [g_1, \dots, g_m]$ will be treated in the next subsection.

EXERCISES

6.40   (i) Write the first 10 monic monomials in $k[x, y]$ in lexicographic order and in degree-lexicographic order.

   (ii) Write all the monic monomials in $k[x, y, z]$ of total degree at most 2 in lexicographic order and in degree-lexicographic order.

6.41 Give an example of a well ordered set X that contains an element u for which $\{x \in X : x \preceq u\}$ is infinite.

6.42 Let $G = [x - y, x - z]$ and $G' = [x - z, x - y]$. Show that the remainder of $x \bmod G$ (degree-lexicographic order) is distinct from the remainder of $x \bmod G'$.

6.43 Let $\preceq$ be a monomial order on $\mathbb{N}^n$, and let $f(X), g(X) \in k[X] = k[x_1, \dots, x_n]$ be nonzero polynomials. Prove that if $f + g \neq 0$, then

$$\mathrm{Deg}(f + g) \preceq \max\{\mathrm{Deg}(f), \mathrm{Deg}(g)\},$$

and that strict inequality can occur only if $\mathrm{Deg}(f) = \mathrm{Deg}(g)$.

6.44 Use the degree-lexicographic order in this exercise.
   (i) Find the remainder of $x^7 y^2 + x^3 y^2 - y + 1 \bmod [xy^2 - x, x - y^3]$.
   (ii) Find the remainder of $x^7 y^2 + x^3 y^2 - y + 1 \bmod [x - y^3, xy^2 - x]$.

6.45 Use the degree-lexicographic order in this exercise.
   (i) Find the remainder of $x^2 y + xy^2 + y^2 \bmod [y^2 - 1, xy - 1]$.
   (ii) Find the remainder of $x^2 y + xy^2 + y^2 \bmod [xy - 1, y^2 - 1]$.

6.46 Let $c_\alpha X^\alpha$ be a nonzero monomial, and let $f(X), g(X) \in k[X]$ be polynomials none of whose terms is divisible by $c_\alpha X^\alpha$. Prove that none of the terms of $f(X) - g(X)$ is divisible by $c_\alpha X^\alpha$.

6.47 An ideal I in $k[X]$ is a **monomial ideal** if it is generated by monomials: $I = (X^{\alpha(1)}, \dots, X^{\alpha(q)})$.
   (i) Prove that $f(X) \in I$ if and only if each term of $f(X)$ is divisible by some $X^{\alpha(i)}$.
   (ii) Prove that if $G = [g_1, \dots, g_m]$ and r is reduced mod G, then r does not lie in the monomial ideal $(\mathrm{LT}(g_1), \dots, \mathrm{LT}(g_m))$.

Gröbner Bases

For the remainder of this section we will assume that $\mathbb{N}^n$ is equipped with some monomial order (the reader may use the degree-lexicographic order), so that $\mathrm{LT}(f)$ is defined and the division algorithm makes sense.

We have seen that the remainder of $f \bmod [g_1, \dots, g_m]$ obtained from the division algorithm can depend on the order in which the g_i are listed. A *Gröbner basis* $\{g_1, \dots, g_m\}$ of the ideal $I = (g_1, \dots, g_m)$ is a basis such that, for any of the m-tuples G formed from the g_i, the remainder of $f \bmod G$ is always the obstruction to whether f lies in I; this will be a consequence of the definition (which is given to make make sure Gröbner bases are sets and not m-tuples).

Definition.   A set of polynomials $\{g_1, \ldots, g_m\}$ is a ***Gröbner basis***[9] of the ideal $I = (g_1, \ldots, g_m)$ if, for each nonzero $f \in I$, there is some g_i with $LT(g_i) \mid LT(f)$.

Example 6.11 shows that

$$\{y^2 + z^2, x^2 y + yz, z^3 + xy\}$$

is not a Gröbner basis of the ideal $(y^2 + z^2, x^2 y + yz, z^3 + xy)$.

Proposition 6.50.   *A set $\{g_1, \ldots, g_m\}$ of polynomials is a Gröbner basis of $I = (g_1, \ldots, g_m)$ if and only if, for each m-tuple $G_\sigma = [g_{\sigma(1)}, \ldots, g_{\sigma(m)}]$ (where $\sigma \in S_m$), every $f \in I$ has remainder 0 mod G_σ.*

Proof.   Assume there is some permutation $\sigma \in S_m$ and some $f \in I$ whose remainder mod G_σ is not 0. Among all such polynomials, choose f of minimal Degree. Since $\{g_1, \ldots, g_m\}$ is a Gröbner basis, $LT(g_i) \mid LT(f)$ for some i; select the smallest $\sigma(i)$ for which there is a reduction $f \overset{g_{\sigma(i)}}{\to} h$, and note that $h \in I$. Since $\mathrm{Deg}(h) \prec \mathrm{Deg}(f)$, by Proposition 6.48, the division algorithm gives a sequence of reductions $h = h_0 \to h_1 \to h_2 \to \cdots \to h_p = 0$. But the division algorithm for f adjoins $f \to h$ at the front, showing that 0 is the remainder of f mod G_σ, a contradiction.

Conversely, let $\{g_1, \ldots, g_m\}$ be a Gröbner basis of $I = (g_1, \ldots, g_m)$. If there is a nonzero $f \in I$ with $LT(g_i) \nmid LT(f)$ for every i, then in any reduction $f \overset{g_i}{\to} h$, we have $LT(h) = LT(f)$. Hence, if $G = [g_1, \ldots, g_m]$, the division algorithm mod G gives reductions $f \to h_1 \to h_2 \to \cdots \to h_p = r$ in which $LT(r) = LT(f)$. Therefore, $r \neq 0$; that is, the remainder of f mod G is not zero, and this is a contradiction.   •

Corollary 6.51.   *Let $\{g_1, \ldots, g_m\}$ be a Gröbner basis of the ideal $I = (g_1, \ldots, g_m)$, and let $G = [g_1, \ldots, g_m]$ be any m-tuple formed from the g_i. If $f(X) \in k[X]$, then there is a unique $r(X) \in k[X]$, which is reduced mod $\{g_1, \ldots, g_m\}$, such that $f - r \in I$; in fact, r is the remainder of f mod G.*

Proof.   The division algorithm gives a polynomial r which is reduced mod $\{g_1, \ldots, g_m\}$, and polynomials $a_1, \ldots, a_m$ with $f = a_1 g_1 + \cdots + a_m g_m + r$; clearly, $f - r = a_1 g_1 + \cdots + a_m g_m \in I$.

---

[9]It was B. Buchberger who, in his dissertation, proved the main properties of Gröbner bases. He named these bases to honor his thesis advisor, W. Gröbner.

To prove uniqueness, suppose that r and r' are reduced mod $\{g_1, \ldots, g_m\}$ and that $f - r$ and $f - r'$ lie in I, so that $(f - r') - (f - r) = r - r' \in I$. Since r and r' are reduced mod $\{g_1, \ldots, g_m\}$, none of their terms is divisible by any $\mathrm{LT}(g_i)$. If $r - r' \neq 0$, then Exercise 6.46 says that no term of $r - r'$ is divisible by any $\mathrm{LT}(g_i)$; in particular, $\mathrm{LT}(r - r')$ is not divisible by any $\mathrm{LT}(g_i)$, and this contradicts Proposition 6.50. Therefore, $r = r'$.   •

The next corollary shows that Gröbner bases resolve the problem of different remainders in the division algorithm arising from different m-tuples.

Corollary 6.52.   *Let* $\{g_1, \ldots, g_m\}$ *be a Gröbner basis of the ideal* $I = (g_1, \ldots, g_m)$, *and let* $G = [g_1, \ldots, g_m]$.

(i) *If* $f(X) \in k[X]$ *and* $G_\sigma = [g_{\sigma(1)}, \ldots, g_{\sigma(m)}]$, *where* $\sigma \in S_m$ *is a permutation, then the remainder of* f mod G *is equal to the remainder of* f mod G_σ.

(ii) *A polynomial* $f \in I$ *if and only if* f *has remainder* 0 mod G.

Proof.   (i) If r is the remainder of f mod G, then Corollary 6.51 says that r is the unique polynomial, reduced mod $\{g_1, \ldots, g_m\}$, with $f - r \in I$; similarly, the remainder r_σ of f mod G_σ is the unique polynomial, reduced mod $\{g_1, \ldots, g_m\}$, with $f - r_\sigma \in I$. The uniqueness assertion in Corollary 6.51 gives $r = r_\sigma$.

(ii) Proposition 6.50 shows that if $f \in I$, then its remainder is 0. For the converse, if r is the remainder of f mod G, then $f = q + r$, where $q \in I$. Hence, if $r = 0$, then $f \in I$.   •

There are several obvious questions. Do Gröbner bases exist and, if they do, are they unique? Given an ideal I in $k[X]$, is there an algorithm to find a Gröbner basis of I?

The notion of *S-polynomial* will allow us to recognize a Gröbner basis, but we first introduce some notation.

Definition.   If $\alpha = (\alpha_1, \ldots, \alpha_n)$ and $\beta = (\beta_1, \ldots, \beta_n)$ are in $\mathbb{N}^n$, define

$$\alpha \vee \beta = \mu,$$

where $\mu = (\mu_1, \ldots, \mu_n)$ is given by $\mu_i = \max\{\alpha_i, \beta_i\}$.

It is easy to see that the least common multiple of monomials X^α and X^β is $X^{\alpha \vee \beta}$.

Definition.   Let $f(X), g(X) \in k[X]$, where $\mathrm{LT}(f) = a_\alpha X^\alpha$ and $\mathrm{LT}(g) = b_\beta X^\beta$. Define

$$L(f, g) = X^{\alpha \vee \beta}.$$

The *S-polynomial* $S(f, g)$ is defined by

$$S(f, g) = \frac{L(f, g)}{\mathrm{LT}(f)} f - \frac{L(f, g)}{\mathrm{LT}(g)} g;$$

that is, if $\mu = \alpha \vee \beta$, then

$$S(f, g) = a_\alpha^{-1} X^{\mu - \alpha} f(X) - b_\beta^{-1} X^{\mu - \beta} g(X).$$

Note that $S(f, g) = -S(g, f)$.

The following technical lemma indicates why S-polynomials are relevant.

Lemma 6.53.   *Given* $g_1(X), \ldots, g_\ell(X) \in k[X]$ *and monomials* $c_j X^{\alpha(j)}$, *let* $h(X) = \sum_{j=1}^{\ell} c_j X^{\alpha(j)} g_j(X)$.

Let δ *be a multidegree. If* $\mathrm{Deg}(h) \prec \delta$ *and* $\mathrm{Deg}(c_j X^{\alpha(j)} g_j(X)) = \delta$ *for all* $j < \ell$, *then there are* $d_j \in k$ *with*

$$h(X) = \sum_j d_j X^{\delta - \mu(j)} S(g_j, g_{j+1}),$$

where $\mu(j) = \mathrm{Deg}(g_j) \vee \mathrm{Deg}(g_{j+1})$, *and for all* $j < \ell$,

$$\mathrm{Deg}\left(X^{\delta - \mu(j)} S(g_j, g_{j+1})\right) \prec \delta.$$

Remark.   The lemma says that if $\mathrm{Deg}(\sum_j a_j g_j) \prec \delta$, where the a_j are monomials, while $\mathrm{Deg}(a_j g_j) = \delta$ for all j, then h can be rewritten as a linear combination of S-polynomials, with monomial coefficents, each of whose terms has multidegree strictly less than δ.   ◄

Proof.   Let $\mathrm{LT}(g_j) = b_j X^{\beta(j)}$, so that $\mathrm{LT}(c_j X^{\alpha(j)} g_j(X)) = c_j b_j X^\delta$. The coefficient of X^δ in $h(X)$ is thus $\sum_j c_j b_j$. Since $\mathrm{Deg}(h) \prec \delta$, we must have $\sum_j c_j b_j = 0$. Define monic polynomials

$$u_j(X) = b_j^{-1} X^{\alpha(j)} g_j(X).$$

There is a telescoping sum

$$h(X) = \sum_{j=1}^{\ell} c_j X^{\alpha(j)} g_j(X)$$

$$= \sum_{j=1}^{\ell} c_j b_j u_j$$

$$= c_1 b_1 (u_1 - u_2) + (c_1 b_1 + c_2 b_2)(u_2 - u_3) + \cdots$$
$$+ (c_1 b_1 + \cdots + c_{\ell-1} b_{\ell-1})(u_{\ell-1} - u_\ell)$$
$$+ (c_1 b_1 + \cdots + c_\ell b_\ell) u_\ell.$$

Now the last term $(c_1 b_1 + \cdots + c_\ell b_\ell) u_\ell = 0$ because $\sum_j c_j b_j = 0$. Since $\text{Deg}(c_j X^{\alpha(j)} g_j(X)) = \delta$, we have $\alpha(j) + \beta(j) = \delta$, so that $X^{\beta(j)} \mid X^\delta$ for all j. Hence, for all $j < \ell$, we have $\text{lcm}\{X^{\beta(j)}, X^{\beta(j+1)}\} = X^{\beta(j) \vee \beta(j+1)} \mid X^\delta$; that is, if we write $\mu(j) = \beta(j) \vee \beta(j+1)$, then $\delta - \mu(j) \in \mathbb{N}^n$. But

$$X^{\delta-\mu(j)} S(g_j, g_{j+1}) = X^{\delta-\mu(j)} \left(\frac{X^{\mu(j)}}{\text{LT}(g_j)} g_j(X) - \frac{X^{\mu(j)}}{\text{LT}(g_{j+1})} g_{j+1}(X) \right)$$

$$= \frac{X^\delta}{\text{LT}(g_j)} g_j(X) - \frac{X^\delta}{\text{LT}(g_{j+1})} g_{j+1}(X)$$

$$= b_j^{-1} X^{\alpha(j)} g_j - b_{j+1}^{-1} X^{\alpha(j+1)} g_{j+1}$$

$$= u_j - u_{j+1}.$$

Substituting this equation into the telescoping sum gives a sum of the desired form, where $d_j = c_1 b_1 + \cdots + c_j b_j$:

$$h(X) = c_1 b_1 X^{\delta-\mu(1)} S(g_1, g_2) + (c_1 b_1 + c_2 b_2) X^{\delta-\mu(2)} S(g_2, g_3) + \cdots$$
$$+ (c_1 b_1 + \cdots + c_{\ell-1} b_{\ell-1}) X^{\delta-\mu(\ell-1)} S(g_{\ell-1}, g_\ell).$$

Finally, since both u_j and u_{j+1} are monic with leading term of multidegree δ, we have $\text{Deg}(u_j - u_{j+1}) \prec \delta$. But we have shown that $u_j - u_{j+1} = X^{\delta-\mu(j)} S(g_j, g_{j+1})$, and so $\text{Deg}(X^{\delta-\mu(j)} S(g_j, g_{j+1})) \prec \delta$, as desired.   •

By Proposition 6.50, $\{g_1, \ldots, g_m\}$ is a Gröbner basis of the ideal $I = (g_1, \ldots, g_m)$ if every $f \in I$ has remainder 0 mod G (where G is any m-tuple formed by ordering the g_i). The importance of the next theorem lies in its showing that it is necessary to compute the remainders of only finitely many polynomials, namely, the S-polynomials, to determine whether $\{g_1, \ldots, g_m\}$ is a Gröbner basis.

Theorem 6.54 (Buchberger).   *A set $\{g_1, \ldots, g_m\}$ is a Gröbner basis of $I = (g_1, \ldots, g_m)$ if and only if $S(g_p, g_q)$ has remainder 0 mod G for all p, q, where $G = [g_1, \ldots, g_m]$.*

Proof.   Clearly, $S(g_p, g_q)$, being a linear combination of g_p and g_q, lies in I. Hence, if $G = \{g_1, \ldots, g_m\}$ is a Gröbner basis, then $S(g_p, g_q)$ has remainder 0 mod G, by Proposition 6.50.

Conversely, assume that $S(g_p, g_q)$ has remainder 0 mod G for all p, q; we must show that every $f \in I$ has remainder 0 mod G. By Proposition 6.50, it suffices to show that if $f \in I$, then $\mathrm{LT}(g_i) \mid \mathrm{LT}(f)$ for some i. Since $f \in I = (g_1, \ldots, g_m)$, we may write $f = \sum_i h_i g_i$, and so

$$\mathrm{Deg}(f) \preceq \max_i \{\mathrm{Deg}(h_i g_i)\}.$$

If there is equality, then $\mathrm{Deg}(f) = \mathrm{Deg}(h_i g_i)$ for some i, and so Proposition 6.47 gives $\mathrm{LT}(g_i) \mid \mathrm{LT}(f)$, as desired. Therefore, we may assume strict inequality: $\mathrm{Deg}(f) \prec \max_i \{\mathrm{Deg}(h_i g_i)\}$.

The polynomial f may be written as a linear combination of the g_i in many ways. Of all the expressions of the form $f = \sum_i h_i g_i$, choose one in which $\delta = \max_i \{\mathrm{Deg}(h_i g_i)\}$ is minimal (which is possible because $\preceq$ is a well order). If $\mathrm{Deg}(f) = \delta$, we are done, as we have seen above; therefore, we may assume that there is strict inequality: $\mathrm{Deg}(f) \prec \delta$. Write

$$f = \sum_{\substack{j \\ \mathrm{Deg}(h_j g_j) = \delta}} h_j g_j + \sum_{\substack{\ell \\ \mathrm{Deg}(h_\ell g_\ell) \prec \delta}} h_\ell g_\ell. \tag{1}$$

Now if $\mathrm{Deg}(\sum_j h_j g_j) = \delta$, then $\mathrm{Deg}(f) = \delta$, a contradiction. Hence, $\mathrm{Deg}(\sum_j h_j g_j) \prec \delta$. But the coefficient of X^δ in this sum is obtained from its leading terms, so that

$$\mathrm{Deg}\left(\sum_j \mathrm{LT}(h_j) g_j\right) \prec \delta.$$

Now $\sum_j \mathrm{LT}(h_j) g_j$ is a polynomial satisfying the hypotheses of Lemma 6.53, and so there are constants d_j and multidegrees $\mu(j)$ so that

$$\sum_j \mathrm{LT}(h_j) g_j = \sum_j d_j X^{\delta - \mu(j)} S(g_j, g_{j+1}), \tag{2}$$

where $\mathrm{Deg}\left(X^{\delta-\mu(j)}S(g_j, g_{j+1})\right) \prec \delta.$[10]

Since each $S(g_j, g_{j+1})$ has remainder $0 \bmod G$, the division algorithm gives $a_{ji}(X) \in k[X]$ with

$$S(g_j, g_{j+1}) = \sum_i a_{ji}g_i,$$

where $\mathrm{Deg}(a_{ji}g_i) \preceq \mathrm{Deg}(S(g_j, g_{j+1}))$ for all j, i. It follows that

$$X^{\delta-\mu(j)}S(g_j, g_{j+1}) = \sum_i X^{\delta-\mu(j)}a_{ji}g_i.$$

Therefore, Lemma 6.53 gives

$$\mathrm{Deg}(X^{\delta-\mu(j)}a_{ji}) \preceq \mathrm{Deg}(X^{\delta-\mu(j)}S(g_j, g_{j+1})) \prec \delta. \qquad (3)$$

Substituting into Eq. (2), we have

$$\sum_j \mathrm{LT}(h_j)g_j = \sum_j d_j X^{\delta-\mu(j)}S(g_j, g_{j+1})$$

$$= \sum_j d_j \left(\sum_i X^{\delta-\mu(j)}a_{ji}g_i\right)$$

$$= \sum_i \left(\sum_j d_j X^{\delta-\mu(j)}a_{ji}\right) g_i.$$

If we denote $\sum_j d_j X^{\delta-\mu(j)}a_{ji}$ by h_i', then

$$\sum_j \mathrm{LT}(h_j)g_j = \sum_i h_i'g_i, \qquad (4)$$

where, by Eq. (3),

$$\mathrm{Deg}(h_i'g_i) \prec \delta \quad \text{for all } i.$$

---

[10]The reader may wonder why we consider all S-polynomials $S(g_p, g_q)$ instead of only those of the form $S(g_i, g_{i+1})$. The answer is that the remainder condition is applied only to those $h_j g_j$ for which $\mathrm{Deg}(h_j g_j) = \delta$, and so the indices viewed as i's need not be consecutive.

Finally, we substitute the expression in Eq. (4) into Eq. (1):

$$
\begin{aligned}
f &= \sum_{\substack{j \\ \mathrm{Deg}(h_j g_j)=\delta}} h_j g_j + \sum_{\substack{\ell \\ \mathrm{Deg}(h_\ell g_\ell) \prec \delta}} h_\ell g_\ell \\
&= \sum_{\substack{j \\ \mathrm{Deg}(h_j g_j)=\delta}} \mathrm{LT}(h_j) g_j + \sum_{\substack{j \\ \mathrm{Deg}(h_j g_j)=\delta}} [h_j - \mathrm{LT}(h_j)] g_j + \sum_{\substack{\ell \\ \mathrm{Deg}(h_\ell g_\ell) \prec \delta}} h_\ell g_\ell \\
&= \sum_i h_i' g_i + \sum_{\substack{j \\ \mathrm{Deg}(h_j g_j)=\delta}} [h_j - \mathrm{LT}(h_j)] g_j + \sum_{\substack{\ell \\ \mathrm{Deg}(h_\ell g_\ell) \prec \delta}} h_\ell g_\ell.
\end{aligned}
$$

We have rewritten f as a linear combination of the g_i in which each term has multidegree strictly smaller than δ, contradicting the minimality of δ. This completes the proof.　•

Here is the main result.

Theorem 6.55 (Buchberger's Algorithm).　*Every ideal $I = (f_1, \ldots, f_s)$ in $k[X]$ has a Gröbner basis[11] which can be computed by an algorithm.*

Proof.　Here is a pseudocode for an algorithm.

> Input : $B = \{f_1, \ldots, f_s\}$　$G = [f_1, \ldots, f_s]$
> Output : a Gröbner basis $B = \{g_1, \ldots, g_m\}$ containing $\{f_1, \ldots, f_s\}$
> $B := \{f_1, \ldots, f_s\}$　$G := [f_1, \ldots, f_s]$
> REPEAT
> 　　　$B' := B$　$G' := G$
> 　　　FOR each pair g, g' with $g \ne g' \in B'$ DO
> 　　　　　$r :=$ remainder of $S(g, g') \bmod G'$
> 　　　　　IF $r \ne 0$
> 　　　　　　THEN $B := B \cup \{r\}$ and $G' = [g_1, \ldots, g_m, r]$
> 　UNTIL $B = B'$

Now each loop of the algorithm enlarges a subset $B \subset I = (g_1, \ldots, g_m)$ by adjoining the remainder mod G of one of its S-polynomials $S(g, g')$. As

---

[11] A nonconstructive proof of the existence of a Gröbner basis can be given using the proof of the Hilbert basis theorem; for example, see Section 2.5 of the book of Cox, Little, and O'Shea (they give a constructive proof in Section 2.7).

$g, g' \in I$, the remainder r of $S(g, g')$ lies in I, and so the larger set $B \cup \{r\}$ is contained in I.

The only obstruction to the algorithm's stopping at some B' is if some $S(g, g')$ does not have remainder 0 mod G'. Thus, if the algorithm stops, then Theorem 6.54 shows that B' is a Gröbner basis.

To see that the algorithm does stop, suppose a loop starts with B' and ends with B. Since $B' \subset B$, we have an inclusion of monomial ideals

$$\big(\mathrm{LT}(g') \colon g' \in B'\big) \leq (\mathrm{LT}(g) \colon g \in B) \,.$$

We claim that if $B' \subsetneqq B$, then there is also a strict inclusion of ideals. Suppose that r is a (nonzero) remainder of some S-polynomial mod B', and that $B = B' \cup \{r\}$. By definition, the remainder r is reduced mod G', and so no term of r is divisible by $\mathrm{LT}(g')$ for any $g' \in B'$; in particular, $\mathrm{LT}(r)$ is not divisible by any $\mathrm{LT}(g')$. Hence, $\mathrm{LT}(r) \notin (\mathrm{LT}(g') \colon g' \in B')$, by Exercise 6.47. On the other hand, we do have $\mathrm{LT}(r) \in (\mathrm{LT}(g) \colon g \in B)$. Therefore, if the algorithm does not stop, there is an infinite strictly ascending chain of ideals in $k[X]$, and this contradicts the Hilbert basis theorem, for $k[X]$ has the ACC.   •

Example 6.12.
The reader may show that $B' = \{y^2 + z^2, x^2 y + yz, z^3 + xy\}$ is not a Gröbner basis because $S(y^2 + z^2, x^2 y + yz) = x^2 z^2 - y^2 z$ does not have remainder 0 mod G'. However, adjoining $x^2 z^2 - y^2 z$ does give a Gröbner basis B because all S-polynomials in B have remainder 0 mod B'.   ◄

Theoretically, Buchberger's algorithm computes a Gröbner basis, but the question arises how practical it is. In very many cases, it does compute in a reasonable amount of time; on the other hand, there are examples in which it takes a very long time to produce its output. The efficiency of Buchberger's algorithm is discussed in Section 2.9 of the book by Cox, Little, and O'Shea.

Corollary 6.56.

(i) *If $I = (f_1, \ldots, f_t)$ is an ideal in $k[X]$, then there is an algorithm to determine whether a polynomial $h(X) \in k[X]$ lies in I.*

(ii) *If $I = (f_1, \ldots, f_t)$ and $I' = (f_1', \ldots, f_s')$ are ideals in $k[X]$, then there is an algorithm to determine whether $I = I'$.*

Proof.   (i) Use Buchberger's algorithm to find a Gröbner basis B of I, and then use the division algorithm to compute the remainder of h mod G (where

G is any m-tuple arising from ordering the polynomials in B). By Corollary 6.52(ii), $h \in I$ if and only if $r = 0$.

(ii) Use Buchberger's algorithm to find Gröbner bases $\{g_1, \ldots, g_m\}$ and $\{g_1', \ldots, g_m'\}$ of I and I', respectively. By part (i), there is an algorithm to determine whether each $g_j' \in I$, and $I' \leq I$ if each $g_j' \in I$. Similarly, there is an algorithm to determine the reverse inclusion, and so there is an algorithm to determine whether $I = I'$. ●

One must be careful here. Corollary 6.56 does not begin by saying "If I is an ideal in $k[X]$;" instead, it specifies a basis: $I = (f_1, \ldots, f_t)$. The reason, of course, is that Buchberger's algorithm requires a basis as input. For example, if $J = (h_1, \ldots, h_s)$, then the algorithm cannot be used to check whether a polynomial $f(X)$ lies in the radical $\sqrt{J}$, for one does not have a basis of $\sqrt{J}$. (There do exist algorithms giving bases of $\sqrt{(f_1, \ldots, f_t)}$; see the book by Becker and Weispfenning.)

A Gröbner basis $B = \{g_1, \ldots, g_m\}$ can be too large. For example, it follows from Proposition 6.50 that if $f \in I$, then $B \cup \{f\}$ is also a Gröbner basis of I; thus, one may seek Gröbner bases that are, in some sense, minimal.

Definition. A basis $\{g_1, \ldots, g_m\}$ of an ideal I is ***reduced*** if

(i) each g_i is monic;

(ii) each g_i is reduced mod $\{g_1, \ldots, \widehat{g_i}, \ldots, g_m\}$.

Exercise 6.53 gives an algorithm for computing a reduced basis for every ideal $(f_1, \ldots, f_t)$. When combined with the algorithm in Exercise 6.55, it shrinks a Gröbner basis to a *reduced* Gröbner basis. It can be proved that a reduced Gröbner basis of an ideal is unique.

In the special case when each $f_i(X)$ is linear, that is,

$$f_i(X) = a_{i1}x_1 + \cdots + a_{in}x_n.$$

then the common zeros $\mathrm{Var}(f_1, \ldots, f_t)$ are the solutions of a homogeneous system of t equations in n unknowns. If $A = [a_{ij}]$ is the $t \times n$ matrix of coefficients, then it can be shown that the reduced Gröbner basis corresponds to the row-reduced echelon form for the matrix A (see Section 10.5 in the book of Becker and Weispfenning).

Another special case occurs when when $f_1, \ldots, f_t$ are polynomials in one variable. The reduced Gröbner basis obtained from $\{f_1, \ldots, f_t\}$ turns out to be their gcd, and so the euclidean algorithm has been generalized to polynomials in several variables.

EXERCISES

Use the degree-lexicographic monomial order in the following exercises.

6.48 Let $I = (y - x^2, z - x^3)$.

   (i) Order $x \prec y \prec z$, and let $\preceq_{\text{lex}}$ be the corresponding monomial order on
       $\mathbb{N}^3$. Prove that $[y - x^2, z - x^3]$ is not a Gröbner basis of I.

   (ii) Order $y \prec z \prec x$, and let $\preceq_{\text{lex}}$ be the corresponding monomial order on
        $\mathbb{N}^3$. Prove that $[y - x^2, z - x^3]$ is a Gröbner basis of I.

6.49 Find a Gröbner basis of $I = (x^2 - 1, xy^2 - x)$.

6.50 Find a Gröbner basis of $I = (x^2 + y, x^4 + 2x^2y + y^2 + 3)$.

6.51 Find a Gröbner basis of $I = (xz, xy - z, yz - x)$. Does $x^3 + x + 1$ lie in I?

6.52 Find a Gröbner basis of $I = (x^y - y, y^2 - x, x^2y^2 - xy)$. Does $x^4 + x + 1$ lie
in I?

6.53 Show that the following pseudocode gives a reduced basis Q of an ideal $I =
(f_1, \dots, f_t)$.

```
Input: P = [f₁, . . . , fₜ]
Output: Q = [q₁, . . . , qₛ]
Q := P
WHILE there is q ∈ Q which is
         not reduced mod Q − {q} DO
     select q ∈ Q which is not reduced mod Q − {q}
     Q := Q − {q}
     h := the remainder of q mod Q
     IF h ≠ 0 THEN
         Q := Q ∪ {h}
     END IF
END WHILE
make all q ∈ Q monic
```

6.54 If G is a Gröbner basis of an ideal I, and if Q is the basis of I obtained from
the algorithm in Exercise 6.53, prove that Q is also a Gröbner basis of I.

6.55 Show that the following pseudocode replaces a Gröbner basis G with a reduced
Gröbner basis H.

Input: $G = \{g_1, \ldots, g_m\}$
Output: H
$H := \varnothing; \quad F := G$
WHILE $F \neq \varnothing$ DO
   select f' from F
   $F := F - \{f'\}$
   IF $\mathrm{LT}(f) \nmid \mathrm{LT}(f')$ for all $f \in F$ AND
       $\mathrm{LT}(h) \nmid \mathrm{LT}(f')$ for all $h \in H$ THEN
     $H := H \cup \{f'\}$
  END IF
END WHILE
apply the algorithm in Exercise 6.53 to H

Hints to Exercises

1.1   The sum is n^2.

1.2   The sum $1 + \sum_{j=1}^{n} j!\, j = (n+1)!$.

1.3   (ii) Either prove this by induction, or use part (i).

1.4   This may be rephrased to say that there is an integer q_n with $10^n = 9q_n + 1$.

1.9   There are $n + 1$ squares on the diagonal, and the triangular areas on either side have area $\sum_{i=1}^{n} i$.

1.10   (i) Compute the area R of the rectangle in two ways.

1.10   (ii) As indicated in Figure 1.3, a rectangle with height $n + 1$ and base $\sum_{i=1}^{n} i^k$ can be subdivided so that the shaded staircase has area $\sum_{i=1}^{n} i^{k+1}$, whereas the area above it is

$$1^k + (1^k + 2^k) + (1^k + 2^k + 3^k) + \cdots + (1^k + 2^k + \cdots + n^k).$$

1.10   (iii) In Alhazen's formula, write $\sum_{i=1}^{n} i = \frac{1}{2} \sum_{i=1}^{n} i^2 + \frac{1}{2} \sum_{i=1}^{n} i$, and then solve for $\sum_{i=1}^{n} i^2$ in terms of the rest.

1.11   (i) In the inductive step, use $n \geq 10$ implies $n \geq 4$.

1.11   (ii) In the inductive step, use $n \geq 17$ implies $n \geq 7$.

1.12   The base step is the product rule for derivatives.

1.13   The inequality $1 + x > 0$ allows one to use Proposition 1.5.

1.14   Model your solution on the proof of Proposition 1.12. Replace "even" by "multiple of 3" and "odd" by "not a multiple of 3."

1.15   What is the appropriate form of induction to use?

1.16   Use Theorem 1.13 and geometric series.

1.17   For the inductive step, try adding and subtracting the same terms.

1.18   If $2 \leq a \leq n + 1$, then a is a divisor of $a + (n + 1)!$. Many proofs do not use induction!

1.21   Check that the properties of addition and multiplication used in the proof for real numbers also hold for complex numbers.

1.23   Consider $f(x) = (1 + x)^n$ when $x = 1$.

1.24   (i) Consider $f(x) = (1 + x)^n$ when $x = -1$.

1.25   Take the derivative of $f(x) = (1 + x)^n$.

1.26   (i) Use the triangle inequality and induction on n.

1.26   (ii) Use the following properties of the dot product of complex numbers: $|u|^2 = u \cdot u$ and $u \cdot v = |u||v| \cos \theta$, where θ is the angle between u and v.

1.27   Only odd powers of i are imaginary.

1.29   (ii) Compare with part (i).

1.31   How many selections of 5 numbers are there?

1.33   Even though there is a strong resemblance, there is no routine derivation of the Leibniz formula from the binomial theorem (there is a derivation using a trick of hypergeometric series).

1.34   (i) $1/z = \bar{z}/z\bar{z}$.

1.36   (i) The polar coordinates of $(8, 15)$ are $(17, 62°)$, and $\sin 31° \approx .515$ and $\cos 31° \approx .857$.

1.36   (ii) $\sin 15.5° \approx .267$ and $\cos 15.5° \approx .967$.

1.37   Use Corollary 1.21.

1.38   Use the portion of the full division algorithm that has already been proved.

1.40   Prove the contrapositive.

1.41 $19 \mid f_7$, but 7 is not the smallest k.

1.43   Write m in base 2.

1.45   (i) Assume $\sqrt{n} = a/b$, where a/b is in lowest terms, and adapt the proof of Corollary 1.36.

1.45 (ii) Assume that $\sqrt[3]{2}$ can be written as a fraction in lowest terms.

1.49 If $ar + bm = 1$ and $sr' + tm = 1$, consider $(ar + bm)(sr' + tm)$.

1.50 If $2s + 3t = 1$, then $2(s + 3) + 3(t - 2) = 1$.

1.51 Use Corollary 1.34.

1.52 Show that if k is a common divisor of ab and ac, then $k \mid a(b, c)$.

1.53 Use the idea in antanairesis.

1.58 (i) If neither a nor b is 0, show that $ab/(a, b)$ is a common multiple of a and b that divides every common multiple c of a and b.

1.61 (ii) Use Corollary 1.42.

1.62 The sets of prime divisors of a and b are disjoint.

1.63 Assume otherwise, cross-multiply, and use Euclid's lemma.

1.67 Cast out 9's.

1.68 $11 \equiv -1 \bmod 10$.

1.69 $100 = 2 \cdot 49 + 2$.

1.72 Use the fact, proved in Example 1.10(i), that if a is a perfect square, then $a^2 \equiv 0, 1$, or 4 mod 8.

1.73 If the last digit of a^2 is 5, then $a^2 \equiv 5 \bmod 10$; if the last two digits of a^2 are 35, then $a^2 \equiv 35 \bmod 100$.

1.75 Use Euclid's lemma.

1.77 Use Euclid's lemma with $21 \mid (a + 1)(a - 1)$.

1.80 (i) Consider the parity of a and of b.

1.81 Try -4 coconuts.

1.82 Easter always falls on Sunday. (There is a Jewish variation of this problem, for Yom Kippur must fall on either Monday, Wednesday, Thursday, or Saturday; secular variants can involve Thanksgiving Day, which always falls on a Thursday, or Election Day, which always falls on a Tuesday.)

1.83 The year $y = 1900$ was not a leap year.

1.84 On what day did March 1, 1896, fall?

1.85 (iii) Either use congruences or scan the 14 possible calendars: there are 7 possible common years and 7 possible leap years, for January 1 can fall on any of the 7 days of the week.

2.1   Use Proposition 2.1.

2.2   Does g have an inverse?

2.3   In more traditional notation, prove that $(g \circ f)' = (g' \circ f) \circ f'$.

2.5   Either find an inverse or show that f is injective and surjective.

2.6   It isn't.

2.7   If f is a bijection, there are m distinct elements $f(x_1), \ldots, f(x_m)$ in Y, and so $m \leq n$; using the bijection f^{-1} in place of f gives the reverse inequality $n \leq m$.

2.8   If $A \subset X$ and $|A| = n = |X|$, then $A = X$; after all, how many elements are in X but not in A?

2.10   (i) Compute composites.

2.12   One of the axioms constraining the $\in$ relation is that the statement

$$a \in x \in a$$

is always false.

2.13   (i) You may use the facts: (1) lines ℓ_1 and ℓ_2 having slopes m_1 and m_2, respectively, are perpendicular if and only if $m_2 m_2 = -1$; (2) the midpoint of the line segment having endpoints (a, b) and (c, d) is $(\frac{1}{2}(a + c), \frac{1}{2}(b + d))$.

2.14   (i) A sequence $x_0, x_1, x_2, \ldots, x_n, \ldots$ is a function $f : \mathbb{N} \to X$, where $f(n) = x_n$.

2.14   (ii) Write the elements of X as a sequence, and write the elements of S as a subsequence.

2.14   (iii) Construct an appropriate subsequence.

2.14   (iv) For each $y \in Y$, choose some $x = x_y \in X$ with $y = f(x)$.

2.15   Form an array whose nth row is a sequence listing all the elements of X_n. Starting in the upper left corner, construct a zigzag path that passes through each term in the array.

2.17   Use the complete factorizations of σ and of σ'.

2.18 There are r cycle notations for any r-cycle.

2.19 (i) If $\alpha = (i_0 \ldots i_{r-1})$, show that $\alpha^k(i_0) = i_k$.

2.19 (ii) Use Proposition 2.8.

2.21 Use induction on $j - i$.

2.23 Not always.

2.24 (i) First show that $\beta\alpha^k = \alpha^k\beta$ by induction on k.

2.25 Place the complete factorization of β over that of γ, and define α to be the downward function. For example, if

$$\beta = (2\ 3\ 1)(4\ 5)(6)$$
$$\gamma = (5\ 6\ 2)(3\ 1)(4),$$

then

$$\alpha = \begin{pmatrix} 1 & 2 & 3 & 4 & 5 & 6 \\ 2 & 5 & 6 & 3 & 1 & 4 \end{pmatrix},$$

and so $\alpha = (1\ 2\ 5)(3\ 6\ 4)$. Note that there are other choices for α as well.

2.27 Let $\tau = (1\ 2)$, and define $f : A_n \to O_n$, where A_n is the set of all even permutations in S_n and O_n is the set of all odd permutations, by

$$f : \alpha \mapsto \tau\alpha.$$

Show that f is a bijection, and conclude that $|A_n| = |O_n|$.

2.28 (i) There are 6 permutations in S_5 commuting with α, only 3 of which are even.

2.28 (ii) There are 8 permutations in S_4 commuting with $(1\ 2)(3\ 4)$, and only 4 of them are even.

2.31 See Proposition 2.11.

2.34 (i) There are 25 elements of order 2 in S_5 and 75 in S_6.

2.34 (ii) You may express your answer as a sum.

2.36 (i) If $h' * h = e$, evaluate $h' * h * h$ in two ways.

2.36 (ii) Consider $(x * x')^2$.

2.36 (iii) Evaluate $x * x' * x$ in two ways.

2.36   (iv) Show that $(e')^2 = e'$.

2.36   (v) Evaluate $x' * x * x''$ in two ways.

2.37   Clearly, $(y')^p = 1$. Use Lemma 2.24 to show that no smaller power of y' is equal to 1.

2.40   (i) Use induction on k.

2.42   Pair each element with its inverse.

2.43   No general formula is known for arbitrary n.

2.46   Let G be the four-group **V**.

2.48   If $x \in H \cap K$, then $x^{|H|} = 1 = x^{|K|}$.

2.50   If $S \neq 1$, choose k to be the smallest positive integer with $a^k \in S$.

2.51   If $G = \langle a \rangle$ and $n = dk$, consider $\langle a^k \rangle$.

2.55   If $\alpha \in S_X$, define $\varphi(\alpha) = f \circ \alpha \circ f^{-1}$. In particular, show that if $|X| = 3$, then φ takes a cycle involving symbols 1, 2, 3 into a cycle involving a, b, c, as in the text.

2.60   Show that the bijection $f : G \to S_3$ is an isomorphism by comparing multiplication tables of G and of S_3.

$$f(x) = 1; \quad f(1 - x) = (2\ 3); \quad f(1/x) = (1\ 2);$$
$$f(x/(x - 1)) = (1\ 3\ 2); \quad f(1/(1 - x)) = (1\ 2\ 3); \quad f((x - 1)/x) = (1\ 3).$$

2.64   Consider

$$\varphi : A = \begin{bmatrix} \cos \alpha & -\sin \alpha \\ \sin \alpha & \cos \alpha \end{bmatrix} \mapsto (\cos \alpha, \sin \alpha).$$

2.65   List the prime numbers $p_0 = 2$, $p_1 = 3$, $p_2 = 5, \ldots$, and define

$$\varphi(e_0 + e_1 x + e_2 x^2 + \cdots + e_n x^n) = p_0^{e_0} \cdots p_n^{e_n}.$$

2.70   Show that squaring is an injective function $G \to G$, and use Exercise 2.8.

2.71   Take $G = S_3$, $H = \langle (1\ 2) \rangle$, and $g = (2\ 3)$.

2.72   Show that if A is a matrix which is not a scalar matrix, then there is some nonsingular matrix that does not commute with A. (There is a proof of this for $n \times n$ matrices given in Proposition 4.18(i).)

2.73 (iii) Consider cases $A^i A^j$, $A^i B A^j$, $B A^i A^j$, and $(B A^i)(B A^j)$.

2.73 (v) Define a function $G \to D_{2n}$ using the unique expression of elements in G in the form $X = B^i A^j$.

2.74 (ii) Note that $A^2 = -E = B^2$.

2.76 Use Exercise 2.58.

2.77 Use Proposition 2.39(ii).

2.78 (ii) See Example 2.15(iv).

2.80 The vertices $X = \{v_1, \ldots, v_n\}$ of π_n are permuted by every motion $\sigma \in \Sigma(\pi_n)$.

2.82 (iii) Define $f : H \times K \to H$ by $f : (h, k) \mapsto h$.

2.83 If $G/Z(G)$ is cyclic, prove that a generator gives an element outside of $Z(G)$ which commutes with each element of G.

2.85 If $H \leq G$ and $|H| = |K|$, what happens to elements of H in G/K?

2.86 Use the fact that $H \subset HK$ and $K \subset HK$.

2.89 $|G| = |G/H||H|$.

2.91 Use Wilson's theorem.

2.92 (ii) Use Exercise 2.90.

2.93 Use a conjugation.

2.95 Use Cauchy's theorem.

2.97 Use Proposition 2.69.

2.98 (i) Recall that A_4 has no element of order 6.

2.98 (ii) Each element $x \in D_{12}$ has a unique factorization of the form $x = b^i a$, where $b^6 = 1$ and $a^2 = 1$.

2.99 (ii) Use the second isomorphism theorem.

2.99 (iii) If $\alpha = (1\ 2\ 3\ 4\ 5)$, then $|C_{S_5}(\alpha)| = 5$ because $24 = \dfrac{120}{|C_{S_5}(\alpha)|}$; hence $C_{S_5}(\alpha) = \langle \alpha \rangle$. What is $C_{A_5}(\alpha)$?

2.101 (i) Show that $(1\ 2\ 3)$ and $(i\ j\ k)$ are conjugate, in two steps: first, if they are not disjoint (so the permutations move at most 5 letters); then, if they are disjoint.

2.102   Use Proposition 2.12, checking the various cycle structures one at a time.

2.103   Use Proposition 2.39(ii).

2.104   (i) Kernels are normal subgroups.

2.104   (ii) Use part (i).

2.105   Show that G has a subgroup H of order p, and use the representation of G on the cosets of H.

2.106   If H is a second such subgroup, then H is normal in S_n and hence $H \cap A_n$ is normal in A_n.

2.107   The parity of n is relevant.

2.110   (i) The group $G = D_{10}$ is acting. Use Example 2.18 to assign to each symmetry a permutation of the vertices, and then show that

$$P_G(x_1, \dots, x_5) = \tfrac{1}{10}(x_1^5 + 4x_5 + 5x_1 x_2^2)$$

and

$$P_G(q, \dots, q) = \tfrac{1}{10}(q^5 + 4q + 5q^3).$$

2.110   (ii) The group $G = D_{12}$ is acting. Use Example 2.18 to assign to each symmetry a permutation of the vertices, and then show that

$$P_G(x_1, \dots, x_6) = \tfrac{1}{12}(x_1^6 + 2x_6 + 2x_3^2 + 3x_2^2 + 4x_2^3)$$

and so

$$P_G(q, \dots, q) = \tfrac{1}{12}(q^6 + 2q + 4q^2 + 4q^3).$$

3.3   See Exercise 1.21.

3.8   (i) You may use some standard facts of set theory:

$$U \cap (V \cup W) = (U \cap V) \cup (U \cap W);$$

if V' denotes the complement of V, then

$$U - V = U \cap V';$$

the ***de Morgan law***:

$$(U \cap V)' = U' \cup V'.$$

3.12   Every subring R of $\mathbb{Z}$ contains 1.

3.13   Use Theorem 1.53.

3.14   (i) Yes.

3.14   (ii) No.

3.22   It is tedious to check the associative law (there are $4^3 = 64$ ordered triples).

3.23   If $R^\times$ denotes the set of nonzero elements of R, prove that multiplication by r is an injection $R^\times \to R^\times$, where $r \in R^\times$.

3.24   Use Corollary 1.21.

3.28   (i) See Example 2.15(iv).

3.29   If x^{-1} exists, what is its degree?

3.30   (i) Compute degrees.

3.32   Use Fermat's theorem.

3.33   This is not a hard exercise, but it is a long one.

3.34   The assumption $x - a \mid f(x)$ is not needed to prove $\Rightarrow$.

3.35   (ii) Generalize the example in part (i).

3.35   (ii) The condition is that there should be a polynomial $g(x) = \sum a_n x^n$ with $f(x) = g(x^p)$; that is, $f(x) = \sum b_n x^{np}$, where $b_n^p = a_n$ for all n.

3.36   (i) The proof for polynomials, Proposition 3.19, works here.

3.36   (iii) If $\sigma = (s_0, s_1, \dots) \in R[[x]]$ is nonzero, define the **order** of σ, denoted by $\mathrm{ord}(\sigma)$, to be the smallest $n \geq 0$ for which $s_n \neq 0$. If R is a domain and $\sigma, \tau \in R[[x]]$ are nonzero, prove that $\mathrm{ord}(\sigma\tau) = \mathrm{ord}(\sigma) + \mathrm{ord}(\tau) \neq 0$, and hence $\sigma\tau \neq 0$.

3.42   This is another of the routine but long calculations.

3.43   First prove that $1 + 1 = 0$, and then show that the nonzero elements form a cyclic group of order 3 under multiplication.

3.44   (i) Use Exercise 2.8.

3.46   (i) Define $\Phi : \mathrm{Frac}(A) \to \mathrm{Frac}(R)$ by $[a, b] \mapsto [\varphi(a), \varphi(b)]$.

3.49   (ii) See Theorem 2.62.

3.50   (i) Show that (r, s) is a unit in $R \times S$ if and only if r is a unit in R and s is a unit in S.

3.51   Define $\varphi : F \to \mathbb{C}$ by $\varphi(A) = a + ib$.

3.52   The answer is $x - 2$.

3.53   (i) Use Frac(R).

3.56   Use Frac(R).

3.57   See Exercise 1.49.

3.58   Mimic the proof of Corollary 1.36 which shows that $\sqrt{2}$ is irrational.

3.59   If I is a nonzero ideal in R, show that $I = (a)$, where $a \in I$ and $\partial(a) \leq \partial(b)$ for all nonzero $b \in I$.

3.61   (ii) The general proof can be generalized from a proof of the special case of polynomials.

3.62   There are $q, r \in R$ with $b^i = qb^{i+1} + r$.

3.63   (i) If I is a nonzero ideal, choose $\tau \in I$ of smallest order. Use Exercise 3.37 to prove that $I = (\tau)$.

3.64   (i) Example 3.9.

3.65   See Exercise 1.58.

3.68   (i) Use a degree argument.

3.69   Show that $\sqrt{x} + 1$ is not a polynomial.

3.70   Adapt the proof of Proposition 3.49.

3.71   Let k be a field and let R be the subring of $k[x]$ consisting of all polynomials having no linear term; that is, $f(x) \in R$ if and only if

$$f(x) = s_0 + s_2 x^2 + s_3 x^3 + \cdots .$$

Show that x^5 and x^6 have no gcd.

3.73   (i) Use Exercise 3.34.

3.73   (ii) Corollary 3.47.

3.74   (i) Use Theorem 3.33.

3.74   (ii) Set $x = a/b$ if $b \neq 0$.

3.75   (vi) Show that $f(x)$ has no roots in $\mathbb{Z}_3$ and that a factorization of $f(x)$ as a product of quadratics would force impossible restrictions on coefficients.

3.75 (vii) Show that $f(x)$ has no rational roots and that a factorization of $f(x)$ as a product of quadratics would force impossible restrictions on coefficients.

3.76 The irreducible quintics in $\mathbb{Z}_2[x]$ are:

$$x^5 + x^3 + x^2 + x + 1 \quad x^5 + x^4 + x^2 + x + 1$$
$$x^5 + x^4 + x^3 + x + 1 \quad x^5 + x^4 + x^3 + x^2 + 1$$
$$x^5 + x^3 + 1 \quad x^5 + x^2 + 1.$$

3.77 Use Corollary 3.64.

3.78 (i) Use the Eisenstein criterion.

3.79 $f(x) \mapsto f^*(x)$, which reverses coefficients, is not a well-defined function $k[x] \to k[x]$.

3.82 Adapt the proof of Exercise 1.51.

3.83 (i) Adapt the proof of Theorem 1.55.

3.83 (ii) See the proof of Theorem 2.62.

3.85 Show that $\mathbb{Z}_p^\times \cong \langle -1 \rangle \times H$, where H is a group of odd order m, say, and observe that either 2 or -2 lies in H because

$$\mathbb{Z}_2 \times \mathbb{Z}_m = (\{1\} \times H) \cup (\{-1\} \times H).$$

Finally, use Exercise 2.70.

3.86 (ii) Equate like coefficients after expanding the right-hand side.

3.86 (iii) In the first case, set $a = 0$ and use b to factor $x^4 + 1$. If $a \neq 0$, then $d = b$ and $b^2 = 1$ (so that $b = \pm 1$); now use a to factor $x^4 + 1$.

3.86 (iv) Use Exercise 3.85.

3.87 See Exercise 3.28.

3.89 Use an irreducible cubic over $\mathbb{Z}_2$.

3.90 (ii) Exercise 2.95.

4.4 The slope of a vector $v = (a, b)$ is $m = b/a$.

4.6 When are two polynomials equal?

4.7   (ii) Rewrite the vectors u, v, and n using coordinates in $\mathbb{R}^3$.

4.8   (ii) Take a basis of $U \cap U'$ and extend it to bases of U and of U'.

4.10   You may assume the following fact of linear algebra: Given a system of linear equations with coefficients in a field,

$$ax + by = p$$
$$cx + dy = q,$$

then there exists a unique solution if and only if $ad - bc \neq 0$.

4.11   (iii) The statement is: If $f : V \to W$ is a linear transformation with $\ker f = U$, then U is a subspace of V and there is an isomorphism $V/U \cong \operatorname{im} f$.

4.12   Prove that if $v_1 + U, \ldots, v_r + U$ is a basis of V/U, then the list $v_1, \ldots, v_r$ is linearly independent.

4.14   Use Corollary 4.13(ii).

4.15   Use Proposition 4.24 to obtain an irreducible polynomial $p(x) \in k[x]$; the polynomial $p(x)$ may factor in $K[x]$.

4.16   (ii) Use Corollary 3.57.

4.18   Apply complex conjugation σ to the equation $f(u) = 0$.

4.19 $r = \cos 3\theta = \cos 3(\theta + 120°) = \cos 3(\theta + 240°)$.

4.20   The roots are -4 and $2 \pm \sqrt{-3}$.

4.21   The roots are 17 and $\frac{1}{2}(-1 \pm \sqrt{-3})$.

4.22   (i) The roots appear in unrecognizable form.

4.22   (ii) The roots are 4 and $-2 \pm \sqrt{3}$.

4.23   The roots are 2 and $-1 \pm \sqrt{3}$.

4.24   This is a tedious calculation. The roots are $-3, -1, 2 \pm \sqrt{6}$.

4.26   No.

4.27   (i) Use Proposition 1.33.

4.27   (iii) Use Exercise 2.8 to prove that the Frobenius $F : k \to k$ is surjective when k is finite.

4.28   (ii) Use Proposition 4.24.

4.28 (iii) Prove that if $\sigma \in G$, then σ is completely determined by $\sigma(\alpha)$, which is a root of the irreducible polynomial of α.

4.28 (iv) Prove that F has order $\geq n$.

4.31 (i) If α is a real root of $f(x)$, then $\mathbb{Q}(\alpha)$ is not the splitting field of $f(x)$.

4.31 (ii) Use the two facts and Theorem 4.45.

4.32 (i) Use Exercise 4.31.

4.32 (ii) Use Exercise 4.31.

4.32 (iii) Use Exercise 4.31.

4.33 (ii) Use Exercise 3.73.

4.35 Consider $t^p - x$.

5.3 (ii) Use part (i).

5.3 (iii) Use part (i).

5.5 There are 14 groups.

5.7 (i) If B is a direct sum of k copies of a cyclic group of order p^n, then how many elements of order p^n are in B?

5.7 (ii) Take G of order p^3.

5.8 If $a, b \in \mathbb{Q}$ are not zero, then there is $c \in \mathbb{Q}$ with $a, b \in \langle c \rangle$.

5.10 Consider $S_3 \times S_3$.

5.11 If $g \in G$, then gPg^{-1} is a Sylow p-subgroup of K, and so it is conjugate to P in K.

5.12 You may use the fact that the only nonabelian groups of order 8 are D_8 and $\mathbf{Q}$.

5.13 It suffices to find a subgroup of S_6 of order 16.

5.14 Use the fact that any other Sylow p-subgroup of G is conjugate to P.

5.15 Compute the order of the subgroup generated by the Sylow subgroups.

5.16 (i) Show that $[G/H : HP/H]$ and $[H : H \cap P]$ are prime to p.

5.16 (ii) Choose a subgroup H of S_4 with $H \cong S_3$, and find a Sylow 3-subgroup P of S_4 with $H \cap P = \{1\}$.

5.18   Some of these are not tricky.

5.19   Adapt the proof of the primary decomposition.

5.21   Show first that both subgroups have order p.

5.22   Remember that odd $\times$ odd $=$ even.

5.23   Use the correspondence theorem.

5.24   If $H \lhd G$, then $|G| = |G/H||H|$.

5.26   (i) If $p = q$, then G is abelian. If $p < q$, then a divisor r of pq for which $r \equiv 1 \bmod q$ must equal 1.

5.26   (ii) If G is not simple, use Proposition 5.35. If $p > q$, then $r \equiv 1 \bmod p$ forces $r = 1$. If $p < q$, then $r = p^2$ and there are more than $p^2 q$ elements in G.

5.27   For sufficiency, choose a "least criminal": a nonsolvable group G of smallest odd order. By hypothesis, G is not simple, and so it has a proper nontrivial normal subgroup.

5.29   Use Schreier's theorem.

5.30   (i) The subgroup ST is a homomorphic image of $S \times T$.

5.35   Consider homomorphisms $G \to S_n$.

5.39   By Cauchy's theorem, G must contain an element a of order p, and $\langle a \rangle \lhd G$ because it has index 2.

6.3   (ii) Let $f : \mathbb{Z} \to \mathbb{Z}_4$ be the natural map, and take $Q = \{0\}$.

6.12   (ii) If I and J are coprime, there are $a \in I$ and $b \in J$ with $1 = a + b$. If $r, r' \in R$, prove that $(d + I, d + J) = (r + I, r' + J) \in R/I \times R/J$, where $d = r'a + rb$.

6.13   When is a Boolean ring a domain?

6.14   (ii) You may assume that every nonunit in a commutative ring lies in some maximal ideal (this result is proved using Zorn's lemma).

6.27   Use the proof of the Hilbert basis theorem, but replace the degree of a polynomial by the order of a power series (where the *order* of a nonzero power series is its first nonzero coefficient).

6.31   Use Exercise 1.3(i).

6.32   If $f^r \in I$ and $g^s \in I$, prove that $(f + g)^{r+s} \in I$.

Bibliography

Baker, A., *Transcendental Number Theory*, Cambridge University Press, 1979.

Becker, T., and Weispfenning, V., *Gröbner Bases: a Computational Approach to Commutative Algebra*, Springer-Verlag, New York, 1993.

Berlekamp, E. R., Conway, J. H., and Guy, R. K., *Winning Ways for Your Mathematical Plays*, Academic Press, Orlando, FL, 1982.

Biggs, N. L., *Discrete Mathematics*, Oxford University Press, 1989.

Birkhoff, G., and Mac Lane, S., *A Survey of Modern Algebra*, 4th ed., Macmillan, New York, 1977.

Burnside, W., *The Theory of Groups of Finite Order*, Cambridge University Press, 1911.

Cajori, F., *A History of Mathematical Notation*, Open Court, 1928; Dover reprint, 1993.

Carmichael, R., *An Introduction to the Theory of Groups of Finite Order*, Ginn, Boston, 1937.

Cox, D., Little, J., and O'Shea, D., *Ideals, Varieties, and Algorithms*, 3d ed., Springer-Verlag, New York, 1992.

Edwards, C. H., Jr., and Penney, D. E., *Calculus and Analytic Geometry*, 3d ed., Prentice-Hall, Upper Saddle River, NJ, 1990.

Eisenbud, D., *Commutative Algebra with a View Toward Algebraic Geometry*, Springer-Verlag, New York, 1995.

Fröhlich, A., and Taylor, M. J., *Algebraic Number Theory*, Cambridge University Press, Cambridge, 1991.

Hadlock, C., *Field Theory and its Classical Problems*, Carus Mathematical Monographs, Mathematical Association of America, 1978.

Herstein, I. N., *Topics in Algebra*, 2d ed., Wiley, New York, 1975.

Jacobson, N., *Basic Algebra* I, Freeman, San Francisco, 1974.

————, *Basic Algebra* II, Freeman, San Francisco, 1980.

Kaplansky, I., *Fields and Rings*, 2d ed., U. Chicago, 1974.

Koblitz, N., *Algebraic Aspects of Cryptography*, Springer-Verlag, New York, 1998.

Li, C. C., *An Introduction to Experimental Statistics*, McGraw-Hill, New York, 1964.

McCoy, N. H., and Janusz, G. J., *Introduction to Modern Algebra*, 5th ed., Wm. C. Brown Publishers, Dubuque, Iowa, 1992.

Niven, I., and Zuckerman, H. S., *An Introduction to the Theory of Numbers*, Wiley, New York, 1972.

Pollard, H., *The Theory of Algebraic Numbers*, Carus Mathematical Monographs 9, Mathematical Association of America, 1950.

Rotman, J. J., *Galois Theory*, 2d ed., Springer-Verlag, New York, 1998.

————, *An Introduction to the Theory of Groups*, 4th ed., Springer-Verlag, New York, 1995.

————, *Journey into Mathematics*, Prentice-Hall, Upper Saddle River, NJ, 1998.

Ryser, H., *Combinatorial Mathematics*, Carus Mathematical Monographs, Mathematical Association of America, 1963.

Stillwell, J., *Mathematics and Its History*, Springer-Verlag, New York, 1989.

Suzuki, M., *Group Theory I*, Springer-Verlag, New York, 1982.

Tignol, J.-P., *Galois' Theory of Equations*, Longman Scientific and Technical, 1988.

van der Waerden, B. L., *Geometry and Algebra in Ancient Civilizations*, Springer-Verlag, New York, 1983.

————, *A History of Algebra*, Springer-Verlag, New York, 1985.

————, *Modern Algebra*, 4th ed., Ungar, 1966.

————, *Science Awakening*, Wiley, New York, 1963.

Zariski, O., and Samuel, P., *Commutative Algebra*, volume II, von Nostrand, Princeton, 1960.

Index